Digital Circuits
and
Logic Design

Samuel C. Lee

Professor of Electrical Engineering
University of Oklahoma

Prentice-Hall, Inc. Englewood Cliffs, New Jersey

Library of Congress Cataloging in Publication Data

LEE, SAMUEL C (date)
    Digital circuits and logic design.

    Includes bibliographies and index.
    1. Digital electronics. 2. Logic design.
3. Integrated circuits. I. Title.
TK7868.D5L4   621.3819′58′35   76–5446
ISBN 0–13–212225–1

10 9 8 7 6 5 4 3 2 1

Printed in the United States of America

PRENTICE-HALL INTERNATIONAL, INC., *London*
PRENTICE-HALL OF AUSTRALIA PTY. LIMITED, *Sydney*
PRENTICE-HALL OF CANADA, LTD., *Toronto*
PRENTICE-HALL OF INDIA PRIVATE LIMITED, *New Delhi*
PRENTICE-HALL OF JAPAN, INC., *Tokyo*
PRENTICE-HALL OF SOUTHEAST ASIA PTE. LTD., *Singapore*

To my parents
Dr. and Mrs. Chou-Ying Lee

Contents

Foreword

We are all the beneficiaries of the tremendous changes that have occurred in the science, technology, and economics of digital logics systems, from the major improvements made in large computing systems and the expanded use of digital logic components in all engineering systems to the ubiquitous pocket calculator which has become a part of all of us. As is always true, these changes have not taken place without some unexpected and frequently unmet costs. In the educational environment, these exceedingly rapid developments have tended to exaggerate the unavoidable lag between the learning experiences of students and the realities of engineering practice that they encounter following their graduation. While every faculty strives to provide classroom and laboratory experiences that will permit graduates to keep pace with the expanding base of knowledge and application in their chosen fields, the achievement of this goal is frequently limited by the lack of texts and monographs that reflect a truly contemporary synthesis of knowledge.

Dr. S. C. Lee is well qualified by both his academic experience and his own professional involvement in this field to produce a current synthesis of understanding application in DIGITAL CIRCUITS AND LOGIC DESIGN. For many years he has taught introductory and advanced courses for undergraduate and graduate students in both electrical engineering and computer sciences. In these activities he has successfully embedded the underlying theory and understanding in context of innovative and advanced applications. Moreover, he has thoroughly evaluated many patterns of student learning in this area to develop the synthesis offered in this text.

Dr. Lee is equally committed to the two primary academic missions of teaching and research: the preparation of each student for a successful and contributive professional career and the contributions of his own research and that of his students. This book is a reflection of these commitments and will serve to extend his influence and leadership in these areas far beyond the geographical limits of the University of Oklahoma. We believe that it is a worthy presentation of his accomplishments and a valuable contribution to engineering education.

WM. R. UPTHEGROVE
Dean, College of Engineering
University of Oklahoma

Preface

In the past five years, digital circuits and logic design have undergone dramatic change. Integrated circuits of increasing complexity have not only become available but also become cost competitive. Many of the classical rules of logic design, such as the minimization of the number of gates or flip-flops, are now useful only in the design of circuit components, such as MSI and LSI devices, and circuits that are not available in chip form. They are not useful in the design of large circuits using MSI and/or LSI devices as components to achieve a low total system cost.

There is a gap between the classical rules of logic design found in textbooks and taught in universities and the practical logic design using currently available integrated circuit components. As a result, after having taken a course on digital circuits and logic design, the student is unable to design any practical, reliable, and low cost digital circuits, not even a very simple one!

This book is an attempt to bridge this gap—the gap between the blackboard and the breadboard in the area of digital circuits and logic design. The intent is to teach the student to relate current products and practices in the industry to theory and design examples in the classroom. It is my believe that it is extremely important for a student to be introduced to the practice of engineering as an integral part of his academic program. This book covers both fundamental theory and practical design guidelines and examples of modern logic design. Students, after taking this course, will be acquainted with both the theoretical and practical aspects of logic design. Consequently, they will be better equipped to enter the job market.

The following approaches have been used in preparing this book:

1. A simple, comprehensive "mathematical language" is always welcomed by the student and the instructor. In this book, I have made an effort to make notations as simple and clear as possible to avoid unnecessary definitions.

2. For methods and theorems that are difficult to explain or prove, examples will first be presented, then followed by the formal description of the procedure or a formal proof. This approach, in the author's opinion, has the following two advantages. Firstly, it will help the reader to know what the methods and theorems are about.

Secondly, it will help the reader to familiarize himself with the mathematical symbols, equations, etc. involved, and consequently understand their formal proofs given later.

3.   It is believed that a picture is worth a thousand words. Tables, block diagrams, and flow charts are often used throughout the book.

A fairly good idea of the contents and scope of the book can be obtained by scanning the Contents listing on p. v. The material presented is illustrated by many examples. A considerable number of exercises are included at the end of almost every section. Material in this book is arranged as follows.

Chapter 1 covers the essentials of Boolean algebra. The fundamentals of switching algebras and basic combinational circuit design procedures are then presented. These include methods for minimizing completely and incompletely specified switching functions, computer-aided switching function minimization, and minimization of multiple-output switching circuits. The basic procedure of designing combinational logic circuits is included.

Chapter 2 introduces various commonly-used number and coding systems and their conversions. Several examples of logic design of commonly-used combinational circuits, such as code converters, binary and decimal addition and subtraction of unsigned and signed numbers are presented. It is shown that even with today's "block approach" the function-minimization approach presented in Chapter 1 is still useful at least in the design of circuit components such as those presented in this chapter.

The mapping from "logic circuit" to "electronic circuit," known as voltage symbolism, is introduced in Chapter 3. The purpose of this chapter is to introduce presently available combinational SSI and MSI chips plus semiconductor memory devices, and the combinational logic design using these components as building blocks. The read-only memory (ROM) and the random-access memory (RAM) are discussed. The combinational logic design using read-only memory (ROM) array is included. The various types and families of semiconductor memories and their fabrication processes are given in Appendix A.

Chapter 4 is a continuation of Chapter 3. It introduces SSI, MSI, and LSI sequential integrated circuits and their applications to modern sequential logic design. These circuits include flip-flops, counters, and shift registers. The MOS/LSI programmable logic array (PLA) is also included.

The sequential machine, a general mathematical model of sequential circuits, is presented in Chapter 5. General analysis and synthesis procedures based on sequential machine theory for synchronous and asynchronous sequential circuits are presented. Undesirable transient phenomena such as races and hazards in asynchronous sequential circuits are discussed and methods for eliminating them are presented. A practical method for designing digital circuits and systems using the sequential machine flow chart is presented in Chapter 6. This method has found great acceptance among practicing engineers.

Because of the lack of internal test points and the increase in number of components per chip, the task of testing methods for IC digital devices and systems (from their external leads) has become an increasingly difficult problem in today's designing, manufacturing, and trouble-shooting of digital systems. Chapters 7 and 8 are devoted to the discussion of this problem. In Chapter 7, several methods for deriving economical fault-detection and fault-location test experiments for combinational circuits are presented. Moreover, both preset- and adaptive-scheduled test experiments for fault detection and fault location are discussed and compared. A discussion of how to apply the methods presented in these chapters to the testing of integrated circuits is included.

There are two distinctly different approaches to the problem of fault detection in sequential circuits which are discussed in Chapter 8. The first approach is the circuit test approach which assumes the experimenter has an exact knowledge of the circuit realization and each fault which can occur. The second approach is the transition checking approach which assumes no knowledge whatsoever of the circuit realization but assumes knowledge only of the desired transition table. This second approach may also be used for state and machine identification purposes. The fault-location of single and multiple faults in synchronous sequential circuits using the first approach is also included.

Chapter 9 introduces the logic design using microprocessors—the state-of-the-art. This chapter begins with the introduction of basic computer functions. The difference between the ordinary (large and mini-) computer and the microprocessor is then pointed out. A detailed design procedure including programming programmable read-only memory (PROM) is presented and illustrated by an example using the popularly-used 4-bit Intel 4040 microprocessor.

At the end of each chapter, a list of references and a bibliographical remarks are included. The references are by no means complete, nor are they intended to be, since compilation of a large number of nonselective references is, in the author's opinion, a waste of effort for both teacher and student. However, in some cases, certain references to the original work or to some excellent papers may have been omitted unintentionally.

This book is compiled from the notes I prepared for a course on digital circuits and logic design at the senior and first-year graduate level for both electrical engineering and computer science students in the past five years. I would like to thank my students who took the course and made valuable contributions in various ways to this book, in particular, T. Fry, H. Frost, Y. Keren-Zvi, and K. Pahlavan.

I would like to acknowledge the support of Prentice-Hall in publishing this book. I am particularly thankful to Drs. J. W. Gault and G. H. Foster for their careful review of the manuscript and many constructive suggestions; to Mr. K. Karlstrom, assistant vice-president of Prentice-Hall, who made many editorial comments and suggestions during the preparation of the manuscript; to Mr. L. Baudean, regional sales manager of Prentice-Hall, who voluntarily typed the first draft of Chapter 6; and to Dr. C. R. Haden, Director of the School of Electrical Engineering, the University of Oklahoma, whose encouragement and support have made the publishing of this book a pleasant endeavor. Finally, it is my privilege to thank Dr. M. E. Van

Valkenburg of the University of Illinois for his encouragement and interest without which this book could have never been written.

I am also deeply indebted to my wife Elisa and my children Vivian and Jennifer for their lost evenings and weekends.

S. C. LEE

1

Switching Algebra and Switching Function

In this chapter *two-element Boolean algebra* (B_2), or switching algebra, is described in detail. Since Boolean algebra is the mathematical foundation of switching algebra, a brief review of the essentials of Boolean algebra is presented first. Boolean functions defined on B_2 are called *switching functions*. Explicit logical equivalent formulas of binary variables and general theorems of switching functions are discussed. Two basic two-level AND–OR and OR–AND realizations are introduced. It is assumed that the input variables of the first-level AND gates and first-level OR gates in these realizations are available in both uncomplemented and complemented forms. To obtain minimal two-level AND–OR and OR–AND realizations, three switching-function-minimization methods, the Karnaugh method, the Quine–McCluskey method, and the iterative consensus method are presented. The first is a graphic method and is convenient for hand computation; the second and the third are tabular methods and are suitable for machine computation. A switching function is said to be minimized or in its minimal sum-of-products (product-of-sums) form if (1) its two-level AND–OR (OR–AND) realization of the function uses a minimum number of AND gates (OR gate), and (2) no AND gate (OR gate) in the realization can be replaced by an AND gate (OR gate) with fewer inputs. These two conditions, when stated in terms of the product (sum) terms of the function, are: (1) it has a minimum number of product (sum) terms, and (2) no product (sum) term of it can be replaced by a product (sum) term with fewer variables.

In the case of multiple-switching-function minimization (i.e., a set of switching functions being minimized simultaneously), our minimization criterion is as follows. In addition to the two conditions required by the above criterion for single-switching-function minimization, we add the following one: Make each minimized function have as many terms in common with those in the other minimized functions as possible. This requirement ensures that the basic two-level AND–OR or OR–AND realization of the multiple transmission functions achieves optimal usage of each of its components.

Finally, a basic procedure for designing combinational logic circuits is presented and illustrated by an example.

1.1 Boolean Algebra

Boolean algebra, first studied by George Boole,† constitutes an area of mathematics that has sprung into prominence with the advent of the digital computer. It is widely used in the design of digital circuits and computers; it is the mathematical foundation of switching theory and logic design.

DEFINITION 1.1.1

A *Boolean algebra* is an algebra $(B; \cdot, +, '; 0, 1)$ consisting of a set B (which contains at least the two elements 0 and 1) together with three operations, the AND (Boolean product) operation $\cdot$, the OR (Boolean sum) operation $+$, and the NOT (complement) operation $'$ defined on the set, such that for any element x, y, and z of B, $x \cdot y$ (the product of x and y), $x + y$ (the sum of x and y), and x' (the complement of x) are in B. For a Boolean algebra, the following axioms hold:

A1 Idempotent: $x \cdot x = x$ $x + x = x$
A2 Commutative: $x \cdot y = y \cdot x$ $x + y = y + x$
A3 Associative: $x \cdot (y \cdot z) = (x \cdot y) \cdot z$ $x + (y + z) = (x + y) + z$
A4 Absorptive: $x \cdot (x + y) = x$ $x + (x \cdot y) = x$
A5 Distributive: $x \cdot (y + z) = (x \cdot y) + (x \cdot z)$ $x + (y \cdot z) = (x + y) \cdot (x + z)$
A6 Zero (null, smallest) and one (universal, largest) elements:

There exists a *unique* element (the one element) $1 \in B$ such that for each $x \in B$,

$$x \cdot 1 = 1 \cdot x = x$$

There exists a *unique* element (the zero element) $0 \in B$ such that for each $x \in B$,

$$x + 0 = 0 + x = x$$

A7 Complement‡:   For each $x \in B$ there exists a *unique* element $x' \in B$, called the *complement* of x, such that

$$x \cdot x' = 0 \qquad\qquad x + x' = 1$$

Note that the operations $x \cdot y$ and $x + y$ are *not* ordinary algebraic operations, and the elements 0 and 1 do not mean the zero and one in ordinary algebra.

Two simple examples of Boolean algebra are given below.

Example 1.1.1

Consider the two-element Boolean algebra $B_2 = (\{0, 1\}; \cdot, +, '; 0, 1)$. The three operations $\cdot$, $+$, and $'$ are defined as follows:

---

†G. Boole, *An Investigation into the Laws of Thought*, Open Court, Chicago, 1854/1940.

‡In this book, a prime (′) and a bar (⁻) will be used interchangeably to denote the complementation of a Boolean variable or a Boolean expression.

·	0	1
0	0	0
1	0	1

+	0	1
0	0	1
1	1	1

′	
0	1
1	0

Next consider the power set $P(S)$ of a set S, which is a singleton. Clearly, $P(S)$ contains only two elements, $\varnothing$ and S. Three tables, similar to those shown above for the set-theoretic intersection, union, and complementation of $\varnothing$ and S, may be constructed:

∩	$\varnothing$	S
$\varnothing$	$\varnothing$	$\varnothing$
S	$\varnothing$	S

∪	$\varnothing$	S
$\varnothing$	$\varnothing$	S
S	S	S

~	
$\varnothing$	S
S	$\varnothing$

Together, these two sets of tables of operations show that the two Boolean algebras are isomorphic† to each other.

Example 1.1.2

Consider the four-element Boolean algebra $B_4 = (\{0, a, b, 1\}; \cdot, +, '; 0, 1)$ The three operations are described by the following tables:

·	0	a	b	1
0	0	0	0	0
a	0	a	0	a
b	0	0	b	b
1	0	a	b	1

+	0	a	b	1
0	0	a	b	1
a	a	a	1	1
b	b	1	b	1
1	1	1	1	1

′	
0	1
a	b
b	a
1	0

Now consider the power set $P(I)$, where $I = \{a, b\}$. The three proper subsets of I are $\{a\}, \{b\}$, and the empty set. Denote them by S_a, S_b, and $\varnothing$, respectively. $P(I)$ is a Boolean algebra with the three operations shown in the following tables:

∩	$\varnothing$	S_a	S_b	I
$\varnothing$	$\varnothing$	$\varnothing$	$\varnothing$	$\varnothing$
S_a	$\varnothing$	S_a	$\varnothing$	S_a
S_b	$\varnothing$	$\varnothing$	S_b	S_b
I	$\varnothing$	S_a	S_b	I

∪	$\varnothing$	S_a	S_b	I
$\varnothing$	$\varnothing$	S_a	S_b	I
S_a	S_a	S_a	I	I
S_b	S_b	I	S_b	I
I	I	I	I	I

~	
$\varnothing$	I
S_a	S_b
S_b	S_a
I	$\varnothing$

†Two algebras are isomorphic if there exists a one-to-one and onto mapping between them.

Again, it is seen that B_4 is isomorphic to $P(I)$ with $I = \{a, b\}$. In fact, it can be shown (Problem 1) that every Boolean algebra is isomorphic to a power set $P(A)$ with the three set operations $\cap$, $\cup$, and $\sim$ as the AND, OR, and NOT operations, respectively. Since the power set $P(A)$ with $A = \{a_1, a_2, \ldots, a_n\}$ has exactly 2^n elements (Problem 2), the number of elements of a Boolean algebra is a power of 2. In other words, for a given Boolean algebra B, there exists a positive integer n such that the number of elements of B is 2^n.

Just like in other algebras, the study of the properties of functions defined on them is always a subject of interest. There is no exception for Boolean algebra. A Boolean function is a mapping (or function) from a Boolean algebra into itself. The type of definition which we shall use is that known as a "recursive" definition. In this definition, two simple types of functions, the constant function and the project function, are defined, and rules are given for constructing all other Boolean functions.

DEFINITION 1.1.2

An element of a Boolean algebra B is called a *constant on B*.

DEFINITION 1.1.3

A symbol that may represent any one of the elements of B is called a (*Boolean*) *variable on B*.

A Boolean function of a Boolean algebra is defined as follows:

DEFINITION 1.1.4

Let $x_1, x_2, \ldots, x_n$ be variables on a Boolean algebra B. A mapping f of B into itself is a *Boolean function* of n variables, denoted by $f(x_1, \ldots, x_n)$ if it can be constructed according to the following rules.

1. Let a denote a constant on B. $f(x_1, \ldots, x_n) = a$ and $f(x_1, \ldots, x_n) = x_i$ are Boolean functions. The former is called the *constant function* and the latter is called the *projection function*.

2. If $f(x_1, \ldots, x_n)$ is a Boolean function, then $(f(x_1, \ldots, x_n))'$ is a Boolean function.

3. If $f_1(x_1, \ldots, x_n)$ and $f_2(x_1, \ldots, x_n)$ are Boolean functions, then $f_1(x_1, \ldots, x_n) + f_2(x_1, \ldots, x_n)$ and $f_1(x_1, \ldots, x_n) \cdot f_2(x_1, \ldots, x_n)$ are Boolean functions.

4. Any function that can be constructed by a finite number of applications of the above rules, and only such a function, is a Boolean function.

Thus a Boolean function is any function that can be constructed from the constant function and projection functions by finitely many uses of operations of $'$, $+$, and $\cdot$. For a function of one variable, the projection function is the *identity* function $f(x) = x$.

Example 1.1.3

The function

$$f(x_1, x_2, x_3) = ax_1 + (bx_2x_3')'(x_2 + x_3') \qquad (1.1.1)$$

is a Boolean function of three variables x_1, x_2, and x_3 defined on B_4.

Although the $x \cdot y$ and $x + y$ are not ordinary algebraic operations, we adopt the following convenient conventions used in ordinary algebra:

1. The dots between variables can be omitted; e.g., $x \cdot y \cdot z$ can be written as xyz.
2. The parentheses around Boolean product terms can be omitted; e.g., $(uvw) + (xyz)$ can be written as $uvw + xyz$.

Every Boolean function has the following two canonical forms:
Canonical sum-of-products form:

$$f(x_1, \ldots, x_n) = \sum f(e_1, \ldots, e_n)x_1^{e_1}x_2^{e_2} \ldots x_n^{e_n} \qquad (1.1.2)$$

where $x_i^{e_i} = 1$, if $x_i = e_i$
$\qquad = 0$, otherwise
Canonical product-of-sums form:

$$f(x_1, \ldots, x_n) = \sum (f(e_1, \ldots, e_n) + x_1^{e_1'} + x_2^{e_2'} + x_n^{e_n'}) \qquad (1.1.3)$$

where $x_i^{e_i'} = 0$, if $x_i = e_i$
$\qquad = 1$, otherwise
The proofs of these two forms are left to the reader as an exercise (Problem 3).

Example 1.1.4

To obtain the canonical sum-of-products form of the function $f(x_1, x_2)$ defined on B_4 whose truth table is described below.

x_1	x_2	$f(x_1, x_2)$
0	0	1
0	a	a
a	b	b
a	1	1
b	0	1
b	a	a
1	b	1
1	1	1
otherwise		0

The canonical sum-of-products form of this function is

$$f(x_1, x_2) = 1 \cdot x_1^0 x_2^0 + a \cdot x_1^0 x_2^a + b \cdot x_1^a x_2^b + 1 \cdot x_1^a x_2^1$$
$$+ 1 \cdot x_1^b x_2^0 + a \cdot x_1^b x_2^a + 1 \cdot x_1^1 x_2^b + 1 \cdot x_1^1 x_2^1 \qquad (1.1.4)$$

Exercise 1.1

1. Prove that every Boolean algebra is isomorphic to a power set $P(A)$ with the three set operations $\cap$, $\cup$, and $\sim$, as the AND, OR, and NOT operations, respectively.

2. Prove that the power set $P(A)$ with $A = \{a_1, a_2, \ldots, a_n\}$ has exactly 2^n elements.

3. Prove the two canonical forms of a Boolean function.

1.2 Switching Algebra

Among all the Boolean algebras, the two-element Boolean algebra B_2 (see Example 1.1.1), known as *switching algebra*, is the most useful. It is the mathematical foundation of the analysis and design of switching circuits that make up digital systems. Switching algebra contains the two extreme elements, the largest number represented by 1 and the smallest number represented by 0. Boolean functions defined on the switching algebra are called *switching functions*. Many properties of switching functions will be studied. Explicit logical equivalent formulas of switching functions will be presented that can be used for direct-switching-function minimization.

The basic properties of switching algebra can be classed in three groups, by the number of variables involved, as shown in Table 1.2.1.

TABLE 1.2.1   Logical Equivalent Formulas

$n = 1$†

1.1. $x_1 + x_1 = x_1$	1.1'. $x_1 \cdot x_1 = x_1$
1.2. $x_1 + x_1' = 1$	1.2'. $x_1 \cdot x_1' = 0$
1.3. $1 + x_1 = x_1 + 1 = 1$	1.3'. $1 \cdot x_1 = x_1$
1.4. $0 + x_1 = x_1 + 0 = x_1$	1.4'. $0 \cdot x_1 = 0$

$n = 2$

2.1. $x_1 + x_2 = x_2 + x_1$	2.1'. $x_1 \cdot x_2 = x_2 \cdot x_1$
2.2. $x_1 + x_1 \cdot x_2 = x_1$	2.2'. $x_1 \cdot (x_1 + x_2) = x_1$
2.3. $(x_1 + x_2') \cdot x_2 = x_1 \cdot x_2$	2.3'. $x_1 \cdot x_2' + x_2 = x_1 + x_2$

$n = 3$

3.1. $x_1 + x_2 + x_3 = (x_1 + x_2) + x_3$ $= x_1 + (x_2 + x_3)$	3.1'. $x_1 \cdot x_2 \cdot x_3 = (x_1 \cdot x_2) \cdot x_3$ $= x_1 \cdot (x_2 \cdot x_3)$
3.2. $x_1 \cdot x_2 + x_1 \cdot x_3 = x_1 \cdot (x_2 + x_3)$	3.2'. $(x_1 + x_2) \cdot (x_1 + x_3) = x_1 + x_2 \cdot x_3$
3.3. $(x_1 + x_2) \cdot (x_2 + x_3)(x_3 + x_1')$ $= (x_1 + x_2) \cdot (x_3 + x_1')$	3.3'. $x_1 \cdot x_2 + x_2 \cdot x_3 + x_3 x_1' = x_1 \cdot x_2 + x_3 \cdot x_1'$
4.4. $(x_1 + x_2) \cdot (x_1' + x_3) = x_1 \cdot x_3 + x_1' x_2$	

†n, number of binary variables.

Next, we shall present several general theorems of switching functions. A very important and interesting one, which has many useful applications in regard to switching functions, is De Morgan's theorem.

THEOREM 1.2.1 (Generalized) De Morgan's Theorem

(a) $(x_1 + x_2 + \ldots + x_n)' = x_1' \cdot x_2' \cdot \ldots \cdot x_n'$
(b) $(x_1 \cdot x_2 \cdot \ldots \cdot x_n)' = x_1' + x_2' + \ldots + x_n'$

The proof of this theorem is left to the reader (Problem 1).

De Morgan's theorem does not convey completely the relation between complementary functions. Shannon has suggested a generalization of the theorem, which is described next.

THEOREM 1.2.2 Shannon's Theorem

$$(f(x_1, x_2, \ldots, x_n, +, \cdot))' = f(x_1', x_2', \ldots, x_n', \cdot, +)$$

In words, this theorem says that the complement of any function is obtained by replacing each variable by its complement and, at the same time, by interchanging the AND and OR operations.

Proof: Since any $(f(x_1, \ldots, x_n, +, \cdot))'$ may be expressed by

$$(f(x_1, \ldots, x_n, +, \cdot))' = (f_1(x_1, \ldots, x_n, +, \cdot) + f_2(x_1, \ldots, x_n, +, \cdot))'$$
$$= (f_1(x_1, \ldots, x_n, +, \cdot))' \cdot ((f_2(x_1, \ldots, x_n, +, \cdot))'$$

and/or

$$(f(x_1, \ldots, x_n, +, \cdot))' = (f_1(x_1, \ldots, x_n, +, \cdot) \cdot f_2(x_1, \ldots, x_n, +, \cdot))'$$
$$= (f_1(x_1, \ldots, x_n, +, \cdot))' + (f_2(x_1, \ldots, x_n, +, \cdot))'$$

where f_1 and f_2 represent two partial functions of f. Repeat the same argument for f_1 and f_2, and so forth, until De Morgan's theorem is applied to every variable of each term of f. Since at each time De Morgan's theorem is applied to f (its partial function), the complement of f (its partial function) is obtained by replacing each partial function (subpartial function) by its complement and, at the same time, by interchanging the AND and OR operations after De Morgan's theorem is applied to every variable of each term of f, the final result should be $f(x_1', x_2', \ldots, x_n', \cdot, +)$. ∎[†]

In applying Shannon's theorem to a switching function, it is advised to keep all the dots between variables and all the parentheses around product terms of the function. The following example illustrates the application of this theorem.

Example 1.2.1

If $f(x_1, x_2, x_3) = (x_1 \cdot x_2)' \cdot x_2 + x_1 \cdot (x_2 + x_3')$, then the complement of this function is

$$f'(x_1, x_2, x_3) = ((x_1 \cdot x_2) + x_3') \cdot (x_1' + (x_2' \cdot x_3))$$

Another interesting theorem of switching algebra is the expansion theorem.

[†]The symbol ∎ denotes the end of a proof.

THEOREM 1.2.3 Expansion Theorem

(a) $f(x_1, x_2, \ldots, x_n) = x_1 \cdot f(1, x_2, \ldots, x_n) + x_1' \cdot f(0, x_2, \ldots, x_n)$

(b) $f(x_1, x_2, \ldots, x_n) = [x_1 + f(0, x_2, \ldots, x_n)][x_1' + f(1, x_2, \ldots, x_n)]$

The function in this form is said to be *expanded with respect to x_1*. Similarly, expressions can be written to represent expansions with respect to any of the variables.

Proof: If we first substitute $x_1 = 1$ and $x_1' = 0$ and then substitute $x_1 = 0$ and $x_1' = 1$ in each equation, they reduce to identities. ∎

Any switching function in n variables may be developed into a series by means of the expansion theorem.

Example 1.2.2

The function $f(x_1, x_2, x_3) = (x_1 x_2)' x_3 + x_1(x_2 + x_3')$ can be expressed by a series as

$$f(x_1, x_2, x_3) = x_1 \cdot (x_2' x_3 + x_2 + x_3') + x_1' \cdot (x_3)$$

or

$$f(x_1, x_2, x_3) = [x_1 + x_3][x_1' + (x_2' x_3 + x_2 + x_3')]$$

From the expansion theorem it follows that

THEOREM 1.2.4

(a) (1) $x_1 \cdot f(x_1, x_2, \ldots, x_n) = x_1 \cdot f(1, x_2, \ldots, x_n)$

(2) $x_1 + f(x_1, x_2, \ldots, x_n) = x_1 + f(0, x_2, \ldots, x_n)$

(b) (1) $x_1' \cdot f(x_1, x_2, \ldots, x_n) = x_1' \cdot f(0, x_2, \ldots, x_n)$

(2) $x_1' + f(x_1, x_2, \ldots, x_n) = x_1' + f(1, x_2, \ldots, x_n)$

Using the formulas in Table 1.2.1 and the above general theorems (Theorems 1.2.1–1.2.4), we can easily simplify all the possible 2^{2^n} switching functions of n variables in canonical sum-of-products form. For $n = 2$, this is demonstrated in Table 1.2.2. For n greater than 2, such tables increase rapidly in size, with 256 ($= 2^{2^3}$) rows for three variables and 65,536 ($= 2^{2^4}$) rows for four variables.

In Table 1.2.2 the first column indicates that there are 16 possible switching functions of two variables in canonical sum-of-products form. The second column shows the number of terms in each form. In the third column, under each of the canonical sum terms, a 0 or a 1 is entered. If a 0 is entered in a column, it means that the term at the head of that column is not included in the canonical sum-of-products form. If a 1 is entered in a column, it means that the term at the head of that column is included in the canonical sum-of-products form. In the fourth column, 16 Boolean functions in canonical sum-of-products form are listed. Each function in the third column may be minimized to its simplest equivalent form, which are shown in the last column of Table 1.2.2. These 16 functions are of course distinct, isomorphic to the 16 canonical sum-of-products forms, and form a lattice, as shown in Fig. 1.2.1.

Functions with three or less variables may be minimized by applying the formulas in Table 1.2.1 and Theorems 1.2.1–1.2.4. For example,

**TABLE 1.2.2 Simplified Forms of All Possible Switching
Functions of Two Variables in Canonical Form**

Number of Possible Functions	Number of Terms in Canonical Sum-of-Products Form	$x_1'x_2'$	$x_1'x_2$	x_1x_2'	x_1x_2	Canonical Sum-of-Products Form	Minimized Function
1	0	0	0	0	0	0	0
2	1	0	0	0	1	x_1x_2	x_1x_2
3	1	0	0	1	0	x_1x_2'	x_1x_2'
4	1	0	1	0	0	$x_1'x_2$	$x_1'x_2$
5	1	1	0	0	0	$x_1'x_2'$	$x_1'x_2'$
6	2	0	0	1	1	$x_1x_2' + x_1x_2$	x_1
7	2	1	1	0	0	$x_1'x_2' + x_1'x_2$	x_1'
8	2	0	1	0	1	$x_1'x_2 + x_1x_2$	x_2
9	2	1	0	1	0	$x_1'x_2' + x_1x_2'$	x_2'
10	2	1	0	0	1	$x_1'x_2' + x_1x_2$	$x_1x_2 + x_1'x_2'$
11	2	0	1	1	0	$x_1'x_2 + x_1x_2'$	$x_1x_2' + x_1'x_2$
12	3	0	1	1	1	$x_1'x_2 + x_1x_2' + x_1x_2$	$x_1 + x_2$
13	3	1	0	1	1	$x_1'x_2' + x_1x_2' + x_1x_2$	$x_1 + x_2'$
14	3	1	1	0	1	$x_1'x_2' + x_1'x_2 + x_1x_2$	$x_1' + x_2$
15	3	1	1	1	0	$x_1'x_2' + x_1'x_2 + x_1x_2'$	$x_1' + x_2'$
16	4	1	1	1	1	$x_1'x_2' + x_1'x_2 + x_1x_2' + x_1x_2$	1

$$
\begin{aligned}
f(x_1, x_2, x_3) &= x_1'x_2x_3' + x_1'x_2x_3 + x_1x_2'x_3' + x_1x_2x_3 \\
&= x_1'x_2(x_3' + x_3) + x_1x_2'x_3' \qquad \text{by formula 1.1 of Table 1.2.1} \\
&\quad + (x_1' + x_1)x_2x_3 \\
&= x_1'x_2 + x_1x_2'x_3' + x_2x_3 \qquad \text{by formula 1.2 of Table 1.2.1}
\end{aligned}
$$

$$(1.2.1)$$

But, for minimizing functions with more than three variables, this method becomes inconvenient. Three general systematic methods of minimizing switching functions of n variables are presented later in the chapter.

Exercise 1.2

1. Prove Theorem 1.2.1.

Simplify the following switching functions by using the formulas given in Table 1.2.1 and Theorems 1.2.1–1.2.4.

2. $f_1(x_1, x_2) = x_1x_2' + x_1'x_2$

3. $f_2(x_1, x_2) = x_1 + x_1'x_2$

4. $f_3(x_1, x_2, x_3) = x_1x_2x_3 + x_1'x_2'x_3 + x_1'x_2x_3' + x_1x_2'x_3' + x_1'x_2'x_3'$

5. $f_4(x_1, x_2, x_3) = x_1x_2 + x_2'x_3' + x_1x_2x_3 + x_1x_2'x_3'$

6. $f_5(x_1, x_2, x_3) = x_1x_2 + x_1'x_3 + x_2x_3$

7. $f_6(x_1, x_2, x_3) = x_1x_3 + x_1x_2x_3 + x_1'x_3 + x_1'x_2'x_3' + x_2x_3$

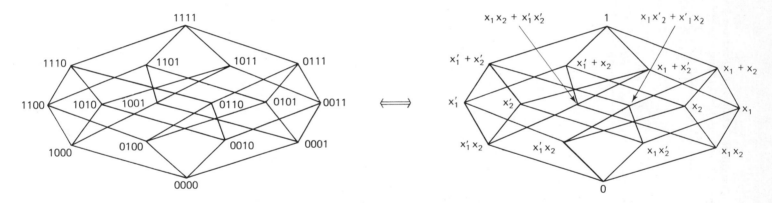

Fig. 1.2.1 Lattice representation of the sixteen simplified functions.

1.3 Completely and Incompletely Specified Switching Functions

Since the value of $f(e_1, e_2, \ldots, e_n)$ of a switching function can *only* be either 0 or 1, Eqs. (1.1.2) and (1.1.3) can be expressed as:

Canonical sum-of-products form of a switching function:

$$f(x_1, \ldots, x_n) = \sum_{\substack{\text{all the combinations} \\ \text{of the values of } x_1, \ldots, x_n, \\ \text{for which } f(e_1, \ldots, e_n) = 1}} x_1^{e_1} x_2^{e_2} \ldots x_n^{e_n} \qquad (1.3.1)$$

Canonical product-of-sums form of a switching function:

$$f(x_1, \ldots, x_n) = \prod_{\substack{\text{all the combinations} \\ \text{of the values of } x_1, \ldots, x_n, \\ \text{for which } f(e_1, \ldots, e_n) = 0}} (x_1^{e_1'} + e_2^{e_2'} + \ldots + x_n^{e_n'}) \qquad (1.3.2)$$

Here, $x_i^{e_i}(x_i^{e_i'})$ is interpreted as $x_i(x_i')$ or $x_i'(x_i)$ accordingly as e_i has the value 1 or 0. For example, for the switching function of Eq. (1.2.1), $f(x_1, x_2, x_3) = x_1'x_2 + x_1x_2'x_3' + x_2x_3$, the canonical forms are

$$f(x_1, x_2, x_3) = x_1'x_2x_3' + x_1'x_2x_3 + x_1x_2'x_3' + x_1x_2x_3 \qquad (1.3.3)$$

and

$$f(x_1, x_2, x_3) = (x_1 + x_2 + x_3)(x_1 + x_2 + x_3')(x_1' + x_2 + x_3')(x_1' + x_2' + x_3) \qquad (1.3.4)$$

Each term of the canonical sum-of-products of Eq. (1.3.1) is called a *minterm*, and each term of the canonical product-of-sums of Eq. (1.3.2) is called a *maxterm*. It is interesting to note that from the definitions of canonical forms of a switching function [Eqs. (1.3.1) and (1.3.2)], the minterms of the canonical sum-of-products form of a switching function can be directly obtained from the rows of the truth table of the function that are mapped into 1, and the maxterms of the canonical product-of-sums form of a switching function can be obtained from the rows that are mapped into 0. A simple example is given in Table 1.3.1. The reason that a row, say 010, corresponds

TABLE 1.3.1

Row Number r	Minterms and Maxterms	x_1 x_2 x_3	$f(x_1, x_2, x_3)$ of Eq. (1.2.1)
0	$M_0 = x_1 + x_2 + x_3$	0 0 0	0
1	$M_1 = x_1 + x_2 + x_3'$	0 0 1	0
2	$m_2 = x_1'x_2x_3'$	0 1 0	1
3	$m_3 = x_1'x_2x_3$	0 1 1	1
4	$m_4 = x_1x_2'x_3'$	1 0 0	1
5	$M_5 = x_1' + x_2 + x_3'$	1 0 1	0
6	$M_6 = x_1' + x_2' + x_3$	1 1 0	0
7	$m_7 = x_1x_2x_3$	1 1 1	1

and only corresponds to the minterm $x_1'x_2x_3'$ is because $x_1'x_2x_3' = 1$ if and only if $x_1 = 0$, $x_2 = 1$, and $x_3 = 0$. The reasons for the other three rows, 011, 100, and 111, describing the minterms $x_1'x_2x_3$, $x_1x_2'x_3'$, and $x_1x_2x_3$ are similar. Thus we find that

RULE 1.3.1

All the minterms of the canonical sum-of-products form of a switching function can be obtained from the rows of the truth table that are mapped into 1. Each 0 value of a variable indicates the complemented form of the variable, and each 1 value of a variable indicates the uncomplemented form of the variable.

Dually, the reason that row 000 corresponds and only corresponds to the maxterm $x_1 + x_2 + x_3$ is because $x_1 + x_2 + x_3 = 0$ if and only if $x_1 = 0$, $x_2 = 0$, and $x_3 = 0$. The reasons for the other three rows, 001, 101, and 110, describing the maxterms $x_1 + x_2 + x_3'$, $x_1' + x_2 + x_3'$, and $x_1' + x_2' + x_3$ are similar. Therefore, we have:

RULE 1.3.2

All the maxterms of the canonical product-of-sums form of a switching function can be obtained from the rows of the truth table that are mapped into 0. Each 0 value of a variable indicates the uncomplemented form of the variable, and each 1 value of a variable indicates the complemented form of the variable.

Because of the one-to-one correspondence between the minterms and maxterms and the rows of the values of the variables of a truth table, we can have a simple representation of minterms and maxterms. It is customary to use m and M to denote minterms and maxterms, respectively. Instead of indicating rows of the truth table by the binary numbers (000, 001, 010, etc.), we use their corresponding decimal numbers (0, 1, 2, etc.). Since the binary-to-decimal and decimal-to-binary conversions are unique in the sense that, given any binary number, there is one and only one decimal-number representation of it, and given any decimal number, its binary-number representation is unique; therefore, every minterm and maxterm can be represented by an m and M letter with a subscript r that is called a *row number*, the decimal-number representation of the binary numbers denoting the row to which it corresponds (see columns 1 and 2 of Table 1.3.1). The above two canonical forms can then be represented by

$$f(x_1, x_2, x_3) = m_2 + m_3 + m_4 + m_7 \qquad (1.3.5)$$

and

$$f(x_1, x_2, x_3) = M_0 \cdot M_1 \cdot M_5 \cdot M_6 \qquad (1.3.6)$$

which can be further simplified by

$$f(x_1, x_2, x_3) = \sum (2, 3, 4, 7)$$

and

$$f(x_1, x_2, x_3) = \prod (0, 1, 5, 6)$$

respectively. Note that the expression $\sum (2, 3, 4, 7)$ indicates that the rows whose row numbers are not in this summation expression, rows 0, 1, 5, and 6, are mapped into 0, and similarly, the expression $\prod (0, 1, 5, 6)$ indicates that the rows whose row numbers are not in this product expression, rows 2, 3, 4 and 7, are mapped into 1. Also, notice that a row, if it describes a minterm (maxterm), cannot also describe a maxterm (minterm). In other words, the set of rows describing the minterms of a completely specified switching function are mutually exclusive, with the set of rows describing the maxterms. These two sets are complementary. Therefore, if the canonical sum-of-products (product-of-sums) form of a switching function is known, the canonical product-of-sums (sum-of-products) form of the function can be obtained from all the rows that are not in the set of rows that correspond to the minterms (maxterms) and application of rule 1.3.2 (rule 1.3.1).

There are six basic electronic gates: AND, OR, NOT, NAND (*NOT AND*), NOR (*NOT OR*), and XOR (EXCLUSIVE-OR), whose symbolic representations are given in Table 1.3.2. These symbols are based on the military services standard sym-

TABLE 1.3.2 Military Service Standard Symbology
MIL-STD-806B.

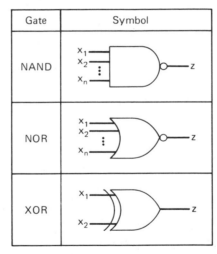

bology MIL-STD-806B. The AND, OR, and NOT gates are the realizations of the and, or, and not operations, respectively. The NAND and NOR gates are the realizations of the complements of the and and or operations, respectively. The XOR gate with two input variables x and y is the realization of the operation $xy' + x'y$. The NAND, NOR, and XOR gates will be discussed in Chapter 3. There are two basic realizations of a switching function using the AND, OR, and NOT gates as building blocks: two-level AND–OR realization and two-level OR–AND realization, which realize a sum-of-products form and product-of-sums form of the function, respectively. In these two realizations, it is assumed that the input variables of the first-level AND gates and first-level OR gates are available in both uncomplemented and comple-

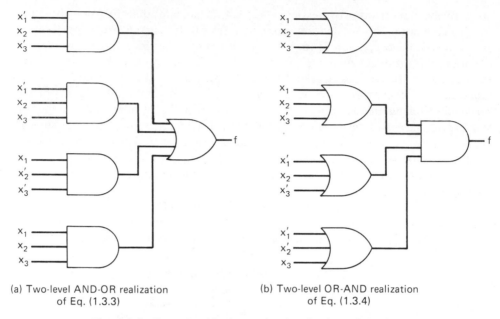

(a) Two-level AND-OR realization   (b) Two-level OR-AND realization
of Eq. (1.3.3)   of Eq. (1.3.4)

Fig. 1.3.1   Example of basic two-level realizations of the functions of Eqs. (1.3.3) and (1.3.4).

mented forms. For example, the basic two-level realizations of the functions of Eqs. (1.3.3) and (1.3.4) are shown in Fig. 1.3.1. Two-level circuits possess the following two general characteristics that are attractive to a logic designer:

1.   Because there are only two layers of gates, the circuit-response time is approximately equal to two times the gate-response time, which is shorter than that of those circuits having more layers of gates.

2.   It will be shown in later chapters that two-level circuits are not only easy to build but also make it easy to detect and locate faults whenever they occur.

The function of Eq. (1.3.3) can be minimized to a function $f(x_1, x_2, x_3) = x_1'x_2 + x_1x_2'x_3' + x_2x_3$ [see Eq. (1.2.1)] whose realization is shown in Fig. 1.3.2(a), and the function of Eq. (1.3.4) can also be minimized.

$$f(x_1, x_2, x_3) = (x_1 + x_2 + x_3)(x_1 + x_2 + x_3')(x_1' + x_2 + x_3')(x_1' + x_2' + x_3)$$

$$= (x_1 + x_2 + x_3x_3')(x_1x_1' + x_2 + x_3')(x_1' + x_2' + x_3)$$

by formulas 1.1′ and 3.2′ of Table 1.2.1

$$= (x_1 + x_2)(x_2 + x_3')(x_1' + x_2' + x_3)$$ by formula 1.2′ of Table 1.2.1

which can be realized by three OR gates and one AND gate, as shown in Fig. 1.3.2(b). Comparing the circuits of Figs. 1.3.1 and 1.3.2, the ones in Fig. 1.3.2 have fewer gates.

So far the switching functions that have been discussed are completely specified. However, incompletely specified functions are quite common in digital system design.

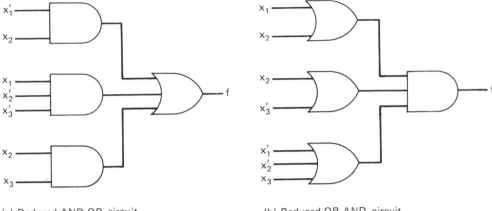

(a) Reduced AND-OR circuit   (b) Reduced OR-AND circuit

Fig. 1.3.2 Reduced two-level AND-OR and OR-AND circuits those of Fig. 1.3.1.

For example, assume that the input to the switching circuit represents a decimal number encoded in binary-coded-decimal (BCD) form. Then only 10 of the four-tuples (x_1, x_2, x_3, x_4) would be specified and the other 6 would correspond to don't-care conditions, denoted by d, which means that we don't care whether these 6 four-tuples are mapped to 0 or 1. Suppose that we want to design this circuit to do the following simple task: to indicate whether the input is an even or odd number by giving, say, a 0 output if it is even, and a 1 output if it is odd. If the 10 decimal numbers $0, 1, \ldots,$ and 9 are coded by their BCD forms, the truth table describing this switching function is shown in Table 1.3.3. Since this switching function has six rows

TABLE 1.3.3

Decimal Number	BCD Form				
r	x_1	x_2	x_3	x_4	$f(x_1, x_2, x_3, x_4)$
0	0	0	0	0	0
1	0	0	0	1	1
2	0	0	1	0	0
3	0	0	1	1	1
4	0	1	0	0	0
5	0	1	0	1	1
6	0	1	1	0	0
7	0	1	1	1	1
8	1	0	0	0	0
9	1	0	0	1	1
10	1	0	1	0	d
11	1	0	1	1	d
12	1	1	0	0	d
13	1	1	0	1	d
14	1	1	1	0	d
15	1	1	1	1	d

whose mappings are unspecified, it is an incompletely specified switching function. The two canonical forms of this function can be represented by

$$f(x_1, x_2, x_3, x_4) = \sum (1, 3, 5, 7, 9) + \sum_d (10, 11, 12, 13, 14, 15) \qquad (1.3.7)$$

and

$$f(x_1, x_2, x_3, x_4) = \prod (0, 2, 4, 6, 8) \cdot \prod_d (10, 11, 12, 13, 14, 15) \qquad (1.3.8)$$

Note that a don't-care minterm of, say, $13_{10} = 1101_2$ in the canonical sum-of-products form is $d \cdot x_1 x_2 x_3' x_4$, and the don't-care maxterm of 13 in the canonical product-of-sums form is $d + x_1' + x_2' + x_3 + x_4'$, where d can be either 0 or 1. It is important to note that by letting all the d's in a sum-of-products function equal to 0 and all the d's in a product-of-sums function equal to 1 does not necessarily result in the simplest functions. As a matter of fact, they are generally not. For example, the functions of Eqs. (1.3.7) and (1.3.8), after letting all the d's be 0 and 1, respectively, are

$$f(x_1, x_2, x_3, x_4) = \sum (1, 3, 5, 7, 9)$$

$$= x_1' x_2' x_3' x_4 + x_1' x_2' x_3 x_4 + x_1' x_2 x_3' x_4 + x_1' x_2 x_3 x_4 + x_1 x_2' x_3' x_4 \qquad (1.3.9)$$

$$= \quad x_1' x_2' x_4 \quad + \quad x_1' x_2 x_4 \quad + \quad x_2' x_3' x_4$$

$$= \quad x_1' x_4 \quad + \quad x_2' x_3' x_4$$

and

$$(f x_1, x_2, x_3, x_4) = \prod (0, 2, 4, 6, 8)$$

$$= (x_1 + x_2 + x_3 + x_4)(x_1 + x_2 + x_3' + x_4)$$

$$\cdot (x_1 + x_2' + x_3 + x_4)(x_1 + x_2' + x_3' + x_4) \qquad (1.3.10)$$

$$\cdot (x_1' + x_2 + x_3 + x_4)$$

$$= (x_1 + x_2 + x_4) \ (x_1 + x_2' + x_4) \ (x_2 + x_3 + x_4)$$

$$= (x_1 + x_4) \ (x_2 + x_3 + x_4)$$

But if we let the don't cares associated with the terms 11,13, and 15 in Eq. (1.3.7) equal to 1 and the rest equal to 0 and we let the don't cares associated with the terms 10, 12, and 14 in Eq. (1.3.8) equal to 0 and the rest equal to 1, we can easily show that

$$f(x_1, x_2, x_3, x_4) = \sum (1, 3, 5, 7, 9) + \sum_{d=1} (11, 13, 15) + \sum_{d=0} (10, 12, 14)$$

$$= \sum (1, 3, 5, 7, 9, 11, 13, 15) = x_4 \qquad (1.3.11)$$

and

$$f(x_1, x_2, x_3, x_4) = \prod_{d=0} (0, 2, 4, 6, 8) \cdot \prod_{d=1} (10, 12, 14) \cdot \prod (11, 13, 15)$$
$$= \prod (0, 2, 4, 6, 8, 10, 12, 14) = x_4$$

(1.3.12)

which are obviously simpler than those of Eqs. (1.3.9) and (1.3.10). From this example it is seen that by judicious assignments of the values 0 and 1 to the don't cares, functions obtained may be simpler than those obtained by letting all *d*'s in the sum-of-products form equal to 0 and all *d*'s in the product-of-sums form equal to 1. The problem of how to assign values of the don't cares to obtain the simplest functions is discussed in the next section.

Exercise 1.3

1. Consider the following functions:

$$f_1(x_1, x_2, x_3) = \Sigma (0, 1, 3, 4, 5, 7)$$
$$f_2(x_1, x_2, x_3) = \prod (0, 2, 3, 4, 6, 7)$$

   (a) Find a minimal AND–OR realization of f_1 and a minimal OR–AND realization of f_2.
   (b) Find a minimal OR–AND realization of f_1 and a minimal AND–OR realization of f_2.

2. Repeat (a) and (b) of problem 1 for the following two functions:

$$f_3(x_1, x_2, x_3) = \Sigma (0, 2, 3) + \sum_d (5, 6)$$
$$f_4(x_1, x_2, x_3) = \prod (4, 5, 6) \cdot \prod_d (0, 3)$$

1.4   Karnaugh Method

In previous sections we introduced the problem of minimization of completely and incompletely specified switching functions and have illustrated the formula-look-up and theorem-application method by several examples. Now we want to formally define the problem and present its solution. The aim of function minimization is to find a minimal form of a given function, which is defined next.

DEFINITION 1.4.1

A minimal sum-of-products (product-of-sums) form of a switching function is a switching function in sum-of-products (product-of-sums) form logically equivalent to the given function meeting the following requirements:

1. The two-level AND–OR (OR–AND) realization of the minimal sum-of-products (product-of-sums) form has a minimum number of AND gates (OR gates).

2.   No AND gate (OR gate) can be replaced by an AND gate (OR gate) with fewer inputs.

In terms of the product (sum) terms of the function, the above two conditions are equivalent, respectively, to

1.   The minimal form has a minimum number of product (sum) terms.
2.   No product (sum) term of it can be replaced by a product (sum) term with fewer variables.

The problem of switching-function minimization may be stated as: Given a switching function, find a minimal form. Three minimization methods, the Karnaugh method, the Quine–McCluskey method, and the iterative consensus method, will be presented. The Karnaugh method is based on a convenient method of representing switching functions graphically. The graph in question is called the "Karnaugh map." The representations of switching functions in canonical sum-of-products and product-of-sums forms by the Karnaugh map and their minimization using the Karnaugh map are quite similar. In the following, we shall first present the construction of the Karnaugh map, the properties of the Karnaugh map, and application of the Karnaugh map to the minimization of switching functions, where the functions are assumed to be in canonical sum-of-products form. We then show the representation, the properties, and the minimization procedure for switching functions when expressed in canonical product-of-sums form.

1.4.1   Construction of the Karnaugh Map

The *Karnaugh map* is a modification of the Venn diagram. Venn diagrams of switching functions with four variables or less in canonical sum-of-products form, for example, are shown in the first column of Fig. 1.4.1. Canonical product-of-sums forms of these switching functions can be represented by the same diagrams except that (1) the · operations between the variables are replaced by the + operations, and (2) the unprimed variables are replaced by prime variables and the prime variables are replaced by unprimed variables. The four Venn diagrams of the first column may be redrawn using a rectangle instead of a circle (or ellipse) to represent a variable that is shown in the second column. A simplification of the diagram in the second column leads to the Karnaugh map. If primed and unprimed variables are represented by 0's and 1's, respectively, in the column or row that it represents, each square of a Karnaugh map may be denoted by a binary number or its decimal-number equivalent, as shown in the last two columns. Suppose that we want to represent the switching function

$$f(x_1, x_2, x_3, x_4) = x_1 x_2 x_3 x_4 + x_1' x_2 x_3 x_4 + x_1' x_2' x_3 x_4 + x_1' x_2' x_3' x_4$$
$$+ x_1' x_2' x_3' x_4' \qquad\qquad (1.4.1a)$$

which may be represented by

$$f(x_1, x_2, x_3, x_4) = \sum (0, 1, 3, 7, 15) \qquad\qquad (1.4.1b)$$

The Venn diagram and the convenient Karnaugh map representations of this function are shown in Fig. 1.4.2. The Venn diagram representation needs no explanation. In the Karnaugh map, we once again use 1 and 0 to indicate whether the term that the square represents is included in the canonical sum-of-products. It is clear that we only need to indicate either all the 1's or all the 0's in a Karnaugh map. Conventionally, we omit 0's; that is, we only put 1's in the squares that correspond to the product terms in the canonical-sum form of the switching function, as shown in Fig. 1.4.2.

1.4.2 Properties of the Karnaugh Map

After having constructed the Karnaugh map, which is a convenient way of representing a switching function with no more than four variables, we now investigate some of its properties, especially those which have direct applications to the minimization of switching functions. For convenience, we shall call each square in the Karnaugh map a *cell*. It is important to note that the following properties depend only on the *relative* positions of the cells marked by 1; they *do not* depend on where they are located at the Karnaugh map.

One obvious advantage of using the Karnaugh map is to display all the adjacencies that are present in a canonical sum-of-products form. When this is done, it no longer becomes necessary to compare all possible pairs of terms in the canonical sum-of-products form in order to eliminate redundant variables. The Karnaugh map described above shows all the adjacencies that exist; therefore, we need only to focus our attention on those groupings of terms that lead immediately to simplification. This leads to the following most basic property of the Karnaugh map.

PROPERTY 1

Any adjacent pair of cells (2^1) marked by 1 in a Karnaugh map can be combined into one term, and one variable is eliminated.

For example, the Karnaugh map of the function of Eq. (1.4.1) was shown in Fig. 1.4.2. The minimization of this function may be worked out directly from the map. The two adjacent cell pairs indicated with 1's may be combined into two terms, with one variable eliminated:

$$x_1' x_2' x_3 x_4 + x_1' x_2 x_3 x_4 = x_1' x_3 x_4 (x_2 + x_2') = x_1' x_3 x_4$$

$$x_1' x_2' x_3' x_4 + x_1' x_2' x_3' x_4' = x_1' x_2' x_3' (x_4 + x_4') = x_1' x_2' x_3'$$

Moreover, since the reason for two adjacent terms can be merged into one is merely because they are different in the value of one variable, they are not restricted to the interior of the map. Since every Boolean expression plus itself is itself, we can use terms to combine with other terms as many times as we want when it is needed. For example, we can use the term $x_1' x_2 x_3 x_4$ once again to combine with $x_1 x_2 x_3 x_4$, giving

$$x_1 x_2 x_3 x_4 + x_1' x_2 x_3 x_4 = x_2 x_3 x_4 (x_1 + x_1')$$

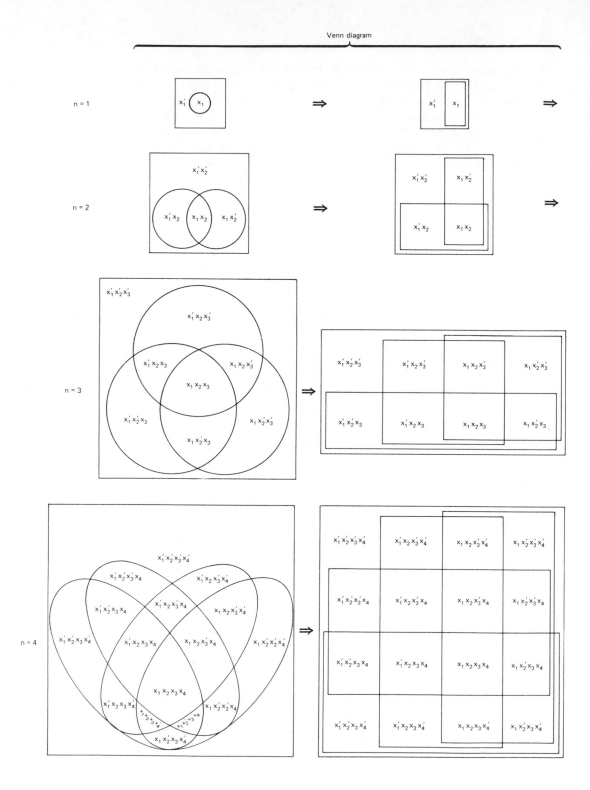

Fig. 1.4.1 The Venn diagram and Karnaugh map representation of switching functions with 1-4 variables.

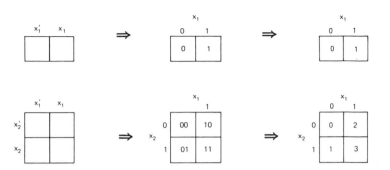

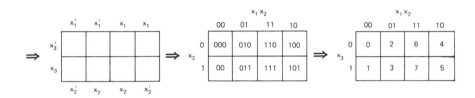

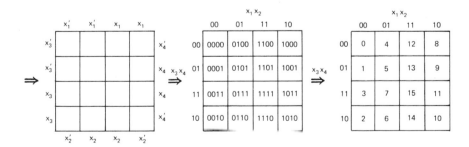

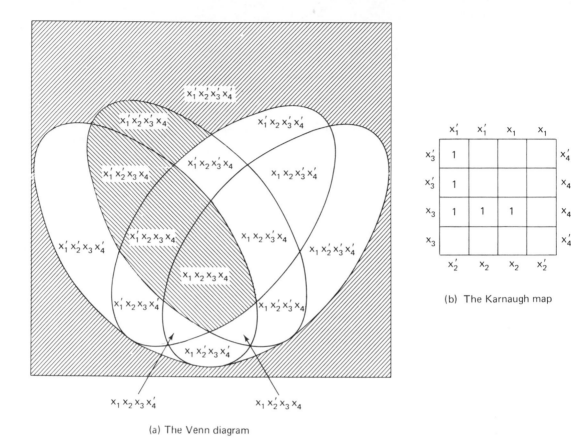

(a) The Venn diagram

(b) The Karnaugh map

Fig. 1.4.2 The Venn diagram and the Karnaugh map representation of the switching function of Eq. (1.4.1).

The sum of the above three terms gives one simplified form of the function of Eq. (1.4.1), which is

$$f(x_1, x_2, x_3, x_4) = x_1'x_3x_4 + x_1'x_2x_3' + x_2x_3x_4 \qquad (1.4.2)$$

The three terms (A, C, and D) are shown in Fig. 1.4.3. It is important to point out that this simplification is *not* unique. For example, another way of combining the adjacent cells of the Karnaugh map of Fig. 1.4.2 is A, B, and D, as shown in Fig. 1.4.3. The simplified function obtained from this combination is

$$f(x_1, x_2, x_3, x_4) = x_2x_3x_4(x_1 + x_1') + x_1'x_2'x_4(x_3' + x_3) + x_1'x_2'x_3'(x_4 + x_4')$$
$$= x_2x_3x_4 + x_1'x_2'x_4 + x_1'x_2'x_3' \qquad (1.4.3)$$

Thus for a given switching function in canonical sum-of-products form, we may obtain *many* "simplest" forms, which are logically equivalent to the given one.

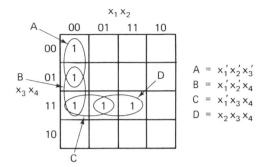

A = $x_1' x_2' x_3'$
B = $x_1' x_2' x_4$
C = $x_1' x_3 x_4$
D = $x_2 x_3 x_4$

Fig. 1.4.3 Another way of combining the adjacent cells to simplify the function of Eq. (1.4.1).

More important, the rule of simplification given above may be repeatedly applied to the simplified terms. This leads to the following two properties of the Karnaugh map.

PROPERTY 2

If four (2^2) cells that are marked by 1 form one of the six patterns shown in Fig. 1.4.4, they can be combined into one term, and two variables can be eliminated.

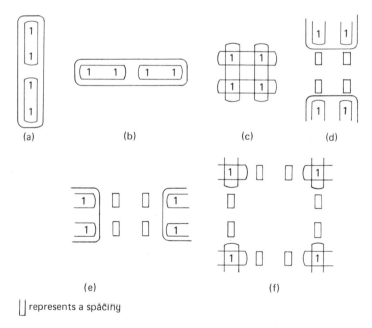

(a)   (b)   (c)   (d)

(e)   (f)

⊔ represents a spacing

Fig. 1.4.4 Six patterns for which two variables can be eliminated.

The proof of this property is very simple. The proofs for all six patterns are similar. The main idea is that for these six patterns, it is possible to combine two adjacent terms in such a way that the simplified terms obtained *again* differ in the value of one variable. The ways of combining the adjacent terms in the six patterns are indicated in Fig. 1.4.4.

Take pattern (a), for instance. Suppose that it appears in the third column of a Karnaugh map. The four terms are

$$x_1x_2x_3x_4 + x_1x_2x_3x_4' + x_1x_2x_3'x_4' + x_1x_2x_3'x_4$$

First simplification: $x_1x_2x_3$ $x_1x_2x_3'$

Second simplification: x_1x_2

As another example, consider pattern (f). The four terms are

$$x_1x_2x_3x_4 + x_1x_2x_3'x_4 + x_1'x_2x_3x_4 + x_1'x_2x_3'x_4$$

First simplification: $x_1x_2x_4$ $x_1'x_2x_4$

Second simplification: x_2x_4

Property 1 can be further extended to

PROPERTY 3

If eight (2^3) cells are marked by one form of the four patterns shown in Fig. 1.4.5, they can be combined into one term and three variables can be eliminated.

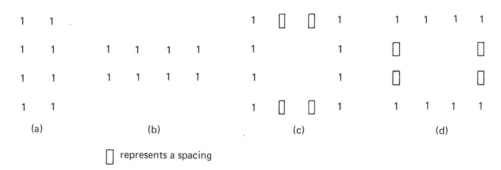

□ represents a spacing

Fig. 1.4.5 Four patterns for which three variables can be eliminated.

For this case we repeatedly use the rule of simplification of property 1 three times. Take pattern (c), for instance. Suppose they are in the second and third columns of the Karnaugh map. The eight terms are

$$x_1x_2'x_3x_4 + x_1'x_2'x_3x_4 + x_1x_2'x_3x_4' + x_1'x_2'x_3x_4' + x_1x_2'x_3'x_4' + x_1'x_2'x_3'x_4' + x_1x_2'x_3'x_4 + x_1'x_2'x_3'x_4$$

First simplification:
$x_2'x_3x_4$ $x_2'x_3x_4'$ $x_2'x_3'x_4'$ $x_2'x_3'x_4$

Second simplification:
$x_2'x_3$ $x_2'x_3'$

Third simplification: x_2'

We repeat that all the patterns discussed above do not depend on where they are located. The crucial point is that their relative positions form such patterns.

1.4.3   Application of Karnaugh Map to the
Minimization of Switching Functions

Much of the convenience and time saving in the use of the Karnaugh map for the minimization of switching functions depends upon our ability to recognize the patterns or subcubes of entities that can be simplified to simpler functions.

Before describing the method for minimizing switching functions using Karnaugh maps, we need the following definitions.

DEFINITION 1.4.2

In the Karnaugh map representation of an *n*-variable function, where $n \leq 4$, every minterm (a cell marked by 1) and a set of 2^i minterms, $i \leq n$, satisfying any one of the patterns described in properties 1–3 are called *implicants*† of the function. The set containing no minterms and the set containing all the minterms are the *trivial implicants* of the function.

DEFINITION 1.4.3

An implicant is called a *prime implicant* if it is *not* a subset of another implicant of the function.

DEFINITION 1.4.4

A prime implicant which includes a 1 cell that is not included in any other prime implicant is called an *essential prime implicant*.

For example, consider the sets A, B, C, and D of the minterms in Fig. 1.4.3. It is easy to verify that they are prime implicants of the function. Among the four prime implicants, only A and D are essential prime implicants, because the terms $x_1'x_2'x_3'x_4'$ and $x_1x_2x_3x_4$ are only covered by A and D, respectively, and no other prime implicants. On the other hand, each term of B and C are covered by more than one prime implicant; hence they are not essential.

Now we are in the position to describe a minimal sum-of-products form of a function that was defined in Definition 1.4.1 in terms of the prime implicants of the functions.

DEFINITION 1.4.5

A set of prime implicants of a function is called a *minimal sum-of-products* form of the function if the following three conditions are satisfied:

1.   It includes all the terms (1's) of the Karnaugh map representation of the function.

---

†The term "implicant" may denote the set of adjacent minterms or the simplified product term obtained from the minterms of the set.

2. Condition (1) will not be satisfied if any one of the prime implicants is removed.

3. None of the prime implicants can be replaced by a simpler prime implicant (i.e., a prime implicant that contains fewer variables).

The procedure for minimizing switching functions by means of Karnaugh maps may then be described as follows:

Step 1 Find all the prime implicants.

Step 2 Find all the essential prime implicants.

Step 3 Find a smallest set of prime implicants that includes (at least) all the essential prime implicants to "cover" all the 1's in the Karnaugh map. Whenever choice exists between two prime implicants, the simpler one should be chosen.

This procedure is best illustrated by the following examples.

Example 1.4.1

Consider the switching function whose Karnaugh map is shown in Fig. 1.4.6. The minimization of this function may be obtained by recognizing the patterns described in properties 1–3 as described below.

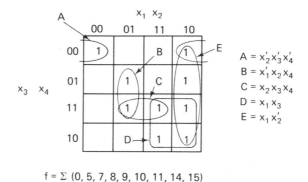

$$f = \Sigma\ (0, 5, 7, 8, 9, 10, 11, 14, 15)$$

Fig. 1.4.6 The Karnaugh map for Example 1.4.1.

Step 1 The prime implicants are $x_2'x_3'x_4'$, $x_1'x_2x_4$, $x_2x_3x_4$, x_1x_3, and x_1x_2'.

Step 2 The essential prime implicants are $x_2'x_3'x_4'$, $x_1'x_2x_4$, x_1x_3, and x_1x_2'. (Note that $x_2x_3x_4$ is not an essential prime implicant.)

Step 3 Since in this example the set of essential prime implicants cover all the terms of the function in the Karnaugh map, the minimal sum-of-products form of the function is

$$f(x_1, x_2, x_3, x_4) = x_1x_3 + x_1x_2' + x_2'x_3'x_4' + x_1'x_2x_4 \qquad (1.4.4)$$

The following is an example of function minimization involving don't-care conditions.

Example 1.4.2

Consider the switching function described by the Karnaugh map of Fig. 1.4.7, which may be represented by

$$f = \sum (0, 2, 5, 9, 15) + \sum_d (6, 7, 8, 10, 12, 13)$$

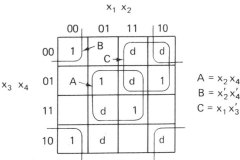

Fig. 1.4.7 The Karnaugh map for Example 1.4.2.

$f = \sum (0, 2, 5, 9, 15) + \sum_d (6, 7, 8, 10, 12, 13)$

In the case where no "don't care" is involved, we shall look for the patterns that contain most terms first and group them; then consider the second largest; and so forth. In the case where the Karnaugh map contains d's, we should try to make use of d's to make as many entities of the map fit the patterns discussed in properties 1–3 (Figs. 1.4.4 and 1.4.5) as we can. In doing so, we should make them as large as possible and consider the largest first. In this example, the largest patterns that we can make are four-element ones, patterns (c) and (f) of Fig. 1.4.4, as shown in Fig. 1.4.7. Any d within the circled lines is set to 1 and outside the circled lines is set to 0. Then the four cells at the center can be reduced to $x_2 x_4$; the four cells at the four corners can be simplified to $x'_2 x'_4$; and the four cells at the upper right corner are combined to $x_1 x'_3$. Therefore, the prime implicants are $x_2 x_4$, $x'_2 x'_4$, and $x_1 x'_3$, which are also essential and cover all the 1's in the map. Hence the minimal sum of the function is

$$f(x_1, x_2, x_3, x_4) = x_2 x_4 + x'_2 x'_4 + x_1 x'_3 \tag{1.4.5}$$

Now one may ask: Will the set of all essential prime implicants of a function always cover all the 1's in the Karnaugh map of the function? The answer is no. This is demonstrated in the following example.

Example 1.4.3

From the Karnaugh map of the function

$$f(x_1, x_2, x_3, x_4) = \sum (0, 1, 2, 3, 6, 7, 9, 13, 14, 15) \tag{1.4.6}$$

which is shown in Fig. 1.4.8, it is found that the prime implicants of the function are $x'_1 x'_2$, $x'_1 x_3$, $x_2 x_3$, $x_1 x_2 x_4$, $x_1 x'_3 x_4$, and $x'_2 x'_3 x_4$, of which only $x'_1 x'_2$ and $x_2 x_3$ are essential. The sum $x'_1 x'_2 + x_2 x_3$, however, does not cover all the 1's in the map. The minimal form of the function is $x'_1 x'_2 + x_2 x_3 + x_1 x'_3 x_4$.

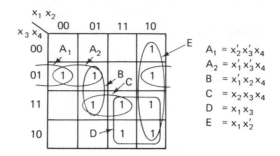

$$f = \Sigma\ (1, 5, 7, 8, 9, 10, 11, 14, 15)$$

Fig. 1.4.8   The Karnaugh map of Example 1.4.3.

Another interesting question one may ask: Is it possible for a function to have more than one minimal form? The answer is yes. A simple example that illustrates this point is given next.

Example 1.4.4

Suppose that the minterm 0 of the function of Fig. 1.4.6 is changed to the minterm 1. The function becomes $f = \Sigma\ (1, 5, 7, 8, 9, 10, 11, 14, 15)$, which is shown in Fig. 1.4.9.

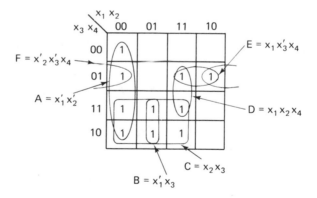

Fig. 1.4.9   The Karnaugh map of Example 1.4.4.

Of the six prime implicants, $x_2'x_3'x_4$, $x_1'x_3'x_4$, $x_1'x_2x_4$, $x_2x_3x_4$, x_1x_3, and x_1x_2', the last two are essential. The sum of these two essential prime implicants $x_1x_3 + x_1x_2'$, again, does not cover all the minterms of the function. The three uncovered minterms are 1, 5, and 7, which can be covered in a number of ways:

$$x_2'x_3'x_4 + x_1'x_2x_4$$

$$x_1'x_3'x_4 + x_1'x_2x_4$$

$$x_1'x_3'x_4 + x_2x_3x_4$$

Since each of these sums has two product terms, each of which has three variables, the amount of hardware needed for realizing them will be the same; thus any one of these sums plus the two essential prime implicants constitutes a minimal form of the function. In other words, this function has three minimal forms.

1.4.4 Five- and Six-Variable Karnaugh Maps

The Karnaugh map discussed above can be extended to five and six variables. First, consider the five-variable case. Let $f(x_1, x_2, x_3, x_4, x_5)$ be a switching function of five variables, x_1, x_2, x_3, x_4, and x_5, in the sum-of-products form. It can be expressed as

$$f(x_1, x_2, x_3, x_4, x_5) = x_1' g_0(x_2, x_3, x_4, x_5) + x_1 g_1(x_2, x_3, x_4, x_5)$$

The function f can therefore be represented by two four-variable Karnaugh maps, as shown in Fig. 1.4.10.

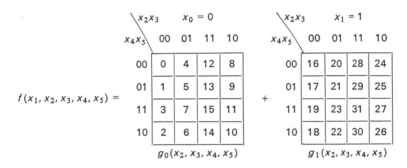

Fig. 1.4.10 Five-variable Karnaugh map.

Similarly, if $f(x_1, \ldots, x_6)$ is a switching function with six variables and is in the sum-of-products form, then

$$f(x_1, \ldots, x_6) = x_1' x_2' g_{00}(x_3, x_4, x_5, x_6) + x_1 x_2' g_{10}(x_3, x_4, x_5, x_6)$$
$$+ x_1' x_2 g_{01}(x_3, x_4, x_5, x_6) + x_1 x_2 g_{11}(x_3, x_4, x_5, x_6)$$

It is seen that a six-variable function can be represented by four four-variable Karnaugh maps, as shown in Fig. 1.4.11.

The basic rules for combining cells on the same four-variable map are exactly the same as those described previously. Two cells that are on different four-variable maps can be combined only if:

1. They occupy the same relative position on their respective four-variable maps.
2. They are in the four-variable maps whose other variables (x_1 and x_2) besides those represented by the maps (i.e., x_3, x_4, x_5, and x_6) have *only one* variable different in value. They are called *adjacent maps*.

$f(x_1, \ldots, x_6) =$

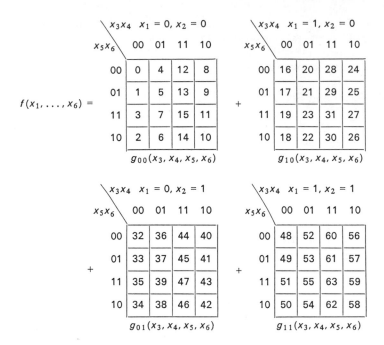

Fig. 1.4.11 Six-variable Karnaugh map.

Taking the six-variable map of Fig. 1.4.11, for example, the cells of the four-variable Karnaugh maps labeled

$$\left.\begin{array}{c} x_1 = 0,\, x_2 = 0 \\ \text{or} \\ x_1 = 1,\, x_2 = 1 \end{array}\right\} \quad \text{can be combined with} \quad \left\{\begin{array}{c} x_1 = 0,\, x_2 = 1 \\ \text{or} \\ x_1 = 1,\, x_2 = 0 \end{array}\right.$$

but the map labeled $x_1 = 0$, $x_2 = 0$ cannot be combined with the map labeled $x_1 = 1$, $x_2 = 1$, nor can the map labeled $x_1 = 0$, $x_2 = 1$ be combined with the map labeled $x_1 = 1$, $x_2 = 0$. However, cells occupying the same relative position on all the four four-variable Karnaugh maps can be combined.

Example 1.4.5

Consider the five-variable function

$$f(x_1, \ldots, x_5) = \Sigma\,(4, 5, 6, 7, 13, 15, 20, 21, 22, 23, 25, 27, 29, 31)$$

The Karnaugh map of f is shown in Fig. 1.4.12. The prime implicants of the function are $x_2'x_3$, x_3x_5, and $x_1x_2x_5$, which are also essential prime implicants (see Fig. 1.4.12). The function after minimization therefore is $x_2'x_3 + x_3x_5 + x_1x_2x_5$.

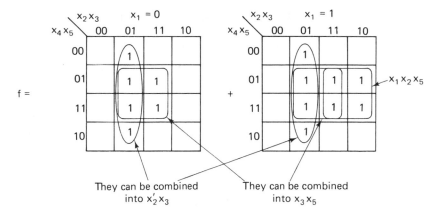

They can be combined   They can be combined
into $x_2' x_3$ into $x_3 x_5$

Fig. 1.4.12 The Karnaugh map of the function of Example 1.4.5.

Example 1.4.6

As an example of a six-variable function, consider

$$f(x_1, \ldots, x_6) = \sum (0, 3, 4, 5, 7, 8, 12, 13, 20, 21, 28, 29, 31, 34, 35, 38, 39, 42, 45,$$
$$46, 50, 54, 58, 61, 62, 63)$$

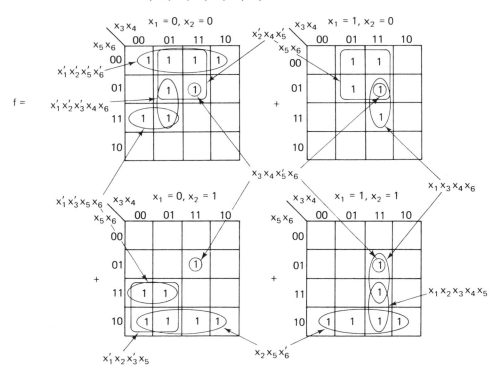

Fig. 1.4.13 The Karnaugh map of the function of Example 1.4.6.

Referring to Fig. 1.4.13, the prime implicants of the function are $x_1'x_2'x_5'x_6'$, $x_1'x_2'x_3'x_4x_6$, $x_1'x_3'x_5x_6$, $x_1'x_2x_3'x_5$, $x_2x_5x_6'$, $x_1x_2x_3x_4x_5$, $x_1x_3x_4x_6$, $x_2'x_4x_5'$, and $x_3x_4x_5'x_6$, of which the first, the third, the fifth, and the last three are essential and cover all the 1's in the map. Hence the minimal sum of the function is

$$f(x_1, \ldots, x_6) = x_1'x_2'x_5'x_6' + x_1'x_3'x_5x_6 + x_2x_5x_6' + x_1x_3x_4x_6 + x_2'x_4x_5' + x_3x_4x_5'x_6$$

1.4.5 Minimization of Switching Function in Canonical Product-of-Sums Form Using the Karnaugh Map

Although all the properties of the Karnaugh map described above were presented in terms of minterms of a function and the methods were illustrated by examples of functions in canonical sum-of-products form, the same properties obtain when minterms are replaced by maxterms and the minimization procedure described above for functions in canonical sum-of-products form is applicable to functions that are expressed in canonical product-of-sums form. For example, the Karnaugh maps of the functions $f(x_1, x_2, x_3, x_4) = \prod(0, 2, 4, 6, 8)$ and $f(x_1, x_2, x_3, x_4) = \prod(0, 2, 4, 6, 8) \cdot \prod_d(10, 11, 12, 13, 14, 15)$ are shown in Fig. 1.4.14(a) and (b), respectively. Notice that the maxterms are now represented by the cells marked by 0's. The results obtained from the Karnaugh maps shown in Fig. 1.4.14 agree completely with those obtained previously [see Eqs. (1.3.10) and (1.3.12)].

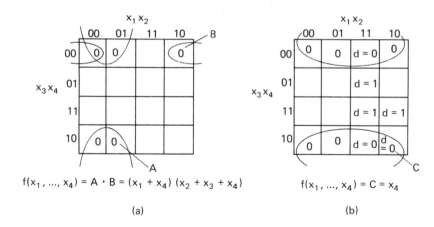

$$f(x_1, \ldots, x_4) = A \cdot B = (x_1 + x_4)(x_2 + x_3 + x_4)$$

(a)

$$f(x_1, \ldots, x_4) = C = x_4$$

(b)

Fig. 1.4.14

Exercise 1.4

1. Minimize the following switching functions using the Karnaugh map. List all prime implicants and essential prime implicants.
    (a) $f_1(x_1, x_2, x_3, x_4) = \sum(0, 1, 2, 3, 5, 6, 8, 10, 15)$
    (b) $f_2(x_1, x_2, x_3, x_4) = x_1x_4 + x_1x_3' + x_2x_4 + x_3 + x_4$

(c) The switching function $f_3(x_1, x_2, x_3, x_4)$ in canonical sum-of-products form whose Karnaugh map is as follows:

$x_1 x_2$

$x_3 x_4$	00	01	11	10
00	1	1	d	d
01	1	d	1	1
11		1	d	
10	d			1

(d) $f_4(x_1, x_2, x_3, x_4) = \prod (0, 3, 4, 6, 7, 8, 10, 13, 14)$

(e) The switching function $f_5(x_1, x_2, x_3, x_4)$ in canonical product-of-sums form whose Karnaugh map is described by

$x_1 x_2$

$x_3 x_4$	00	01	11	10
00	0		d	
01			0	d
11	d	d	0	
10			d	d

2. Find the Boolean sum of the two functions by adding the values of the corresponding cells of their maps.

$x_1 x_2$

$x_3 x_4$	00	01	11	10
00	1		1	
01		d		d
11	1		1	
10		d		1

$+$

	00	01	11	10
00	1		d	
01	d	1		
11			1	
10	1	d		1

$=$

	00	01	11	10
00				
01				
11				
10				

3. Find the Boolean product of the two functions by multiplying the values of the corresponding cells of their maps.

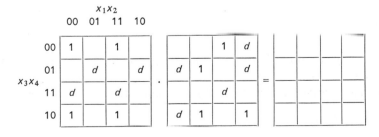

$x_1 x_2$

$x_3 x_4$	00	01	11	10
00	1		1	
01		d		d
11	d		d	
10	1		1	

$\cdot$

	00	01	11	10
00			1	d
01	d	1		d
11			d	
10	d	1		1

$=$

	00	01	11	10
00				
01				
11				
10				

4. Minimize the following functions using the Karnaugh map.
   (a) $f(x_1, \ldots, x_5) = \Sigma\, (0, 1, 4, 5, 6, 7, 9, 12, 13, 14, 15, 16, 17, 20, 21, 22, 23, 25, 29)$
   (b) $f(x_1, \ldots, x_5) = \Sigma\, (0, 1, 2, 3, 4, 5, 6, 7, 16, 17, 20, 21) + \Sigma_d\, (24, 25, 27, 28, 30)$
   (c) $f(x_1, \ldots, x_5) = \prod\, (0, 1, 2, 3, 8, 9, 16, 17, 20, 21, 24, 25, 28, 29, 30, 31) \cdot$
        $\cdot \prod_d (13, 14, 19)$
   (d) $f(x_1, \ldots, x_6) = \Sigma\, (4, 5, 20, 21, 36, 37, 38, 39, 48, 49, 52, 53)$
   (e) $f(x_1, \ldots, x_6) = \prod\, (0, 1, 2, 3, 4, 5, 8, 9, 12, 13, 16, 17, 18, 19, 24, 25, 36, 37, 38, 39,$
        $52, 53, 60, 61)$

5. Minimize the following four-variable functions using sum-of-products Karnaugh maps.

$$f(x_1, x_2, x_3, x_4) = f_1(x_1, x_2, x_3, x_4) \oplus f_2(x_1, x_2, x_3, x_4)$$

where

$$f_1 = x_1' x_4 + x_2 x_3 + x_2' x_3' x_4' + \sum_d (1, 2, 11, 13, 14, 15)$$

and

$$f_2 = \prod\, (0, 2, 4, 8, 9, 10, 14) \cdot \prod_d (1, 7, 13, 15)$$

Note that f_2 is in canonical product-of-sums form. The exclusive-or operation is defined as:

$\oplus$	0	1	d
0	0	1	d
1	1	0	d
d	d	d	d

1.5 Quine–McCluskey Method

We have seen that the minimization of switching functions with four variables or less using the Karnaugh map is not only convenient but also effective. This method, which is based on a graphic display of switching functions, becomes less and less convenient and effective as the number of variables increases. In practice we often encounter logic design problems that involve variables considerably more than four; in these situations, the Karnaugh map method described in the previous section will be difficult to apply.

In this section, another switching-function-minimization method, known as the *Quine–McCluskey method,* is presented. It is particularly attractive when

1. A digital computer is used to minimize switching functions with more than four variables.

2. Function minimization of multiple-output circuits is based on certain cost criteria.

The basic Quine–McCluskey tabular minimization procedure is as follows:

1.   Find the prime implicants of the function.

2.   Construct the prime-implicant table and find the essential prime implicants (essential rows) of the function.

3.   Include the essential prime implicants in the minimal sum.

4.   After all essential prime implicants are deleted from the prime-implicant table, determine the dominated rows and dominating columns in the table, delete all dominated rows and dominating columns, and find the (secondary) essential prime implicants.

5.   Repeat 3 and 4 as many times as they are applicable until a minimal cover of the function is found.

In this section, only minimization of single switching functions is considered. The function minimization of multiple-output circuits is discussed in the next section. The Quine–McCluskey method is best illustrated by examples. For the sake of simplicity, functions with four and five variables will be used. The first example illustrates the finding of prime implicants, the construction of a prime-implicant table, and the determination of essential prime implicants using the Quine–McCluskey method.

Example 1.5.1

Consider the minimization of the switching function of Example 1.4.1 using the Quine–McCluskey method.

(a)   Find the prime implicants of the function.

(1)   Represent each term (minterm) of the canonical sum-of-products form by a binary code. For example, the term $x_1x_2'x_3x_4'$ is represented by 1010.

(2)   Find the decimal number for each binary code.

(3)   Define the number of 1's in a binary number as the *index* of the number. Group all the the binary numbers of the same index into a group. List all the groups in a column in the index ascending order. Within each group, the binary numbers are listed in the ascending order of their decimal-number equivalent. For the function of Example 1.4.1, such a list of all terms is shown in the first column of Fig. 1.5.1(a).

(4)   Start with the terms in the set of lowest index; compare them with those, if any, in the set whose index is 1 greater, and eliminate all redundant variables as described in property 1. In our example there are 10 such reductions, which are listed in the second column of the table of Fig. 1.5.1(a). For instance, the first term in the second column is obtained from

$$
\begin{array}{ll}
0 & 0000 \\
8 & \underline{1000} \\
0,8 & -000
\end{array}
$$

where the dash indicates that the variable at that position is eliminated. Note that −000 represents $x_2'x_3'x_4'$ and that the numbers 0 and 8 on the left of the reduced term −000 indicate that the term −000 *covers* the terms represented by 0 and 8, which are 0000 and 1000 in the canonical-sum-of-products form.

	Decimal Number	Binary Representation of Each Term		Decimal Numbers	First Reduction		Decimal Numbers	Second Reduction	
Index 0 {	0	0000	✓	0, 8	–000	E	8, 9, 10, 11	10– –	B
Index 1 {	8	1000	✓	8, 9	100–	✓	10, 11, 14, 15	1–1–	A
Index 2 {	5	0101	✓	8, 10	10–0	✓			
	9	1001	✓	5, 7	01–1	D			
	10	1010	✓	9, 11	10–1	✓			
Index 3 {	7	0111	✓	10, 11	101–	✓			
	11	1011	✓	10, 14	1–10	✓			
	14	1110	✓	7, 15	–111	C			
Index 4	15	1111	✓	11, 15	1–11	✓			
				14, 15	111–	✓			

Note that a check mark placed at the right of a term means that the term is covered by a larger term and therefore is not a prime implicant.

(a) Table for obtaining all the prime implicants

Prime Implicants \ Minterms	0	5	7	8	9	10	11	14	15
*A						×	×	⊗	×
*B				×	⊗	×	×		
C			×						×
*D		⊗	×						
*E	⊗			×					

(b) Prime-implicant table for finding essential prime implicants

Minimized function: $f(x_1, x_2, x_3, x_4) = A + B + D + E = x_1 x_3 + x_1 x_2' + x_1' x_2 x_4 + x_2' x_3' x_4'$

Fig. 1.5.1 Quine–McCluskey method of minimizing the switching function of Example 1.4.1.

(5) Check off all the terms that entered into the combination. The ones that are left are *prime implicants*, from which a minimum sum is to be selected.

(6) Repeat steps 4 and 5 until no further reduction is possible; thereby we obtain the set of *all* prime implicants each of which is designated by a capital English letter. In this example, there are five prime implicants:

$$1-1- \quad A$$
$$10-- \quad B$$
$$-111 \quad C$$
$$01-1 \quad D$$
$$-000 \quad E$$

(b) Construct the prime-implicant table and find the essential prime implicants of the function.

    (1) Construct a table like the one shown in Fig. 1.5.1(b). Each column carries a decimal number at the top which corresponds to one of the minterms in the canonical sum-of-products form in the given function. The columns are assigned by such a number in ascending order. Each row corresponds to one of the prime implicants designated by A, B, C, D, E at the left.

    (2) Make a cross under each decimal number that is a term contained in the prime implicant represented by that row. For example, for prime implicant B, which covers 8, 9, 10, and 11, we make a cross under each of these numbers in that row, as indicated in Fig. 1.5.1(b).

    (3) Find all the columns that contain a *single* cross and circle them; place an asterisk at the left of those rows in which you have circled a cross. The rows marked with an asterisk are the essential prime implicants.

(c) Include the essential prime implicants in the minimal sum.

In this example, since the sum of the essential prime implicants covers all the minterms of the function, it is the minimal form. Thus no further steps need to be carried out.

In general, the sum of the essential prime implicants does not necessarily cover all the minterms of the function. In fact, there are cases where a function does not even have any essential prime implicants. Another thing that we would like to point out is that the minimal form of a function in general depends upon the cost criterion used. Here we assume that the cost associated with each prime implicant (row of a prime-implicant table) is directly proportional to the number of variables that the prime implicant has.

We assign a "cost" to each implicant according to the number of variables that it contains. A prime implicant having four variables, for example, would have a higher cost compared with a prime implicant having three variables. A prime implicant occupying a larger block of 1's in a Karnaugh map would have a lower cost than a prime implicant occupying a smaller block of 1's, because the former will result in fewer variables than the later.

DEFINITION 1.5.1

Two rows (columns) I and J of a prime-implicant table which have $\times$'s in exactly the same rows (columns) are said to be *equal* (written $I = J$).

DEFINITION 1.5.2

Let columns i and j be two columns of a prime-implicant table. Column i is said to *dominate* column j (written $i \supset j$) if $i = j$ or column i has $\times$'s in all the rows in which column j has $\times$'s. Columns i and j are called the *dominating* and *dominated columns*, respectively.

DEFINITION 1.5.3

Let rows I and J be two rows of a prime-implicant table. Row I is said to *dominate* row J (written $I \supset J$) if $I = J$ or row I has $\times$'s in all the columns in which row J has $\times$'s. Rows I and J are called the *dominating* and *dominated rows*, respectively.

All the dominating columns and dominated rows of a prime-implicant table can be removed without affecting the table for obtaining a minimal sum. This is because a dominating column is guaranteed to be covered by the row that covers its dominated column. Similarly, the columns of a dominated row are guaranteed to be covered by its dominating row included in the minimal sum. Since, by our definition of row dominance (Definition 1.5.3), the cost of a dominated row must not be less than that of its dominating row, the removal of the dominated row will not increase the cost of the minimal sum of the function. The simplification of a prime-implicant table by removing dominating columns and dominated rows of the table is an important step in this minimization procedure. The following example illustrates this point.

Example 1.5.2

Suppose that a switching function with five variables whose canonical sum-of-products form is

$$f(x_1, \ldots, x_5) = \sum (0, 1, 2, 8, 9, 15, 17, 21, 24, 25, 27, 31) \qquad (1.5.1)$$

The first two tables of Fig. 1.5.2 are constructed like those in Example 1.5.1. One is for obtaining the prime implicants, and the other is for obtaining the essential prime implicants. From the first table, we obtain 11 prime implicants, which are designated $A, B, \ldots, K$. From the prime-implicant table, we see that the columns 2, 15, and 21 are, respectively, only covered by rows J, E, and H; thus they are essential prime implicants (indicated by an asterisk) which must be selected for inclusion in the minimal sum. After removing the three essential prime implicants [Fig. 1.5.2(c)], the following dominance relations in the remaining columns and rows are observed:

Column dominance:	$9 \supset 8$	Row dominance:	$A \supset G$
	$25 \supset 24$		$F \supset D$
			$B \supset I$
			$B \supset K$

Thus columns 9 and 25 can be removed. Moreover, since all the dominated rows G, D, I, and K have a cost that is not lower than their respective dominating row, they can also be removed from the table. The resulting table is shown in Fig. 1.5.2(d), in which we see that columns 24 and 27 are only covered by rows A and F, respectively; thus they must be included in the minimal sum. Rows A and F are sometimes referred to as the *secondary essential prime implicants* (indicated by two asterisks). After rows A and F are removed from the table [Fig. 1.5.2(e)], all but minterm 1 has not been covered; however, it can be covered by row B or row C. Suppose that row B is selected; the minimal sum of the function is $E + H + J + A + F + B$.

When a switching function has "don't cares," we *set all the don't cares equal to 1 and find the prime implicants. In determining essential prime implicants and minimum cover of the function, we set all the don't cares equal to 0; that is, the decimal numbers that denote the don't-care terms do not appear in the prime-implicant table.* This is illustrated by the following examples.

	Decimal Number	Binary Representation of Each Minterm	Decimal Numbers	First Reduction	Decimal Numbers	Second Reduction
Index 0 {	0	00000 ✓	0, 1	0000– K	0, 8, 1, 9	0–00– C
	1	00001 ✓	0, 2	000–0 J	1, 9, 17, 25	––001 B
Index 1 {	2	00010 ✓	0, 8	0–000 ✓	8, 9, 24, 25	–100– A
	8	01000 ✓	1, 9	0–001 ✓		
	9	01001 ✓	1, 17	–0001 I		
Index 2 {	17	10001 ✓	8, 9	0100– ✓		
	24	11000 ✓	8, 24	–1000 ✓		
Index 3 {	21	10101 ✓	9, 25	–1001 ✓		
	25	11001 ✓	17, 21	10–01 H		
Index 4 {	15	01111 ✓	17, 25	1–001 ✓		
	27	11011 ✓	24, 25	1100– G		
Index 5 {	31	11111 ✓	25, 27	110–1 F		
			15, 31	–1111 E		
			27, 31	11–11 D		

(a) Table for obtaining all the prime implicants

	0	1	2	8	9	15	17	21	24	25	27	31
A				×	×				×	×		
B		×			×		×			×		
C	×	×		×	×							
D											×	×
*E						⊗						×
F										×	×	
G									×	×		
*H							×	⊗				
I		×					×					
*J	×		⊗									
K	×	×										

(b) Prime-implicant table with essential prime implicants E, H, and J indicated by an asterisk

Fig. 1.5.2 Function minimization of Example 1.5.3 using the Quine–McCluskey method.

	1	8	9	24	25	27	
A		×	×	×	×		Column dominance:
B	×		×		×		9 ⊃ 8
C	×	×	×				25 ⊃ 24
D						×	Row dominance:
F				×	×		A ⊃ G
G			×		×		F ⊃ D
I	×						B ⊃ I
K	×						B ⊃ K

(c)   Prime-implicant table of (b) after rows E, H, and J are removed

	1	8	24	27
**A		×	⊗	
B	×			
C	×	×		
**F				⊗

(d)   Table of (c) after the dominating columns 8 and 24 and the dominated rows G, D, I, and K are deleted

	1
B	×
C	×

(e)   Table of (d) after rows A and F (secondary essential prime implicants) are removed

Fig. 1.5.2   (continued)

Example 1.5.3

Consider the same function as that of Example 1.5.1 except that the minterm 0 is changed to a don't care as

$$f(x_1, x_2, x_3, x_4) = \sum (5, 7, 8, 9, 10, 11, 14, 15) + \sum_a (0) \qquad (1.5.2)$$

By setting the don't-care term equal to 1, this function becomes the function of Example 1.5.1 whose prime implicants were found and given in Fig. 1.5.1. Modifying the table of Fig. 1.5.1(b) by not including the minterm 0 (since it is a don't-care minterm), we find that the prime implicant E of this function is no longer an essential prime implicant. The essential prime implicants of this function are A, B, and D. Since the sum of these three essential prime implicants covers all the minterms of the function [not necessary to include the don't-care minterm(s)], it is the minimal sum of the function. Because the don't-care minterm 0 is contained only in the prime implicant E, which is not included in the minimal sum, the don't-care minterm 0 is not included in the minimal sum. This implies that this don't-care minterm should be set equal to zero in order to achieve the simplest minimal sum.

Example 1.5.4

As another example, consider the incompletely specified function

$$f(x_1, x_2, x_3, x_4, x_5) = \sum (0, 2, 4, 5, 6, 7, 8, 10, 14, 17, 18, 21, 29, 31) + \sum_d (11, 20, 22)$$

$$(1.5.3)$$

The table of Fig. 1.5.3(a) shows the table for obtaining all the prime implicants of the func-

	Decimal Number	Binary Representation of Each Term	Decimal Numbers	First Reduction	Decimal Numbers	Second Reduction
Index 0 {	0	00000 ✓	0, 2	000–0 ✓	0, 2, 4, 6	00– –0 E
	2	00010 ✓	0, 4	00–00 ✓	0, 2, 8, 10	0–0–0 F
Index 1 {	4	00100 ✓	0, 8	0–000 ✓	2, 6, 10, 14	0– –10 G
	8	01000 ✓	2, 6	00–10 ✓	2, 6, 18, 22	–0–10 H
	5	00101 ✓	2, 10	0–010 ✓	4, 5, 6, 7	001– – I
	6	00110 ✓	2, 18	–0010 ✓	4, 5, 20, 21	–010– J
	10	01010 ✓	4, 5	0010– ✓	4, 20, 6, 22	–01–0 K
Index 2 {	17	10001 ✓	4, 6	001–0 ✓		
	18	10010 ✓	4, 20	–0100 ✓		
	20	10100 ✓	8, 10	010–0 ✓		
	7	00111 ✓	5, 7	001–1 ✓		
	11	01011 ✓	5, 21	–0101 ✓		
Index 3 {	14	01110 ✓	6, 7	0011– ✓		
	21	10101 ✓	6, 14	0–110 ✓		
	22	10110 ✓	6, 22	–0110 ✓		
Index 4 {	29	11101 ✓	10, 14	01–10 ✓		
Index 5 {	31	11111 ✓	10, 11	0101– A		
			17, 21	10–01 B		
			18, 22	10–10 ✓		
			20, 21	1010– ✓		
			20, 22	101–0 ✓		
			21, 29	1–101 C		
			29, 31	111–1 D		

(a) Table for obtaining all the prime implicants with don't cares = 1

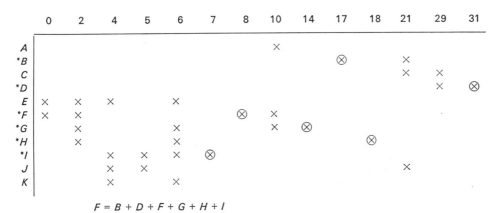

	0	2	4	5	6	7	8	10	14	17	18	21	29	31
A								×						
*B										⊗		×		
C												×	×	
*D													×	⊗
E	×	×	×		×									
*F	×	×					⊗	×						
*G		×			×			×	⊗					
*H		×			×						⊗			
*I			×	×	×	⊗								
J			×	×								×		
K			×		×									

$$F = B + D + F + G + H + I$$

$$= x_1 x_4' x_3' x_5 + x_1 x_2 x_3 x_5 + x_1' x_3' x_5' + x_1' x_4 x_5' + x_2' x_4 x_5' + x_1' x_2' x_3$$

(Note that the don't-care minterms 11, 20, and 22 do not appear in this table.)

(b) Table for finding a minimal cover

Fig. 1.5.3 Minimization of the switching function of Eq. (1.5.3) which has don't cares.

tion. In constructing this table, all the don't cares are set to equal to 1. It is found that there are 11 prime implicants. From the prime-implicant table of Fig. 1.5.3(b) in which all the don't-care minterms 11, 20, and 22 do not appear, it is seen that there are six essential prime implicants and the sum of them covers the function.

It is important to note that it is not always true that removing essential prime implicants, dominated rows, and dominating columns will suffice for "solution" of a prime-implicant table. It can happen that each column of an original prime-implicant table, or a table after all its essential rows, dominated rows, and dominating columns are removed, contains at least two ×'s, and there does not exist any dominance relation among the rows and the columns of the table and therefore no rows and columns can be removed.

DEFINITION 1.5.4

A prime-implicant table is said to be *semicyclic* if (1) it does not have any essential prime implicant (essential row), that is, each column contains at least two ×'s; (2) no dominance relation exists among rows and columns; and (3) the costs of the rows are not the same.

To solve a semicyclic prime-implicant table, we can *select a row with the lowest cost* (i.e., the prime implicant represented by this row has the fewest variables) to include in the (possible) minimal sum and then using the reduction techniques to remove rows and columns from the table, which results after removal of this row. *This entire process must then be repeated for each row that could replace the original selected row, and the final minimal sum is obtained by comparing the costs of the expressions that result from each arbitrary choice of a selected row.* This is illustrated by the following example.

Example 1.5.5

Consider the function whose Karnaugh map is shown in Fig. 1.5.4(a). It is seen that this function has 11 prime implicants, denoted by the letters $A, B, \ldots, K$, which are shown in the

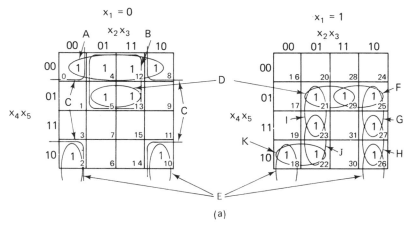

(a)

Fig. 1.5.4 Example 1.5.5.

(b)

	0	2	4	5	8	10	12	13	18	21	22	23	25	26	27	29
A	×		×		×		×									
B			×	×			×	×								
C	×	×			×	×										
D				×				×		×						×
E		×				×			×					×		
F													×			×
G													×		×	
H														×	×	
I										×		×				
J											×	×				
K									×		×					

(c)

		2	5	10	13	18	21	22	23	25	26	27	29
	B		×		×								
	C	×		×									
	D		×		×		×						×
	E	×		×		×							
$D \supset B$	F								×				×
$E \supset C$	G								×			×	
	H										×	×	
	I						×		×				
	J							×	×				
	K					×		×					

(d)

	2	5	10	13	18	21	22	23	25	25	27	29
*D		⊗		⊗		×						×
*E	⊗		⊗		×					×		
F								×				×
G								×			×	
H										×	×	
I						×		×				
J							×	×				
K					×		×					

(e)

	22	23	25	27
F			×	
G			×	×
H				×
I		×		
J	×	×		
K	×			

$G \supset F$
$G \supset H$
$J \supset I$
$J \supset K$

(f)

	22	23	25	27
**G			⊗	⊗
**J	⊗	⊗		

Fig. 1.5.4 (continued)

rows of the prime-implicant table of Fig. 1.5.4(b). This table has no essential rows and no dominance relation among the rows and the columns, and the costs of the rows are not the same. It is a semicyclic table. There are two cost-group rows, $\{A, B, C, D, E\}$ and $\{F, G, H, I, J, K\}$. The rows in the first group cost less than the ones in the second. We can arbitrarily select a row from the first group, say row A, and remove it from the table. The resulting table is shown in Fig. 1.5.4(c). It is observed that there are still no essential rows and no dominance relation among the columns. But there exist dominance relations among the rows: $D \supset B$ and $E \supset C$; thus rows B and C can be removed from the table. After these two rows are removed [Fig. 1.5.4(d)], rows D and E become essential. The table of Fig. 1.5.4(e) shows that after rows D and E are removed from the table of Fig. 1.5.4(d), $G \supset F$, $G \supset H$, $J \supset I$, and $J \supset K$. By removing dominated rows F, H, I, and K, rows G and J become essential [Fig. 1.5.4(f)] and cover all the four remaining columns. Thus one possible minimal sum is $S_1 = A + D + E + G + J$.

Now suppose that in the semicyclic table of Fig. 1.5.4(b), instead of removing row A, we remove one of the other rows in the lower-cost group (i.e., row B, C, D, or E). The following are the other possible minimal sums:

$$S_2 = B + C + D + E + G + J$$
$$S_3 = B + C + F + H + I + K$$

By comparing the costs of S_1, S_2, and S_3, we find that S_1 has the lowest cost and thus is the minimal sum of the function.

DEFINITION 1.5.5

A semicyclic prime-implicant table is said to be *cyclic* if the costs of all its rows are the same.

Note that the class of semicyclic tables includes the class of cyclic tables as a special case. The minimization involving a cyclic prime-implicant table is similar to that involving a semicyclic prime-implicant table described above. Since the costs of all the rows of a cyclic table are the same, we can select *any one* of the rows to include in the possible minimal sum. This is illustrated by the following example.

Example 1.5.6

Consider the minimization of the function

$$f(x_1, x_2, x_3, x_4) = \sum (0, 2, 4, 10, 12, 14) \qquad (1.5.4)$$

The prime implicants of this function are

$$\begin{aligned} A &= x_1' x_2' x_4' & D &= x_2 x_3' x_4' \\ B &= x_1' x_3' x_4' & E &= x_1 x_3 x_4' \\ C &= x_2' x_3 x_4' & F &= x_1 x_2 x_4' \end{aligned} \qquad (1.5.5)$$

None of them is an essential prime implicant! This can be easily seen from the Karnaugh map of the function, which is shown in Fig. 1.5.5(a). Every one of these prime implicants has three variables and can be implemented by a three-input AND gate; their costs are the same. The table of Fig. 1.5.5(b) shows that it is a cyclic table. We can arbitrarily select one of the rows A, B, C, D, E, or F, say row A, to include in a minimum sum. After it is removed from the table [see Fig. 1.5.5(c)], the resulting table has two $\times$'s under each column. But this

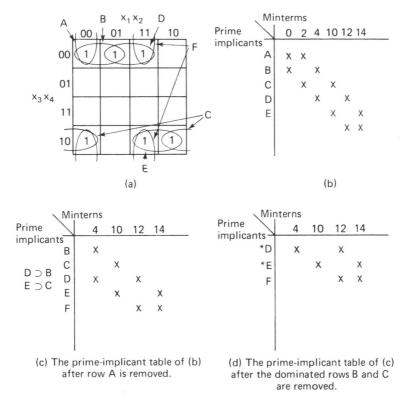

(a)

Prime implicants	Minterms 0	2	4	10	12	14
A	X	X				
B	X		X			
C		X		X		
D			X		X	
E				X		X
					X	X

(b)

Prime implicants	Minterns 4	10	12	14
B	X			
C		X		
D	X		X	
E		X		X
F			X	X

$D \supset B$
$E \supset C$

(c) The prime-implicant table of (b) after row A is removed.

Prime implicants	Minterns 4	10	12	14
*D	X		X	
*E		X		X
F			X	X

(d) The prime-implicant table of (c) after the dominated rows B and C are removed.

Fig. 1.5.5 An example of function minimization involving a cyclic table.

table is not cyclic, because rows B and C are dominated by rows D and E, respectively, and thus can be removed from the table. The remaining table is shown in Fig. 1.5.5(d), in which rows D and E are essential and cover the remaining columns. Hence a possible minimal sum is $S_1 = A + D + E$. Another possible minimal sum can be obtained from the table of Fig. 1.5.5(b) by selecting row B, C, or F for inclusion in the possible minimal sum. This will result in a second possible minimal sum, $S_2 = B + C + F$. These are the only possible minimal sums of the function. By comparing their costs, we find that they are identical; therefore, both of them are minimal sums of the function.

Example 1.5.7

As another example of a function involving a cyclic table, consider

$$f(x_1, x_2, x_3, x_4) = \Sigma \, (0, 1, 3, 4, 7, 12, 13, 15)$$

which has eight prime implicants [see Fig. 1.5.6(a)] which are denoted by A, B, C, D, E, F, G, and H. None of them is essential [see Fig. 1.5.6(b)]. There are two possible minimal sums that can be obtained from the cyclic table of Fig. 1.5.6(b). They are: $S_1 = A + C + E + G$ (by initially selecting A, C, E, or G) and $S_2 = B + D + F + H$ (by initially selecting B, D, F, or H). Both of them have the same cost; thus they are minimal sums of the function.

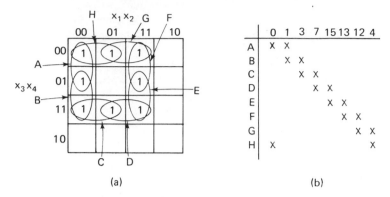

Fig. 1.5.6   Example 1.5.7.

In summary, we can describe the Quine–McCluskey method by a flow chart shown in Fig. 1.5.7. As a final remark, with slight modification, the method described in this section can be applied to the minimization of a switching function in the canonical product-of-sums form. The derivation of this procedure is left to the reader as an exercise (Problem 6).

Exercise 1.5

1. Complete the following minimization procedure for minimizing the function

$$f = \sum (0, 1, 2, 3, 6, 7, 8, 9, 10, 13, 14, 15)$$

Decimal Number	Binary Representation of Each Minterm	Decimal Numbers	First Reduction
0	0 0 0 0 ✓	0, 1	0 0 0 –
1	0 0 0 1 ✓	0, 2	0 0 – 0
2	0 0 1 0 ✓	0, 8	– 0 0 0
8	1 0 0 0 ✓	1, 3	0 0 – 1
3	0 0 1 1 ✓	1, 9	– 0 0 1
6	0 1 1 0 ✓	2, 3	0 0 1 –
9	1 0 0 1 ✓	2, 6	0 – 1 0
10	1 0 1 0 ✓	2, 10	– 0 1 0
7	0 1 1 1 ✓	8, 9	1 0 0 –
13	1 1 0 1 ✓	8, 10	1 0 – 0
14	1 1 1 0 ✓	3, 7	0 – 1 1
15	1 1 1 1 ✓	6, 7	0 1 1 –
		6, 14	– 1 1 0
		9, 13	1 – 0 1
		10, 14	1 – 1 0
		7, 15	– 1 1 1
		13, 15	1 1 – 1
		14, 15	1 1 1 –

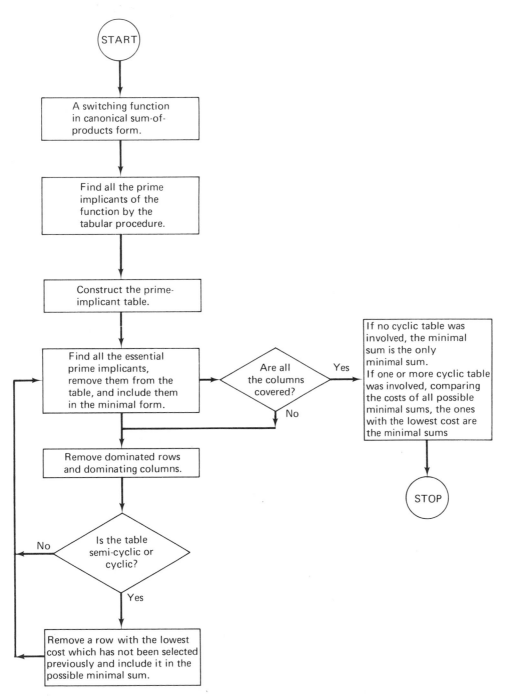

Fig. 1.5.7 A flow chart describing the Quine-McCluskey method.

2. Minimize the function of problem 1 with minterms 8 and 10 changed to don't cares.

3. Find the prime implicants, essential prime implicants, and a minimum sum-of-products for the functions of problems 1(a), (b), and (c) of Exercise 1.4. Use the Quine–McCluskey method and compare your results with those obtained by the Karnaugh method.

4. Minimize the following functions which involve cyclic tables. Use the Quine–McCluskey method.

(a) $f(x_1, x_2, x_3, x_4) = \sum (4, 5, 6, 13, 14, 15)$

(b) $f(x_1, x_2, x_3, x_4) = \sum (4, 5, 6, 8, 9, 10, 13, 14, 15)$

(c) $f(x_1, x_2, x_3, x_4) = \sum (4, 5, 6, 13, 14, 15) + \sum_d (8, 9, 10)$

(d) $f(x_1, x_2, x_3, x_4, x_5) = \sum (0, 4, 8, 10, 20, 26, 28, 30)$

5. Minimize the following functions using the Quine–McCluskey method.

(a) $f(x_1, \ldots, x_5) = \sum (0, 2, 4, 9, 11, 14, 15, 16, 17, 19, 23, 25, 29, 31)$

(b) $f(x_1, \ldots, x_5) = \sum (1, 3, 4, 5, 7, 8, 10, 12, 13, 14, 17, 19, 20, 21, 22, 23, 28, 29, 30, 31)$

(c) $f(x_1, \ldots, x_5) = \sum (0, 1, 2, 3, 5, 7, 9, 11, 17, 19, 29, 31) + \sum_d (4, 6, 13, 15, 21, 23, 25, 27)$

(d) $f(x_1, \ldots, x_6) = \sum (0, 1, 3, 4, 8, 12, 23, 28, 30, 39, 43, 48, 49, 63)$

(e) $f(x_1, \ldots, x_6) = \sum (1, 3, 5, 7, 9, 11, 13, 15, 20, 21, 22, 23, 28, 29, 30, 31, 36, 37, 38, 39, 44, 45, 46, 47, 49, 51, 53, 55, 57, 59, 61, 63)$

(f) $f(x_1, \ldots, x_6) = \sum (5, 7, 13, 15, 16, 18, 21, 23, 24, 26, 29, 31, 32, 34, 40, 42, 45, 47, 53, 55, 56, 57, 58, 59)$

(g) $f(x_1, \ldots, x_6) = \sum (0, 1, 4, 7, 8, 10, 12, 17, 19, 21, 23, 33, 35, 37, 39, 41, 43, 52, 54, 60, 62) + \sum_d (2, 3, 9, 15, 28, 29, 30, 31, 45, 47, 53, 55, 61, 63)$

6. (a) Derive the Quine–McCluskey tabular procedure for minimizing a switching function in the product-of-sums form.

(b) Apply this procedure to minimize the following functions:

(1) $f(x_1, \ldots, x_5) = \prod (0, 1, 2, 3, 8, 9, 10, 11, 17, 19, 21, 23, 25, 27, 29, 31)$

(2) $f(x_1, \ldots, x_5) = \prod (0, 2, 4, 6, 8, 10, 12, 14, 16, 18, 20, 22, 25, 27, 29, 31) + \prod_d (5, 7, 13, 15, 24, 26, 28, 30)$

(3) $f(x_1, \ldots, x_6) = \prod (4, 5, 6, 7, 8, 9, 10, 11, 16, 19, 24, 25, 41, 43, 45, 47, 49, 51, 53, 55) \cdot \prod_d (1, 3, 12, 14, 17, 27, 48, 50, 57, 59, 61, 63)$

1.6 Function Minimization of Multiple-Output Circuits

So far we have discussed the minimization of the single switching function by the Karnaugh method and the Quine–McCluskey method. In practice, many logic design problems involve designing switching circuits to have more than one output. For these problems, which are too large or complicated to be solved by the Karnaugh method, it is necessary to turn to a tabular method for either hand computation or digital-computer usage. The multiple-switching-function minimization criterion is as follows:

1. Each minimized function should have as many terms in common with those in other minimized functions as possible.

2. Each minimized function should have a minimum number of product (sum) terms and no product (sum) term of it can be replaced by a product (sum) term with few variables.

The procedure for minimizing a set of m switching functions according to the above minimization criterion using the Quine–McCluskey method is described below.

Step 1   Find the prime implicants of the m switching functions. In finding the prime implicants of the m switching functions, we use a single table to find the prime implicants of the m functions simultaneously by using the tag. Suppose that we want to minimize the set of three output functions

$$f_1(x_1, x_2, x_3, x_4) = \sum (1, 2, 3, 5, 7, 8, 9, 12, 14) \qquad (1.6.1)$$
$$f_2(x_1, x_2, x_3, x_4) = \sum (0, 1, 2, 3, 4, 6, 8, 9, 10, 11) \qquad (1.6.2)$$
$$f_3(x_1, x_2, x_3, x_4) = \sum (1, 3, 5, 7, 8, 9, 12, 13, 14, 15) \qquad (1.6.3)$$

The single table that can be used to find the prime implicants of the three functions simultaneously is shown in Fig. 1.6.1(a). Notice that the first column of this table includes the minterms (represented in decimal numbers) of all three functions. They are grouped into groups as usual according to the number of 1's in their binary representations, which are shown in the second column. The third column consists of a tag for each minterm (row) for indicating which of the output functions include the minterm. Each symbol of the tag corresponds to one of the output functions and is either 0 or 1, depending on whether the corresponding minterm is not included in the output function or is included in the output function. For example, the tag of the first row is

f_1	f_2	f_3
0	1	0

which indicates that the minterm 0 is included in the output function f_2 but is not included in the output functions f_1 and f_3.

The first, second, and third reductions of the implicants to form prime implicants are the same as those in the single-output technique. The rule for obtaining the tag of each row of the tables of Fig. 1.6.1(b), (c), and (d) is as follows:

Symbol of the Tag of Row i in the $(k - 1)$th Reduction Table	Symbol of the Tag of Row j in the $(k - 1)$th Reduction Table	Symbol of the Tag of the Row in the kth Reduction Table Obtained from Rows i and j of the $(k - 1)$th Reduction Table
0	0	0
0	1	0
1	0	U
1	1	1

Notice that there is a 1 in a new tag *only* where there are 1's in *both* the original implicants. The tag of the implicant (0, 1) in the first row of the table in Fig. 1.6.1(b) is

f_1	f_2	f_3
0	1	0

This is because only the symbols of the tags of the minterms 0 and 1 under f_2 are both 1. The reasoning behind this tag-formation rule is as follows. The new 2^i-cell implicant corresponds to a product term which is included in those functions which include *both* the 2^{i-1}-cell implicants used in forming the product term. It should be noted that when the tag of a new 2^i-cell implicant is (0 0 . . . 0), which means that none of the output functions includes this new implicant, this new implicant should, therefore, not be included. For example, in the table of Fig. 1.6.1(a), minterms 4 and 5 are combinable. Since their tags are (0 1 0) and (1 0 1), the resulting tag is (0 0 0); thus implicant (4, 5) does not appear in the table of Fig. 1.6.1(b). For the same reason, implicant (11, 15) is not included in the same table.

It is important to point out that there is a difference between single-function minimization and multiple-function minimization in determining prime implicants in

	Decimal Number	Binary Representation of Each Term	$f_1\ f_2\ f_3$
Index 0 {	0	0 0 0 0	0 1 0 ✓
Index 1	1	0 0 0 1	1 1 1 ✓
	2	0 0 1 0	1 1 0 ✓
	4	0 1 0 0	0 1 0 ✓
	8	1 0 0 0	1 1 1 ✓
Index 2	3	0 0 1 1	1 1 1 ✓
	5	0 1 0 1	1 0 1 ✓
	6	0 1 1 0	0 1 0 ✓
	9	1 0 0 1	1 1 1 ✓
	10	1 0 1 0	0 1 0 ✓
	12	1 1 0 0	1 0 1 ✓
Index 3	7	0 1 1 1	1 0 1 ✓
	11	1 0 1 1	0 1 0 ✓
	13	1 1 0 1	0 0 1 ✓
	14	1 1 1 0	1 0 1 ✓
Index 4 {	15	1 1 1 1	0 0 1 ✓
			Tag

(a)

Decimal Numbers	First Reduction	$f_1\ f_2\ f_3$
0, 1	0 0 0 –	0 1 0 ✓
0, 2	0 0 – 0	0 1 0 ✓
0, 4	0 – 0 0	0 1 0 ✓
0, 8	– 0 0 0	0 1 0 ✓
1, 3	0 0 – 1	1 1 1 A
1, 5	0 – 0 1	1 0 1 ✓
1, 9	– 0 0 1	1 1 1 B
2, 3	0 0 1 –	1 1 0 C
2, 6	0 – 1 0	0 1 0 ✓
2, 10	– 0 1 0	0 1 0 ✓
4, 6	0 1 – 0	0 1 0 ✓
8, 9	1 0 0 –	1 1 1 D
8, 10	1 0 – 0	0 1 0 ✓
8, 12	1 – 0 0	1 0 1 E
3, 7	0 – 1 1	1 0 1 ✓
3, 11	– 0 1 1	0 1 0 ✓
5, 7	0 1 – 1	1 0 1 ✓
5, 13	– 1 0 1	0 0 1 ✓
9, 11	1 0 – 1	0 1 0 ✓
9, 13	1 – 0 1	0 0 1 ✓
10, 11	1 0 1 –	0 1 0 ✓
12, 13	1 1 0 –	0 0 1 ✓
12, 14	1 1 – 0	1 0 1 F
7, 15	– 1 1 1	0 0 1 ✓
13, 15	1 1 – 1	0 0 1 ✓
14, 15	1 1 1 –	0 0 1 ✓
		Tag

(b)

Fig. 1.6.1 Example of minimizing a set of three output functions using the Quine–McCluskey method.

Decimal Numbers	Second Reduction	$f_1\ f_2\ f_3$	
0, 1, 2, 3	0 0 – –	0 1 0	✓
0, 1, 8, 9	– 0 0 –	0 1 0	✓
0, 2, 4, 6	0 – – 0	0 1 0	G
0, 2, 8, 10	– 0 – 0	0 1 0	✓
1, 3, 5, 7	0 – – 1	1 0 1	H
1, 3, 9, 11	– 0 – 1	0 1 0	✓
1, 5, 9, 13	– – 0 1	0 0 1	I
2, 3, 10, 11	– 0 1 –	0 1 0	✓
8, 9, 10, 11	1 0 – –	0 1 0	✓
8, 9, 12, 13	1 – 0 –	0 0 1	J
5, 7, 13, 15	– 1 – 1	0 0 1	K
12, 13, 14, 15	1 1 – –	0 0 1	L

Tag

(c)

Decimal Numbers	Third Reduction	$f_1\ f_2\ f_3$	
0, 1, 2, 3, 8, 9, 10, 11	– 0 – –	0 1 0	M

Tag

(d)

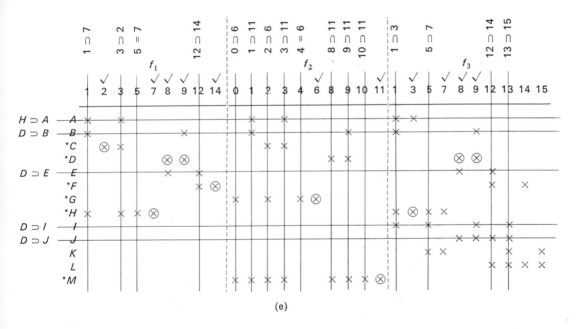

(e)

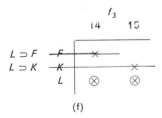

(f)

Fig. 1.6.1   (continued)

the function reduction table. For example, in the first reduction column of Fig. 1.6.1(b), the implicants (0, 2) and (1, 3) are combined into the implicant (0, 1, 2, 3) whose tag is (0 1 0) which is the same as that of the implicant (0, 2), but *not* that of the implicant (1, 3); hence the implicant (0, 2) is "checked out", but the implicant (1, 3) is not. The rule for checking out implicants in a multiple-function reduction table is: an implicant is checked out ($\checkmark$) if and only if its tag is identical to that of the newly-formed implicant.

Step 2 Construct the multiple-output prime-implicant table as shown in Fig. 1.6.1(e).

The row dominance, row $I \supset$ row J, in a multiple-output prime-implicant table is defined as: row $I \supset$ row J for every output function. The definition of column dominance for the multiple-output case is the same as that for the single-output case.

Step 3 Remove all the dominating columns from the table.

Step 4 Remove all the dominated rows from the table.

Step 5 Remove the essential prime implicants from the table and include them in their respective function. For over example, with reference to Fig. 1.6.1(e), we find that $f_1 = C + D + F + H$, $f_2 = G + M$, and two terms (essential prime implicants) of f_3, which are D and H. The remaining terms of f_3 are to be determined from the remaining prime-implicant table shown in Fig. 1.6.1(f).

Step 6 Repeat Steps 3–5 as many times as needed until every minterm of each function is covered by an essential prime implicant. Fig. 1.6.1(f) shows that $L \supset F$ and $L \supset K$, and thus rows F and K can be removed from the table. Therefore, the set of minimized output functions are

$$f_1(x_1, x_2, x_3, x_4) = C + D + F + H = x_1'x_2'x_3 + \underline{x_1x_2'x_3'} + x_1x_2x_4' + \underline{x_1'x_4} \quad (1.6.4)$$

$$f_2(x_1, x_2, x_3, x_4) = G + M = x_1'x_4' + x_2' \quad (1.6.5)$$

$$f_3(x_1, x_2, x_3, x_4) = D + H + L = \underline{x_1x_2'x_3'} + \underline{x_1'x_4} + x_1x_2 \quad (1.6.6)$$

A multiple-output prime-implicant table is *semicyclic*† if it is semicyclic with respect to each of the output functions. The criterion in selecting a row of the table to include in the minimal sum is to select a low-cost row that is contained in several output functions. The reasoning behind this is that in the final realization, the gate that realizes this row "will be shared" by the realizations of several minimized output functions; hence the complexity of the circuitry will be reduced. After such a row is selected and included in the minimal sum, we can use the reduction techniques described above to remove rows and columns from the table. Just as in the single-output case, this entire process must be repeated for each row that could replace the original selected row, and the final minimal sum is obtained by comparing the costs of the expressions that result from each arbitrary choice of a selected row.

It is important to point out that the cost of the minimized multiple-output circuit obtained by the method described in this section is in general much lower compared

---

†The definition of a semicyclic table for the single output case was given in Definition 1.5.4, which includes that of a cyclic table as a special case.

to that of the circuit obtained by applying the single-function tabular minimization method described in Section 1.5 to the set of output functions individually (Problem 1).

By slight modification, the minimization procedure described in this section can be applied to the minimization of a set incompletely specified functions.

Exercise 1.6

1. Show that the cost of the minimized multiple-output circuit constructed from Eqs. (1.6.4), (1.6.5), and (1.6.6) is lower than that of the circuit obtained by applying a single-function minimization method to the functions of Eqs. (1.6.1), (1.6.2), and (1.6.3), individually.

2. Minimize the following sets of functions using the Quine–McCluskey method.
   (a) $f_1(x_1, x_2, x_3) = x_1 x_2 + x_2' x_4' + x_3 x_4' + x_4$
       $f_2(x_1, x_2, x_3) = x_2' x_4' + x_3 x_4' + x_1 x_3 + x_2' x_3' x_4'$
   (b) $f_1(x_1, x_2, x_3, x_4) = x_1 + x_4 + x_1' x_3' x_4$
       $f_2(x_1, x_2, x_3, x_4) = x_2' x_4' + x_1 x_2 x_4 + x_2 x_3' x_4$
       $f_3(x_1, x_2, x_3, x_4) = x_2' x_4' + x_1 x_2 x_3 x_4 + x_1' x_2' x_3'$
   (c) $f_1(x_1, x_2, x_3, x_4) = x_1 + x_1' x_3' x_4 + x_4$
       $f_2(x_1, x_2, x_3, x_4) = x_1 x_2 x_4 + x_2' x_4' + x_2 x_3' x_4$
       $f_3(x_1, x_2, x_3, x_4) = x_1 x_2 x_3 x_4 + x_2' x_4' + x_1' x_2' x_3'$
   (d) $f_1(x_1, x_2, x_3, x_4) = \sum (2, 3, 4, 5, 6, 7, 11, 14) + \sum_d (9, 10, 13, 15)$
       $f_2(x_1, x_2, x_3, x_4) = \sum (0, 1, 3, 4, 5, 7, 11, 14) + \sum_d (8, 10, 12, 13)$
   (e) $f_1(x_1, x_2, x_3, x_4) = \prod (3, 4, 5, 7, 11, 13, 15) \cdot \prod_d (6, 8, 10, 12)$
       $f_2(x_1, x_2, x_3, x_4) = \prod (2, 7, 9, 10, 11, 12, 14, 15) \cdot \prod_d (0, 4, 6, 8)$
   (f) $f_1(x_1, \ldots, x_5) = \sum (0, 2, 4, 6, 9, 10, 13, 14, 15, 16, 17, 21, 25, 28, 30, 31)$
       $f_2(x_1, \ldots, x_5) = \sum (0, 1, 3, 8, 9, 13, 14, 15, 16, 17, 19, 25, 27, 31)$

3. Minimize the set of three incompletely specified functions using the Quine-McCluskey method.
   $f_1(x_1, x_2, x_3, x_4) = \sum (1, 2, 3, 5, 7, 8, 9) + \sum_d (12, 14)$
   $f_2(x_1, x_2, x_3, x_4) = \sum (0, 1, 2, 3, 4, 6, 8, 9) + \sum_d (10, 11)$
   $f_3(x_1, x_2, x_3, x_4) = \sum (1, 3, 5, 7, 8, 9, 12, 13) + \sum_d (14, 15)$

1.7   Iterative Consensus Method

In the Quine–McCluskey tabular method, if a function to be minimized is not in a canonical form, we must first convert it into a canonical form before we apply the minimization procedure. This expanded form usually contains a large number of terms that have to be handled in the process. The iterative consensus method for obtaining the prime implicants overcomes these disadvantages. It can begin with a function that is not in a canonical form. This method consists of the following two basic processes:

1. *Term generation.*   This is based on the equality

$$XY + \bar{X}Z = XY + \bar{X}Z + YZ \qquad (1.7.1)$$

Whenever two terms in an expression can be represented by XY and $\bar{X}Z$, a new term YZ (if it is not zero) is added to the expression. For example, suppose that in an expression we have two terms, $x_2' x_3' x_4$ and $x_1' x_2 x_3'$. Let $X = x_2$. Then $Y = x_1 x_3'$ and

$Z = x_3' x_4$. By Eq. (1.7.1), we can add the term $YZ = x_1 x_3' x_4$ to the expression without changing the function.

2. *Term elimination.* This is based on the following two equalities:

$$X + XY = X \qquad\qquad (1.7.2)$$

$$X + X = X \qquad\qquad (1.7.2a)$$

The second equality can be considered as a special case of the first one, with $Y = 1$. Whenever two terms in an expression can be represented by X and XY, the latter *subsumes* the former and can be eliminated from the expression with changing the function. For example, if both terms $x_1 x_2 x_3'$ and $x_2 x_3'$ are present in an expression, the terms $x_1 x_2 x_3'$ can be deleted.

The method can be stated as follows. We apply the term generation process to all pairs of terms of a given sum-of-products function (not in canonical form) systematically to all pairs of terms to obtain all possible included terms, which are added to the expression. The pairing continues as the terms are added. At the same time, the term-elimination process is applied to delete terms whenever possible. The process is continued until no more terms can be formed. The existing terms at this point comprise all the prime implicants of the given function.

Example 1.7.1

Minimize the function

$$f(x_1, x_2, x_3, x_4) = x_2' x_3' x_4 + x_1' x_2 x_3' + x_1 x_2 x_3' + x_1 x_2 x_3 \qquad (1.7.3)$$

using the iterative consensus method.

Solution:

Start: $\boxed{x_2' x_3' x_4} + x_1' x_2 x_3' + x_1 x_2 x_3' + x_1 x_2 x_3$

First cycle: $x_2' x_3' x_4 + \boxed{x_1' x_2 x_3'} + x_1 x_2 x_3' + x_1 x_2 x_3 + x_1' x_3' x_4 + x_1 x_3' x_4$
$\qquad\qquad\qquad\qquad\qquad\qquad\qquad\qquad\qquad\qquad\qquad \underbrace{\qquad\qquad\qquad\qquad}_{\text{added terms}}$

Second cycle: $x_2' x_3' x_4 + \cancel{x_1' x_2 x_3'} + \cancel{x_1 x_2 x_3'} + \boxed{x_1 x_2 x_3} + x_1' x_3' x_4 + x_1 x_3' x_4 + x_2 x_3' + x_2 x_3' x_4$
$\qquad\qquad\qquad\qquad \text{(because they subsume } x_2 x_3') \qquad\qquad\qquad\qquad\qquad\qquad\qquad \underbrace{\qquad\qquad}_{\text{added terms}}$

Third cycle: $x_2' x_3' x_4 + \cancel{x_1 x_2 x_3} + \boxed{x_1' x_3' x_4} + x_1 x_3' x_4 + x_2 x_3' + \cancel{x_1 x_2 x_4} + \cancel{x_1 x_2}$
$\qquad\qquad\qquad\qquad\qquad\qquad\qquad\qquad\qquad\qquad\qquad\qquad\qquad \underbrace{\qquad\qquad}_{\text{added terms}}$

Fourth cycle: $\cancel{x_2' x_3' x_4} + \cancel{x_1' x_3' x_4} + \cancel{x_1 x_3' x_4} + x_2 x_3' + x_1 x_2 + \underbrace{x_3' x_4 + x_2 x_3' x_4}_{\text{added terms}}$

This procedure is terminated at the fourth cycle, since no more included terms can be formed. Therefore, the prime implicants of the function are $x_2 x_3'$, $x_1 x_2$, and $x_3' x_4$.

This process is more commonly carried out in tabular form. This is described as follows.

Representation of a Term Each term of the function is represented by a row consisting of 1's, 0's, and dashes. A 1, 0, and dash (–) denote that the corresponding variable is in true form, complemented form, or missing, respectively. For example, the term $x_2'x_3'x_4$ of Eq. (1.7.3) is represented by – 001.

All pairs of rows—original rows, and rows that may be added to the table—are systematically compared for consensus [Eq. (1.7.1)] and subsumption [Eq. (1.7.2)]. The rules for term generation and term elimination are described below.

Rule for Term Generation Two rows generate a consensus row if, *in one column only*, one row has a 1 and the other has a 0. The consensus row has a dash in that column. In the remaining columns, the entries are determined by the following table:

¢	1	0	–
1	1	x	1
0	x	0	0
–	1	0	–

The entries marked by x mean that they should not occur if the consensus row is nonzero. For example, in the example above, the consensus term of $x_2'x_3'x_4$ and $x_1'x_2x_3'$ can be found by applying this rule as

	x_1	x_2	x_3	x_4	
A	–	0	0	1	
B	0	1	0	–	
$A \notin B$	0	–	0	1	$(x_1'x_3'x_4)$

The consensus row is added to the table only if it does not subsume a row already in the table.

Rule for Term Elimination A row subsumes another row if it has a 1 in every column in which other row has a 1, and if it has a 0 in every column in which the other row has a 0. Any row that subsumes another is eliminated. For example, the first three rows in the following table subsume the fourth row and therefore can all be eliminated.

	x_1	x_2	x_3	x_4	
A	0	1	0	–	$(x_1'x_2x_3')$
B	1	1	0	–	$(x_1x_2x_3')$
C	–	1	0	1	$(x_2x_3'x_4)$
D	–	1	0	–	(x_2x_3')

(see the second cycle of the minimization
procedure in Example 1.7.1)

This tabular method for obtaining prime implicants of a function is illustrated by the following example.

Example 1.7.2

Minimize the function of Eq. (1.7.3) by use of the iterative consensus tabular method.

Solution:

		x_1	x_2	x_3	x_4	
Initial sum-	A	–	0	0	1	
of-products	B	0	1	0	–	
	C	1	1	0	–	
	D	1	1	1	–	
First cycle	A	–	0	0	1	
	B	0	1	0	–	
	C	1	1	0	–	
	D	1	1	1	–	
	$E = A ¢ B$	0	–	0	1	
	$F = A ¢ C$	1	–	0	1	
Second cycle	A	–	0	0	1	
	B	0	1	0	–	✓
	C	1	1	0	–	✓
	D	1	1	1	–	
	E	0	–	0	1	
	F	1	–	0	1	
	$G = B ¢ C$	–	1	0	–	
	$H = B ¢ F$	–	1	0	1	✓
Third cycle	A	–	0	0	1	
	D	1	1	1	–	✓
	E	0	–	0	1	
	F	1	–	0	1	
	G	–	1	0	–	
	$I = D ¢ F$	1	1	–	1	✓
	$J = D ¢ G$	1	1	–	–	
Fourth cycle	A	–	0	0	1	✓
	E	0	–	0	1	✓
	F	1	–	0	1	✓
	G	–	1	0	–	$(x_2 x_3')$
	J	1	1	–	–	$(x_1 x_2)$
	$K = E ¢ F$	–	–	0	1	$(x_3' x_4)$

Note that a check mark placed at the right of a row means that the row is deleted.

After the prime implicants of a function are obtained by the iterative consensus method, the selection of an optimum set of these prime implicants is then accomplished as described earlier in the chapter.

With some modification, the iterative consensus method can be applied to the obtaining of prime implicants of (1) product-of-sums functions and (2) functions of a multiple-output circuit.

Exercise 1.7

Find the prime implicants of the following functions. Use the iterative consensus tabular method.

1. $f(x_1, x_2, x_3, x_4) = x_1 x_2 x_3' + x_1' x_2 x_3 + x_1 x_2' x_3' + x_1 x_3 x_4$

2. $f(x_1, x_2, x_3, x_4) = x_1' x_3' x_4 + x_1 x_2' x_4 + x_1 x_2 x_3' x_4 + x_1' x_2' x_3 x_4 + x_1 x_2 x_3' x_4'$

3. The set of functions of a multiple-output circuit:

$$f_1(x_1, x_2, x_3, x_4) = x_2 x_3' x_4 + x_1' x_2' x_3' + x_2 x_3 x_4 + x_1' x_2 x_3' x_4' + x_1 x_2' x_3' x_4$$
$$f_2(x_1, x_2, x_3, x_4) = x_1' x_2 x_4 + x_2' x_3 x_4' + x_1 x_2 x_4 + x_1' x_2' x_3' + x_1 x_2' x_3' x_4'$$

1.8 Basic Procedure for Designing Combinational Logic Circuits

The basic procedure of designing combinational logic circuits based on the gate-minimization criterion to optimize a design is presented in this section. This procedure consists of five steps and is illustrated by the following design example.

Example 1.8.1

Design a combinational circuit that is to count the coins placed into an automatic toll collector used on toll highways to speed traffic flow. Suppose that the toll is 15 cents and that the machine accepts nickels and dimes only. When 15 cents is received by the collector, the go light is flashed on and a change-collect signal is sent out to collect the coins. Otherwise, the stop light is to remain on.

The step-by-step procedure of designing combinational circuits is described below and is illustrated by this design example.

Step 1 Construct a truth table that describes the given design problem.

Examining the problem statement, we see that there are two input signals and one output signal. They are:

$$N = \text{number of nickels deposited}$$
$$D = \text{number of dimes deposited}$$
$$z = \text{command to the signal light and collection control}$$

These signals will take on the following integer and logical values:

$$0 \leq N \leq 3$$
$$0 \leq D \leq 2$$
$$z = 0 \quad \text{15 cents has not been deposited}$$
$$z = 1 \quad \text{15 cents has been deposited}$$

It is assumed that the information concerning the number of coins deposited shows up at the same time, so that the circuit does not have a "memory." Then the operations of this collector are defined by the following truth table:

N	D	z
0	0	0
0	1	0
1	0	0
2	0	0
0	2	1
1	1	1
3	0	1
1	2	1
2	1	1
2	2	1
3	1	1
3	2	1

Step 2   Code the input and output symbols of the system in binary codes and express the truth table obtained in step 1 in coded form.

It is easy to see that the minimum number of input variables needed to code the numbers of nickels and dimes deposited is 4. We shall use the input variables x_1, x_2 and x_3, x_4 to code the number of nickels deposited and the number of dimes deposited, respectively.

N	x_1	x_2
0	0	0
1	0	1
2	1	0
3	1	1

D	x_3	x_4
0	0	0
1	0	1
2	1	0

The output signals in this example are already in binary, so we do not need to code them. Thus the preceding truth table, expressed in term of x_1, x_2, x_3, x_4, and z, is as follows:

x_1	x_2	x_3	x_4	z
0	0	0	0	0
0	0	0	1	0
0	0	1	0	1
0	0	1	1	d
0	1	0	0	0
0	1	0	1	1
0	1	1	0	1
0	1	1	1	d
1	0	0	0	0
1	0	0	1	1
1	0	1	0	1
1	0	1	1	d
1	1	0	0	1
1	1	0	1	1
1	1	1	0	1
1	1	1	1	d

Step 3 Obtain the switching function in the canonical sum-of-products from the coded truth table.

$$z = f(x_1, x_2, x_3, x_4) = \sum (2, 5, 6, 9, 10, 12, 13, 14) + \sum_d (3, 7, 11, 15)$$

Step 4 Obtain the minimal sum of the switching function. From the Karnaugh map of Fig. 1.8.1(a), we find that the minimal sum is

$$z = f(x_1, x_2, x_3, x_4) = A + B + C + D = x_3 + x_1 x_2 + x_2 x_4 + x_1 x_4$$

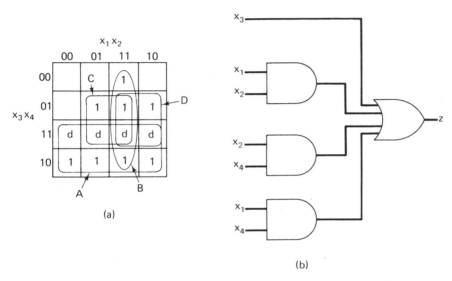

Fig. 1.8.1 Logic design of Example 1.8.1.

Step 5 Obtain the logic-circuit realization of the minimized switching function. The AND–OR two-level realization of the minimized function obtained in step 4 is depicted in Fig. 1.8.1 (b).

Exercise 1.8

1. Use the procedure described in this section to design a majority circuit M, which is as follows. Referring to the circuit of Fig. P1.8.1, C_1, C_2, and C_3 are three copies of the desired circuit. The outputs of these three circuits are fed to a fourth circuit M, which is constructed of components of very high reliability. The output signal z of circuit M agrees with the majority of the input signals y_1, y_2, and y_3 to the circuit. The overall circuit will thus have the correct output, even though the output of one of the copies of the desired circuit is in error.

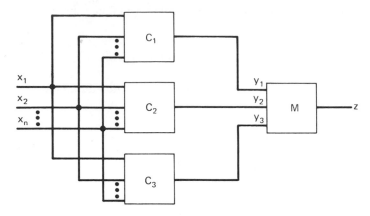

Fig. P1.8.1

2. It is desired to design an indicator circuit for a room that has two swinging doors, D_1 and D_2. Two switches, s_{i1} and s_{i2}, are associated with each door D_i. The s_{i1} (s_{i2}) switch is closed only when the corresponding door D_i is open in (open out). An indicator lamp is to be lit whenever there is a clear path through the room (one door open in and the other door open out).

Bibliographical Remarks

All the standard textbooks (references 1–11) have extensive discussions on the subjects of switching algebra and switching function. Excellent treatments on switching algebra (Section 1.2) and switching function (Section 1.3) are contained in references 1 and 6, respectively. A detailed discussion of switching-function minimization can be found in reference 8. Reference 12 is the original paper of the Karnaugh method (Section 1.4), references 13–15 are the original papers of the Quine–McCluskey method (Section 1.5), and references 16 and 17 are the original papers on the function minimization of multiple-output switching circuits (Section 1.6). They are recommended reading. Minimization of incompletely specified switching functions can be found in references 18 and 19. Two computer algorithms for generating prime implicants are contained in references 20–22. Several computer algorithms for finding a minimal prime-implicant cover of a given switching function from its prime-implicant table can be found in references 23–26. An extensive treatment on the iterative consensus method (Section 1.7) can be found in reference 8.

References

Books

1. S. H. CALDWELL, *Switching Circuits and Logical Design*, Wiley, New York, 1958.

2. D. D. GIVONE, *Introduction to Switching Theory*, McGraw-Hill, New York, 1970.

3. M. A. HARRISON, *Introduction to Switching and Automata Theory*, McGraw-Hill, New York, 1965.

4. F. J. HILL and G. R. PETERSON, *Introduction to Switching Theory and Logical Design*, Wiley, New York, 1968.

5. Z. KOHAVI, *Switching and Finite Automata Theory*, McGraw-Hill, New York, 1970.

6. R. R. KORFHAGE, *Logic and Algorithms*, Wiley, New York, 1966.

7. M. KRIEGER, *Basic Switching Circuit Theory*, McGraw-Hill, New York, 1970.

8. E. J. McCLUSKEY, JR., *Introduction to the Theory of Switching Theory and Logical Design*, Wiley, New York, 1968.

9. R. E. MILLER, *Switching Theory*, Vol. 1, Wiley, New York, 1968.

10. H. C. TORNG, *Switching Circuits Theory and Logic Design*, Addison-Wesley, Reading, Mass., 1972.

11. P. E. WOOD, JR., *Switching Theory*, McGraw-Hill, New York, 1968.

Papers
Function Minimization (Original Papers)

12. M. KARNAUGH, "The Map Method of Synthesis of Combinational Logic Circuits," *Commun. Electron.*, No. 9, 1953, pp. 593–599.

13. E. J. McCLUSKEY, JR., "Minimization of Boolean Functions," *Bell System Tech. J.*, Vol. 35, No. 6, 1956, pp. 1417–1444.

14. W. V. QUINE, "The Problem of Simplifying Truth Functions," *Amer. Math. Monthly*, Vol. 59, No. 8, 1952, pp. 521–531.

15. W. V. QUINE, "A Way to Simplify Truth Function," *Amer. Math. Monthly*, Vol. 62, No. 9, 1955, pp. 627–631.

Function Minimization

16. S. R. DAS, "An Approach for Simplifying Switching Functions by Utilizing the Cover Table Representation," *IEEE Trans. Computers*, Vol. C-20, Mar. 1971, pp. 355–359.

17. S. R. DAS, and A. K. CHOUDHURY, "Maxterm Type Expression of Switching Functions and Their Prime Implicants," *IEEE Trans. Electron. Computers*, Vol. EC-14, Dec. 1965, pp. 920–923.

18. E. J. McCLUSKEY, JR., "Minimal Sums for Boolean Functions Having Many Unspecified Fundamental Products," *AIEE Trans. Commun. Electron.*, Vol. 81, Nov. 1962, pp. 387–392.

19. K. K. ROY, A. K. CHOUDHURY, and S. R. DAS, "Simplification of Switching Functions Involving a Very Large Number of 'Don't Care' States," *Int. J. Contr.*, Vol. 3, Jan. 1966, pp. 17–28.

20. J. R. SLAGLE, C.-L. CHANG, and R. C. T. LEE, "A New Algorithm for Generating Prime Implicants," *IEEE Trans. Computers*, Vol. C-19, No. 4, 1970, pp. 304–310.

21. S. R. DAS, "Comments on 'A New Algorithm for Generating Prime Implicants'," *IEEE Trans. Computers*, Vol. C-20, Dec. 1971, pp. 1614–1615.

22. S. R. DAS and N. S. KHABRA, "Clause-Column Table Approach for Generating All the Prime Implicants of Switching Functions," *IEEE Trans. Computers*, Vol. C-21, No. 11, 1972, pp. 1239–1246.

Function-Minimization Computer Programs

23. V. BUBENIK, "Weighting Method for Determination of the Irredundant Set of Prime Implicants," *IEEE Trans. Computers*, Vol. C-21, Dec. 1972, pp. 1449–1451.

24. A. COBHAM, R. FRIDSHA, and J. H. NORTH, "An Application of linear Programming to the Minimization of Boolean Functions," *Proc. 2nd Ann. Symp. Switching Circuit Theory and Logical Design*, AIEE Spec. Publ. S-134, pp. 3–9, 1961.

25. J. F. GIMPEL, "A Reduction Technique for Prime Implicant Tables," *IEEE Trans. Electron. Computers*, Vol. EC-14, Aug. 1965, pp. 535–541.

26. I. B. PYNE and E. J. MCCLUSKEY, JR., "The Reduction of Redundancy in Solving Prime Implicant Tables," *IRE Trans. Electron. Computers*, Vol. EC-11, Aug. 1962, pp. 473–482.

2

Combinational Logic Design

The objective of this chapter is threefold. One is to introduce the binary-number system and various binary-coded decimal systems and their conversion from one system to another, and to show how signed binary and binary-coded decimal numbers are presented in digital systems and how the addition of signed numbers is performed. These are considered to be fundamentals in logic design.

The second objective is to present several examples of logic design of commonly used combinational circuits, such as various kinds of code converters and binary and decimal adders and subtractors of unsigned and signed numbers. After the reader learns their functions and how they are designed, he or she may apply them to the designs of larger and more sophisticated circuits which use them as components. Many such examples are presented in Chapter 3.

The third objective is to show that even with today's "block approach," the function-minimization-oriented approach presented in Section 1.8 is still useful, at least in designing circuit components such as those presented in this chapter.

The digital-circuit implementations of code converters and adders using integrated circuits will be discussed in Chapter 3.

2.1 Number Systems and Their Conversion

The ordinary number system is decimal, but the number system most commonly used in computers is binary. The popularity of binary-number systems is essentially due to the simplicity of the manner in which the binary digit 0 and 1 relate to physical implementation. Binary digits 0 and 1 can easily be represented by a digital component voltage being low or high (this will be discussed in detail in Section 3.1), and the digital system can only process binary numbers or binary-coded numbers of other systems, such as decimal numbers. Because of this inherent restriction of the digital (binary) system, numbers given in other forms must be converted to binary form before they can be processed by a digital system. At the end of the process, the result (in binary form) may be converted back to its original-number-system form.

Any integral number N of n digits of base r can be expressed as

$$N_r = a_n r^n + a_{n-1} r^{n-1} + \ldots + a_1 r^1 + a_0 r^0 = \sum_{k=0}^{n} a_k r^k \qquad (2.1.1)$$

where a_i, $i = 0, 1, 2, \ldots, n$, is the digit in the $(i + 1)$th position from the right. For base r, $a_i \in \{0, 1, 2, \ldots, r - 1\}$. Decimal, binary, and octal number systems, the three most commonly used number systems, are outlined as follows:

Number System	Base	General Form	Example
Decimal	$r = 10$	$\sum_{k=0}^{n} a_k 10^k$	1975_{10}
Binary	$r = 2$	$\sum_{k=0}^{n} b_k 2^k$	$1975_{10} = 11110110111_2$
Octal	$r = 8$	$\sum_{k=0}^{n} c_k 8^k$	$1975_{10} = 3667_8$

2.1.1   Decimal-to-Binary Conversion: The Dibble-Dabble Method

There are many ways to convert a decimal number to a binary number and a binary number to its decimal equivalent, but the most popular is the dibble-dabble method. This method is used to convert a decimal to a binary number by repeatedly dividing the decimal number by 2, yielding a succession of remainders of 0 or 1. The remainders, when read in reverse order, yield the binary equivalent to the decimal number. The conversion from binary to decimal can be accomplished by the same method, in reverse order. This is illustrated by the following examples.

Example 2.1.1

Convert 1975_{10} to binary by use of the dibble-dabble method.

$$
\begin{array}{rl}
2\underline{|1975} & remainder \\
2\underline{|987} & 1 \\
2\underline{|493} & 1 \\
2\underline{|246} & 1 \\
2\underline{|123} & 0 \\
2\underline{|61} & 1 \quad \text{read in reverse order} \\
2\underline{|30} & 1 \\
2\underline{|15} & 0 \\
2\underline{|7} & 1 \\
2\underline{|3} & 1 \\
2\underline{|1} & 1 \\
0 & 1
\end{array}
$$

$$1975_{10} = 11110110111_2$$

Example 2.1.2

Convert 11110110111_2 to decimal.
(a)   Use Eq. (2.1.1).

$$11110110111_2 = 1 \times 2^{10} + 1 \times 2^9 + 1 \times 2^8 + 1 \times 2^7 + 1 \times 2^5 + 1 \times 2^4$$
$$+ 1 \times 2^2 + 1 \times 2^1 + 1 \times 2^0$$
$$= 1975_{10}$$

(b)   Use the dabble-dibble method. Begin with leftmost bit as

$$1 \times 2 = 2$$

Add the next bit:

$$2 + 1 = 3$$

Multiply and repeat:

$3 \times 2 = 6$	$6 + 1 = 7$
$7 \times 2 = 14$	$14 + 1 = 15$
$15 \times 2 = 30$	$30 + 0 = 30$
$30 \times 2 = 60$	$60 + 1 = 61$
$61 \times 2 = 122$	$122 + 1 = 123$
$123 \times 2 = 246$	$246 + 0 = 246$
$246 \times 2 = 492$	$492 + 1 = 493$
$493 \times 2 = 986$	$986 + 1 = 987$
$987 \times 2 = 1974$	$1974 + 1 = 1975$

Thus

$$11110110111_2 = 1975_{10}$$

2.1.2   Octal-to-Binary Conversion

The conversion of a binary number to an octal number can be accomplished easily by simply dividing the binary number into groups of 3 bits. For example,

$$11 \mid 110 \mid 110 \mid 111_2 = 3667_8$$

Obviously, the conversion of an octal number to binary can be accomplished by converting each octal bit of the number to 3-bit binary numbers and then cascading them.

2.1.3   Decimal-to-Octal Conversion

This conversion can be accomplished by using the dibble-dabble method. The conversion from decimal to octal can also be done by first converting it to binary, then from binary to octal, using the methods described in Sections 2.1.1 and 2.1.2.

2.2 Basic Binary Adders and Subtractors

The arithmetic operations (addition, subtraction, multiplication, and division) of binary numbers of several digits can be carried out in the same manner as that of decimal numbers. For example, the addition and subtraction of the binary numbers 10101_2 and 111_2 are

$$
\begin{array}{ll}
 & \text{borrow}\quad 111 \\
21_{10} = 10101_2 & 21_{10} = \overset{\frown}{10101}_2 \\
+\ 7_{10} = \underset{\frown}{111}_2 & -\ 7_{10} = \quad 111_2 \\
\hline
\text{carry}\quad 111 & \\
28_{10} = 11100_2 & 14_{10} = \quad 1110_2
\end{array}
$$

The basic building blocks in the arithmetic unit of a digital computer to carry out these operations are basic adders and basic subtractors. We shall first discuss the basic adder. A *half-adder* circuit adds two bits. It has two inputs and two outputs. The two inputs are the two 1-bit numbers A and B, and the two outputs are the sum S of A and B and the carry bit, denoted by C. The sum S of A and B may be either 0 or 1, as shown in Fig. 2.2.1(b). In the same figure it is seen that the carry bit C is 1 only when both A and B are 1. From the truth table it is evident that $S = A'B + AB' = (A + B)(A' + B') = A \oplus B$ and $C = AB$. An AND–OR gate realization of the half-adder is shown in Fig. 2.2.1(c-1). In this realization no circuits are common to the sum and carry operations. However, few components are often required if circuits are shared between operations. Figure 2.2.1(c-2) is a NOR-gate realization that has some circuits common to both outputs. A third realization, presented in Fig. 2.2.1(c-3), is a realization using an OR gate, a NOT gate, and two AND gates which is obtained based on the formulas $S = (A + B)(AB)'$ and $C = AB$.

A half-adder itself can hardly be of any practical use. A third input is needed for carries. An adder with three inputs is a full adder, which is depicted in Fig. 2.2.2(a), in which C_{in} denotes the carry from the prior addition and C_{out} denotes the new carry. The truth table is shown in Fig. 2.2.2(b). From this table and the Karnaugh maps of S and C_{out} [Fig. 2.2.2(c)], we find that

$$
S = A'B'C_{in} + A'BC'_{in} + AB'C'_{in} + ABC_{in}
$$
$$
= (A + B + C_{in})(A + B' + C'_{in})(A' + B + C'_{in})(A' + B' + C_{in})
$$
$$
C_{out} = AB + AC_{in} + BC_{in} = (B + C_{in})(A + B)(A + C_{in})
$$

Although the sum-of-products and the product-of-sums expressions for the sum and output are in expanded form, they cannot be minimized further. The full-adder can therefore be implemented using basic logic elements as shown in Fig. 2.2.2(d). It can also be realized by using two half-adders and an OR gate, as given in Fig. 2.2.2(e).

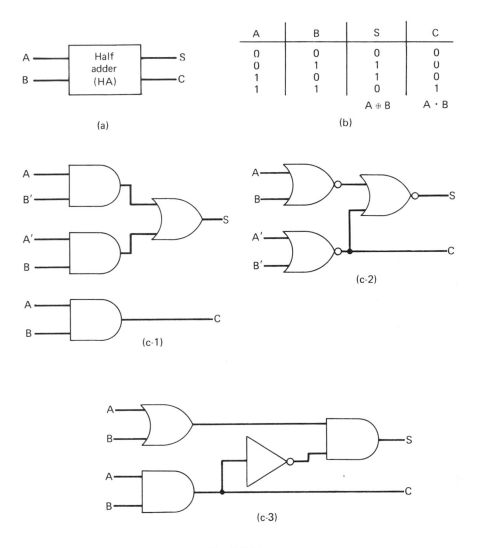

A	B	S	C
0	0	0	0
0	1	1	0
1	0	1	0
1	1	0	1
		$A \oplus B$	$A \cdot B$

(a)   (b)

(c-1)   (c-2)

(c-3)

Fig. 2.2.1

Now suppose that the numbers A and B have n-bits binary numbers, and we want to construct an n-bit adder to perform the addition of A and B. Let $A = A_n A_{n-1} \ldots A_1$ and $B = B_n B_{n-1} \ldots B_1$, where A_i's and B_i's denote n bits of A and B, respectively. The simplest n-bit combinational adder is the parallel adder shown in Fig. 2.2.3(a). This adder uses n full adders in parallel with n bits of the input numbers A and B to the n adders at the same time. The carry generated by the ith adder is fed as an input to the $(i + 1)$th adder. The sum $\sum$ of A and B is $\sum_{n+1} \sum_n \ldots \sum_1$. In this circuit arrangement, the carry must be allowed time to propagate through each

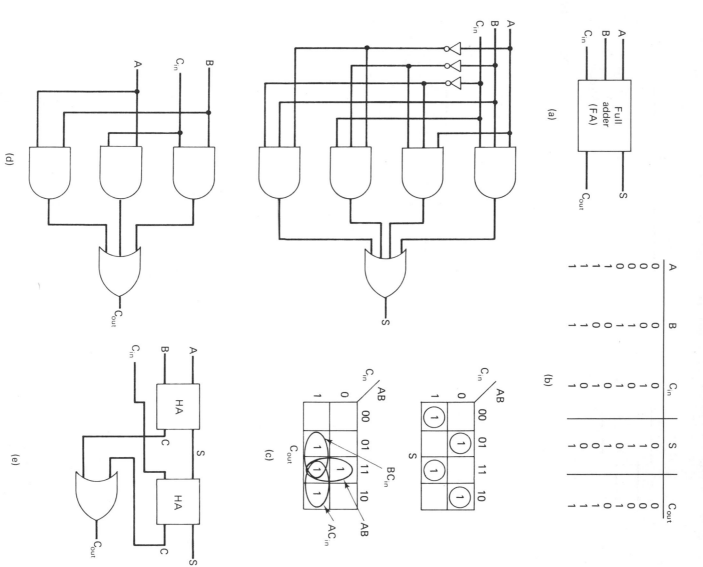

Fig. 2.2.2   Full-adder.

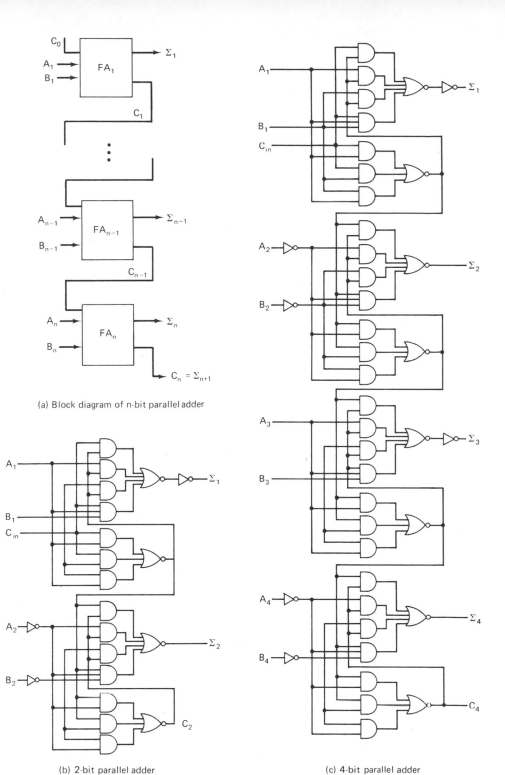

(a) Block diagram of n-bit parallel adder

(b) 2-bit parallel adder

(c) 4-bit parallel adder

Fig. 2.2.3 Parallel-adder.

of the adders before the addition can be considered complete. Two examples of parallel adder are given in Figs. 2.2.3(b) and (c).

As can be summarized from our prior adder discussion, a half-subtractor is a two-input and two-output circuit, depicted in Fig. 2.2.4(a), in which the input A designates a digit in the minuend, and the input B, the corresponding digit in the subtrahend. The two outputs are D, the difference bit, and β, the borrow bit. The truth table of the subtraction B from A is shown in Fig. 2.2.4(b). From this table we find that $D = A'B + AB' = (A + B)(A' + B') = A \oplus B$ and $\beta = A'B$. It should be noted that the expression for the difference bit is the EXCLUSIVE-OR of the inputs, which is the same expression as for the sum bit of a half-adder. One of the ways the half-subtractor can be realized is shown in Fig. 2.2.4(c).

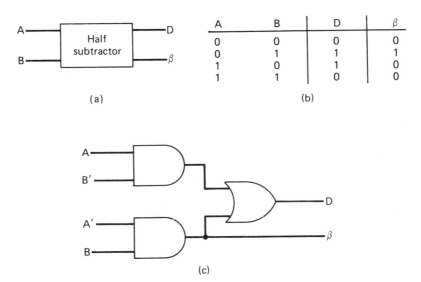

A	B	D	β
0	0	0	0
0	1	1	1
1	0	1	0
1	1	0	0

(a)   (b)

(c)

Fig. 2.2.4 Half-subtractor.

Similar to the full-adder, a full-subtractor has three inputs. Besides the minuend A and the subtrahend B, the third input is the borrow β_{in} from the prior stage of subtraction. The block diagram of a full subtractor is shown in Fig. 2.2.5(a), in which β_{out} is the borrow bit generated by the present subtraction operation. From the truth table in Fig. 2.2.5(b) and the Karnaugh maps of D and β_{out}, it is found that

$$D = A'B'\beta_{in} + A'B\beta'_{in} + AB'\beta'_{in} + AB\beta_{in}$$
$$= (A + B' + \beta'_{in})(A' + B + \beta'_{in})(A + B + \beta_{in})(A' + B' + \beta_{in})$$

and

$$\beta_{out} = A'\beta_{in} + B\beta_{in} + A'B = (A' + B)(A' + \beta_{in})(\beta_{in} + B)$$

Note that the expression for the difference bit of the full-subtractor is the same as

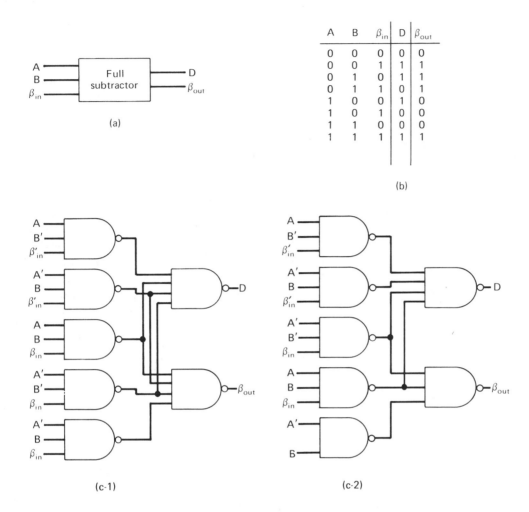

A	B	β_{in}	D	β_{out}
0	0	0	0	0
0	0	1	1	1
0	1	0	1	1
0	1	1	0	1
1	0	0	1	0
1	0	1	0	0
1	1	0	0	0
1	1	1	1	1

(b)

(a)

(c-1)   (c-2)

Fig. 2.2.5   Full-subtractor.

that for the sum bit of the full-adder. We could implement the difference expression and the minimized borrow expression easily enough, but a better realization can be obtained. From a glance at the truth table or the Karnaugh map, we see that the expanded expressions for the difference and borrow bits have three terms in common. Thus, in a realization of the expanded expressions, considerable components can be shared, as in Fig. 2.2.5(c-1).

The subtractor of Fig. 2.2.5(c-1) can be improved, economically speaking, by using only two NAND gates common to both outputs. This economy can be achieved by rearranging and simplifying the borrow expression: $\beta_{out} = A'B'\beta_{in} + AB\beta_{in} + (A'B\beta_{in} + A'B\beta'_{in}) = A'B'\beta_{in} + AB\beta_{in} + A'B$. The $A'B\beta_{in}$ input is still needed for the difference bit D, but it is not needed for the borrow bit β_{out} and can be combined with $A'B\beta'_{in}$ to simplify to $A'B$. As a result of this simplification, the final NAND gate for the borrow-bit output need have only three inputs. The total saving for the improved circuit shown in Fig. 2.2.5(c-2) is two inputs.

2.3 Carry Look-Ahead Adders

As the length of a typical carry-propagating parallel adder increases, the time required to complete an addition increases by the delay time per stage for each bit added. *The carry look-ahead adder* reduces carry delay by reducing the number of gates through which a carry signal must pass. The truth table for the full-adder is repeated in Table 2.3.1, but this time emphasizes the conditions under which carry generation occurs. Entries 1, 2, 7, and 8 show instances where the output carry C_i is independent of C_{i-1}. In entries 1 and 2, the output carry is always zero, and in entries 7 and 8, it is always unity. These are known as *carry-generate* combinations. Entries 3, 4, 5, and 6 show input combinations where the output carry depends upon the input carry. In other words, C_i is 1 only when C_{i-1} is 1. These are *carry-propagate* combinations. Suppose that G_i denotes the unity carry-generate condition of the ith stage of a parallel adder (see Fig. 2.2.3) and P_i the carry-propagate condition of the same stage.

TABLE 2.3.1

Entry	A_i	B_i	C_{i-1}	C_i	Condition
1	0	0	0	0	No carry generate
2	0	0	1	0	
3	0	1	0	0	
4	0	1	1	1	Carry propagate
5	1	0	0	0	
6	1	0	1	1	
7	1	1	0	1	Carry generate
8	1	1	1	1	

Without loss of generality, consider the addition of two 4-bit binary numbers

$$A = A_4 A_3 A_2 A_1$$

and

$$B = B_4 B_3 B_2 B_1$$

From Table 2.3.1, the carry-generate and carry-propagate (switching) functions in terms of A_i and B_i, $i = 1, 2, 3,$ and 4, are found to be

$$G_i = A_i B_i$$
$$P_i = A_i + B_i = A_i \oplus B_i$$

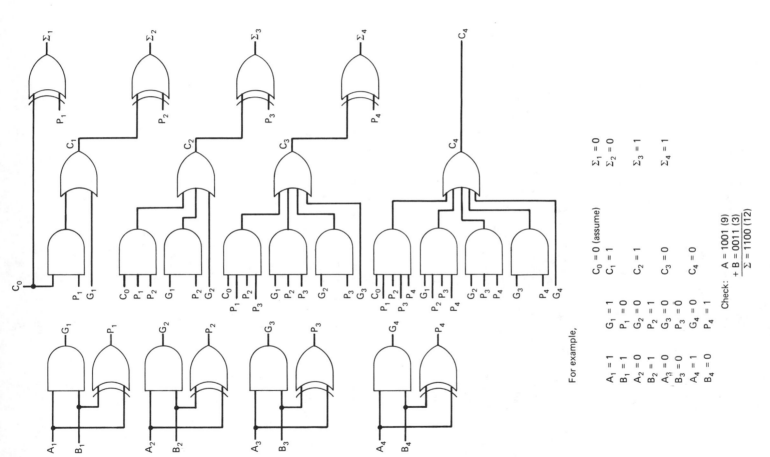

For example,

$A_1 = 1$	$G_1 = 1$	$C_0 = 0$ (assume)	$\Sigma_1 = 0$
$B_1 = 1$	$P_1 = 0$	$C_1 = 1$	$\Sigma_2 = 0$
$A_2 = 0$	$G_2 = 0$	$C_2 = 1$	$\Sigma_3 = 1$
$B_2 = 1$	$P_2 = 1$	$C_3 = 0$	
$A_3 = 0$	$G_3 = 0$	$C_4 = 0$	$\Sigma_4 = 1$
$B_3 = 0$	$P_3 = 0$		
$A_4 = 1$	$G_4 = 0$		
$B_4 = 0$	$P_4 = 1$		

Check: $A = 1001\ (9)$
$+\ B = 0011\ (3)$
$\overline{\Sigma = 1100\ (12)}$

Fig. 2.3.1 A 4-bit carry look-ahead adder.

The unity output carry of the ith stage can be expressed in terms of G_i, P_i, and C_{i-1}, which is the unity output carry of the $(i-1)$th stage, as

$$C_i = G_i + P_i \cdot C_{i-1}$$

For example, for $i = 1, 2, 3$, and 4, the C_i's are

$$C_1 = G_1 + P_1 C_0$$
$$C_2 = G_2 + P_2 C_1 = G_2 + P_2 G_1 + P_2 P_1 C_0$$
$$C_3 = G_3 + P_3 C_2 = G_3 + P_3 G_2 + P_3 P_2 G_1 + P_3 P_2 P_1 C_0$$
$$C_4 = G_4 + P_4 C_3 = G_4 + P_4 G_3 + P_4 P_3 G_2 + P_4 P_3 P_2 G_1 + P_4 P_3 P_2 P_1 C_0$$

The sum $\sum$ of A and B is $\sum = C_4 \sum_4 \sum_3 \sum_2 \sum_1$, where

$$\sum_i = A_i \oplus B_i \oplus C_{i-1}$$

for $i = 1, 2, 3$, and 4. A realization of a 4-bit carry look-ahead adder is shown in Fig. 2.3.1. It is seen that the addition of two n-bit binary numbers for any finite n can be easily accomplished by a carry look-ahead adder in four-gate propagation time.† The price to pay for achieving this time saving is the need for a considerable amount of excessive hardware. An example of adding the two binary numbers 1001_2 and 11_2 by this adder is also given in Fig. 2.3.1.

It should be noted that if no restrictions are imposed on the maximum allowable number of inputs (fan-in) to a gate, the carry look-ahead adder can be realized by a two-level AND–OR (or OR–AND) circuit, which means that the addition of two n-bit numbers can be accomplished in three-gate propagation time†† if all the bits of the two numbers are available simultaneously at the time when the addition is performed. This is because the C_i's can be expressed directly in terms of the inputs A_i's, B_i's, and C_0, which implies that the $\sum_i$'s can be expressed directly in terms of the inputs A_i's, B_i's, and C_0. Since the $\sum_i$'s are switching functions, they can be expressed in the sum-of-products (or product-of-sums) form. By applying a function minimization technique for multiple-output circuits, such as the one presented in Section 1.6, we can obtain a minimal two-level AND–OR (or OR–AND) realization of the carry look-ahead adder.

2.4 Representation and Addition of Signed Binary Numbers

An n-bit signed binary number consists of two parts: a part denoting the sign of the number and a part denoting the magnitude. The first bit of a number is called the *sign bit*, which denotes the sign of the number, and the convention is that 0 and 1

---

†Assume that the gate propagation times for an AND, OR, and XOR are nearly the same.

††One gate propagation time is for inverting variables from their true form to their complemented form when needed.

denote "the number is positive" and "the number is negative," respectively. The remaining $n-1$ bits denote the magnitude of the number. There are several ways to represent the magnitude of a signed number in a digital system. Three forms of signed

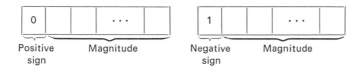

Positive   Magnitude   Negative   Magnitude
sign   sign

(binary) numbers are popular: signed magnitude, signed-1's complement, and signed-2's complement. Number systems using these forms are referred to as *signed-magnitude number system*, *signed-1's-complement number system*, and *signed-2's-complement number system*, respectively. Because the forms of the magnitude part of a signed number for these number systems are different, so are the methods for arithmetic operations for them.

In the following, for each of the three number systems, we shall present (1) the forms of a positive and a negative number and the form of zeros, (2) the method for adding two numbers, and (3) the proof that the method works. It should be pointed out that since the subtraction of a number N_2 from N_1 is the same as the addition of $N_1 + (-N_2)$, every subtraction of a number (subtrahend) from a number (minuend) can be converted into addition of the two numbers (with the sign of the subtrahend changed). Consequently, the addition operation for signed number can perform both addition and subtraction. Figure 2.4.1 shows the data flow through a typical digital

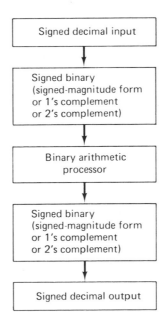

Fig. 2.4.1 Data flow in a typical digital arithmetic system.

system that processes arithmetic data, receiving information from and returning information to a human user.

2.4.1 Signed-Magnitude Number Systems

In the signed-magnitude form, a positive or negative number is represented by a sign bit followed by the magnitude in binary. For example, for 6 bits,

+19 is represented by

0	1	0	0	1	1

sign bit 2^4 $+2^1 + 2^0$ $= + (16 + 2 + 1) = +19$

−10 is represented by

1	0	1	0	1	0

sign bit 2^3 $+2^1$ $= -(8 + 2) = -10$

There is a plus zero and a minus zero, which for 6 bits are 000000 and 100000, respectively. Let N_1 and N_2 denote two signed binary numbers, and $|N_1|$ and $|N_2|$ denote the magnitudes of N_1 and N_2, respectively. Let N_1' and $|N_1|'$ be the complements of N_1 and $|N_1|$, respectively. For example, if $N_1 = 010011$ and $N_2 = 101010$, then

$$|N_1| = 10011$$
$$|N_2| = 01010$$

and

$$N_1' = 101100 \qquad |N_1|' = 01100$$
$$N_2' = 010101 \qquad |N_2|' = 10101$$

Now consider the addition of two numbers of this system.

CASE 1

N_1 and N_2 have the same sign.

RULE 1

When N_1 and N_2 have the same sign, an adder adds the magnitude bits and makes the sign bit of the sum agree with that of the numbers.

For example, the addition of +19 and +10 and the addition of −19 and −10 will be obtained from

$$
\begin{array}{ll}
10011 & |N_1| \\
+01010 & |N_2| \\
\hline
11101 & |N_1| + |N_2|
\end{array}
$$

which are 011101 (+29) and 111101 (−29), respectively.

CASE 2

N_1 and N_2 have different signs.

RULE 2

When N_1 and N_2 have different signs, one of the numbers (*either* N_1 or N_2) is complemented before the addition. After the complement is taken, the magnitude of the uncomplemented number is added to the magnitude of this complement.

(a) If there is an overflow, called an *end-around carry* (EAC), it is added to the least significant digit of the sum, and the sign of the sum will be the same as the sign of the number that was *not* complemented.

(b) If there is *no* end-around carry, the sum is complemented and the sign bit of the sum will be the same as that of the number that was complemented.

For example, consider the addition of $+19$ and -10, in which the two numbers have different signs. Suppose that we complement the magnitude of the second number. Then the addition is performed as

$$
\begin{array}{ll}
10011 & |N_1| \\
+10101 & |N_2|' \\
\hline
101000 & \\
\ \ \hookrightarrow +1 & \text{end-around carry} \\
\hline
01001 & ||N_1| + |N_2||'
\end{array}
$$

Since there is an end-around carry, the sign bit of the sum is that of the number that was *not* complemented. The result is 001001 ($+9$).

If, in the above addition, we complement the magnitude of the first number, then

$$
\begin{array}{ll}
01100 & |N_1|' \\
+01010 & |N_2| \\
\hline
10110 & ||N_1|' + |N_2||
\end{array}
$$

Because there is no end-around carry, the result is complemented to 01001 and the sign of the complemented number is used. Thus the sum is 101001 (-9).

Proof of Rules 1 and 2:   The proof for rule 1 should be obvious. The proof of rule 2 is as follows. Suppose that N_1 and N_2 are two n-bit numbers. Let $M = 2^{n-1}$. Then the complement of the magnitude of a number N_1 is

$$|N_1|' = M - |N_1| - 1 \tag{2.4.1}$$

For example, if $n = 6$ and $N_1 = 010011$ ($+19$), the complement of $|N_1|$ is

$$|N_1|' = 2^5 - 19 - 1 = 12$$

which is equal to the complement of $|N_1| = 10011$ [i.e., 01100 (12)]. Now $|N_1|'$ is added to $|N_2|$.

$$\begin{array}{r} M - |N_1| - 1 \\ + |N_2| \\ \hline M + |N_2| - |N_1| - 1 \end{array}$$

If $|N_2|$ is greater than $|N_1|$, the sum is equal to or greater than M, and an end-around carry is generated that is similar to an overflow but which is really the modulus M. When this carry is added to the least significant digit, it cancels the -1:

$$\begin{array}{r} |N_2| - |N_1| - 1 \\ + 1 \quad \text{from } M \\ \hline |N_2| - |N_1| \end{array}$$

Obviously, the sign of the sum should be that having magnitude $|N_2|$, the number that was *not* complemented.

If $|N_1|$ is greater than $|N_2|$, then $M + |N_2| - |N_1| - 1$ is less than M, and the sum fits into the specified number of bits without an overflow or end-around carry being generated. If this sum is complemented, by Eq. (2.4.1), the result is

$$M - (M + |N_2| - |N_1| - 1) - 1 = |N_1| - |N_2|$$

and the correct sign for this result is that of the complemented number.

If N_1 and N_2 are equal in magnitude but opposite in sign, then $M + |N_2| - |N_1| - 1$ $= M - 1$ (all 1's). Since in this case there is no end-around carry, the result is complemented to all 0's. Thus the sum is $000\ldots0$ or $100\ldots0$, which depends on which number was complemented. If the complemented number was positive, the sum is the plus zero $000\ldots0$. If the complemented number was negative, the sum is the minus zero $100\ldots0$.

It is possible, of course, to have an overflow when two positive numbers or two negative numbers are added. This is a true overflow and should not be confused with the end-around carry that may occur if the signs of the number differ. ∎

2.4.2 Signed-1's-Complement Number System

Positive numbers in the signed-1's-complement system are the same as in the signed-magnitude number system, but negative numbers differ, for they are represented in 1's-complement form. For example, the 1's-complement form of -19 for a 6-bit digital system is the complement of 010011 $(+19)$, which is 101100 (-19). Also, since plus zero is 000000, minus zero for the signed-1's-complement number system is therefore 111111.

The main difference between the addition of two signed-magnitude numbers described above and the addition of two signed-1's-complement numbers that will be described below is that in the latter, the sign bits are added along with the magnitude bits. In other words, the sign bits are added as if they were magnitude bits.

CASE 1

N_1 and N_2 are positive.

RULE 1

When N_1 and N_2 are positive, add the signed numbers (the signs and the magnitudes). If the sign bit shows a 1, it indicates an overflow.

For example, consider the addition of 19 and 10. In signed-1's-complement form, they are 010011 and 001010, whose sum is

$$
\begin{array}{rl}
010011 & (+19) \\
+001010 & (+10) \\
\hline
011101 & (+29)
\end{array}
$$

The 19 plus 19 is

$$
\begin{array}{rl}
010011 & (+19) \\
+010011 & (+19) \\
\hline
100110 & (+38)
\end{array}
$$

indication of
an overflow

since the sign bit is 1, indicating an overflow.

CASE 2

N_1 and N_2 are negative.

RULE 2

If two negative numbers are added, an end-around carry always occurs, which is produced by the two 1 sign bits of the numbers being added. This carry is added to the least signficant bit position.

(a)   If the sign bit of the resulting number is a 1, it indicates that the answer is correct.

(b)   If the sign bit of the resulting number is a 0, it indicates an overflow.

For example, the sum of -19 and -10 is

$$
\begin{array}{rl}
101100 & (-19) \\
+110101 & (-10) \\
\hline
1100001 & \\
\longrightarrow +1 & \\
\hline
100010 & (-29)
\end{array}
$$

The number 100010 is -29 in signed-1's-complement form because the complement of 100010 is 011101, which in decimal numbers is 29. The sum of -19 and -19 is

$$101100$$
$$+101100$$
$$\overline{1011000}$$

$$\uparrow$$

indication of
an overflow

Observe that the sign bit is 0, which indicates an overflow.

CASE 3

N_1 and N_2 have different signs.

RULE 3

When two numbers, N_1 and N_2, of unlike signs are added and the positive number is larger, an end-around carry occurs that must be added to the least significant digit. If the negative number is larger, there will be no such carry.

For example, the sum of 19 and −10 and the sum of −19 and 10 are

$$
\begin{array}{ll}
010011 & (+19) \\
+110101 & (-10) \\
\hline
1001000 & \\
\:\longmapsto +1 & \\
\hline
001001 & (+9)
\end{array}
\quad \text{and} \quad
\begin{array}{ll}
101100 & (-19) \\
+001010 & (+10) \\
\hline
110110 & (-9)
\end{array}
$$

Proof of Rules 1, 2, and 3: The proofs of these rules are quite similar to those of rules 1 and 2 of Section 2.4.1. First we need to consider the form of the negative numbers in more detail. In the signed 1's complement number system, a negative number N of magnitude $|N|$ has the form $M + (M - |N| - 1) = 2M - |N| - 1$, where $M = 2^{n-1}$ and n is the number of bits. Note that the first M is for the sign bit and $M - |N| - 1$ is the 1's complement of the magnitude of the number. For example, $N = 19$ in signed-1's-complement form is

$$N = 2 \times 2^{6-1} - |-19| - 1 = 44 = 101100\,[= (\underbrace{010011}_{19})']$$

The validity of rule 1 is obvious. Now consider the sum of two negative numbers.

$$
\begin{array}{llll}
 & 2M - & |N_1| & -1 \\
+ & 2M - & |N_2| & -1 \\
\hline
& 4M - (|N_1| + |N_2|) - 2
\end{array}
$$

In this partial result, $2M$ of the $4M$ produces an end-around carry that, when added to the remainder, produces

$$
\begin{array}{lr}
 & 2M - (|N_1| + |N_2|) - 2 \\
+ & +1 \quad \text{from } 2M \\
\hline
& 2M - (|N_1| + |N_2|) - 1
\end{array}
$$

This $2M - (|N_1| + |N_2|) - 1$ is thus the 1's complement of $|N_1| + |N_2|$ and hence is the correct result.

In the case where $|N_1| + |N_2|$ is greater than M, an overflow occurs. Note that $2M - (|N_1| + |N_2|) - 1 = M + [M - (|N_1| + |N_2|) - 1]$. When $|N_2| + |N_2| \geq M$, $2M - (|N_1| + |N_2|) - 1 < M$, which implies that a 0 is in the sign-bit position. Since the sum of two negative numbers must be negative, there should not be a 0 in the sign-bit position of the sum. This discrepancy shows the existence of an overflow. Rule 2 is thus proved.

Finally, let us prove rule 3. When a positive number N_1 is added to a negative number N_2, the first step of the addition is

$$
\begin{array}{r}
|N_1| \\
+\ 2M - |N_2| - 1 \\
\hline
2M + |N_1| - |N_2| - 1
\end{array}
$$

If $|N_1| > |N_2|$, $2M + |N_1| - |N_2| - 1 \geq 2M$; hence the addition generates an end-around carry and the result is $[|N_1| - |N_2| - 1] + 1$ (from $2M$) $= |N_1| - |N_2|$, which is positive. On the other hand, if $|N_1| \leq |N_2|$, $2M + |N_1| - |N_2| - 1 \leq 2M$, for which no end-around carry would generate. The result is the sum of $N_1 + N_2$, since $2M + |N_1| - |N_2| - 1$ may be written $2M - (|N_2| - |N_1|) - 1$, in which $|N_2| - |N_1|$ is positive. ∎

Both signed-magnitude and signed-1's-complement representation are adequate for parallel computation systems. But they are not suitable for serial computation systems. This is mainly because considerably more computation time will be required to generate the end-around carries by a serial adder than by a parallel adder. A method of avoiding the need for end-around carries in the computation is presented next.

2.4.3   Signed-2's-Complement Number System

In this system a positive number is represented in the same form as in the other two systems. However, the negative numbers are in 2's-complement form. For example, -10 in a 6-bit word is 110110. This is obtained form

$$-10 = -32 + 22$$

$$\underset{110110}{\searrow \ \downarrow}$$

There is only plus zero, which is all 0's; no minus zero exists.

The addition of two positive numbers will not be discussed because this addition is the same as the 1's-complement system.

CASE 1

N_1 and N_2 are negative.

RULE 1

When two negative numbers are added, a carry must be disregarded. This carry resulted from the sum of the two 1 sign bits. Moreover, the sign bit of the sum must be 1, since it is negative. If a 0 is shown positive in the sign bit, an overflow is indicated.

For example, the sum of -19 and -10 and the sum of -19 and -19 are

$$
\begin{array}{ll}
101101 & (-19) \\
+110110 & (-10) \\
\hline
1100011 & (-29)
\end{array}
\qquad \text{and} \qquad
\begin{array}{ll}
101101 & (-19) \\
+101101 & (-19) \\
\hline
1011010 &
\end{array}
$$

$\uparrow$ $\uparrow$

disregard   indication of overflow

CASE 2

N_1 and N_2 have opposite signs.

RULE 2

A carry is generated if the sum is positive. In this case, the carry is ignored. No carry is generated if the sum is negative.

For example,

$$
\begin{array}{ll}
010011 & (+19) \\
+110110 & (-10) \\
\hline
1001001 & (+9)
\end{array}
\qquad\qquad
\begin{array}{ll}
101101 & (-19) \\
+001010 & (+10) \\
\hline
110111 & (-9)
\end{array}
$$

$\uparrow$

disregard

Proof of Rules 1 and 2: From the 1's complement it should be obvious that a negative number of magnitude N has the form $2M - N$ in the 2's-complement number system. For example,

$$-19 = 2 \times 2^{6-1} - |-19| = 45 = 101101$$

$$32 - 13 = 19$$

When two negative numbers are added,

$$
\begin{array}{l}
2M - |N_1| \\
+ 2M - |N_2| \\
\hline
4M - (|N_1| + |N_2|)
\end{array}
$$

In this sum, $2M$ of the $4M$ produces a carry that is ignored. Hence the result is $2M - (|N_1| + |N_2|)$. However, when $|N_1| + |N_2| > M$, $2M - (|N_1| + |N_2|) < M$, which implies the sign bit being 0. This overflow can thus be detected by the presence of a 0 at the sign-bit position. This proves rule 1.

TABLE 2.4.1 Comparison of Three Signed-Number Systems

Number System / Topic	Signed Magnitude	Signed 1's Complement	Signed 2's Complement
1. Representation of a positive number	0 □□□□□□□ (sign / magnitude)	0 □□□□□□□ (sign / magnitude)	0 □□□□□□□ (sign / magnitude)
2. Representation of a negative number	1 □□□□□□□ (sign / magnitude)	$2M - |N| - 1$	$2M - |N|$
3. Form of zero	Plus zero = 00 . . . 0 Minus zero = 10 . . . 0	Plus zero = 00 . . . 0 Minus zero = 11 . . . 1	Plus zero = 00 . . . 0 No minus zero
4. Negative numbers in complement form?	No	Yes	Yes
5. Sign bit involved in addition or subtraction	No	Yes	Yes
6. Use end-around carry?	Yes	Yes	No
7. Method for detecting overflow in adding two positive numbers	Carry from magnitude bits	Negative sum	Negative sum
8. Method for detecting overflow in adding two negative numbers	Carry from magnitude bits	Positive sum	Positive sum

Now consider the case where two numbers of unlike sign are added. If a positive number N_1 of magnitude $|N_1|$ is added to a negative number N_2 of magnitude $|N_2|$, the result is

$$
\begin{array}{r}
|N_1| \\
+\ 2M - |N_2| \\
\hline
2M + |N_1| - |N_2|
\end{array}
$$

If $|N_1|$ is greater than $|N_2|$, the sum is greater than $2M$, in which case a carry is generated. Ignoring this carry, we obtain $|N_1| - |N_2|$. If N_2 has the larger magnitude, then $2M + |N_1| - |N_2| < 2M$, the sum fits into n bits and no carry is generated. The final possibility is that the magnitudes of N_1 and N_2 are equal. In this case, the result is $2M$, which is the carry that is ignored. What is left (all 0's) is positive zero, the only zero in the system. ∎

It should be noted that the 2's complement of a positive binary number can also be generated in the following two convenient ways. One way is to convert the number to its 1's complement and then to add a 1. For example, the 2's complement of

$10_{10} = \boxed{0\ 0\ 1\ 0\ 1\ 0}_2$ can be obtained from

1's complement of 10_{10}	$\boxed{1\ 1\ 0\ 1\ 0\ 1}$
Add a 1	$+\qquad\qquad\quad 1$
	$\boxed{1\ 1\ 0\ 1\ 1\ 0}$

Another way to obtain the 2's complement of a binary number is to start at the right (least significant) bit and proceed to the left while copying down all zeros and the first 1 encountered, then inverting every bit to the left of the first 1. For example, the 2's complement of 10_{10} can be obtained from

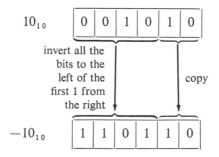

A comparison of the three signed-number systems is shown in Table 2.4.1.

The 1's complement is usually achieved by the true/complement and zero/one device, as shown in Fig. 2.4.2. Adders/subtractors using signed-magnitude, 1's-complement, and 2's-complement notations are shown in Fig. 2.4.3(a), (b), and (c), respectively. Each of these adders is 4-bit, including the sign bit.

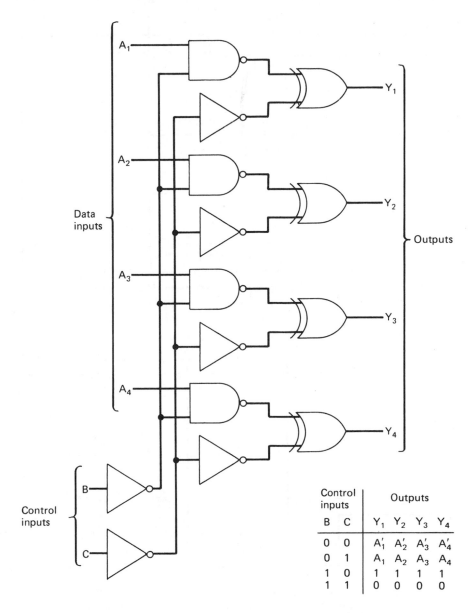

Control inputs		Outputs			
B	C	Y_1	Y_2	Y_3	Y_4
0	0	A_1'	A_2'	A_3'	A_4'
0	1	A_1	A_2	A_3	A_4
1	0	1	1	1	1
1	1	0	0	0	0

Fig. 2.4.2 True/complement and zero/one device.

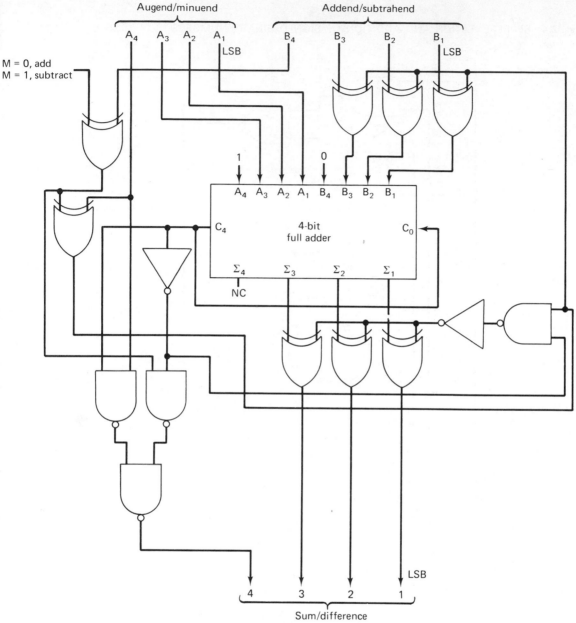

Note that: 1. NC denotes no connection.
2. Although only three bits are needed, a 4-bit adder is used for comparison with Fig. 2.4.3(b) and (c). A 1 and 0 are placed on B_4 and A_4 so that a carry-out of bit 3 will propagate to C_4.

(a) Adder/subtractor using
signed-magnitude number system

Fig. 2.4.3 Adders/subtractors using signed-magnitude, 1's complement, and 2's complement number systems.

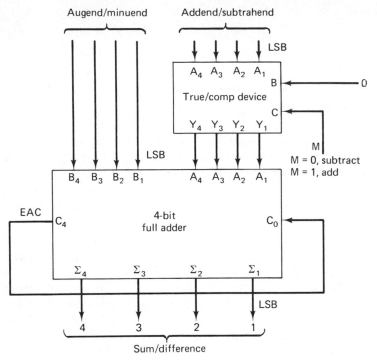

(b) Adder/subtractor using 1's
complement number system

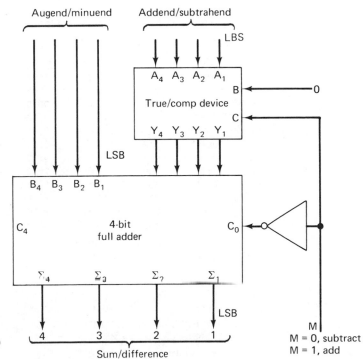

Fig. 2.4.3 (continued)

(c) Adder/subtractor using 2's
complement number system

2.5 Codes for Decimal Numbers and Their Conversions

In the foregoing sections we have discussed the addition and subtraction of binary numbers. However, in some instances it is desirable to perform the arithmetic operation directly, using decimal numbers. This need often occurs in systems where the result of the arithmetic operation is to be displayed directly and it is desired to avoid conversions. Most numerical display systems interface directly with decimal-code representations.

The most common scheme to code a decimal number is to code each digit individually. Each digit has the values 0–9; hence a minimum of four binary variables is needed. Since four variables can actually assume $2^4 = 16$ values, there are always six redundancies in this type of encoding. The four-variable representation of a decimal digit is commonly called *binary-coded decimal* (BCD). The most common BCD code is 8421 BCD. This code represents each decimal value by its 4-bit binary equivalent. This code is tabulated in the first column of Table 2.5.1. In the second column of the same table, another common BCD code, known as the *excess-3* code, is presented which is generated from the 8421 BCD code by adding a 3 to the binary

TABLE 2.5.1 Various BCD Codes

Decimal	8421 BCD	Excess-3	Gray	Modified Gray	
0	0000	0011	0000	0010	
1	0001	0100	0001	0110	
2	0010	0101	0011	0111	
3	0011	0110	0010	0101	one
4	0100	0111	0110	0100	decimal
5	0101	1000	0111	1100	digit and
6	0110	1001	0101	1101	its codes
7	0111	1010	0100	1111	
8	1000	1011	1100	1110	
9	1001	1100	1101	1010	
10	1010		1111		
11	1011		1110		
12	1100		1010		
13	1101		1011		
14	1110		1001		
15	1111		1000		

complete 4-bit
binary and Gray codes

value of the 8421 BCD code. This BCD code has certain desirable arithmetic properties which will be discussed in the next section.

Both 8421 BCD codes and the excess-3 code are not particularly suitable for electrical or electro-optical encoder systems (angular position-shaft encoders, etc.) because a movement from one state to the next often results in more than one bit change; for example, from seven to eight, the binary code changes from 0111 to 1000. Such bit changes can never really be simultaneous, so the encoder always generates erroneous transient codes when switching between certain positions. This problem is avoided with a Gray code, which is shown in the third column of Table 2.5.2. The Gray code is a reflected digital code with the special property that two adjacent Gray code numbers differ by only one bit. This type of encoding scheme is called a *unit-distance code*. A fourth useful BCD coding system is the modified Gray code, which is shown in

TABLE 2.5.2 Gray Codes to Encoding Angular Positions

Quadrant (deg)	2-Bit Gray Code	Half-Quadrant (deg)	3-Bit Gray Code	Quarter-Quadrant (deg)	4-Bit Gray Code
0– 89	00	0–44	000	0–22	0000
90–179	01	45–89	001	22.5–44.5	0001
80–269	11	90–134	011	45–67	0011
270–359	10	135–179	010	67.5–89.5	0010
		180–224	110	90–112	0110
		225–269	111	112.5–134.5	0111
		270–314	101	135–157	0101
		315–359	100	157.5–179.5	0100
				180–202	1100
				202.5–224.5	1101
				225–247	1111
				247.5–269.5	1110
				270–292	1010
				292.5–314.5	1011
				315–337	1001
				337.5–359.5	1000

the fourth column of Table 2.5.1. The attractive feature of this code is that there is only one bit change at a time, even on a 9-to-0 transition.

Code Conversion   From Table 2.5.1 it is seen that any code of a coding system can be converted to a code of another by a multiple-input and multiple-output combinational network. This is illustrated by the following examples.

Example 2.5.1

Convert a 10-line decimal into a 4-line BCD.

The 10-line decimal-to-4-line-BCD conversion table is given in Fig. 2.5.1(a). A 10-line decimal-to-BCD converter using NOR and NAND gates is shown in Fig. 2.5.1(b). Note that

Inputs										Outputs			
9	8	7	6	5	4	3	2	1	0	B_3	B_2	B_1	B_0
0	0	0	0	0	0	0	0	0	0	0	0	0	0
0	0	0	0	0	0	0	0	1	0	0	0	0	1
0	0	0	0	0	0	0	1	0	0	0	0	1	0
0	0	0	0	0	0	1	0	0	0	0	0	1	1
0	0	0	0	0	1	0	0	0	0	0	1	0	0
0	0	0	0	1	0	0	0	0	0	0	1	0	1
0	0	0	1	0	0	0	0	0	0	0	1	1	0
0	0	1	0	0	0	0	0	0	0	0	1	1	1
0	1	0	0	0	0	0	0	0	0	1	0	0	0
1	0	0	0	0	0	0	0	0	0	1	0	0	1

(a) Conversion table

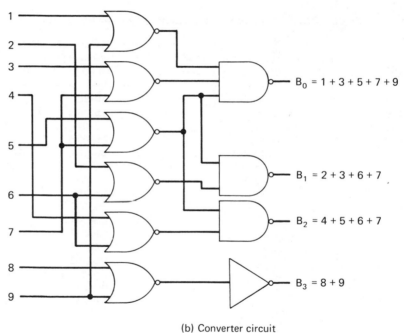

(b) Converter circuit

Fig. 2.5.1 Ten-line decimal-to-8421 BCD converter.

in this converter, only nine input lines are needed, since the 8421 BCD zero (i.e., 0000) is generated when all the inputs are zero.

Example 2.5.2

Convert the 4-bit binary code into the 4-bit Gray code.
The 4-bit binary code and the corresponding Gray code are shown in Fig. 2.5.2. The

	Binary (input)				Gray (output)			
	B_3	B_2	B_1	B_0	G_3	G_2	G_1	G_0
0	0	0	0	0	0	0	0	0
1	0	0	0	1	0	0	0	1
2	0	0	1	0	0	0	1	1
3	0	0	1	1	0	0	1	0
4	0	1	0	0	0	1	1	0
5	0	1	0	1	0	1	1	1
6	0	1	1	0	0	1	0	1
7	0	1	1	1	0	1	0	0
8	1	0	0	0	1	1	0	0
9	1	0	0	1	1	1	0	1
10	1	0	1	0	1	1	1	1
11	1	0	1	1	1	1	1	0
12	1	1	0	0	1	0	1	0
13	1	1	0	1	1	0	1	1
14	1	1	1	0	1	0	0	1
15	1	1	1	1	1	0	0	0

(a)

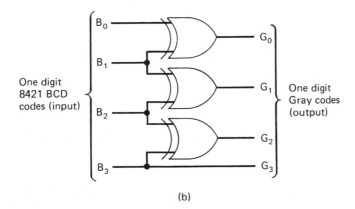

One digit
8421 BCD
codes (input)

One digit
Gray codes
(output)

(b)

Fig. 2.5.2 Binary-to-Gray converter.

Karnaugh maps of $G_i(B_0, B_1, B_2, B_3)$, $i = 0, 1, 2, 3$, are shown in Fig. 2.5.3. The minimal forms of these functions are found to be

$$G_0 = B_0 B_1' + B_0' B_1 = B_0 \oplus B_1$$
$$G_1 = B_1 B_2' + B_1' B_2 = B_1 \oplus B_2$$
$$G_2 = B_2 B_3' + B_2' B_3 = B_2 \oplus B_3$$
$$G_3 = B_3$$

A binary-to-Gray converter is shown in Fig. 2.5.2(b).

Note that a 10-line decimal-to-Gray converter may be obtained by cascading the above two converters.

The preceding examples illustrate the general procedure of constructing a code converter. In other words, the conversion of a code from any system into another can be obtained in the same manner. Several examples of code conversions are shown in Table 2.5.3.

TABLE 2.5.3　Examples of Code Conversions

10-Line-Decimal Active 1's	Binary (8421)	Gray	Excess-3	Minimized Conversion Functions
⟶				$B_0 = 1 + 3 + 5 + 7 + 9$ $B_1 = 2 + 3 + 6 + 7$ $B_2 = 4 + 5 + 6 + 7$ $B_3 = 8 + 9$
	⟶			$G_0 = B_0 \oplus B_1$ $G_1 = B_1 \oplus B_2$ $G_2 = B_2 \oplus B_3$ $G_3 = B_3$
	⟶			$E_0 = B_0'$ $E_1 = B_0 B_1 + B_1' B_0' = (B_0 \oplus B_1)'$ $E_2 = B_0 B_2' + B_1 B_2' + B_0' B_1' B_2$ $E_3 = B_0 B_2 + B_1 B_2 + B_3$
	⟵			$B_0 = E_0'$ $B_1 = E_0 E_1' + E_0' E_1 = E_0 \oplus E_1$ $B_2 = E_0' E_2' + E_0 E_1 E_2 + E_0 E_1' E_3$ $B_3 = E_0 E_1 E_3 + E_2 E_3$
	⟵			$B_0 = G_0 \oplus G_1 \oplus G_2 \oplus G_3$ $B_1 = G_1 \oplus G_2 \oplus G_3$ $B_2 = G_2 \oplus G_3$ $B_3 = G_3$
⟵				$0 = B_0' B_1' B_2' B_3' \qquad 5 = B_0' B_1 B_2' B_3$ $1 = B_0' B_1' B_2' B_3 \qquad 6 = B_0' B_1 B_2 B_3'$ $2 = B_0' B_1' B_2 B_3' \qquad 7 = B_0' B_1 B_2 B_3$ $3 = B_0' B_1' B_2 B_3 \qquad 8 = B_0 B_1' B_2' B_3'$ $4 = B_0' B_1 B_2' B_3' \qquad 9 = B_0 B_1' B_2' B_3$

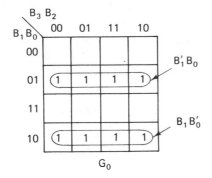

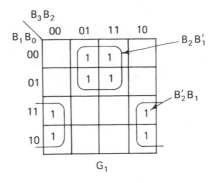

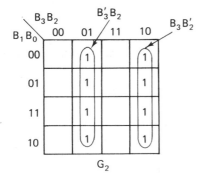

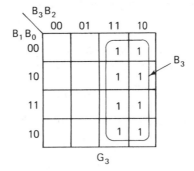

Fig. 2.5.3 Minimization of $G_i(B_3, B_2, B_1, B_0)$, $i = 0, 1, 2, 3$, using Karnaugh maps.

2.6   Decimal Adders/Subtractors

Using binary adders as building blocks, we can design decimal adders (i.e., adders for one digit of a coded decimal number). Two codes will be considered here: the 8421 BCD code and the excess-3 code. For example, the numbers 495_{10} and 278_{10} are represented in these codes as follows:

	8421 BCD	Excess-3
Decimal	10^2 10^1 10^0	10^2 10^1 10^0
495	0100/1001/0101	0111/1100/1000
278	0010/0111/1000	0101/1010/1011

It should be mentioned that conversion between 8421 BCD and excess-3 is quite simple. The excess-3 code can be obtained from the 8421 BCD code by adding a 3 [see Fig. 2.6.1(a)], and the 8421 BCD code can be obtained from the excess-3 code by subtracting a 3. The later is, however, equivalent to adding a 13 (which is the 2's complement of the binary representation of 3) to the excess-3 code [see Fig. 2.6.1(b)].

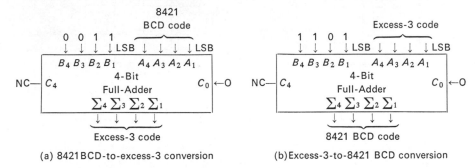

(a) 8421 BCD-to-excess-3 conversion   (b) Excess-3-to-8421 BCD conversion

Fig. 2.6.1

We shall discuss separately the addition and subtraction of BCD numbers for use with these two codes.

2.6.1 8421 BCD Addition

Let us consider the addition of the numbers in 8421 BCD codes.

Example 2.6.1

	10^2	10^1	10^0
495	0100	1001	0101
+278	0010	0111	1000
773	0110	10000	1101 $(= 13_{10})$

$(= 16_{10})$ correction required

The BCD sums 10000 and 1101, which are not valid 8421 BCD codes, therefore need to be corrected as follows:

Uncorrected BCD Sum		Corrected BCD Sum
1101	to be corrected to	10011
		3_2
		a carry-out
10000	to be corrected to	10110
		6_2
		a carry-out

This example illustrates that when 8421 BCD is performed with a possible carry-in, 20 different sums can be produced. Of these, only 10 will be correct; the remainder will require correction. For a sum equal to or greater than 10_{10}, subtraction of 10_{10} will give the correct result for the digit in question, and a carry to the next decade will also be required, as can be seen in Table 2.6.1. The required subtraction of 10_{10} can be achieved by adding the 2's complement of the BCD representation of 10_{10} (1010_2), which is 0110_2 in BCD or decimal 6. A second four-bit adder stage is used for the addition. A decoding scheme is required to generate a carry for the following digit, and to control when 6 is added to the sum for correction. The Karnaugh map in Table 2.6.1 represents the decoding of the uncorrected sums. This result is combined with the carry-out signal (C_4) to detect a possible sum of 10_{10} through 19_{10}, thus giving

$$C_n = C_4 + \Sigma_4\,\Sigma_3 + \Sigma_4\,\Sigma_2$$

Note that C_n bears the following important information: "When $C_n = 0$, it means that the BCD sum is between 0 and 9; no correction is required. When $C_n = 1$, it means that the BCD sum is greater than 9; correction is required, which can be accomplished by adding 0110_2 to the sum."

TABLE 2.6.1   Results of BCD Addition with Corrections Indicated

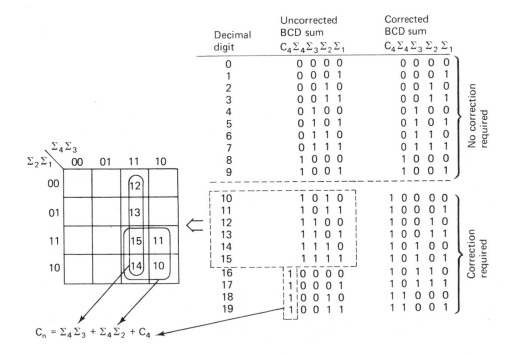

Decimal digit	Uncorrected BCD sum $C_4\,\Sigma_4\,\Sigma_3\,\Sigma_2\,\Sigma_1$	Corrected BCD sum $C_4\,\Sigma_4\,\Sigma_3\,\Sigma_2\,\Sigma_1$	
0	0 0 0 0	0 0 0 0	
1	0 0 0 1	0 0 0 1	
2	0 0 1 0	0 0 1 0	
3	0 0 1 1	0 0 1 1	
4	0 1 0 0	0 1 0 0	No correction required
5	0 1 0 1	0 1 0 1	
6	0 1 1 0	0 1 1 0	
7	0 1 1 1	0 1 1 1	
8	1 0 0 0	1 0 0 0	
9	1 0 0 1	1 0 0 1	
10	1 0 1 0	1 0 0 0 0	
11	1 0 1 1	1 0 0 0 1	
12	1 1 0 0	1 0 0 1 0	
13	1 1 0 1	1 0 0 1 1	
14	1 1 1 0	1 0 1 0 0	Correction required
15	1 1 1 1	1 0 1 0 1	
16	1 1 0 0 0 0	1 0 1 1 0	
17	1 1 0 0 0 1	1 0 1 1 1	
18	1 1 0 0 1 0	1 1 0 0 0	
19	1 1 0 0 1 1	1 1 0 0 1	

$$C_n = \Sigma_4\,\Sigma_3 + \Sigma_4\,\Sigma_2 + C_4$$

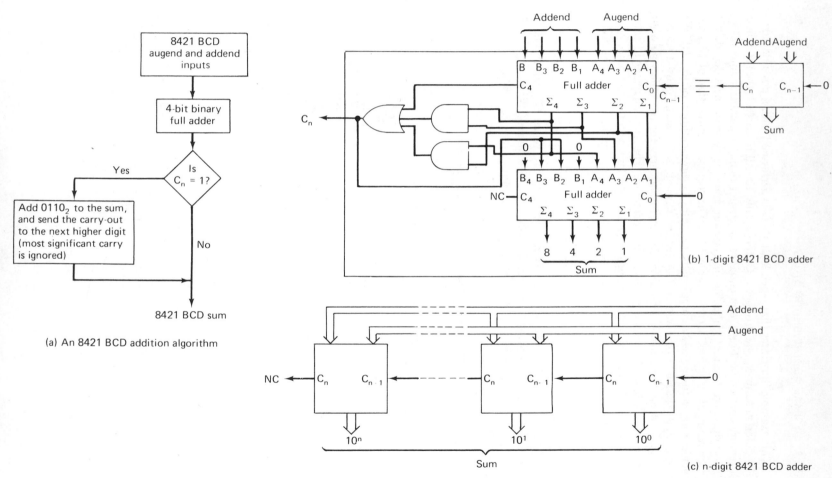

(a) An 8421 BCD addition algorithm

(b) 1-digit 8421 BCD adder

(c) n-digit 8421 BCD adder

Fig. 2.6.2 8421 BCD adder.

The one-digit 8421 BCD addition algorithm may be described by a flow chart as shown in Fig. 2.6.2(a). An implementation of this algorithm for one decimal digit is given in Fig. 2.6.2(b), and an n-bit 8421 BCD adder is depicted in Fig. 2.6.2(c).

Now let us go back to our example.

Example 2.6.1   (Continued)

$$
\begin{array}{r}
495 \\
+278 \\
\hline
773
\end{array}
$$

	10^2	10^1	10^0
	0100	1001	0101
	0010	0111	1000
	0110	10000	1101

(Compute C_n:
$C_n = \Sigma_4 \Sigma_3 + \Sigma_4 \Sigma_2 + C_4$)

$C_n = 0$ no correction $\quad$ $C_n = 1$ add 0110 $\quad$ $C_n = 1$ add 0110

0110 $\quad$ 10110 $\quad$ 10011

$\qquad\quad$ 1 $\quad\quad$ 1

0111 $\quad\quad$ 0111

"true" results

2.6.2 8421 BCD Subtraction

One method for performing 8421 BCD subtraction is the addition of the 9's complement of the subtrahend to the minuend. The 9's complement of the 8421 BCD code is described in Table 2.6.2. There are two commonly used methods to generate the 8421 BCD 9's complement. One is using a 4-bit binary adder to add 1010_2 ($= 10_{10}$)

TABLE 2.6.2 9's Complement of 8421 BCD

	True					9's Complement			
		BCD					BCD		
	8	4	2	1		8	4	2	1
Decimal	A	B	C	D	Decimal	W	X	Y	Z
0	0	0	0	0	9	1	0	0	1
1	0	0	0	1	8	1	0	0	0
2	0	0	1	0	7	0	1	1	1
3	0	0	1	1	6	0	1	1	0
4	0	1	0	0	5	0	1	0	1
5	0	1	0	1	4	0	1	0	0
6	0	1	1	0	3	0	0	1	1
7	0	1	1	1	2	0	0	1	0
8	1	0	0	0	1	0	0	0	1
9	1	0	0	1	0	0	0	0	0

to the binary 1's complement of the 8421 BCD code and ignoring the carry. For example, the 9's complement of 278_{10} is generated by

True 278_{10}			9's complement of 278_{10}		
10^2	10^1	10^0	10^2	10^1	10^0
0010	0111	1000	1101	1000	0111 (1's complement)
			+ 1010	+ 1010	+ 1010
			10111	10010	10001

ignore these carries

The hardware implementation of this method for generating the 9's complement is shown in Fig. 2.6.4(a).

The second method for generating the 9's complement is through the use of a combinational circuit [see Fig. 2.6.4(b)] which realizes the following minimized switching functions:

$$W = B'_3 B'_2 B'_1$$
$$X = B_2 \oplus B_1$$
$$Y = B_1$$
$$Z = B'_0$$

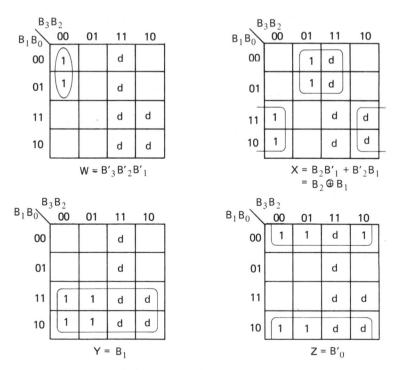

Fig. 2.6.3 Minimization of the functions W, X, Y, and Z.

These functions are obtained from the Karnaugh maps of the functions W, X, Y, and Z (see Fig. 2.6.3). A 8421 BCD subtractor made up of 8421 BCD adders and 9's-complement generators is shown in Fig. 2.6.5. This decimal subtraction is illustrated by the following two examples. The 9's-complement subtraction for a positive result produces a carry-out of the MSD (most significant digit), and the digit results are represented as their true values. A negative result will not produce a carry from the

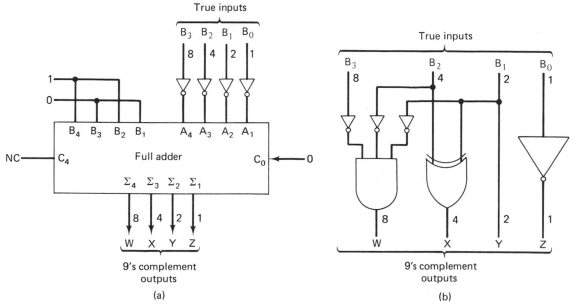

Fig. 2.6.4 Two 9's-complement generators

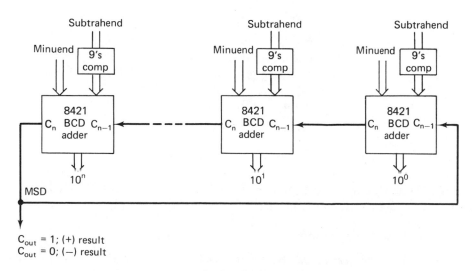

Fig. 2.6.5 8421 BCD subtractor.

MSD, and the digits will be represented in the 9's-complement form. If necessary, 9's-complement generator stages can be used to convert a negative result to the true form.

Example 2.6.2

Subtract 278_{10} from 495_{10}.

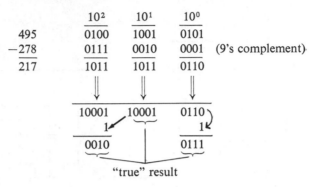

(This is a positive number, since EAC = 1.)

Example 2.6.3

Subtract 495_{10} from 278_{10}.

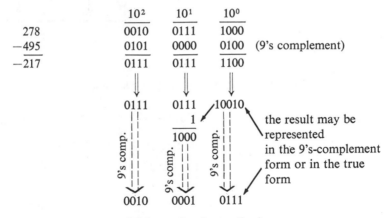

(This number is negative.)

Just like in the binary signed subtraction where the signed-2's-complement number system is used to avoid the end-around-carry, the 10's-complement number system (instead of the 9's-complement number system) may be used in the 8421 BCD subtraction to avoid the end-around-carry (Problem 15).

2.6.3 Excess-3 Addition

Excess-3 sum correction can be directly decided by whether C_4 is 0 (a no-carry) or 1 (a carry-out). If $C_4 = 0$, a 3 must be subtracted from the sum. This will be accomplished by adding 13, which is the 2's complement of the binary representation of 3; the carry-out for this addition is disregarded. For a carry-out, the effect is a carry-out of magnitude 16, which is 6 too much (see Table 2.6.3). A magnitude of 6 must be added and 3 must be subtracted to obtain an excess-3 for this decade; the net effect is an addition of 3.

TABLE 2.6.3 Comparison of Excess-3 Sums with the 8421 BCD Sums

Decimal Digit	8421 BCD Sum $C_4 \sum_4 \sum_3 \sum_2 \sum_1$						Excess-3 $C_4 \sum_4 \sum_3 \sum_2 \sum_1$				
0		0	0	0	0			0	0	1	1
1		0	0	0	1			0	1	0	0
2		0	0	1	0			0	1	0	1
3		0	0	1	1			0	1	1	0
4		0	1	0	0			0	1	1	1
5		0	1	0	1			1	0	0	0
6		0	1	1	0			1	0	0	1
7		0	1	1	1			1	0	1	0
8		1	0	0	0			1	0	1	1
9		1	0	0	1			1	1	0	0
10		1	0	1	0			1	1	0	1
11		1	0	1	1			1	1	1	0
12 do not have		1	1	0	0			1	1	1	1
13 a carry-out		1	1	0	1		1	0	0	0	0
14		1	1	1	0		1	0	0	0	1
15		1	1	1	1		1	0	0	1	0	have a
16	1	0	0	0	0		1	0	0	1	1	carry-out
17	1	0	0	0	1		1	0	1	0	0
18	1	0	0	1	0		1	0	1	0	1
19	1	0	1	1	1		1	0	1	1	0

8421 BCD + 6

Example 2.6.4

Repeat the addition performed in Example 2.6.1. Use the excess-3 code and the above algorithm.

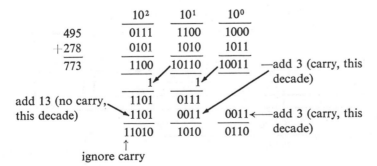

	10^2	10^1	10^0
495	0111	1100	1000
+278	0101	1010	1011
773	1100	10110	10011

—add 3 (carry, this decade)

add 13 (no carry, this decade) 1101 0111
1101 0011 0011← —add 3 (carry, this decade)
11010 1010 0110

↑
ignore carry

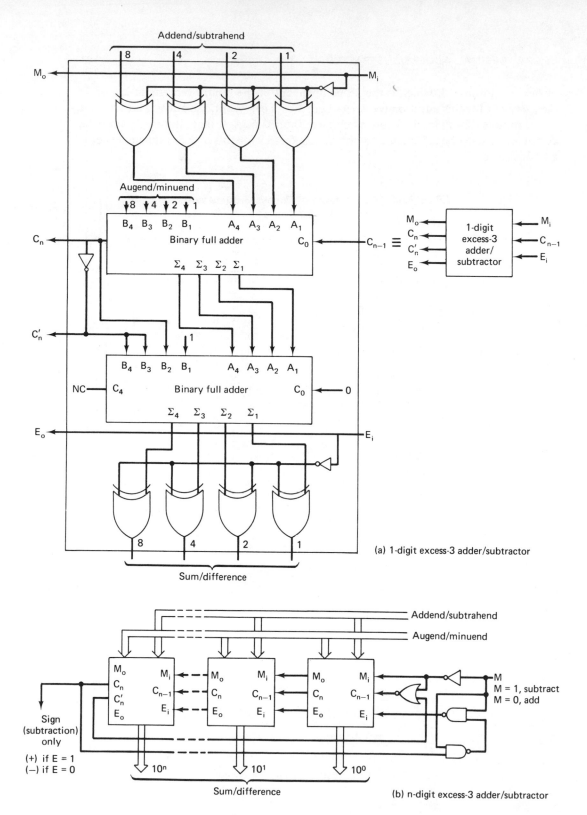

(a) 1-digit excess-3 adder/subtractor

Sum/difference

(b) n-digit excess-3 adder/subtractor

Fig. 2.6.6 Excess-3 adder/subtractor.

2.6.4 Excess-3 Subtraction

The second attractive feature of excess-3 code for decimal arithmetic is its self-complementing property; no special complement-generator circuit is needed. The complement of the excess-3 code can be obtained using the ordinary 1's-complement circuit, for example, the true/complement device similar to the one described in Fig. 2.4.2. An excess-3 decimal adder/subtractor is shown in Fig. 2.6.6.

Note that there is no carry in the 8421 BCD sum from 10 to 15; hence a detecting circuit is needed to detect whether C_n is 1, which determines whether the sum needs to be corrected or not. But if excess-3 code is used, these sums will have a carry-out, as indicated.

Example 2.6.5

Subtract 278_{10} from 495_{10} using the excess-3 subtractor.

	10^2	10^1	10^0
495	0111	1100	1000
−278	1010	0101	0100
217	10001	10001	1100
	1		1
	0010		1101
	0011	0011	1101
	0101	0100	11010

ignore carry

Example 2.6.6

Subtract 495_{10} from 278_{10}. Use the excess-3 subtractor.

	10^2	10^1	10^0
278	0101	1010	1011
−495	1000	0011	0111
−217	1101	1101	10010
		1	
		1110	
	1101	1101	0011
	11010	11011	0101

ignore carry

Exercises

1. (a) Perform the following number conversions:
    (1) $1000_{10} = ?_2 = ?_8$
    (2) $110110111_2 = ?_8 = ?_{10}$
    (3) $7777_8 = ?_{10} = ?_2$

(b) Perform the following binary arithmetic operations:
    (1) $11111110_2 + 10101010_2$
    (2) $11111110_2 - 10101010_2$

2. The hexadecimal number system is a number system with a base of 16. The 16 hexadecimal symbols are usually represented by

    Decimal: 0 1 2 3 4 5 6 7 8 9 10 11 12 13 14 15

    Hexadecimal: 0 1 2 3 4 5 6 7 8 9 A   B   C   D   E   F

    (a) Convert the following decimal numbers to hexadecimal. Use the dibble-dabble method:
        (1) 1000_{10} (2) 1976_{10} (3) 2000_{10} (4) 2468_{10}
    (b) Convert the following hexadecimal numbers to decimal using the dabble-dibble method:
        (1) 789_{16} (2) ADD_{16} (3) ABC_{16} (4) $EF56_{16}$
    (c) Derive a simple algorithm for the hexadecimal-to-binary conversion.

3. (a) Construct an octal full-adder using binary adders.
    (b) Construct a hexadecimal full-adder using binary adders.

4. Repeat problem 3 for full subtractors.

5. Derive the expressions of C_k, $k = 1, 2, \ldots, 8$, in terms of P_k and G_k for an eight-bit carry look-ahead adder.

6. Find a two-level AND–OR minimal realization of a 4-bit carry look-ahead adder using the minimization technique described in Section 1.6.

7. Design a 10-line decimal-to-4-line-BCD converter for active 0's on input lines.

8. (a) Design a 10-line decimal-to-excess-3 converter for active 1's on input lines using NAND and NOR gates.
    (b) Design a Gray-to-excess-3 converter and an excess-3-to-Gray converter using NAND and NOR gates.

9. Referring to the circuit of Fig. P2.9, show that it is a 4-bit controllable 8421 BCD-to-Gray/Gray-to-8421 BCD converter. When the control signal MC (mode control) is set to 1 and 0, it acts as a 8421 BCD-to-Gray converter and a Gray-to-8421 BCD converter, respectively.

10. Perform the following decimal additions for use with (1) the 8421 BCD code and (2) the excess-3 code:
    (a) $386_{10} + 756_{10}$
    (b) $123_{10} + 987_{10}$

11. Perform the following decimal subtractions for use with (1) the 8421 BCD code and (2) the 9's-complement subtraction algorithm:
    (a) $546_{10} - 429_{10}$
    (b) $429_{10} - 546_{10}$

12. Repeat problem 11 for use with the excess-3 code.

13. (a) Design a decimal half adder for use with the 8421 BCD code.
    (b) Construct a 8421 BCD full adder from decimal half adders.

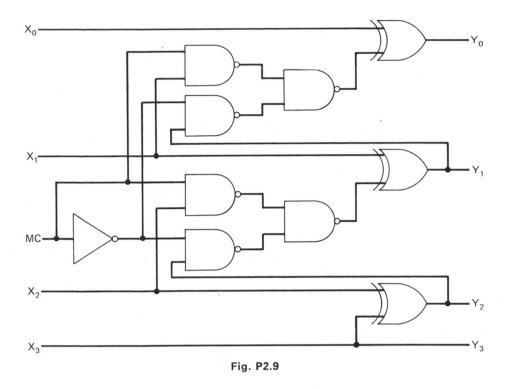

Fig. P2.9

14. Table P2.14 shows two self-complementing BCD codes. Construct a decimal subtractor for use with each of these codes.

TABLE P2.14

| Decimal | 6,3,1,−1 Code | | | | 2421 Code | | | |
	6	3	1	−1	2	4	2	1
0	0	0	1	1	0	0	0	0
1	0	0	1	0	0	0	0	1
2	0	1	0	1	0	0	1	0
3	0	1	1	1	0	0	1	1
4	0	1	1	0	0	1	0	0
5	1	0	0	1	1	0	1	1
6	1	0	0	0	1	1	0	0
7	1	0	1	0	1	1	0	1
8	1	1	0	1	1	1	1	0
9	1	1	0	0	1	1	1	0

15. Derive the 10's-complement subtraction algorithms for use with (1) the 8421 BCD code and (2) the excess-3 code. Apply them to the decimal subtraction $495_{10} - 278_{10}$.

Bibliographical Remarks

The subjects covered in this chapter are standard and can be found in the following references.

References

1. T. C. Bartee, *Digital Computer Fundamentals*, 3rd ed., McGraw-Hill, New York, 1972.

2. Y. Chu, *Digital Computer Design Fundamentals*, McGraw-Hill, New York, 1962.

3. D. Eadie, *Introduction to the Basic Computer*, Prentice-Hall, Englewood Cliffs, N.J., 1968.

4. I. Flores, *The Logic of Computer Arithmetic*, Prentice-Hall, Englewood Cliffs, N.J., 1963.

5. H. W. Gschwind, *Design of Digital Computers*, Springer-Verlag, New York, 1967.

6. M. F. Heilweil and G. A. Maley, *Introduction to Digital Computers*, Prentice-Hall, Englewood Cliffs, N.J., 1968.

7. P. A. Ligomenides, *Information-Processing Machines*, Holt, Rinehart and Winston, New York, 1969.

8. R. L. Morris and J. R. Miller, (ed.) *Designing with TTL Integrated Circuits*, McGraw-Hill, New York, 1971.

9. L. Nashelsky, *Digital Computer Theory*, Wiley, New York, 1966.

10. J. O'Malley, *Introduction to the Digital Computer*, Holt, Rinehart and Winston, New York, 1972.

11. C. M. Pease, "Matrix Inversion Using Parallel Processing," *J. ACM*, Vol. 14, No. 4, 1967, pp. 757–764.

12. V. T. Rhyne, *Fundamentals of Digital Systems Design*, Prentice-Hall, Englewood Cliffs, N.J., 1973.

13. S. Winogard, "On the Time Required to Perform Addition," *J. ACM*, Vol. 12, No. 2, 1965, pp. 277–285.

14. S. Winogard, "On the Time Required to Perform Addition," *J. ACM*, Vol. 12, No. 2, 1967, pp. 793–802.

3

Combinational Logic Design
Using Integrated Circuits

There are two types of digital circuits: combinational and sequential. The combinational digital circuit, whose outputs are functions of its inputs, is described by one or a set of switching functions, whereas the sequential digital circuit whose outputs are functions of its inputs and its state is described by a sequential machine. The latter which is an extension of the former will be discussed in later chapters.

The progress in integrated-circuit technology has made it possible to produce the inexpensive, ready-made SSI (*small-scale-integration*) gating circuits, MSI (*medium-scale-integration*) functional-level devices, and LSI (*large-scale-integration*) systems [such as the single-chip integrated-circuit pocket calculator, single-chip computer memories, single-chip computer central process unit (CPU), etc.], which have now become the basic "building blocks" of the design process. The linking of the logic circuit to its electronic-circuit realization is accomplished by a mapping from logic levels 0 and 1 of the former to two circuit-voltage levels of the latter. This mapping is referred to as *logic assignment*. It is clear that different logic assignments to a electronic circuit define different logic operations. In other words, the same circuit may perform several different logic operations, depending upon the logic assignment being used. A very convenient way to indicate the logic assignment at every part of a circuit is to use the voltage symbolism. Several important remarks about logic assignment and voltage symbolism are given.

There are eight major families of logic: resistor–transistor logic (RTL), diode–transistor logic (DTL), transistor–transistor logic (TTL), complementary-transistor logic (CTL), emitter-coupled logic (ECL), metal-oxide semiconductor (MOS), complementary metal-oxide semiconductor (CMOS), and integrated injection logic (IIL). They all have quite different characteristics. Possible characteristics to be considered when choosing a logic circuit are listed. The comparison of the characteristics of the eight major digital logic families are presented. Among the eight types, TTL, ECL, MOS, and CMOS are more popularly used. TTL, ECL, and CMOS are suitable for SSI and MSI, and MOS and IIL are particularly attractive for LSI.

The design of combinational digital circuits using SSI gates and MSI devices is presented. Available TTL/SSI and TTL/MSI of the most commonly used series,

series 54/74, are used in illustrative design examples. Considerations such as parallel connection of gates, propagation delay of a circuit, the restriction on circuit structure, and the time requirements on the input pulse signals are discussed.

Two basic catagories of semiconductor memories—read-only memories (ROM's) and random-access memories (RAM's)—are then discussed. Examples of commercially available MSI/TTL and LSI/MOS ROM's and RAM's are given. Finally, combinational logic design using programmable ROM arrays is presented.

A carefully prepared summary of semiconductor memories, including their basic circuitries, characteristics, and fabrication processes, is given in Appendix A. Readers interested in semicoductor memories are advised to read it.

3.1 Integrated Circuit Gates, Logic Assignment, and Voltage Symbolism

The advent of microelectronic integrated-circuit technology has changed many circuit-design criteria and much of the methodology used in the design of discrete-component circuits. For example, transistors in discrete form were far more expensive than passive components, but with the present integrated-circuit technology, transistors can be built at a cost that is much less than that of a resistor or a capacitor. The design criterion for discrete-component circuits was to minimize the number of active elements, which, for the reason above, have now been changed to minimize the total area required for constructing all the passive components of the design. As for design methodology, it is very seldom necessary today to build one's own gates, because commerically available IC gates are reliable, easy to use, and cost less. All we need to do is look at the IC gates in manufacturers' catalogs to find those that meet the design specifications, and then interconnect them to make the circuit.

IC chips are classified as SSI, MSI, or LSI, according to their complexity,

DEFINITION 3.1.1

A *gate-equivalent circuit* is a basic unit of measure of relative digital-circuit complexity. The number of gate-equivalent circuits is that number of individual logic gates that would have to be interconnected to perform the same function.

DEFINITION 3.1.2

LSI is a concept whereby a complete major subsystem function is fabricated as a single microcircuit. In this context a major subsystem or system, whether logical or linear, is considered to be one that contains 100 or more equivalent gates or circuitry of similar complexity.

DEFINITION 3.1.3

MSI is a concept whereby a complete subsystem or system function is fabricated as a single microcircuit. The subsystem or system is smaller than for LSI but, whether digital or linear, is considered to be one that contains 12 or more equivalent gates or circuitry of similar complexity.

DEFINITION 3.1.4

SSI circuits are integrated circuits of less complexity than MSI.

It is important to distinguish between the logic design and the circuit design of a digital system. The former is concerned only with the logical aspect of the design problem, whereas the latter concerns the electronic-circuit realization of the logic design after it is obtained. Linking the logic design with the circuit design is accomplished by a mapping from logic levels 0 and 1 to two circuit-voltage levels of an IC chip, which may be SSI, MSI, or LSI. This mapping is determined by circuit and logic considerations, and it is of the form

$$\text{logic } 0 \longleftrightarrow V(0) = V_L \text{ or } V_H$$
$$\text{logic } 1 \longleftrightarrow V(1) = V_H \text{ or } V_L$$

where the subscripts L and H represent the low and high voltage levels, respectively. In particular, logic 0 and logic 1 are mapped to two distinct circuit-voltage intervals:

$$\text{logic } 0 \longleftrightarrow V(0)_U > V > V(0)_L$$
$$\text{logic } 1 \longleftrightarrow V(1)_U > V > V(1)_L$$

where the U and L subscripts indicate the upper and lower limits, respectively. For a logic gate to be useful, it must satisfy several requirements.

　1.　The input logic voltage levels and the output logic voltage levels should be the same.

　2.　It must be possible to cascade several stages without degrading performance.

　3.　Each gate must be capable of driving several other gates.

　4.　The logic levels should be separated by a wide-enough difference that determination of a given logic level is easy and noise will have a minimal effect.

In addition, for circuits to be implemented by the monolithic technology, the following constraints apply:

　5.　The circuit should require small amounts of power.

　6.　Only diodes, resistors, and transistors should be required.

There are many different types of modern voltage-sensitive gating circuits: resistor–transistor logic (RTL), diode–transistor logic (DTL), transistor–transistor logic (TTL), complementary-transistor logic (CTL), emitter-coupled logic (ECL), metal-oxide semiconductor (MOS), complementary metal-oxide semiconductor (CMOS), and integrated injection logic (IIL). The circuits of the basic gates of these eight major families of logic are given in Fig. 3.1.1. The voltage truth tables of these circuits are shown in the second column of Table 3.1.1.

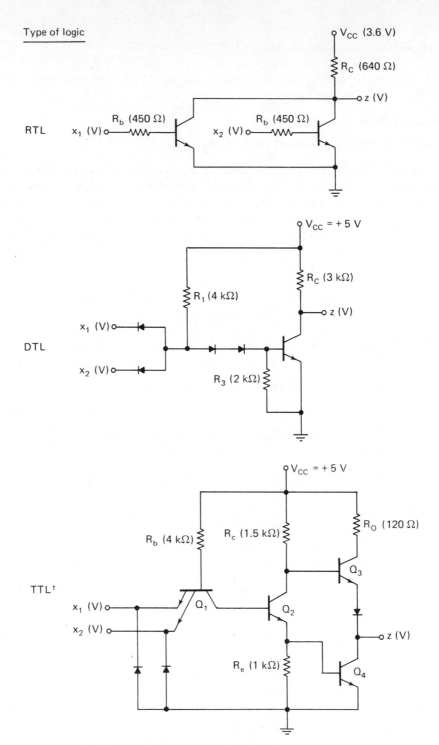

†TTL NOR gate is also standard in the sense that its fabrication is essentially of the same order of complexity of the TTL NAND gate

Fig. 3.1.1 Basic gate circuits of the families of logic.

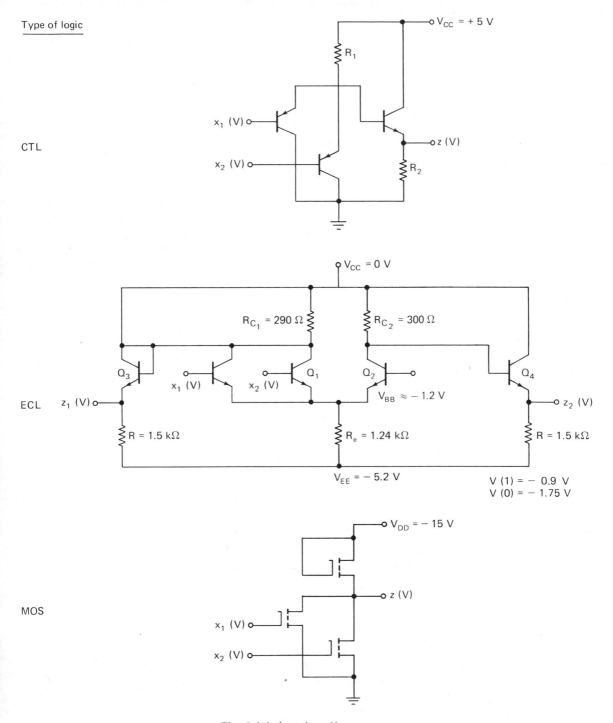

CTL

ECL

MOS

Fig. 3.1.1 (continued)

Type of logic

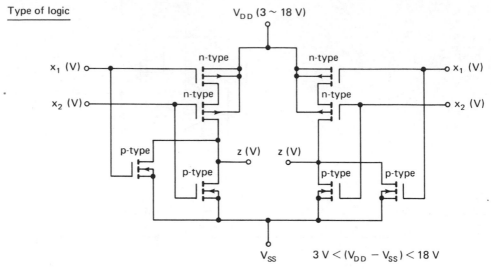

CMOS

Basic CMOS circuit 1 (positive NOR)

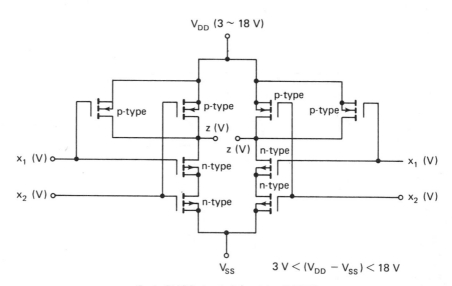

Basic CMOS circuit 2 (positive NAND)

Fig. 3.1.1 (continued)

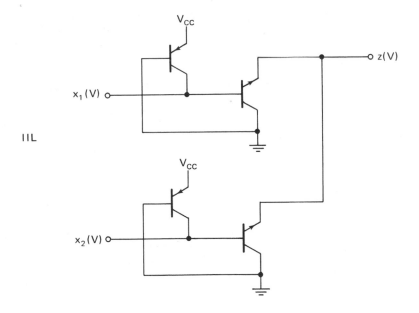

Fig. 3.1.1 (continued)

DEFINITION 3.1.5

$V(1)$ and $V(0)$ denote the voltages (or voltage intervals) that map to logic 1 and logic 0, respectively. Any voltage-logic mapping such that $V(1) > V(0)$ $[V(1) < V(0)]$ for the entire circuit (i.e., for all the inputs and the outputs of a digital circuit) is said to be a *positive (negative) logic assignment*. A gate obtained from a positive (negative) logic assignment of a digital circuit is called a *positive (negative) gate*. If the logic assignment of a part of a digital circuit is positive and that of the other part is negative, the logic assignment of the entired circuit is called a *mixed-logic assigiment*. A gate obtained from a mixed-logic assignment is called a *mixed gate*.

With a positive assignment to the basic gate circuits we obtain their logic truth tables and the basic gates they represent, which are shown in the third and fourth columns of Table 3.1.1, respectively. For example, with a positive logic assignment to the basic gate circuit of TTL, the logic truth table of the circuit shows that the circuit would perform a NAND operation, and thus the circuit is a positive NAND gate.

A very convenient way to indicate the mapping of the voltage levels (high and low) to the logic levels (logic 1 and 0) of an input or an output variable of a gate is to use the *voltage symbolism*. The use of voltage symbolism involves the placing of a plus or minus sign at each input and output point in the logic circuit. It specifies the relationship between the electronic characteristics of the gating circuit and the logical function that it is to perform. The use of the symbol follows:

TABLE 3.1.1 Voltage Truth Tables of the Circuits of Fig. 3.1.1

Type of logic	Truth table		Gate symbol with voltage symbols indicated
	Voltage truth table	Logic truth table	
RTL	x_1 (V) x_2 (V) z (V) 0 V 0 V + 3.6 V 0 V + 3.6 V 0 V + 3.6 V 0 V 0 V + 3.6 V + 3.6 V 0 V	V (1) = 3.6 V V (0) = 0 V x_1 x_2 z 0 0 1 0 1 0 1 0 0 1 1 0	x_1 x_2 z positive NOR
DTL	x_1 (V) x_2 (V) z (V) + 0.5 V + 0.5 V + 5 V + 0.5 V + 5 V + 5 V + 5 V + 0.5 V + 5 V + 5 V + 5 V + 0.5 V	V (1) = 5 V V (0) = 0.5 V x_1 x_2 z 0 0 1 0 1 1 1 0 1 1 1 0	x_1 x_2 z positive NAND
TTL	x_1 (V) x_2 (V) z (V) 0 V 0 V + 5 V 0 V + 5 V + 5 V + 5 V 0 V + 5 V + 5 V + 5 V 0 V	V (1) = 5 V V (0) = 0 V x_1 x_2 z 0 0 1 0 1 1 1 0 1 1 1 0	x_1 x_2 z positive NAND
CTL	x_1 (V) x_2 (V) z (V) 0 V 0 V 0 V 0 V + 5 V 0 V + 5 V 0 V 0 V + 5 V + 5 V + 5 V	V (1) = 5 V V (0) = 0 V x_1 x_2 z 0 0 0 0 1 0 1 0 0 1 1 1	x_1 x_2 z positive AND
ECL	x_1 (V) x_2 (V) z_1 (V) z_2 (V) − 1.75 V − 1.75 V − 0.9 V − 1.75 V − 1.75 V − 0.9 V − 1.75 V − 0.9 V − 0.9 V − 1.75 V − 1.75 V − 0.9 V − 0.9 V − 0.9 V − 1.75 V − 0.9 V	V (1) = − 0.9 V V (0) = − 1.75 V x_1 x_2 z_1 z_2 0 0 1 0 0 1 0 1 1 0 0 1 1 1 0 1	x_1 x_2 z_1 positive NOR x_1 x_2 z_2 positive OR

$$\frac{x}{+} \text{ means: } \begin{cases} \text{more positive voltage appears at } x \longleftrightarrow \text{ the logic value of } x \text{ is 1} \\ \text{more negative voltage appears at } x \longleftrightarrow \text{ the logic value of } x \text{ is 0} \end{cases}$$

TABLE 3.1.1 continued

MOS	x_1 (V) $\quad$ x_2 (V) $\quad$ z (V) 0 V $\qquad$ 0 V $\qquad$ − 15 V 0 V $\qquad$ − 15 V $\qquad$ 0 V − 15 V $\qquad$ 0 V $\qquad$ 0 V − 15 V $\qquad$ − 15 V $\qquad$ 0 V	V (1) = 0 V V (0) = − 15 V x_1 $\quad$ x_2 $\quad$ z 1 $\quad$ 1 $\quad$ 0 1 $\quad$ 0 $\quad$ 1 0 $\quad$ 1 $\quad$ 1 0 $\quad$ 0 $\quad$ 1	positive NAND
CMOS	Basic circuit 1: x_1 (V) $\qquad$ x_2 (V) $\qquad$ z (V) 0 V $\qquad$ 0 V $\qquad$ (0.45 ~ 0.55) x V_{DD} 0 V $\qquad$ (0.45 ~ 0.55) x V_{DD} $\qquad$ 0 V (0.45 ~ 0.55) x V_{DD} $\qquad$ 0 V $\qquad$ 0 V (0.45 ~ 0.55) X V_{DD} $\qquad$ (0.45 ~ 0.55) X V_{DD} $\qquad$ 0 V 3 V < V_{DD} < 18 V Basic circuit 2: x_1 (V) $\qquad$ x_2 (V) $\qquad$ z (V) 0 V $\qquad$ 0 V $\qquad$ (0.45 ~ 0.55) X V_{DD} 0 V $\qquad$ (0.45 ~ 0.55) x V_{DD} $\qquad$ 0 V (0.45 ~ 0.55) x V_{DD} $\qquad$ 0 V $\qquad$ 0 V (0.45 ~ 0.55) x V_{DD} $\qquad$ (0.45 ~ 0.55) x V_{DD} $\qquad$ 0 V 3 V < V_{DD} < 18 V	V (1) = (0.45 ~ 0.55) x V_{DD} V (0) = 0 V x_1 $\quad$ x_2 $\quad$ z 0 $\quad$ 0 $\quad$ 1 0 $\quad$ 1 $\quad$ 0 1 $\quad$ 0 $\quad$ 0 1 $\quad$ 1 $\quad$ 0 V (1) = (0.45 ~ 0.55) x V_{DD} V (0) = 0 V x_1 $\quad$ x_2 $\quad$ z 0 $\quad$ 0 $\quad$ 1 0 $\quad$ 1 $\quad$ 1 1 $\quad$ 0 $\quad$ 1 1 $\quad$ 1 $\quad$ 0	positive NOR positive NAND
IIL	x_1 (V) $\quad$ x_2 (V) $\quad$ z (V) 0 V $\qquad$ 0 V $\qquad$ + 0.85 V 0 V $\qquad$ + 0.85 V $\qquad$ 0 V + 0.85 V $\qquad$ 0 V $\qquad$ 0 V + 0.85 V $\qquad$ + 0.85 V $\qquad$ 0 V	V (1) = 0.85 V V (0) = 0 V x_1 $\quad$ x_2 $\quad$ z 0 $\quad$ 0 $\quad$ 1 0 $\quad$ 1 $\quad$ 0 1 $\quad$ 0 $\quad$ 0 1 $\quad$ 1 $\quad$ 0	positive NOR

$$\frac{x}{-} \text{means:} \begin{cases} \text{more positive voltage appears at } x \longleftrightarrow \text{ the logic value} \\ \text{of } x \text{ is } 0 \\ \text{more negative voltage appears at } x \longleftrightarrow \text{ the logic value} \\ \text{of } x \text{ is } 0 \end{cases}$$

The voltage symbols of the inputs and the output of the basic gates of the eight logic families are indicated in the fourth column of Table 3.1.1.

It is important to point out the difference between a logic gate and an electronic digital circuit gate. A comparison of the meanings of a logic gate symbol and an electronic digital circuit gate symbol is shown in Table 3.1.2. A logic gate is represented by a logic gate symbol whose input and output values are logic 0 and 1. Its function is described by a logic truth table. An electronic digital circuit gate with type of logic family specified is represented by a logic gate indicating its logic function with voltage

TABLE 3.1.2 Comparison of the Meanings of a Logic Gate Symbol and an Electronic Digital Circuit Gate Symbol

Type	Logic gate	Electronic digital circuit gate with type of logic family specified	Electronic digital circuit gate with type of logic family unspecified
Symbol	 (No logic assignment)	 (Positive logic assignment)	 (Mixed logic assignment)
Input and output values	Logic 0 and 1	0 V and + 5 V	LOW and HIGH voltage levels (intervals), V_L and V_H
Function description	Logic truth table	Voltage truth table	Voltage truth table
Truth table	x_1 x_2 z 0 0 1 0 1 1 1 0 1 1 1 0	x_1 x_2 z 0 V 0 V + 5 V 0 V + 5 V + 5 V + 5 V 0 V + 5 V + 5 V + 5 V 0 V	x_1 x_2 z V_L V_H V_L V_L V_L V_L V_H V_H V_L V_H V_L V_H

symbols indicated at each input and output variable of the gate. The type of logic family specifies the values of the low and high voltage levels of the gate. The circuit input/output relation is described by a voltage truth table. The representation of an electronic digital circuit gate with type of logic family unspecified shown in the third column of Table 3.1.2 is similar to that described in the second column of the same, table, except that the low and high voltage level are unspecified and denoted by V_L and V_H, respecitvely.

It is worth noting that each basic gate circuit of the eight logic families shown in Fig. 3.1.1 can perform several different logic operations, depending upon the logic assignment being used. Taking the TTL basic gate circuit for example, it performs:

1. A NAND operation if a positive logic assignment is used,
2. A NOR operation if a negative logic assignment is used,

and 3.   An AND or OR operation if a mixed logic assignment is used.

They are shown in Table 3.1.3.

Several important remarks about logic assignment and voltage symbolism are in order.

TABLE 3.1.3

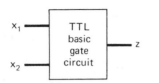

x_1 (V)	x_2 (V)	z (V)
0 V	0 V	+5 V
0 V	+5 V	+5 V
+5 V	0 V	+5 V
+5 V	+5 V	0 V

Block diagram representation   The voltage truth table
of TTL basic gate circuit   of the circuit

Logic assignment (LS)	Truth table	Gate
(positive logic assignment)	x_1 x_2 z 0 0 1 0 1 1 1 0 1 1 1 0	(positive NAND)
(negative logic assignment	x_1 x_2 z 1 1 0 1 0 0 0 1 0 0 0 1	(negative NOR)
(mixed logic assignment)	x_1 x_2 z 0 0 0 0 1 0 1 0 0 1 1 1	
(mixed logic assignment)	x_1 x_2 z 1 1 1 1 0 1 0 1 1 0 0 0	

1. Most commercial ready-made SSI gates are positive (logic assignment) gates.

2. The positive logic assignment is used throughout the circuit in most MSI and LSI devices.

3. In industry, voltage symbols are not used. Positive and negative gates are represented by their logic gate symbol with the type of logic assignment specified. For example.

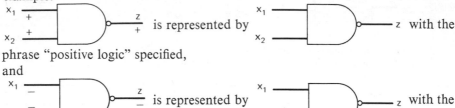

phrase "positive logic" specified,

and

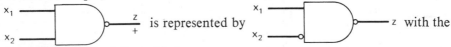

phrase "negative logic" specified.

For mixed logic assignment gates, circles are used to denote the negative voltage symbols. For example,

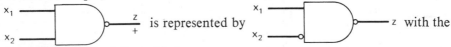

phrase "positive logic" specified,

which indicates that $z = x_1 \uparrow x_2'$. From now on, we shall use this convention to denote voltage symbols of a gate which is most generally used by industry. By doing so, the mixed logic assignment case is eliminated.

DEFINITION 3.1.6

Two electronic digital circuit gates are said to be *electrically equivalent* if they have the same voltage truth table. This equivalent relation is denoted by the symbol $\equiv$.

Several examples of electrically equivalent electronic digital circuit gates are given in Fig. 3.1.2. In this figure we see that a positive OR gate, for example, is electrically equivalent to a negative AND gate. This indicates that an electronic digital circuit that realizes a positive OR will also realize a negative AND, and vice versa.

Example 3.1.1

The voltage truth table given in Fig. 3.1.3(a) describes a gating circuit. Using the positive logic assignment, find all the logic gates that this circuit can be used for.

The two gates, an OR gate and (a) NAND gate with the required input and output voltage symbols, are shown in Fig. 3.1.3(b). Note that the positive logic is assumed and the low and high voltage levels of this circuit are -10 V and 0 V, respectively.

Example 3.1.2

(a) Construct the voltage table for each of the gates in the circuit diagram of Fig. 3.1.4(a). Assume that $V_H = 5$ V and $V_L = 0$ V.

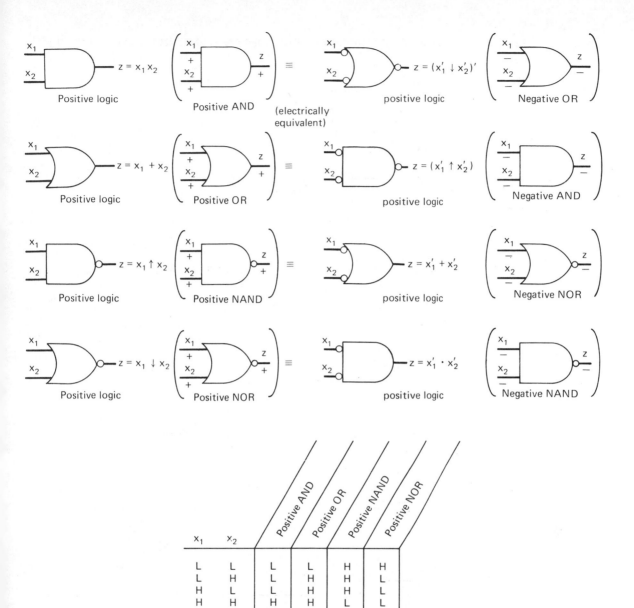

L denotes the low voltage level
H denotes the high voltage level

Fig. 3.1.2 Several examples of electrically equivalent gates.

A(V)	B(V)	C(V)
0	0	0
0	− 10	− 10
− 10	0	0
− 10	− 10	0

(a) Voltage truth table

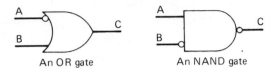

An OR gate   An NAND gate

(b) Two positive logic gates with the
required input and output voltage
symbols indicated.   **Fig. 3.1.3** Example 3.1.1.

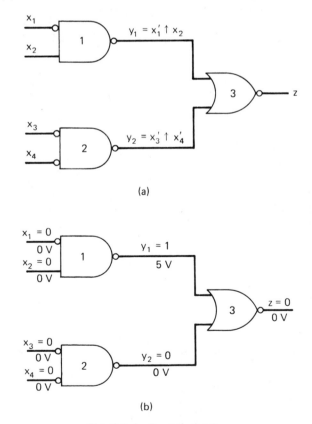

(a)

(b)

Fig. 3.1.4 Example 3.1.2.

(b)   Indicate the voltage on each branch of the circuit of Fig. 3.1.4(a) for $x_1 = x_2 = x_3 = x_4 = 0$ (logic 0).

The voltage truth tables of the three gates are as follows:

Gate 1

x_1 (V)	x_2 (V)	y_1 (V)
5	0	5
5	5	5
0	0	5
0	5	0

Gate 2

x_3 (V)	x_4 (V)	y_2 (V)
5	5	5
5	0	5
0	5	5
0	0	0

Gate 3

y_3 (V)	y_4 (V)	f (V)
0	0	5
0	5	0
5	0	0
5	5	0

The voltage on each branch of the circuit is shown in Fig. 3.1.4(b).

It should also be remarked that besides the sum-of-products (AND–OR) and the product-of-sums (OR–AND) canonical forms, other canonical forms exist. They are:

$$\begin{array}{ll} \text{NAND–NAND} & \text{AND–NOR} \\ \text{OR–NAND} & \text{NAND–AND} \\ \text{NOR–OR} & \text{NOR–NOR} \end{array}$$

For example, consider a function $f(x_1, x_2) = x_1 x_2 + x_1' x_2'$. The canonical forms of this function are

$$
\begin{aligned}
f(x_1, x_2) &= x_1 x_2 + x_1' x_2' & \text{(AND–OR)} \\
&= ((x_1 x_2 + x_1' x_2')')' \\
&= ((x_1 x_2)' \cdot (x_1' x_2')')' = (x_1 \uparrow x_2) \uparrow (x_1' \uparrow x_2') & \text{(NAND–NAND)} \\
&= ((x_1' + x_2') \cdot (x_1 + x_2))' = (x_1' + x_2') \uparrow (x_1 + x_2) & \text{(OR–NAND)} \\
&= (x_1' + x_2')' + (x_1 + x_2)' = (x_1' \downarrow x_2') + (x_1 \downarrow x_2) & \text{(NOR–OR)}
\end{aligned}
$$

The other four canonical forms may be obtained from $f'(x_1, x_2)$ in sum-of-products form:

$$
\begin{aligned}
f'(x_1, x_2) &= x_1 x_2' + x_1' x_2 \\
f(x_1, x_2) &= (x_1 x_2' + x_1' x_2)' = (x_1 x_2') \downarrow (x_1' x_2) & \text{(AND–NOR)} \\
&= (x_1 x_2')' \cdot (x_1' x_2)' = (x_1 \uparrow x_2') \cdot (x_1' \uparrow x_2) & \text{(NAND–AND)} \\
&= (x_1' + x_2) \cdot (x_1 + x_2') & \text{(OR–AND)} \\
&= (((x_1' + x_2) \cdot (x_1 + x_2'))')' \\
&= ((x_1' + x_2)' + (x_1 + x_2')')' = (x_1' \downarrow x_2) \downarrow (x_1 \downarrow x_2') & \text{(NOR–NOR)}
\end{aligned}
$$

The negation of a variable can be accomplished by using a NOT gate which can be obtained from a NAND gate or a NOR gate, as shown in Fig. 3.1.5. Note that a NOT gate can also be realized by other ways (problem 5).

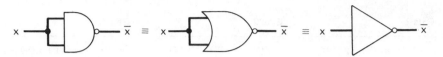

Fig. 3.1.5 NAND and NOR gates acting as NOT gates

Example 3.1.3

Consider the NAND-NAND realization of the function

$$z = f(x_1, x_2, x_3) = x_1 x_2' + x_1' x_2 x_3 + x_1 x_3' \qquad (3.1.1)$$

which can be expressed in the NAND-NAND canonical form as:

$$z = (x_1 \uparrow x_2') \uparrow (x_1' \uparrow x_2 \uparrow x_3) \uparrow (x_1 \uparrow x_3') \qquad (3.1.2)$$

The NAND–NAND realization of this function is shown in Fig. 3.1.6.

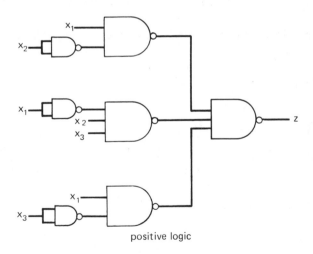

positive logic

Fig. 3.1.6 Example 3.1.3.

Exercise 3.1

1. Show that with appropriate logic assignment to the inputs and the output of the gate circuit of Fig. P3.1.1(a), whose voltage truth table is described in Fig. P3.1.1(b), the circuit can act as a NAND gate.

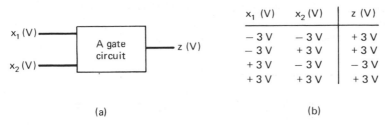

x_1 (V)	x_2 (V)	z (V)
− 3 V	− 3 V	+ 3 V
− 3 V	+ 3 V	+ 3 V
+ 3 V	− 3 V	− 3 V
+ 3 V	+ 3 V	+ 3 V

(a)   (b)

Fig. P3.1.1

2. Find the eight canonical forms of the following functions.
   (a) $f(x_1, x_2, x_3) = x_1 x_2 + x_2' x_3$
   (b) $f(x_1, x_2, x_3, x_4) = (x_1 + x_2')(x_3 + x_2 x_4')$
   (c) $f(x_1, x_2, x_3, x_4) = (x_1' + x_2')(x_3' + x_1' x_4)(x_2 + x_4)$

3. Use TTL gates to realize
   (a) the full-adder described in Fig. 2.2.2(d).
   (b) the full-subtractor described in Fig. 2.2.5(c-2).
   Assume that positive voltage symbolism is used throughout the circuit.

4. Indicate the voltage value on each branch of the circuit realization of Eq. (3.1.2) shown in Fig. 3.1.6 for $x_1 = x_3 = $ logic 1 and $x_2 = $ logic 0. Assume that TTL gates are used.

5. (a) Verify the eight relationships between the voltage symbolism and the not operation shown in Fig. P3.1.5.

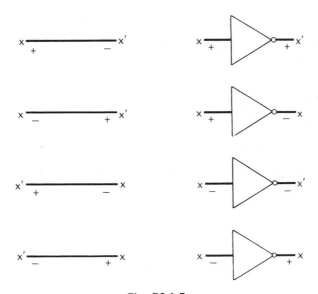

Fig. P3.1.5

   (b) Show that an n-input NAND(NOR) gate can be converted into a NOT gate by connecting its $n - 1$ inputs to logic 1 (logic 0).

6. Realize the logic circuit of Fig. P3.1.6 using TTL gates. Assume that
    (a) positive logic at the circuit inputs and outputs are required.
    (b) negative logic at the circuit inputs and outputs are required.

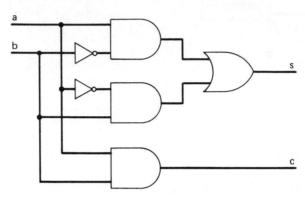

Fig. P3.1.6

7. Realize the two-level AND-OR logic circuit of Fig. P3.1.7 by (a) a positive TTL NAND-NAND realization and (b) a negative TTL NOR-NOR realization.

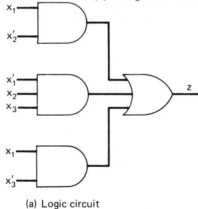

(a) Logic circuit   **Fig. P3.1.7**

8. Repeat problem 7 for the two-level OR-AND logic circuit of Fig. P3.1.8.

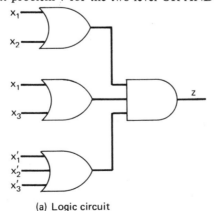

(a) Logic circuit   **Fig. P3.1.8**

3.2 Characteristics and Comparison of the Major Integrated IC Logic Families

Before going on to describe the various families of IC logic gates, it will be appropriate to look at their general characteristics, so that the differences between the types can be better appreciated.

1. *Threshold voltage.* The voltage level at the input of a circuit, at which the circuit changes from one state to the other, is called the *threshold voltage*. One approximation of this is the voltage at the midpoint of the transition between the two states.

2. *Operating speed.* The time delay between the application of a level change at the input and the change of state at the output of a circuit is called the *propagation delay* of the circuit. Generally, the propagation delay of an IC gate is in the range 2† to 50 nanoseconds (a nanosecond ns, being 10^{-9} second). The total propagation delay time of a logic system will be the delay per gate multiplied by the number of gates in series.

3. *Power dissipation.* The power dissipation of a logic circuit is usually defined as the supply power required for the gate to operate with a 50 per cent duty cycle at a specific frequency. (i.e., equal times in the 0 and 1 states). The power dissipation of a typical logic-integrated circuit ranges from a few microwatts (μW) to about 50 mW per gate, depending on the type of circuit. Hence, in any given system there may be many hundreds of gates, and the power required by each gate must be considered. In general, an increase in speed requires higher power; hence a compromise between power dissipation and speed is usually required.

4. *Noise margin.* The difference between the operating input–logic–voltage level and the threshold voltage is called the *noise margin* of the circuit [i.e., the maximum amount of deviation from the nominal values of $V(0)$ and $V(1)$ that the circuit can tolerate without changing states]. To give the most stable operation for a given total logic–voltage swing, it is desirable to have approximately equal logical 0 and logical 1 noise margins. It is relatively straightforward to specify the dc noise margin of a circuit. But it is much more difficult to specify the *ac noise margin* or the *ac noise immunity*, which is defined as the ability of a circuit to withstand short, transient voltage pulses that appear at the input to a logic circuit. The ability of the gate to operate reliably in a noise environment is important in many applications.

5. *Logic voltage levels.* The values for the voltage levels corresponding to logic 1 and logic 0 affect several other specifications for a system such as power dissipation, speed, and noise immunity. Within a given system, a common family of logic will be used that will ensure compatibility of logic levels. The choice of logic levels does, however, become important when a system must be driven by or must drive an existing system. The problem of interfacing with these systems then becomes much simpler if a family of gates is chosen that has the same logic levels as the system to which they must interface.

6. *Fan-in and fan-out.* The *fan-in* of a logic gate is the number of inputs it is designed to have. It is a measurement of how much that input will load a driving

†May be as low as 0.5 ns for ECL.

source, usually referenced to the standard input schematic for a given logic family. The *fan-out* of a logic gate is the number of gates that can be reliably driven by the gate.

7. *Operating temperature.* All IC gates are semiconductor devices that are temperature-sensitive. They must be designed to guarantee a satisfactory performance over a wide range of temperature. Integrated circuits intended for military applications are usually specified to operate over the ambient range -55 to $+125°C$, and those for industrial use from 0 to $+70°C$.

The general characteristics of the eight IC logic families are tabulated in Table 3.2.1. Values quoted are representative on a comparison basis. For any one family, the values may vary somewhat.

RTL logic was the first family of logic circuits established as a standard catalog line. It offered high performance but insufficient noise margins. DTL logic, slower but easier to use, offered improved noise margins and fan-out. It has been gradually superseded by its faster cousin, TTL. TTL logic offers higher speed, better noise immunity and drive capability, by far the largest number of standard SSI gates and MSI complex devices, and various speed/power trade-offs. CTL logic offers higher speed at moderate cost and power dissipation. ECL logic offers the highest speed, drives terminated transmission lines, and is the ultimate choice for very fast systems. MOS logic offers greatly increased complexity and low power consumption but significantly less speed. CMOS logic offers extremely low power consumption when operated at low speed. Its complexity is comparable to TTL, and its speed is between those of MOS and TTL. Finally, IIL logic, a recently (since 1970) developed bipolar technology, offers bipolar performance with a packing density greater than MOS and simpler processing than all technologies except p-MOS (see Appendix A, Section A.1.3). The IIL gate requires no components except lateral PNP and inverted NPN transistors, permitting extremely efficient utilization of silicon real estate. The speed-power product for IIL is constant and supply current may be traded for switching speed over a range of 4–5 decades. This speed-power product for production IIL is as low as that for CMOS (the lowest of all previous logic families) and current laboratory IIL is an order of magnitude better than this. A summary of advantages and disadvantages of the eight logic families is given in Table 3.2.2.

It should be remarked that p-MOS IC's require only one-third of the process steps needed for the standard double-diffused bipolar IC. But the most significant feature is the large number of semiconductor circuit elements that can be put on a small chip. This high circuit density means large-scale integration and permits us to put up to 5,000 devices on a silicon chip only 150 by 150 mils square. Each transistor in the MOS/LSI array requires as little as 1 square mil of chip area, a great reduction over bipolar transistors, which require 49 to 50 square mils. Hence it is predominately used in large regular arrays such as serial and random-access memories, custom LSI circuits for high-volume production—notably calculators and microprocessors that implement logic with programming steps rather than hard-wired connections. The device pictured in Fig. 3.2.1 is a complete 8-bit parallel processor on a 215- by 225-mil chip. It includes an 8-bit parallel ALU with hardware, parity flags, an 8-bit accumulator with file

TABLE 3.2.1   General Characteristics of IC Logic Families

Logic Family	Basic Logic Type	Propagation Time per Gate (ns)	Power Dissipation per Gate (mW)	Typical Noise Margin (V)	Typical Fan-in	Maximum Fan-out	Relative Cost per Gate
RTL	Positive NOR	50	10	0.2	3	4	Medium
DTL	Positive NAND	25	15	0.7	8	8	Medium
TTL	Positive NAND	10	20	0.4	8	12	Low
CTL	Positive AND	5	50	0.4	5	25	High
ECL	Positive OR/NOR	2	50	0.4	5	25	High
MOS	Positive NAND	250	<1	2.5	10	5	Very low
CMOS	Positive NOR and Positive NAND	30	0.05 μW	Depends on V_{DD}. (Typically 45% of V_{DD})	10	100	Low
IIL	Positive NOR	40	<1	0.35		8	Very low

**TABLE 3.2.2 Advantages and Disadvantages of the
Eight Logic Families**

Logic Family	Advantages	Disadvantages
RTL	Low power dissipation	(1) Insufficient noise margins (2) Low speed
DTL	Low power dissipation	(1) Poor noise margins (2) Low speed
TTL	(1) Low power dissipation (2) High speed (3) High fan-out capability (4) High compatibility with existing systems (5) Low cost	(1) Tight V_{CC} tolerance (2) Susceptible to power transients
CTL	Higher speed	(1) High cost (2) Requiring more experience on the part of the user because the logic levels are nonrestoring
ECL	(1) Very high speed (2) High fan-out capabilities (3) Low noise generation (4) Complementary output	(1) Difficult to interface with other logic families (2) High cost (3) Requiring careful circuit layout
MOS	(1) Increased circuit complexity per package (2) Low cost per circuit function (3) Fewer subsystem to test (4) Fewer parts to assemble and inspect (5) Lower power drain per function (6) A choice of standard or custom products	(1) Very low speed (2) Requiring multiple power sources to properly operate and interface to other logic families (3) Reliability generally only fair
CMOS	(1) Extremely low power consumption when operated at low speed (2) Relative insensitive to variations in V_{DD} (3) High noise immunity (4) Compatable (through appropriate buffers) to most existing logic families	(1) Low speed (2) Susceptible to static discharge damage
IIL	(1) Speed-power product constant and low (2) Very high packing density	

register, 8-bit program and memory address registers, and a multilevel program address stack. With suitable random-access memory system, the capabilities of a full 8-bit microcomputer can be achieved.

It is important to point out that within a logic family, there may be several variations. Taking TTL, for instance, the basic type gates are the 5400/7400 series gates. However, the basic 5400/7400 series gates are also manufactured in a high-speed version (54H00/74H00 series gates), a low-power-dissipation version (54L00/74L00 series gates), a very high speed version (54S00/74S00 series gates), and a low power and very high speed version (54LS00/74LS00 series gates). The available choices range

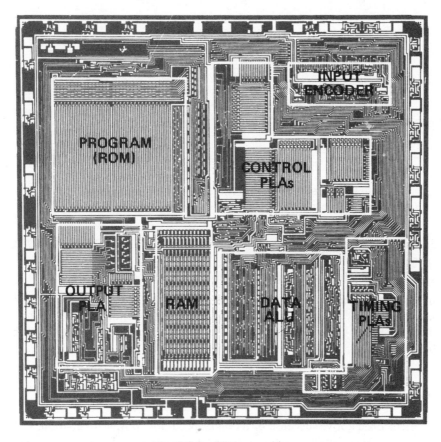

Fig. 3.2.1 CPU on a chip.

from the very high performance of the Schottky-clamped† functions for systems operating typically up to 125 megahertz (MHz) to low-power functions with power consumption of only 1 mW per gate. Typical characteristics of the five TTL series offiered are shown in Table 3.2.3. Because of the general advantages offered by the TTL logic (see Table 3.2.2) and a full spectrum of performance ranges that permits the digital system designer to optimize system cost and performance, TTL/SSI and TTL/MSI have been widely used as building blocks in designing modern digital systems. In addition, the TTL 54/74 series offers the designer the following advantages:

1. A broad range of functions are offered in each series.
2. All series are designed for a single 5-V power supply. (However, the supply must be relatively well regulated).
3. All series provide 1 V or greater typical dc noise margins.
4. Power dissipation relatively insensitive to operating frequency.

†See p. 21 of reference 5.

**TABLE 3.2.3 Typical Characteristics of the Five
TTL 54/74 Series†**

Series	Propagation Time per Gate (ns)	Power Dissipation (mW)
54/74	10	10
54H/74H	6	22
54L/74L	33	1
54S/74S	3	19
54LS/74LS	9.5	2

†There is a new 54SS/74SS (super Schottky) series whose propagation time and power dissipation are <2 ns and 22 mW, respectively.

5. Compatible with most logic families, such as DTL, MOS, CMOS†, and IIL.

6. All TTL types are compatible and may be mixed in a system.

In summary, when one is looking for SSI gates and MSI devices, TTL is recommended for high-speed applications, and ECL is recommended for very high speed applications. But if LSI components or systems are being sought, MOS will be the best bet.

3.3 Combinational Logic Design Using SSI Gates

As mentioned above, because of its high speed, low power dissipation, sufficient noise margins, high fan-in and fan-out, and low cost (see Table 3.2.1), TTL logic offers by far the largest number of standard SSI gates and MSI complex devices. The most commonly used TTL/SSI and TTL/MSI integrated-circuit series is the SN54/SN74 series (SN stands for semiconductor network), or simply the 54/74 series. The 54 series was introduced in 1964 primarily for the military market, where size, power consumption, and reliability requirements were paramount. The operating characteristics of the 54 series were guaranteed in the temperature range −55 to 125°C. Soon the low-cost industrial versions of the 54 series, called the 74 series, were introduced, which were primarily for commercial use, with guaranteed operating characteristics over a 0°C to 70°C range. Not only are individual members of the series 54/74 TTL families compatible, but they have the following typical characteristics in common: (1) supply voltage = 5.0 V, (2) logical 0 output voltage $V_{0L} \leq 0.4$ V, (3) logical 1 output voltage $V_{0H} \geq 2.4$ V, and (4) noise immunity = 0.4 V.

With some modification of the basic TTL gate circuit of Fig. 3.1.1, a complete line of ready-made TTL/SSI positive gates, including NAND, NOT, AND, NOR,

---

†TTL will drive CMOS, but CMOS may not drive the TTL gate. However there exist special buffer chips to provide the proper interface.

OR, and AND–OR–INVERT gates, which are available to the designer are shown in Table 3.3.1.

TABLE 3.3.1 Some Available Combinational TTL/SSI Integrated Circuits

Gate	Function	Device
NAND–NOR	Quad 2-input NAND	54/7400, 54/74L00, 54H/74H00
	Quad 2-input NAND, open collector	54/7401, 54H/74H01
	Quad 2-input NOR	54/7402
	Quad 2-input NAND, open collector	54/7403
	Hex inverter	54/7404, 54/74L04, 54H/74H04
	Hex inverter, open collector	54H/74H05
	Hex inverter buffer/driver, open collector	54/7406
	Hex buffer/driver, open collector	54/7407
	Quad 2-input AND	54/7408
	Quad 2-input AND, open collector	54/7409
	Triple 3-input NAND	54/7410, 54/74L10, 54H/74H10
	Triple 3-input AND	54H/74H11
	Hex inverter buffer/driver, open collector	54/7416
	Hex buffer/driver, open collector	54/7417
	Dual 4-input NAND	54/7420, 54/74L20, 54H/74H20
	Dual 4-input AND	54H/74H21
	Dual 4-input NAND, open collector	54H/74H22
	Quad 2-input NAND, open collector	54/7426
	8-input NAND	54/7430, 54/74L30, 54H/74H30
	Dual 4-input NAND buffer	54/7440, 54H/74H40
AND-OR-INVERT	Expandable dual 2-wide 2-input AND-OR-INVERT	54/7450, 54H/74H50
	Dual 2-wide 2-input AND-OR-INVERT	54/7451, 54/74L51, 54H/74H51
	Expandable 4-wide 2-2-2-3-input AND-OR	54H/74H52
	Expandable 4-wide 2-2-2-3-input AND-OR-INVERT	54/7453, 54H/74H53
	2-2-2-3-input AND-OR-INVERT	54/7454, 54/74L54, 54H/74H54
	Expandable 2-wide 4-input AND-OR-INVERT	54/74L55, 54H/74H55
Expander	Dual 4-input	54/7460, 54H/74H60
	Triple 3-input	54H/74H61
	3-2-2-3-input AND-OR	54H/74H62

The following example illustrates the design of combinational circuits using TTL/SSI gates.

Example 3.3.1

Consider the electronic-circuit realization of the automatic toll collector design problem discussed in Section 1.8. The first five design steps that led to the minimal logic circuit realization (Fig. 1.8.1(b)) were presented in Example 1.8.1.

Step 6 Obtain the electronic-circuit realization of the logic circuit, using ready-made IC components. Suppose that we choose TTL 54/74 series gates to design this circuit.

Several ways are possible for constructing this circuit, depending upon the gates available. For example:
1.   Use one SN74H52 AND–OR gate.
2.   Use one SN7454 AND–OR–INVERT gate.
The electronic-circuit realizations using these gates are shown in Fig. 3.3.1. Note that both these realizations use only one IC component.

In designing combinational circuits, besides the consideration of selecting an appropriate type of logic family to meet design specifications, the following restrictions and practical considerations are also very important:

1.   The circuit should be loop-free (i.e., contains no feedback loops). This is a necessary condition for any circuit to be combinational.

2.   The values of the output signals of a combinational circuit must depend only on the current values of the input signals. This is a necessary and sufficient condition for a circuit to be combinational. Note that every gate has a delay time between its inputs and its output which is of the order of 10^{-9} second (ns). To ensure that this requirement be met, the input signals must be kept at their values until all the input signals have reached every output of the circuit. To be more specific, if we let τ_c and τ_p be the time to hold the input signal at a constant value and the time delay between the input and the output of a path of the circuit, it is necessary that

$$\tau_c \geq \max_{\substack{\text{for all paths from} \\ \text{every input to every} \\ \text{output of the circuit}}} \{\tau_p\}$$

In words, τ_c must be equal to or greater than the delay time of the path from input to output with the most propagational delay.

3.   Most SSI gates are nonexpandable unless otherwise stated in the data book. Take TTL/SSI expandable AND–OR gates SN74H52 and TTL/SSI triple three-input expanders SN74H61. They can be interconnected to realize 3-3-2-2-2-input AND–OR gates, as shown in Fig. 3.3.2.

4.   All major logic family circuitries have the wired logic operation property. For example, the ECL circuit in Fig. 3.3.3 exhibits a wired-OR logic operation at the wired interesection before the output z_1 and a wired-AND logic operation at the wired intersection before the output z_2. As another example, consider the TTL gated full-adder SN7480 shown in Fig. 3.3.4. It is seen that there are two wired-AND gates in this circuitry. A third example of wired-AND and wired-OR circuits is shown in Fig. 3.3.5.

5.   All the circuits described in Chapter 2 can be realized using the TTL/SSI gates described in Table 3.3.1. However, it will be much simpler as well as more economical to use TTL/MSI devices in these circuits realizations, if they are available. This is discussed in the next section.

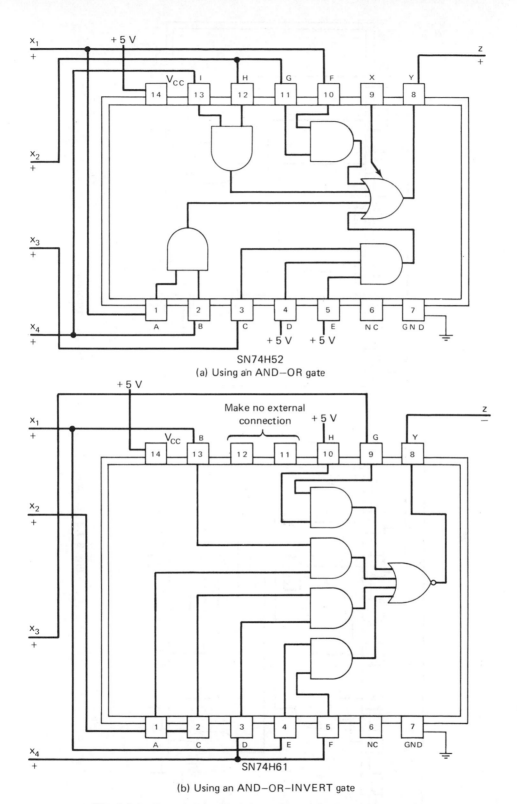

SN74H52
(a) Using an AND−OR gate

SN74H61
(b) Using an AND−OR−INVERT gate

Fig. 3.3.1 Two electric circuit realizations using IC ready-made gates.

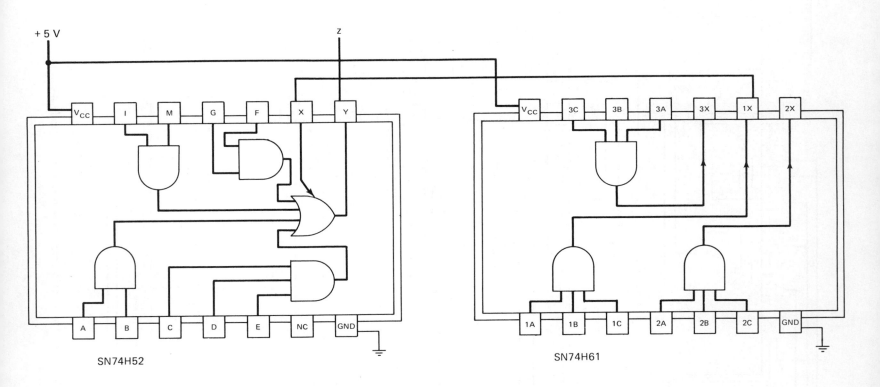

Fig. 3.3.2 Example of expandable gates and their expanders.

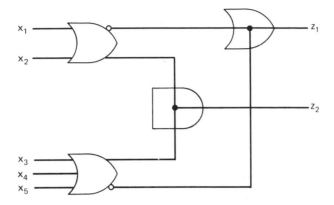

Fig. 3.3.3 Wired-OR and wired-AND logic operations at wired intersections of an ECL circuit.

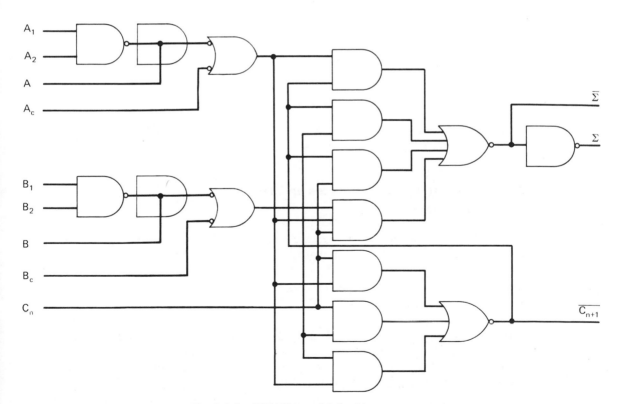

Fig. 3.3.4 SN7480 gated full-adder.

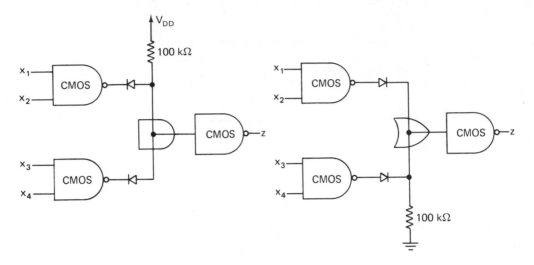

Fig. 3.3.5 CMOS driving CMOS circuits.

Exercise 3.3

1. Use TTL/SSI to realize the following logic diagrams:
    (a) Logic diagrams of Fig. 2.2.1(c-1), (c-2), and (c-3).
    (b) Logic diagram of Fig. 2.2.2(d).
    (c) Logic diagram of Fig. 2.2.4.
    (d) Logic diagrams of Fig. 2.2.5(c-1) and (c-2).
    (e) Logic diagram of Fig. 2.5.1(b).
    (f) Logic diagram of Fig. 2.5.2(b).
    (g) Logic diagram of Fig. 2.6.4(b).

3.4 Combinational MSI Devices and Their Applications

Before the advent of MSI devices, the logic design criterion was to achieve a circuit with a minimum number of components, using such established function-minimization techniques as those described in Chapter 1. System design, logic design, and component selection were independent, requiring little interaction on the part of the designers. But since the advent of MSI and LSI, this relationship has been changed; system design, logic design, and component selection become heavily interdependent and each influences and is influenced by the others. Thus in today's digital system design,† it is no longer sufficient—or even important—to minimize the number of gates (and flip-flops, which will be introduced in Chapter 4) to optimize a design using ready-made MSI and LSI IC chips. It is far more important to select the proper complex integrated circuit which can perform the desired function most economically.

†By digital system design here we mean digital design at a system level, *not* at a circuit level.

It may even be appropriate to redefine subsystems to accommodate more sophisticated and more cost-effective components. Logic design is no longer an isolated art. It has left its ivory tower and is more demanding, but at the same time far more stimulating and rewarding.

One may ask: Is the basic procedure for designing combinational logic circuits described in Section 1.8 still useful? The answer is, of course, yes. It is useful in digital design at a circuit level such as in designing MSI and LSI (both combinational and sequential) components and digital circuits that are not available in chip form. The designs of MSI code converters and adders/subtractors presented in Chapter 2 are examples of applications of this function-minimization-oriented design procedure.

Advantages of MSI Complex MSI devices offer the following advantages over SSI gates:

1. *Simplified mechanical construction.* Since the packing density of MSI is greatly increased (compared to that of SSI), more functions on a printed circuit board can be constructed, which simplifies the mechanical construction of the system.

2. *Reduced interconnections.* As a direct result of the above advantage, the numbers of solder joints, back-plane wiring, and connectors in a system using MSI devices as building blocks would be in general much less compared to those using SSI gates.

3. *Reduced power consumption and total heat generation.* However, the increased density possible with MSI can result in higher power and heat densities.

4. *Low cost.* MSI/SSI cost comparisons must consider true total cost, not only the purchase price of the integrated circuits. An unavoidable overhead cost is associated with each IC, attributed to testing, handling, insertion, and soldering plus the appropriate share of connectors, printed circuit boards, power supplies, cabinets, and so on. When this cost is added to the semiconductor cost, MSI offers more economical solutions, even in cases when the MSI components are more expensive.

5. *Decreased design, debugging, and servicing costs and time.* Because MSI devices are functional subsystems, it is much easier and faster to design a system with them. Functional partitioning also simplifies debugging and service.

6. *Improved system reliability.* Mean time until failure of an MSI is roughly the same as that of an SSI device, so a reduction in package count increases system mean time before failure. Moreover, the reduced interconnections improve reliability.

Because of these obvious advantages, MSI is generally accepted in the regular and repetitive portions of digital designs.

In the following some commonly used combinational TTL/MSI devices and their applications are presented. Inputs and outputs of these combinational MSI devices are labeled with mnemonic letters, as illustrated in Table 3.4.1. Note that an active LOW function labeled outside the logic symbol is given a bar over the label; the same function inside the symbol is labeled without the bar. It should be noted that MSI devices may have several different types, and each type, made by different manufacturers, may have different input/output pin assignments.

TABLE 3.4.1 Input and Output Labels for Combinational MSI and Their Meanings

Label	Meaning	Example
A,B,C,...	Inputs	
Y,W	Outputs	
$\overline{G},\overline{E}$	Strobe (enable) active LOW on all TTL/MSI	
S	Select	

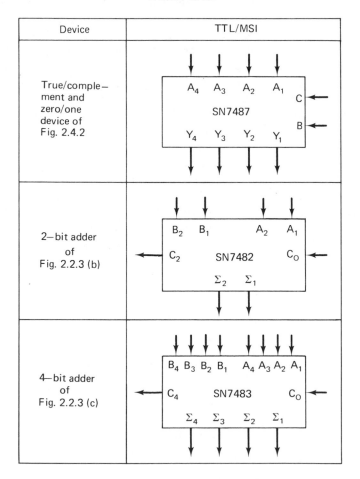

TABLE 3.4.2

Device	TTL/MSI
True/complement and zero/one device of Fig. 2.4.2	
2–bit adder of Fig. 2.2.3 (b)	
4–bit adder of Fig. 2.2.3 (c)	

Table 3.4.2 shows several examples of combinational TTL/MSI devices. The functions and circuit diagrams of these devices were presented in Chapter 2. Table 3.4.3 shows a list of standard 54/74 series combinational TTL/MSI IC's that are now available. They include: data selectors/multiplexers, encoders, code converters, decoders/demultiplexers, digital display systems, arithmetic devices, and comparators.

The TTL/MSI IC chips shown in Table 3.4.3 may be used to design large and complex digital systems. Several examples to illustrate some of the applications of these chips are given next.

TABLE 3.4.3 Some Available Combinational TTL/MSI Integrated Circuits

MSI	Function	Device
Data selector/multiplexer	16-bit	54/74150
	8-bit, with strobe	54/74151
	8-bit	54/74152
	Dual 4-line-to-1-line	54/74153
	Quad 2-input	54/74157
Encoder	Full BCD priority encoder	54/74147
	Cascadable octal priority encoder	54/74148
Code converter	6-line-BCD to 6-line	⎫
	Binary, or 4-line to 4-line	⎬ 54/74184
	BCD 9's/BCD 10's converters	⎭
	6-bit-binary to 6-bit-BCD converter	54/74185A
Decoder/demultiplexer	BCD-to-decimal (4-to-10 lines)	54/7442
	Excess-3-to-decimal (4-to-10 lines)	54/7443
	Excess-3-Gray-to-decimal (4-to-10 lines)	54/7444
	BCD-to-decimal decoder/driver (4-to-10 lines)	54/7445
	BCD-to-7 segment decoder/driver	54/7446
	BCD-to-7 segment decoder/driver	54/7447
	BCD-to-7 segment decoder/driver	54/7448
	BCD-to-7 segment decoder/driver	54/7449
	BCD-to-decimal decoder/driver	74141
	BCD-to-decimal decoder/driver	54/74145
	4-line-to-16-line decoder/demultiplexer	54/74154
	Dual 2-line-to-4 line decoder/demultiplexer	54/74155
	Dual 2-line-to-4 line decoder/demultiplexer, open collector	54/74156
Digital display circuit	BCD-to-decimal decoder/driver	54/7445, 54/74145, 54/74141
	BCD-to-7-segment decoder/driver	54/7446A, 54/7447A, 54/74L46 54/74L47, 54/7448, 54/7449
Arithmetic device	Gated full-adder	54/7480
	2-bit full-adder	54/7482
	4-bit full-adder	54/7483
	Quad 2-input EXCLUSIVE-OR gate	54/7486
	4-bit arithmetic logic unit	54/74181
	Look-ahead carry generator	54/74182
Comparator	4-bit magnitude comparator	54/7485, 54/74L85, 54/74S85

Example 3.4.1   Data Selectors/Multiplexers and Their Applications

Data selectors/multiplexers are combinational devices controlled by a selector address that routes one of many input signals to the output. They can be considered semiconductor equivalents to multiposition switches or stepping switches. There are three main usefulness of data selectors/multiplexers: (1) data selection, (2) time-division multiplexing, and (3) function generation.

Referring to the function table of SN74157 in Fig. 3.4.1(a),

$$Y = \begin{cases} A & \text{if } G = L \text{ and } S = L \\ B & \text{if } G = L \text{ and } S = H \end{cases}$$

Hence it can be used for data selecting. A simple example of using it for time-division multiplexing is shown in Fig. 3.4.1(b).

In most digital systems there are areas, usually in the control section, where a number of inputs generate an output in a highly irregular way. In other words, an

Input				Output
G	S	A	B	Y
H	X	X	X	L
L	L	L	X	L
L	L	H	X	H
L	H	X	L	L
L	H	X	H	H

L: Low voltage
H: High voltage
X: Either low voltage or high
  voltage

(a) Function table of SN74157

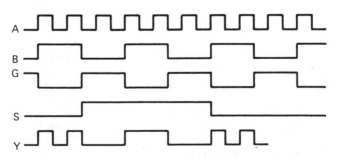

(b) An example of using SN74157 for time-division multiplexing.

Fig. 3.4.1   Example 3.4.1.

unusual function must be generated which is apparently not available as an MSI building block. In such cases many designers tend to return to classical methods of logic design with NAND and NOR gates, using Boolean algebra and Karnaugh maps for logic minimization. Surprisingly enough, multiplexers can simplify these designs.

1. The SN74157 quad two-input multiplexer can generate any *4* of the *16 different functions* of two variables.

2. The SN74153 dual four-input multiplexers can generate any *2* of the *256 different functions* of three variables.

3. The SN74151 and SN74152 eight-input multiplexers can generate any *1* of the *65,536 different functions* of four variables.

4. The SN74150 16-input multiplexer can generate any *1* of the over *4 billion different functions* of five variables.

Example 3.4.2 Seven-Segment Display Circuits

Suppose that we want to design a combinational circuit to command a seven-segment display of 10 numerals 0–9, as shown in Fig. 3.4.2(a).

The desired operations of the circuit are described by the truth table shown as Table 3.4.4. We could design this device following the procedure described in Examples 1.8.1 and

TABLE 3.4.4 Truth Table of Example 3.4.2

x_0	x_1	x_2	x_3	x_4	x_5	x_6	x_7	x_8	x_9	z_1	z_2	z_3	z_4	z_5	z_6	z_7	Symbol
1	0	0	0	0	0	0	0	0	0	1	1	1	1	1	1	0	0
0	1	0	0	0	0	0	0	0	0	0	1	1	0	0	0	0	1
0	0	1	0	0	0	0	0	0	0	1	1	0	1	1	0	1	2
0	0	0	1	0	0	0	0	0	0	1	1	1	1	0	0	1	3
0	0	0	0	1	0	0	0	0	0	0	1	1	0	0	1	1	4
0	0	0	0	0	1	0	0	0	0	1	0	1	1	0	1	1	5
0	0	0	0	0	0	1	0	0	0	0	0	1	1	1	1	1	6
0	0	0	0	0	0	0	1	0	0	1	1	1	0	0	0	0	7
0	0	0	0	0	0	0	0	1	0	1	1	1	1	1	1	1	8
0	0	0	0	0	0	0	0	0	1	1	1	1	0	0	1	1	9

3.3.1 namely, using SSI gates as building blocks. However, a more practical (less laborous, cheaper, and more reliable) approach is to use ready-made MSI functional-level circuits as building blocks. Suppose that we choose TTL to design this device. At the present time no single TTL/MSI circuit which realizes the truth table of Table 3.4.4 is commercially available. However, we could realize it by using a decimal-to-binary converter in cascade with a BCD-to-seven-segment decoder/driver, which is shown in Fig. 3.4.2(b). This corresponds to converting the truth table of Table 3.4.4 into two truth tables (Table 3.4.5). The first table can be realized by a TTL/MSI SN74147 10-line to 4-line priority encoder, and the second table can be realized by a TTL/MSI SN7446A BCD-to-seven-segment decoder/driver except that the outputs of the encoder are opposite to what is required for the inputs of the BCD-to-seven-segment decoder/driver. To match the outputs of the encoder to the inputs of the decoder/driver, four inverters (one for each line) must be used to convert the voltage levels. This can be accomplished by using TTL/SSI SN7404 hex inverters. The block diagram and the circuit diagram of this design are shown in Fig. 3.4.2(c) and (d), respectively. Note that the positive logic assignment is used throughout this design.

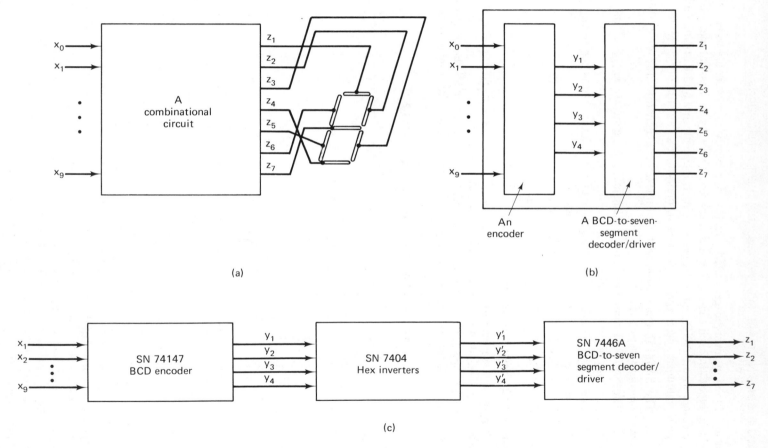

(a)

(b)

(c)

Fig. 3.4.2 Example 3.4.2.

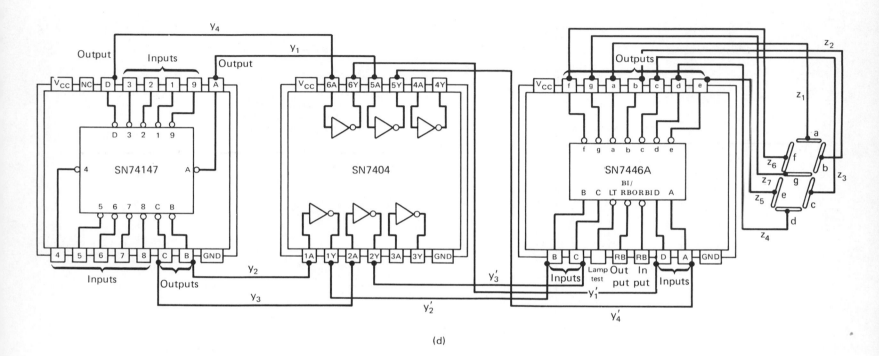

(d)

Note: LT denotes lamp test
BI denotes blanking input
RBO denotes ripple-blanking output
RBI denotes ripple-blanking input

Fig. 3.4.2 (continued)

TABLE 3.4.5 Converted Truth Table of Table 3.4.4

x_0	x_1	x_2	x_3	x_4	x_5	x_6	x_7	x_8	x_9	y_1	y_2	y_3	y_4	z_1	z_2	z_3	z_4	z_5	z_6	z_7	Symbol
1	0	0	0	0	0	0	0	0	0	0	0	0	0	1	1	1	1	1	1	0	0
0	1	0	0	0	0	0	0	0	0	0	0	0	1	0	1	1	0	0	0	0	1
0	0	1	0	0	0	0	0	0	0	0	0	1	0	1	1	0	1	1	0	1	2
0	0	0	1	0	0	0	0	0	0	0	0	1	1	1	1	1	1	0	0	1	3
0	0	0	0	1	0	0	0	0	0	0	1	1	1	0	1	1	0	0	1	1	4
0	0	0	0	0	1	0	0	0	0	0	1	0	1	1	0	1	1	0	1	1	5
0	0	0	0	0	0	1	0	0	0	0	1	1	0	0	0	1	1	1	1	1	6
0	0	0	0	0	0	0	1	0	0	0	1	1	1	1	1	1	0	0	0	0	7
0	0	0	0	0	0	0	0	1	0	1	0	0	0	1	1	1	1	1	1	1	8
0	0	0	0	0	0	0	0	0	1	1	0	0	1	1	1	1	0	0	1	1	9

Truth table 1 which can be realized by a decimal-to-binary converter.

Truth table 2 which can be realized by a seven-segment decoder/driver.

Example 3.4.3   Realization of the Sum-of-Products Form of a Switching Function Using Decoders/Demultiplexers

Among other uses of decoders/demultiplexers, such as code conversions used in the above example, they can also be used to realize switching functions. For example, take the SN74156 (dual two-line-to-four-line decoder/demultiplexer), whose chip descriptions and function tables are described in Fig. 3.4.3(a). It is seen that each of the two decoders can realize

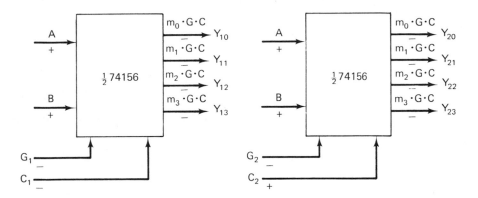

Inputs			Outputs			
Strobe	Data	Select				
G_1	C_1	B	A	Y_{10}	Y_{11}	Y_{12}	Y_{13}
H	X	X	X	H	H	H	H
L	L	L	L	L	H	H	H
L	L	L	H	H	L	H	H
L	L	H	L	H	H	L	H
L	L	H	H	H	H	H	L
X	H	X	X	H	H	H	H

Inputs			Outputs			
Strobe	Data	Select				
G_2	C_2	A	B	Y_{20}	Y_{21}	Y_{22}	Y_{23}
H	X	X	X	H	H	H	H
L	H	L	L	L	H	H	H
L	H	H	L	H	L	H	H
L	H	L	H	H	H	L	H
L	H	H	H	H	H	H	L
X	L	X	X	H	H	H	H

(a)

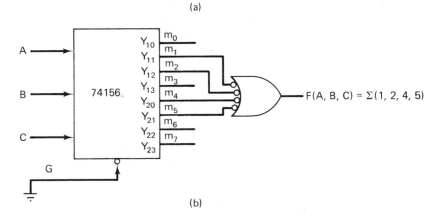

(b)

Fig. 3.4.3   Example 3.4.3.

the (canonical) sum-of-products form of any two-variable function by connecting appropriate $m_i \cdot G \cdot C$ leads (which realize the minterm m_i) to an OR gate. The G and C signals act as enabling signals for the decoder outputs. Note that the C input is positive logic to one decoder and is negative logic to the other. Thus, if the A, B, C, and G inputs to the two decoders are paralleled, the devices can be used as a three-to-eight decoder. A, B, and C then serve as three single-rail Boolean inputs and G serves as an enabling signal. The decoder whose C input is positive logic produces m_4 through m_7, since it is enabled when the C input is HIGH. The composite diagram for the three-to-eight decoder is shown in Fig. 3.4.3(b). The decoder can realize the sum-of-products form of any three-variable switching function. For example, the realization of $\sum (1, 2, 4, 5)$ using this decoder plus an OR gate is shown in Fig. 3.4.3(b).

Example 3.4.4 Two n-Bit Word Magnitude Comparators

A 4-bit word magnitude comparator, SN7485, is available in chip form (see Fig. 3.4.4). It compares straight binary and straight BCD (8–4–2–1) codes. Three fully decoded decisions

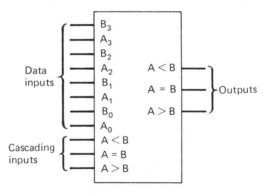

Function Table

Comparing inputs				Cascading inputs			Outputs		
A3.B3	A2.B2	A1.B1	A0.B0	A > B	A < B	A = B	A > B	A < B	A = B
A3 > B3	X	X	X	X	X	X	H	L	L
A3 < B3	X	X	X	X	X	X	L	H	L
A3 = B3	A2 > B2	X	X	X	X	X	H	L	L
A3 = B3	A2 < B2	X	X	X	X	X	L	H	L
A3 = B2	A2 = B2	A1 > B1	X	X	X	X	H	L	L
A3 = B3	A2 = B2	A1 < B1	X	X	X	X	L	H	L
A3 = B3	A2 = B2	A1 = B1	A0 > B0	X	X	X	H	L	L
A3 = B3	A2 = B2	A1 = B1	A0 < B0	X	X	X	L	H	L
A3 = B3	A2 = B2	A1 = B1	A0 = B0	H	L	L	H	L	L
A3 = B3	A2 = B2	A1 = B1	A0 = B0	L	H	L	L	H	L
A3 = B3	A2 = B2	A1 = B1	A0 = B0	L	L	H	L	L	H

H = high level, L = low level, X = irrelevant

Fig. 3.4.4 Two 4-bit word magnitude comparator SN7485.

about two 4-bit words (A, B) are made and externally available at three outputs. These devices are fully expandable to any number of bits without external gates. Words of greater length may be compared by connecting comparators in cascade. The $A > B$, $A < B$, and $A = B$ outputs of a stage that is handling less significant bits are connected to the corresponding $A > B$, $A < B$, and $A = B$ inputs of the next stage, which handles more significant bits. A two-level two-bit word magnitude comparator using six SN7485's is shown in Fig. 3.4.5.

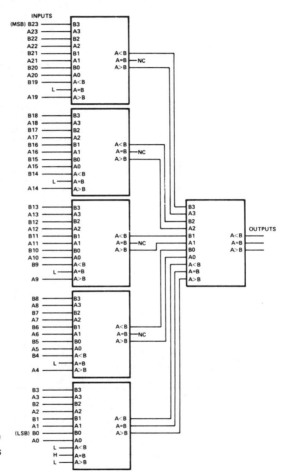

Fig. 3.4.5 Two 24-bit word magnitude comparator using six SN7485. NC denotes no connection.

Example 3.4.5   BCD-to-Binary and Binary to-BCD Converters

A 6-bit BCD-to-binary converter is available in MSI form. This chip is the SN74184, whose chip description and function table when used as BCD-to-binary converter are shown in Fig. 3.4.6. The chip that performs 6-bit binary-to-BCD conversion is the SN74185A, whose chip description and function table are shown in Fig. 3.4.7. By using these chips as building blocks, BCD-to-binary and binary-to-BCD converters for large numbers can be constructed. Several examples are shown in Fig. 3.4.8.

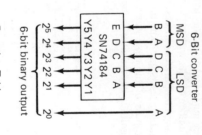

Function Table
BCD-to-Binary Converter

BCD Words	Inputs E	D	C	B	A	G	Outputs Y5	Y4	Y3	Y2	Y1
0-1	L	L	L	L	L	L	L	L	L	L	L
2-3	L	L	L	L	H	L	L	L	L	L	H
4-5	L	L	L	H	L	L	L	L	L	H	L
6-7	L	L	L	H	H	L	L	L	L	H	H
8-9	L	L	H	L	L	L	L	L	H	L	L
10-11	L	H	L	L	L	L	L	L	H	L	H
12-13	L	H	L	L	H	L	L	L	H	H	L
14-15	L	H	L	H	L	L	L	L	H	H	H
16-17	L	H	L	H	H	L	L	H	L	L	L
18-19	L	H	H	L	L	L	L	H	L	L	H
20-21	H	L	L	L	L	L	L	H	L	H	L
22-23	H	L	L	L	H	L	L	H	L	H	H
24-25	H	L	L	H	L	L	L	H	H	L	L
26-27	H	L	L	H	H	L	L	H	H	L	H
28-29	H	L	H	L	L	L	L	H	H	H	L
30-31	H	H	L	L	L	L	L	H	H	H	H
32-33	H	H	L	L	H	L	H	L	L	L	L
34-35	H	H	L	H	L	L	H	L	L	L	H
36-37	H	H	L	H	H	L	H	L	L	H	L
38-39	H	H	H	L	L	L	H	L	L	H	H
Any	X	X	X	X	X	H	H	H	H	H	H

H = high level, L = low level, X = irrelevent

Fig. 3.4.6

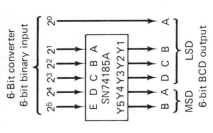

6-Bit converter
6-bit binary input

2^5 2^4 2^3 2^2 2^1 → E D C B A — SN74185A — Y5 Y4 Y3 Y2 Y1 — 2^0 → A

MSD — B A
LSD — D C B A
6-bit BCD output

Function Table

Binary words	Inputs Binary select					Enable	Outputs							
	E	D	C	B	A	G	Y8	Y7	Y6	Y5	Y4	Y3	Y2	Y1
0-1	L	L	L	L	L	L	H	H	H	H	H	H	H	H
2-3	L	L	L	L	H	L	H	H	H	H	H	H	H	L
4-5	L	L	L	H	L	L	H	H	H	H	H	H	L	H
6-7	L	L	L	H	H	L	H	H	H	H	H	H	L	L
8-9	L	L	H	L	L	L	H	H	H	H	H	L	H	H
10-11	L	L	H	L	H	L	H	H	H	H	L	H	H	H
12-13	L	L	H	H	L	L	H	H	H	H	L	H	H	L
14-15	L	L	H	H	H	L	H	H	H	H	L	H	L	H
16-17	L	H	L	L	L	L	H	H	H	H	L	H	L	L
18-19	L	H	L	L	H	L	H	H	H	H	L	L	H	H
20-21	L	H	L	H	L	L	H	H	H	L	H	H	H	H
22-23	L	H	L	H	H	L	H	H	H	L	H	H	H	L
24-25	L	H	H	L	L	L	H	H	H	L	H	H	L	H
26-27	L	H	H	L	H	L	H	H	H	L	H	H	L	L
28-29	L	H	H	H	L	L	H	H	H	L	H	L	H	H
30-31	L	H	H	H	H	L	H	H	H	L	L	H	H	H
32-33	H	L	L	L	L	L	H	H	H	L	L	H	H	L
34-35	H	L	L	L	H	L	H	H	H	L	L	H	L	H
36-37	H	L	L	H	L	L	H	H	H	L	L	H	L	L
38-39	H	L	L	H	H	L	H	H	H	L	L	L	H	H
40-41	H	L	H	L	L	L	H	H	L	H	H	H	H	H
42-43	H	L	H	L	H	L	H	H	L	H	H	H	H	L
44-45	H	L	H	H	L	L	H	H	L	H	H	H	L	H
46-47	H	L	H	H	H	L	H	H	L	H	H	H	L	L
48-49	H	H	L	L	L	L	H	H	L	H	H	L	H	H
50-51	H	H	L	L	H	L	H	H	L	H	L	H	H	H
52-53	H	H	L	H	L	L	H	H	L	H	L	H	H	L
54-55	H	H	L	H	H	L	H	H	L	H	L	H	L	H
56-57	H	H	H	L	L	L	H	H	L	H	L	H	L	L
58-59	H	H	H	L	H	L	H	H	L	H	L	L	H	H
60-61	H	H	H	H	L	L	H	H	L	L	H	H	H	H
62-63	H	H	H	H	H	L	H	H	L	L	H	H	H	L
All	X	X	X	X	X	H	H	H	H	H	H	H	H	H

H = high level, L = low level, X = irrelevant

Fig. 3.4.7

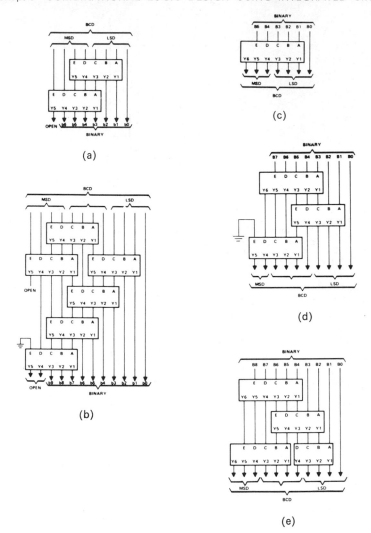

Fig. 3.4.8   (a) BCD-to-binary converter for two BCD decades.
(b) BCD-to-binary converter for three BCD decades.
(c) 6-bit binary-to-BCD converter.
(d) 8-bit binary-to-BCD converter.
(e) 9-bit binary-to-BCD converter.

Example 3.4.6   Carry Look-Ahead Adders

The SN74181 arithmetic logic unit/function generator shown in Fig. 3.4.9 is a parallel
4-bit MSI device that can perform 16 arithmetic and 16 possible logic operations on two
4-bit parallel words. The significant arithmetic operations are ADD, SUBTRACT, PASS,

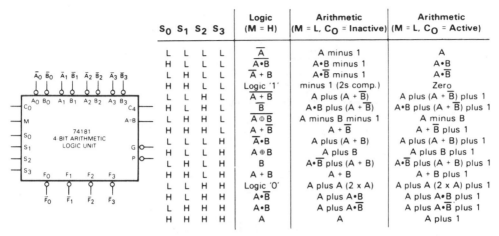

S_0 S_1 S_2 S_3	Logic (M = H)	Arithmetic (M = L, C_O = Inactive)	Arithmetic (M = L, C_O = Active)
L L L L	$\overline{A}$	A minus 1	A
H L L L	$\overline{A \cdot B}$	A•B minus 1	A•B
L H L L	$\overline{A} + B$	A•$\overline{B}$ minus 1	A•$\overline{B}$
H H L L	Logic '1'	minus 1 (2s comp.)	Zero
L L H L	$\overline{A + \overline{B}}$	A plus (A + $\overline{B}$)	A plus (A + $\overline{B}$) plus 1
H L H L	$\overline{B}$	A•B plus (A + $\overline{B}$)	A•B plus (A + $\overline{B}$) plus 1
L H H L	$\overline{A \oplus B}$	A minus B minus 1	A minus B
H H H L	$A + \overline{B}$	A + $\overline{B}$	A + $\overline{B}$ plus 1
L L L H	$\overline{A} \cdot \overline{B}$	A plus (A + B)	A plus (A + B) plus 1
H L L H	$A \oplus B$	A plus B	A plus B plus 1
L H L H	B	A•$\overline{B}$ plus (A + B)	A•$\overline{B}$ plus (A + B) plus 1
H H L H	A + B	A + B	A + B plus 1
L L H H	Logic '0'	A plus A (2 x A)	A plus A (2 x A) plus 1
H L H H	A•$\overline{B}$	A plus A•B	A plus A•B plus 1
L H H H	A•B	A plus A•$\overline{B}$	A plus A•$\overline{B}$ plus 1
H H H H	A	A	A plus 1

Fig. 3.4.9 Arithmetic logic unit/function generator SN74181.

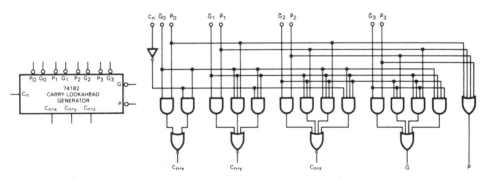

Fig. 3.4.10 Carry lookahead generator SN74182.

INCREMENT, DECREMENT, INVERT, and DOUBLE. The operation is selected by four select lines S_1–S_4 and a mode control line M, which is low for arithmetic operations and high for logic operations. The device has a CARRY IN, a CARRY OUT for ripple carry† cascading of units, and two look-ahead auxiliary carry functions, CARRY GENERATE and CARRY PROPAGATE, for use with the carry look-ahead generator SN74182, which is shown in Fig. 3.4.10. A complete 16-bit carry look-ahead adder can be achieved by using four SN74181 and one SN74182 as shown in Fig. 3.4.11.

Exercise 3.4

1. Find TTL/MSI chips that realize the following logic diagrams:
   (a) Logic diagrams of Fig. 2.4.3(a), (b), and (c).

†See Section 4.3.

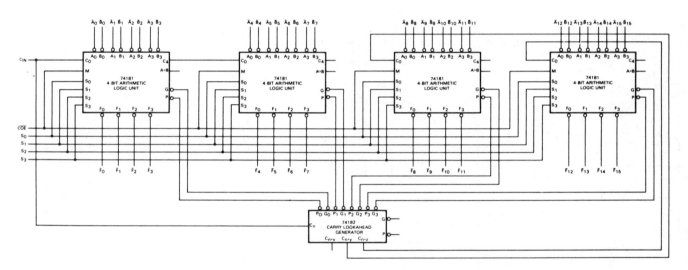

Fig. 3.4.11   16-bit carry lookahead adder.

    (b) Logic diagrams of Fig. 2.6.2(b) and (c).
    (c) Logic diagram of Fig. 2.6.4.
    (d) Logic diagram of Fig. 2.6.5.
    (e) Logic diagrams of Fig. 2.6.6(a) and (b).

2. Design a BCD-to-binary converter for six BCD decades using a SN74184.

3. Design a 12-bit binary-to-BCD converter using a SN74185A and a 16-bit binary-to-BCD converter.

4. Design a 32-bit parallel comparator array using SN7485 chips.

5. (a) Design a 24-bit carry look-ahead adder using one SN74182 and no more than six SN74181's.
    (b) Design a 32-bit carry look-ahead adder using two SN74182's and eight SN74181's.

3.5 Semiconductor Memories

Semiconductor memories fall into two basic catagories: read-only memories (ROM's) and random-access read/write memories (RAM's). In a ROM, the data content is fixed, readout is nondestructive, and the data are retained indefinitely even when the power is shut off. A RAM, however, is capable of read and write operations, the mode of operation being defined by the "write enable" input. In the write mode, the information at the data input is written into the array of latches† selected by the address. In the read mode, the contents of the selected array of latches is fed to the data output. The RAM also has nondestructive readout, but data in the memory are lost when the power is shut off. For reference convenience, a brief summary of semiconductor memories—the state of the art, including their basic circuitries, characteristics, and fabrication processes—is given in Appendix A. Some available MSI/TTL and LSI/MOS ROM's and RAM's are given in Table 3.5.1.

The advent of large-scale integrated arrays has aroused interest in the old and basic idea of using complete arrays to execute combinational logic functions. Previously the arrays had been considered highly inefficient solutions, even though the design was simple, primarily because of the great waste of circuits. However, as the cost per function and the size have decreased, the ease of design has again become appealing. The arrays, most often called read-only memories, have become almost a panacea for logic design problems.

The ROM can take a digital code at its terminals and provide a unique digital code on its output terminals. The relationship between the input and output codes is relatively fixed, usually alterable only by relatively slow techniques, and for this reason we term it "read only." The difference between a read-only memory and a read/write memory is the level of difficulty of changing the stored information. The ROM is generally a random-access type and permits us to access information stored within the memory with a random addressing code.

---

†Latches and how to use them to store information (0's and 1's) are discussed in Section 4.2.

**TABLE 3.5.1   Some Available MSI/TTL and LSI/MOS
ROM's and RAM's**

Size of Integration	Function	Example†
MSI/TTL ROM	256-bit	54/7488A
	256-bit programmable	74188A
	512-bit programmable	54/74186
	1,024-bit	54/74187
LSI/MOS ROM	2,560-bit dynamic	TMS 2300 JC, NC
	Row output character generator	TMS 2400 JC, NC
	2,560-bit static	TMS 2500 JC, NC
	2,048-bit static	TMS 2600 JC, NC
	USACII-to-Selectric/ Selectric-to-USASCII code converter	TMS 2602 JC, NC
	EBCDIC-to-USASCII code converter	TMS 2603 JC, NC
	USASCII-to-EBCDIC/ Selectric-to-EBDIC code generator	TMS 2604 JC, NC
	USASCII, Baudot, Selectric EBCDIC code generator	TMS 2605 JC, NC
	3,072-bit static	TMS 2700 JC, JM, NC
	1,024-bit static	TMS 2800 JC, NC
	1,280-bit static	TMS 2900 JC, NC
	Series character generator	TMS 4100 JC, NC
	4,096-bit static	TMS 4400 JC, NC
MSI/TTL RAM	16-bit register file	54/74170
	16-bit multiple-port register file	74172
	16-bit read/write memory	54/7481A, 54/7484A
	64-bit read/wirte memory	7489
	256-bit read/write memory	74200, 74S200, 74S206
LSI/MOS RAM	256-bit	TMS 1101 JC, NC
	Fully-decoded 1,024-bit	TMS 1103 NC
	High-speed content-addressable	TMS 4000 JC, NC
	256-bit	TMS 4003 JR, NC
	2,048-bit dynamic	TMS 4020 NC
	1,024-bit	TMS 4023 NC
	2,048 dynamic	TMS 4025 NC

†J and N denote ceramic dual-in-line and plastic dual-in-line packages, respectively. C, M, and R denote temperature ranges −25 to +85°C, −55 to +125°C, and −55 to +85°C.

The ROM consists of two sections, the address decoder part and the memory part, as shown in Fig. 3.5.1. The memory part is generally arranged in a rectangular matrix. This two-dimensional matrix is convenient for parallel binary address (W lines) and parallel binary output (B lines). The W and B lines are commonly referred to as *word* and *bit lines*, respectively. ROM's are available in 256-bit, 512-bit, up to 4,096-bit. A typical ROM is shown in Fig. 3.5.2. It is a 16-pin 1,024-bit ROM with

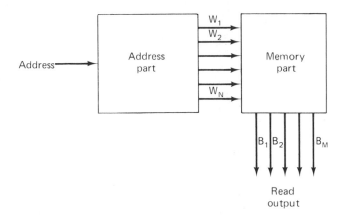

Fig. 3.5.1 Two parts of a ROM.

memory cells organized in a 32 by 32 memory matrix. Word selection of this memory is accomplished in a conventional 8-bit positive-logic binary code with the A select input being the least significant bit, progressing alphabetically through the select inputs to H, which is the most significant bit. The word-select table of this ROM is described in Fig. 3.5.2.

A very significant, and perhaps the best known, application of MOS ROM is for character generation. This includes display control for moving billboards, light-emitting diode arrays, Hollerith card punching, paper-tape coding, digital communications control, and, most commonly, CRT display drivers. For example, the TMS 4100 is a family of MOS ROM's, each with a capacity of 2,240 bits, mounted in a 28-pin package. Two organizations are available:

1. 64 words of 35 bits (5 by 7).
2. 32 words of 70 bits (5 by 14).

The memory is organized to function primarily as a character generator. The seven outputs represent a column in a 5 by 7 dot matrix. The output word appears as a five-word sequence on the output lines. Sequence is controlled by five strobe lines (column select), which feed directly into the buffer section of the memory (see Fig. 3.5.3). By enabling the first strobe line, the first group of 7 bits (first column) is obtained at the output. Then the second, third, fourth, and fifth strobe lines are enabled. The column select can remain fixed while the character address changes, or the character address can remain fixed while the column select changes. The decoder will accept a 7-bit parallel input. Because only 6 bits are required to decode the 64 input words, the seventh bit may be used as a chip enable. If the memory is organized as 32 words of 70 bits, it is possible to have two chip-enable lines.

Another widely used application of MOS ROM is for code generation. In Section 2.5 some most frequently used codes for decimal numbers and their conversions were discussed. However, the use of combinations of Boolean variables to represent items

WORD-SELECT TABLE

WORD	INPUTS							
	H	G	F	E	D	C	B	A
0	L	L	L	L	L	L	L	L
1	L	L	L	L	L	L	L	H
2	L	L	L	L	L	L	H	L
3	L	L	L	L	L	L	H	H
4	L	L	L	L	L	H	L	L
5	L	L	L	L	L	H	L	H
6	L	L	L	L	L	H	H	L
7	L	L	L	L	L	H	H	H
8	L	L	L	L	H	L	L	L
Words 9 thru 250 omitted								
251	H	H	H	H	H	L	H	H
252	H	H	H	H	H	H	L	L
253	H	H	H	H	H	H	L	H
254	H	H	H	H	H	H	H	L
255	H	H	H	H	H	H	H	H

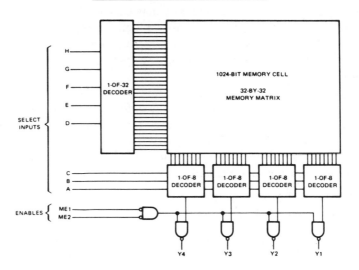

Fig. 3.5.2 Block diagram of MSI/TTL SN74187 1,024-bit ROM.

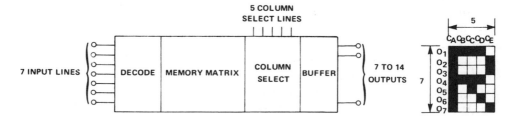

Fig. 3.5.3 Block diagram of MOS/LSI TMS 4100 character generator.

of information is not limited to numerical values. A common requirement of data processing and storage is the representation of the alphabetic, numerical, and punctuation symbols that make up English text. Such codes are called *alphameric codes*, since they represent both *alpha*betic and nu*meric* information. The four most commonly used codes are:

1. The Selectric code. This code is used on the Selectric typewriter.
2. The extended binary-coded-decimal interchange code (EBCDIC).
3. The Baudot code. This code is the teletype (TTY) code, which uses only five bits per character. Since there are 26 letters and 10 numerals, each code word will require at least six bits. Thus, in the Baudot code, some words are used to represent more than one character. This code is extended to represent more than 32 different characters by using the special "mode-change" character. The transmitter and receiver that handle the code must begin in the same mode, usually the alphabetic mode. Changes in mode are inserted into the sequence of code words whenever they are required. The operation is similar to the up-shift/down-shift mechanism by typewriters.
4. The American standard code (ASCII or USASCII).

The off-the-shelf TMS 2605 device, for example, is a 24-pin multiple-code generator which generates the above four types of codes. It is designed to exercise and test keyboards, data communication links, and typing mechanisms. Six inputs of the TMS 2605 are fed from a 6-bit binary count. The seventh and eighth inputs are used as code selects. The message can be generated in the following four codes:

I_7	I_8	Code generated
0	0	Selectric (7 bits plus parity)
1	0	EBCDIC (8 bits)
0	1	Baudot (5 bits)
1	1	ASCII (7 bits plus parity)

For normal operation of this device, Chip Enable must be at logic 1, I_9 at logic 1, and Mode Control at logic 0. The outputs of the ROM correspond as follows:

ROM Output	Selectric	EBCDIC	Baudot	ASCII
O_1	I	D_7	1	B_1
O_2	2	B_6	2	B_2
O_3	4	B_5	3	B_3
O_4	8	B_4	4	B_4
O_5	A	B_3	5	B_5
O_6	B	B_2	—	B_6
O_7	S	B_1	—	B_7
O_8	Parity	B_0	—	Parity

The truth table of the TMS 2606 is given in Table 3.5.2, for the four selected codes:

Address	*Code*	I_7	I_8
0–63	Selectric	0	0
64–127	EBCDIC	1	0
128–191	Baudot	0	1
192–255	ASCII	1	1

TABLE 3.5.2 Truth Table of the TMS 2605 Code Generator

SELECTRIC

Address	Char	O_8 PARITY Y	O_7 S	O_6 B	O_5 A	O_4 8	O_3 4	O_2 2	O_1 1
0	IL1	0	1	0	1	0	0	0	0
1	NL	1	1	0	1	0	0	1	0
2	IL1	0	1	0	1	0	0	0	0
3	IL1	0	1	0	1	0	0	0	0
4	T	1	0	1	0	1	1	1	1
5	H	1	0	1	0	0	1	1	0
6	E	0	0	1	0	1	0	1	1
7	SP	1	1	1	1	1	1	1	1
8	Q	0	0	0	0	1	0	0	1
9	U	1	0	1	0	1	1	0	0
10	I	1	0	0	1	1	0	0	1
11	C	0	0	1	0	1	0	0	0
12	K	1	0	1	0	1	0	0	1
13	SP	1	1	1	1	1	1	1	1
14	B	0	0	1	0	0	1	0	0
15	R	1	0	0	1	1	0	1	0
16	O	0	0	0	1	0	1	1	1
17	W	0	0	0	1	0	1	0	0
18	N	0	0	1	0	1	1	0	1
19	SP	1	1	1	1	1	1	1	1
20	F	0	0	0	0	1	1	0	0
21	O	0	0	0	1	0	1	1	1
22	X	0	0	1	0	1	1	1	0
23	SP	1	1	1	1	1	1	1	1
24	J	0	0	0	0	1	1	1	1
25	U	1	0	1	0	1	1	0	0
26	M	0	0	0	1	1	1	1	0
27	P	1	0	0	0	1	0	1	1
28	S	1	0	0	1	0	1	1	0
29	SP	1	1	1	1	1	1	1	1
30	O	0	0	0	1	0	1	1	1
31	V	1	0	0	1	1	1	0	0
32	E	0	0	1	0	1	0	1	1
33	R	1	0	0	1	1	0	1	0
34	SP	1	1	1	1	1	1	1	1
35	T	1	0	1	0	1	1	1	1
36	H	1	0	1	0	0	1	1	0
37	E	0	0	1	0	1	0	1	1
38	SP	1	1	1	1	1	1	1	1
39	L	0	0	1	0	0	1	1	1
40	A	0	0	0	1	1	0	0	0
41	Z	0	0	1	1	0	1	0	1
42	Y	0	0	0	0	0	1	1	0
43	SP	1	1	1	1	1	1	1	1
44	D	1	0	1	0	1	0	1	0
45	O	0	0	0	1	0	1	1	1
46	G	1	0	0	0	1	1	1	0
47	SP	1	1	1	1	1	1	1	1
48	1	0	1	1	1	1	1	1	0
49	2	0	1	1	1	1	1	0	1
50	3	1	1	1	1	1	1	0	0
51	4	0	1	1	1	0	1	1	1
52	5	0	1	1	1	1	0	1	1
53	6	1	1	1	1	1	0	0	1
54	7	1	1	1	1	1	0	1	0
55	8	0	1	1	1	1	0	0	0
56	9	0	1	1	1	0	1	0	0
57	0	1	1	1	1	0	1	1	0
58	SP	1	1	1	1	1	1	1	1
59	D	1	0	1	0	1	0	1	0
60	E	0	0	1	0	1	0	1	1
61	SP	1	1	1	1	1	1	1	1
62	DEL	1	1	0	0	0	0	0	0
63	DEL	1	1	0	0	0	0	0	0

EBCDIC

Char	Address	O_8 B_0	O_7 B_1	O_6 B_2	O_5 B_3	O_4 B_4	O_3 B_5	O_2 B_6	O_1 B_7
NULL	64	1	1	1	1	1	1	1	1
CR	65	1	1	1	1	0	0	1	0
CA	66	1	1	1	1	0	0	1	0
LF	67	1	1	0	1	1	0	1	0
T	68	0	0	0	1	1	1	0	0
H	69	0	0	1	1	0	1	1	1
E	70	0	0	1	1	1	0	1	0
SP	71	1	0	1	1	1	1	1	1
Q	72	0	0	1	0	0	1	1	1
U	73	0	0	0	1	1	0	1	1
I	74	0	0	1	1	0	1	1	0
C	75	0	0	1	1	1	1	0	0
K	76	0	0	1	0	1	1	0	1
SP	77	1	0	1	1	1	1	1	1
B	78	0	0	1	1	1	1	0	1
R	79	0	0	1	0	0	1	1	0
O	80	0	0	1	0	1	0	0	1
W	81	0	0	0	1	1	0	0	1
N	82	0	0	1	0	1	0	1	0
SP	83	1	0	1	1	1	1	1	1
F	84	0	0	1	1	1	0	0	1
O	85	0	0	1	0	1	0	0	1
X	86	0	0	0	1	1	0	0	0
SP	87	1	0	1	1	1	1	1	1
J	88	0	0	1	0	1	1	1	0
U	89	0	0	0	1	1	0	1	1
M	90	0	0	1	0	1	0	1	1
P	91	0	0	1	0	1	0	0	0
S	92	0	0	0	1	1	1	0	1
SP	93	1	0	1	1	1	1	1	1
O	94	0	0	1	0	1	0	0	1
V	95	0	0	0	1	1	0	1	0
E	96	0	0	1	1	1	0	1	0
R	97	0	0	1	0	0	1	1	0
SP	98	1	0	1	1	1	1	1	1
T	99	0	0	0	1	1	1	0	0
H	100	0	0	1	1	0	1	1	1
E	101	0	0	1	1	1	0	1	0
SP	102	1	0	1	1	1	1	1	1
L	103	0	0	1	0	1	1	0	0
A	104	0	0	1	1	1	1	1	0
Z	105	0	0	0	1	0	1	1	0
Y	106	0	0	0	1	0	1	1	1
SP	107	1	0	1	1	1	1	1	1
D	108	0	0	1	1	1	0	1	1
O	109	0	0	1	0	1	0	0	1
G	110	0	0	1	1	1	0	0	0
SP	111	1	0	1	1	1	1	1	1
1	112	0	0	0	0	1	1	1	0
2	113	0	0	0	0	1	1	0	1
3	114	0	0	0	0	1	1	0	0
4	115	0	0	0	0	1	0	1	1
5	116	0	0	0	0	1	0	1	0
6	117	0	0	0	0	1	0	0	1
7	118	0	0	0	0	1	0	0	0
8	119	0	0	0	0	0	1	1	1
9	120	0	0	0	0	0	1	1	0
0	121	0	0	0	0	1	1	1	1
SP	122	1	0	1	1	1	1	1	1
D	123	0	0	1	1	1	0	1	1
E	124	0	0	1	1	1	0	1	0
SP	125	1	0	1	1	1	1	1	1
DEL	126	1	1	1	1	1	0	0	0
DEL	127	1	1	1	1	1	0	0	0

BAUDOT

Address	Char	O_8 –	O_7 –	O_6 –	O_5 5	O_4 4	O_3 3	O_2 2	O_1 1
128	CR	1	1	1	1	0	1	1	1
129	CR	1	1	1	1	0	1	1	1
130	LF	1	1	1	1	1	1	0	1
131	LTR	1	1	1	0	0	0	0	0
132	T	1	1	1	0	1	1	1	1
133	H	1	1	1	0	1	0	1	1
134	E	1	1	1	1	1	1	1	0
135	SP	1	1	1	1	1	0	1	1
136	Q	1	1	1	0	1	0	0	0
137	U	1	1	1	1	1	0	0	0
138	I	1	1	1	1	1	0	0	1
139	C	1	1	1	1	0	0	0	1
140	K	1	1	1	1	0	0	0	0
141	SP	1	1	1	1	1	0	1	1
142	B	1	1	1	0	0	1	1	0
143	R	1	1	1	1	0	1	0	1
144	O	1	1	1	0	0	1	1	1
145	W	1	1	1	0	1	1	0	0
146	N	1	1	1	1	0	0	1	1
147	SP	1	1	1	1	1	0	1	1
148	F	1	1	1	1	0	0	1	0
149	O	1	1	1	0	0	1	1	1
150	X	1	1	1	0	0	0	1	0
151	SP	1	1	1	1	1	0	1	1
152	J	1	1	1	1	0	1	0	0
153	U	1	1	1	1	1	0	0	0
154	M	1	1	1	0	0	0	1	1
155	P	1	1	1	0	1	0	0	1
156	S	1	1	1	1	1	0	1	0
157	SP	1	1	1	1	1	0	1	1
158	O	1	1	1	0	0	1	1	1
159	V	1	1	1	0	0	0	0	1
160	E	1	1	1	1	1	1	1	0
161	R	1	1	1	1	0	1	0	1
162	SP	1	1	1	1	1	0	1	1
163	T	1	1	1	0	1	1	1	1
164	H	1	1	1	0	1	0	1	1
165	E	1	1	1	1	1	1	1	0
166	SP	1	1	1	1	1	0	1	1
167	L	1	1	1	0	1	1	0	1
168	A	1	1	1	1	1	1	0	0
169	Z	1	1	1	0	1	1	1	0
170	Y	1	1	1	0	1	0	1	0
171	SP	1	1	1	1	1	0	1	1
172	D	1	1	1	1	0	1	1	0
173	O	1	1	1	0	0	1	1	1
174	G	1	1	1	0	0	1	0	1
175	SP	1	1	1	1	1	0	1	1
176	FIG	1	1	1	0	0	1	0	0
177	1	1	1	1	0	1	0	0	0
178	2	1	1	1	0	1	1	0	0
179	3	1	1	1	1	1	1	1	0
180	4	1	1	1	1	0	1	0	1
181	5	1	1	1	0	1	1	1	1
182	6	1	1	1	0	1	0	1	0
183	7	1	1	1	1	1	0	0	0
184	8	1	1	1	1	1	0	0	1
185	9	1	1	1	0	0	1	1	1
186	0	1	1	1	0	1	0	0	1
187	SP	1	1	1	1	1	0	1	1
188	LTR	1	1	1	0	0	0	0	0
189	D	1	1	1	1	0	1	1	0
190	E	1	1	1	1	1	1	1	0
191	SP	1	1	1	1	1	0	1	1

USASCII

Address	Char	O_8 PARITY Y	O_7 B_7	O_6 B_6	O_5 B_5	O_4 B_4	O_3 B_3	O_2 B_2	O_1 B_1
192	NULL	0	0	0	0	0	0	0	0
193	CR	0	1	1	1	0	0	1	0
194	CR	0	1	1	1	0	0	1	0
195	LF	1	1	1	1	0	1	0	1
196	T	0	0	1	0	1	0	1	1
197	H	1	0	1	1	0	1	1	1
198	E	0	0	1	1	1	0	1	0
199	SP	0	1	0	1	1	1	1	1
200	Q	0	0	1	0	1	1	1	0
201	U	1	0	1	0	1	0	1	0
202	I	0	0	1	1	0	1	1	0
203	C	0	0	1	1	1	1	0	0
204	K	1	0	1	1	0	1	0	0
205	SP	0	1	0	1	1	1	1	1
206	B	1	0	1	1	1	1	0	1
207	R	0	0	1	0	1	1	0	1
208	O	0	0	1	1	0	0	0	0
209	W	0	0	1	0	1	0	0	0
210	N	1	0	1	1	0	0	0	1
211	SP	0	1	0	1	1	1	1	1
212	F	0	0	1	1	1	0	0	1
213	O	0	0	1	1	0	0	0	0
214	X	0	0	1	0	0	1	1	1
215	SP	0	1	0	1	1	1	1	1
216	J	0	0	1	1	0	1	0	1
217	U	1	0	1	0	1	0	1	0
218	M	1	0	1	1	0	0	1	0
219	P	1	0	1	0	1	1	1	1
220	S	1	0	1	0	1	1	0	0
221	SP	0	1	0	1	1	1	1	1
222	O	0	0	1	1	0	0	0	0
223	V	1	0	1	0	1	0	0	1
224	E	0	0	1	1	1	0	1	0
225	R	0	0	1	0	1	1	0	1
226	SP	0	1	0	1	1	1	1	1
227	T	0	0	1	0	1	0	1	1
228	H	1	0	1	1	0	1	1	1
229	E	0	0	1	1	1	0	1	0
230	SP	0	1	0	1	1	1	1	1
231	L	0	0	1	1	0	0	1	1
232	A	1	0	1	1	1	1	1	0
233	Z	1	0	1	0	0	1	0	1
234	Y	1	0	1	0	0	1	1	0
235	SP	0	1	0	1	1	1	1	1
236	D	1	0	1	1	1	0	1	1
237	O	0	0	1	1	0	0	0	0
238	G	1	0	1	1	1	0	0	0
239	SP	0	1	0	1	1	1	1	1
240	1	0	1	0	0	1	1	1	0
241	2	0	1	0	0	1	1	0	1
242	3	1	1	0	0	1	1	0	0
243	4	0	1	0	0	1	0	1	1
244	5	1	1	0	0	1	0	1	0
245	6	1	1	0	0	1	0	0	1
246	7	0	1	0	0	1	0	0	0
247	8	0	1	0	0	0	1	1	1
248	9	1	1	0	0	0	1	1	0
249	0	1	1	0	0	1	1	1	1
250	SP	0	1	0	1	1	1	1	1
251	D	1	0	1	1	1	0	1	1
252	E	0	0	1	1	1	0	1	0
253	SP	0	1	0	1	1	1	1	1
254	DEL	0	0	0	0	0	0	0	0
255	DEL	0	0	0	0	0	0	0	0

Code-to-code conversion is often required when various types of digital equipment are tied together to make a data-processing system. The conversion techniques given in Section 2.5 are sufficient for designing the necessary converters between any alphameric codes. Off-the-shelf MOS/LSI ROM devices for performing the code conversions among the above four codes are available. For example.

TMS 2602 code converter: USASCII-to-Selectric/Selectric-to-USASCII
TMS 2603 code converter: EBCDIC-to-USASCII
TMS 2604 code converter: USASCII-to-EBCDIC/Selectric-to-EBCDIC

For a device that is not an off-the-shelf item, we can ask the manufacturer to program one for us. All we need to do is to specify the ROM chip to be programmed on and to describe the desired input/output function table (truth table) on a sequenced deck of standard 80-column computer cards according to a specific format. The number of cards and the data-card format depend upon the type of the chip and the manufacturer. They are described in the manufactures's products catalogs.

After having discussed the ROM and its applications, we now turn to another type of semiconductor memory—the RAM, to introduce some of the commonly used chips and to show how they function.

The RAM represents a memory function that has acquired an unfortunate name. The RAM memory refers to a memory circuit in which one can both read and write the desired data with facility and speed. A better designator for the RAM memory would be "read/write memory." Unfortunately, the designator "RAM" does not adequately differentiate the read/write memory from the ROM discussed above. In any event, it is not the intent of this text to revise accepted nomenclature, and we will continue to refer to read/write memory and RAM memory as one and the same. Any location within the RAM can be reached without regard to any other location. At the selected location, data can be written (stored) in the memory and retrieved from it. A RAM is sometimes also called a *direct-access memory* because any location can be addressed in approximately the same time interval as any other location. Thus for receipt of the information at the output, there is only a small variation in access time from any of the storage locations.

Most commonly used RAM's are 64-bit, 256-bit, 1,024-bit, and 4,096-bit. The trend is toward putting more bits on a chip so that the cost of the memory per bit will be decreased. There are two types of RAM's: a one-word (M-bit) input RAM with an X selector and a one-bit input RAM with X and Y selectors. Block diagrams of these two types of RAM's are shown in Fig. 3.5.4. The SN7489 64-bit RAM and the SN74200 256-bit RAM are examples of these two types of RAM's, which are shown in Fig. 3.5.5(a) and (b), respectively. These memory devices are organized to be compatible with a standard 16-pin package. The write and read operations of these memory devices are described below.

SN7489 Write operation: Information present at the data inputs is written into the memory by addressing the desired word and holding both the memory enable and write enable low. Since the internal output of the data input gate is common to the

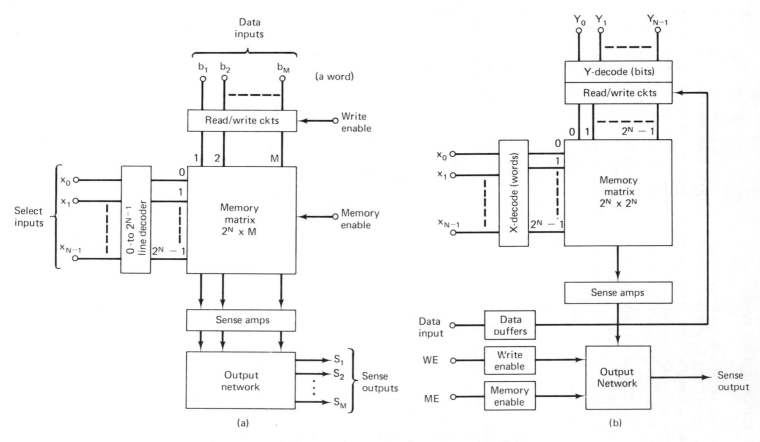

Fig. 3.5.4 (a) Block diagram of the one-word (M-bit) input RAM with the X selector; (b) block diagram of the one-bit input RAM with the X and Y selectors.

input of the sense amplifier, the sense output will assume the opposite state of the information at the data input when the write enable is low.

Read operation: The complement of the information that has been written into the memory is nondestructively read out at the four sense outputs. This is accomplished by holding the memory enable low, the write enable high, and selecting the desired address.

The voltage-function table of SN7489 is as follows:

ME†	WE†	Operation	Condition of Outputs
L	L	Write	Complement of data inputs
L	H	Read	Complement of selected word
H	L	Inhibit storage	Complement of data inputs
H	H	Do nothing	High

†H, high level; L, low level.

SN74200 Write cycle: The complement of the information at the data input is written into the selected location when all memory-enable inputs and write-enable input are low. While the write-enable input is low, the output is in the high-impedance state. When a number of outputs are bus-connected, this high-impedance output state will neither load nor drive the bus line, but it will allow the bus line to be driven by another active output.

Read cycle: The stored information (complement of information applied at the data input during the write cycle) is available at the output when the write-enable input is high and the three memory-enable inputs are low. When any one of the memory enable input is high, the output will be in the high-impedance state.

The voltage-function table of SN74200 is as follows:

Function	Inputs†		Output
	Memory Enable (ME)	Write Enable (WE)	
Write (Store complement of data)	L	L	High impedance
Read	L	H	Stored data
Inhibit	H	X	High impedance

†H, high level; L, low level; X, irrelevant. For memory enable: L, all ME inputs low; H, one or more ME inputs high.

RAM's are utilized in the computer as scratch-pad, buffer, and main-frame memories. At the onset of the 1970 decade, these functions were almost exclusively performed by magnetic devices. In the next generation of computers, MOS/LSI will

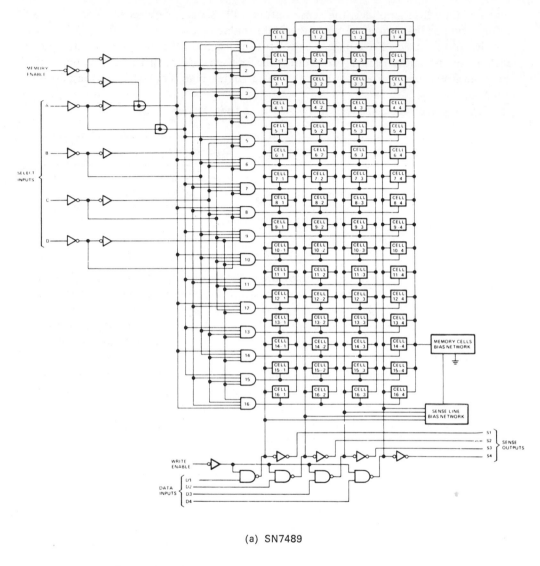

(a) SN7489

Fig. 3.5.5 (a) Block diagram of the SN7489 64-bit RAM; (b) block diagram of the SN74200 256-bit RAM.

be speedily adopted for buffer and mainframe memories. Following is a summary of the advantages of MOS/LSI for performing the RAM function over magnetic devices.

1. *Nondestructive readout.*   Readout of a MOS RAM does not affect the content stored. Readout of magnetic core is destructive and hence affects the content stored. Therefore, with MOS RAM it is unnecessary to perform a write after every read operation as must be done with core memory.

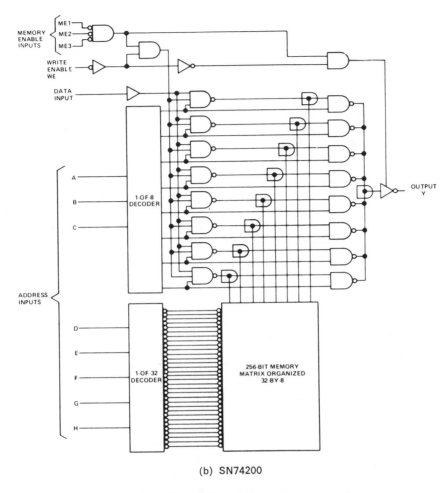

(b) SN74200

Fig. 3.5.5 (continued)

2. *Fast operating speed.* Access time can be as low as 150 ns, with on-chip decoding.

3. *Low power dissipation.* Power dissipation is typically less than 1.0 mW per bit for static design and less than 0.5 mW per bit for dynamic.

4. *Compatibility.* Semiconductor memory systems are entirely self-compatible because they enjoy common interface and technology between sensing and decoding circuitry and the storage element itself.

5. *Economy.* MOS memories are presently more economical than magnetic core for small and medium-sized systems.

The TMS 4023 is a typical example of MOS/LSI RAM's. It is a 1,024-bit RAM, organized as 1,024 1-bit words and is designed as a low-cost main-frame memory for large-storage high-performance operations. The block diagram of the TMS 4023 is

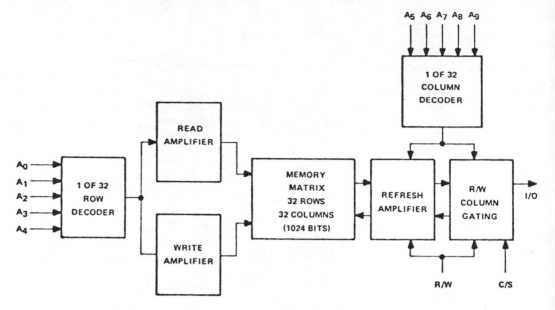

Fig. 3.5.6   Block diagram of the TMS 4023 1024-bit MOS/LSI
RAM.

shown in Fig. 3.5.6. It is seen that a 10-bit address code selects any one of the 1,024
bits for either the Read or Write operation. The memory is inhibited with the application of a logic high to the chip-select line. A Read or Write operation may be performed with the application of a logic low on the chip-enable line.

Read operation: Application of a logic high to the Read/Write line will result
in a Read operation. This may be presented simultaneously with or before application
of the address code. Read-out is nondestructive.

Write operation: A logic low applied to the Read/Write and chip enable will
result in a Write operation. Duration of the Write command and of the input data
must be at least 710 ns to ensure that the data is written into memory.

This memory is mounted in a 24-pin package, as most memory devices of this
size are.

3.6 Combinational Logic Design Using Read-Only
Memory (ROM) Array

In the previous section we discussed the ROM and its applications to generating
characters, codes, and so on. These applications use the ROM as a source of systematically arranged data that can be accessed quickly. Each of these applications for the
ROM involves data retrieval. In this section we introduce another important application of the ROM, the use of the ROM in the implementation of random logic.

The actual logic (NAND, NOR, and so forth) implemented by the ROM depends upon the specific coupling element used. When a diode or resistor is used, the ROM, without any inverters added, perform positive AND logic. When the inverter is cascaded into each ROM bit line, the overall logic function performed by the ROM is positive NAND. For our designs we will refer to ROM circuits that provide NAND logic. The reader should remember that the inverter may or may not be required in the actual ROM structure, depending upon the technology used. For example, if we use the basic MOS multiple-input positive NAND gate circuit in Fig. 3.6.1(a) as the building block for a ROM matrix, inverters are not needed. This NAND gate can be rearranged into ROM format in Fig. 3.6.1(b). A N by N ROM matrix for implementing $y_1 = \overline{x_1 x_2}$ and $y_2 = \overline{\bar{x}_1 \bar{x}_2}$ is shown in Fig. 3.6.2(a). The standard positive-logic equivalent of the circuit diagram of Fig. 3.6.2(a) is shown in Fig. 3.6.2(b). A simplified representation of a ROM is depicted in Fig. 3.6.2(c). When two ROM's are in cascade, as shown in Fig. 3.6.2(d), the resulting positive-logic Boolean functions are Q_1 and Q_2:

$$z_1 = \overline{y_1 y_2} = x_1 x_2 + \bar{x}_1 \bar{x}_2$$
$$z_2 = \bar{y}_1 = x_1 x_2$$

$$(3.6.1)$$

Note that the ROM matrices coupled together as in Fig. 3.6.2(d) provide the possibility for implementing two-level, random, and static logic. These two coupled ROM's will be referred to as an AND–OR ROM pair [see Fig. 3.6.2(e)]. The upper ROM matrix acts as a logical product generator and the lower ROM matrix acts as a logical summer. The logical product generator ROM and the logical summer ROM are called the *AND ROM* and the *OR ROM*, respectively. The AND–OR ROM pair can be used to realize any switching functions or truth table. When the ROM array is used to implement combinational logic, the gate reduction of the design is not important.

Example 3.6.1

Design a switching circuit that converts the 4-bit binary code into the 4-bit Gray code using ROM array.

The switching functions that describe the conversion from 4-bit binary code into the 4-bit Gray code are (see Example 2.5.2)

$$G_0 = B_0 B_1' + B_0' B_1$$
$$G_1 = B_1 B_2' + B_1' B_2$$
$$G_2 = B_2 B_3' + B_2' B_3$$
$$G_3 = B_3$$

$$(3.6.2)$$

They can be conveniently realized by an AND–OR ROM pair as in Fig. 3.6.3(a). This converter can also be realized by directly from the truth table in Fig. 2.5.2(a). This realization is shown in Fig. 3.6.3(b).

Commercially available programmable ROM arrays will be discussed in Section 4.5.

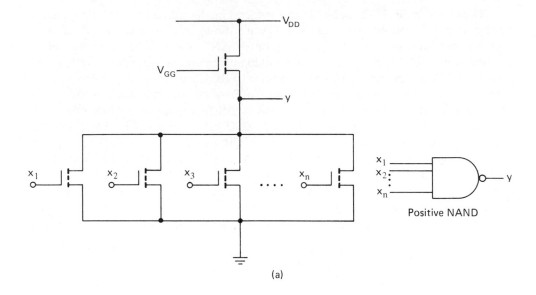

(a)

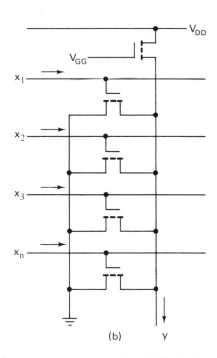

(b)   y

Fig. 3.6.1   (a) Basic MOS multiple-input positive NAND gate circuit; (b) same basic gate for ROM martix.

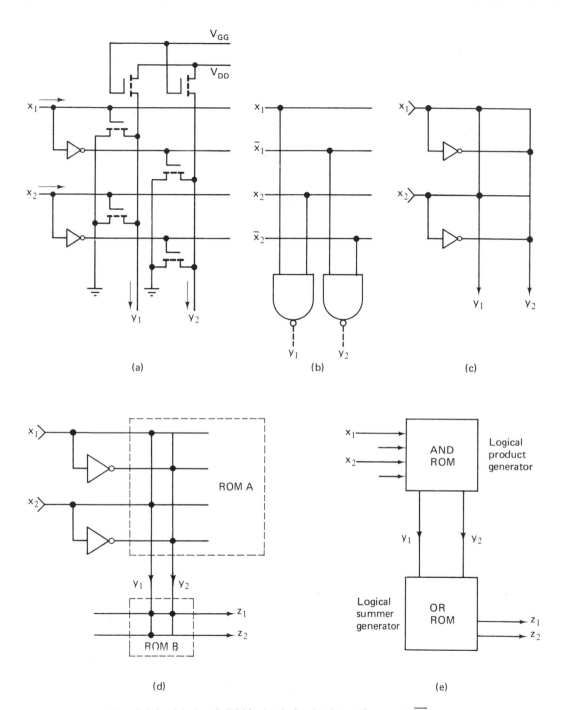

Fig. 3.6.2 (a) 2×2 ROM circuit for implementing $y_1 = x_1 \overline{x_2}$ and $y_2 = \overline{x_1} \overline{x_2}$; (b) standard positive logic equivalent; (c) simplified representation of a ROM; (d) two ROM's in cascade; (e) AND-OR ROM pair.

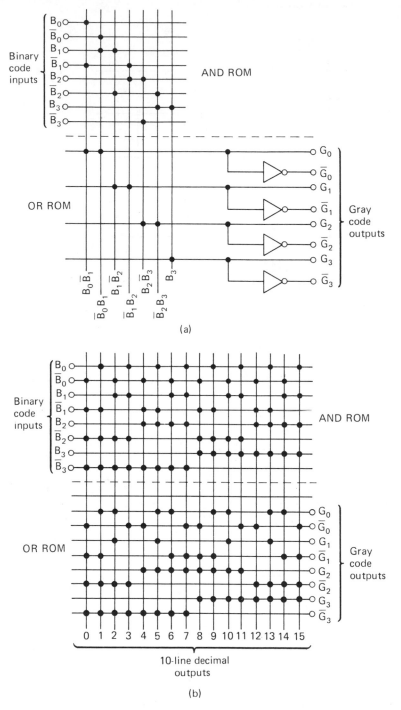

Fig. 3.6.3 Example 3.6.1.

Exercise 3.6

1. Use ROM array to realize the following code conversions:
   (a) From Gray to excess-3.
   (b) From 10-line decimal to binary.
   (c) From binary to excess-3.

2. Use ROM array to realize the function described in Table 3.4.4.

3. Realize the carry look-ahead generator described in Fig. 3.4.10 by an AND–OR ROM-pair circuit.

Bibliographical Remarks

References 1–3 provide detailed information about several TTL integrated circuits manufactured by three major manufactures. A good treatment on the subject of voltage symbolism can be found in reference 6. References 4 and 5 discuss Texas Instruments' series 54/74 integrated circuits and their applications, and reference 7 introduces Fairchild Semiconductor's series TTL/MSI and applications. References 8 and 9 provide design techniques and applications of semiconductor memory in general and MOS/LSI in particular, respectively. References 10, 11, and 12 are recent articles on IIL technology. Finally, a brief summary of up-to-date information about the basic circuitries, characteristics, and fabrication processes of various types of semiconductor memories is provided in Appendix A.

References

1. *The TTL Data Book*, Texas Instruments, Inc.

2. *TTL Data Book*, Fairchild Semiconductor.

3. *Signetics Digital 54/7400 Data Book*, Signetics Corporation.

4. R. G. HIBBERD, *Integrated Circuits*, McGraw-Hill, New York, 1969.

5. R. L. MORRIS and J. R. MILLER (ed.) *Designing with TTL Integrated Circuits*, McGraw-Hill, New York, 1971.

6. V. T. RHYNE, *Fundamentals of Digital Systems Design*, Prentice-Hall, Englewood Cliffs, N.J., 1973.

7. *The TTL Applications Handbook*, Fairchild Semiconductor, Aug. 1973.

8. G. LUECKE, J. P. MIZE, and W. N. CARR, *Semiconductor Memory Design and Application*, McGraw-Hill, New York, 1973.

9. W. N. CARR, and J. P. MIZE, *MOS/LSI Design and Application*, McGraw-Hill, New York, 1972.

10. F. M. KLAASSEN, "Device Physics of Integrated Injection Logic," *IEEE Transactions on Electroni Devices*, March 1975, pp. 145–152.

11. R. L. HORTON, J. ENGLADE, and G. MCGEE, "I^2L Takes Bipolar Integration a Significant Step Forward," *Electronics*, January 23, 1975, pp. 83–89.

12. J. ALTSTEIN, "I^2L: Today's Versatile Vehicle for Tomorrow's Custom LSI," *EDN*, February 20, 1975, pp. 34–38.

4

Sequential Integrated Circuits

In Chapter 3 some practical combinational integrated circuits and their applications to combinational logic design were discussed. This chapter, a continuation of the previous one, will introduce some commonly used TTL/SSI, TTL/MSI, and MOS/LSI sequential integrated circuits and their applications to sequential logic design. These circuits will include flip-flops, counters, shift registers, and programmable logic arrays.

4.1 Differences Between Combinational Circuits and Sequential Circuits

Before discussing sequential circuits, we first want to find out the differences between combinational circuits and sequential circuits.

A *combinational circuit* can be defined from the behavioral point of view as a circuit whose output is dependent only on the inputs at the same time instant, or can be defined from the constructional point of view as a circuit that does not contain any memory element.

A *sequential circuit*, on the other hand, can be defined from the behavioral point of view as a circuit whose output depends not only on the present inputs, but also on the past history of inputs, or can be defined from the constructional point of view as a circuit that contains at least one memory element.

It should be remarked here that, although a circuit with no memory element is never dependent on the past values of inputs, it is still possible to have a sequential circuit with memory elements independent of the past values of inputs. Such a sequential circuit is, of course, a trivial special case. However, strictly speaking, this class of circuits causes a discrepancy between definitions from the behavioral point of view and those from the constructional point of view. From the behavioral point of view, such a circuit should belong to the class of combinational circuits, whereas from the constructional point of view, such a circuit should belong to the class of sequential circuits. Since such a circuit is trivial in that there is no justification to construct a combinational circuit with useless memory elements, this slight discrepancy in the definitions is of little consequence.

Like any other physical system, the whole past history of a sequential circuit can be represented or manifested by a *state* of the sequential circuit. It is customary to use symbols x_i and z_i to denote the input and output variables of a combinational circuit and a sequential circuit. The states of a sequential circuit are represented by a set of *state variables*, denoted by y_i. The x_i's, y_i's, and z_i's are, of course, functions of time. For a sequential circuit, the values of the variables are usually specified at certain discrete time instants rather than over the whole continuous time. Let the time instants be $t_1, t_2, \ldots, t_k, \ldots$ and let $x_i(t_k)$, $y_i(t_k)$, and $z_i(t_k)$ be the variables at t_k. Both the behavioral and the constructional differences between a combinational circuit and a sequential circuit are best displayed by the systematic diagrams shown in Fig. 4.1.1.

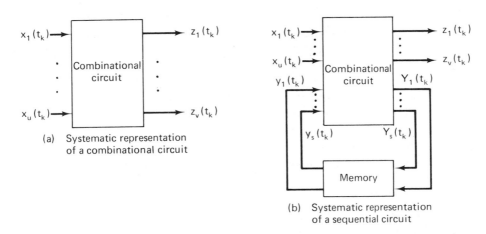

(a) Systematic representation
of a combinational circuit

(b) Systematic representation
of a sequential circuit

Fig. 4.1.1

From these two systematic diagrams, we see that a combinational circuit can be totally described by a set of equations of the form

$$z_j(t_k) = \lambda_j(x_1(t_k), \ldots, x_u(t_k)), \qquad j = 1, 2, \ldots, v$$

But a total description of a sequential circuit requires two sets of equations:

$$Y_i(t_k) = \delta_i(x_1(t_k), \ldots, x_u(t_k); \ y_1(t_k), \ldots, y_s(t_k)), \qquad i = 1, 2, \ldots, s$$

and

$$z_j(t_k) = \lambda_j(x_1(t_k), \ldots, x_u(t_k); \ y_1(t_k), \ldots, y_s(t_k)). \qquad j = 1, 2, \ldots, v$$

The first set of equations is called the *next-state functions*, and the second set is called the *output functions*. A summary of the differences between a combinational circuit and a sequential circuit is given in Table 4.1.1.

**TABLE 4.1.1   Differences Between a Combinational
Circuit and a Sequential Circuit**

Combinational Circuit	Sequential Circuit
It contains no memory elements.	It contains memory elements.
The present values of its outputs are determined solely by the present values of its inputs.	The present values of its outputs are determined by the present values of its inputs and its present state.
Its behavior is described by the set of output functions.	Its behavior is described by the set of next-state functions and the set of output functions.

There are two types of sequential circuits: synchronous and asynchronous sequential circuits. Before presenting their definitions, the concept of "state time" must first be introduced. One of the simplest sequential circuits is a unit-time delay device, as shown in Fig. 4.1.2(a), in which $x(k)$ and $y(k)$ are the discrete input and output of the device, where k denotes the discrete time 1, 2, 3, $\cdots$. This delay device is characterized by *an ability to hold the information existing during a short stable period for the entire state time that follows.* For example, Fig. 4.1.2(b) gives the time-line representation of the logic on the input and output of the delay device of Fig. 4.1.2(a). The darkened regions before each state time indicate the required stable period in the state time. Usually the stable period may represent a larger portion of the state time than shown. The following remarks about the diagram of Fig. 4.1.2(b) are in order.

1. Whenever an input such as (1) in Fig. 4.1.2(b) is a V_H in the stable region, the next-state output is a V_H, as indicated by the arrow (2).

2. Short pulses such as (3) are ignored when they occur in the transition period.

3. The transition (4) occurs during the required stable period, and this makes the next state undefined [either V_H or V_L until the next stable period, (5)]. Such a signal, which may be changing during the stable period, is called *asynchronous* in respect to the state-time system. The opposite definition is a *synchronous* signal.

DEFINITION 4.1.1

A sequential circuit that is based on an equal state time or a state time defined by external means (such as a clock [see Fig. 4.1.2(c)]) is called a *synchronous sequential circuit.* A circuit whose state time depends solely upon the internal logic circuit delays is called an *asynchronous sequential circuit.*

Several examples of synchronous and asynchronous sequential circuits are presented in the next section.

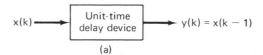

(a)

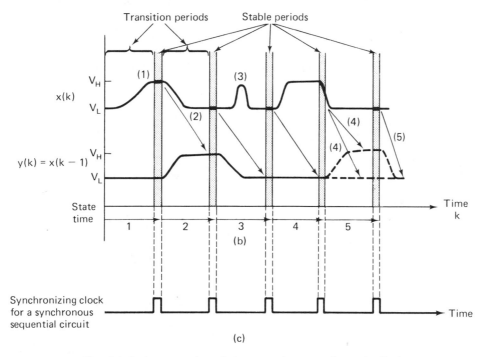

(b)

(c)

Fig. 4.1.2 Interpretation of the state time as a "snapshot" of logic behavior during stable periods.

4.2 Flip-Flops

A device that exhibits two different stable states is extremely useful as a memory element in a binary system. Any electrical circuit that has this characteristic falls into the category of devices commonly known as *flip-flops* (or *bistable multivibrators*).

S-R Latch There are many different implementations of a flip-flop. Among them, the implementation using gates is the simplest. The simplest type of flip-flop is the *set–reset*, or *S-R, latch*, which is a basic building block for other types of flip-flops. It has two inputs: the *S* (set) and *R* (reset) inputs, and two outputs, *which are complementary to each other*, denoted by Q and $\bar{Q}$. The S–R latch can be constructed by cross-coupling two NOR gates as shown in Fig. 4.2.1(a). It has two states: $Q = 0$ and $Q = 1$. We shall use Q_n and Q_{n+1} to denote the present state and the next state of the latch, respectively. It is evident that four possible inputs exist for the S–R latch. The outputs for these four possible inputs are described below.

1. For $S = 0$ and $R = 0$, the value of Q remains unchanged; that is, $Q_{n+1} = Q_n$.
2. For $S = 0$ and $R = 1$, regardless of the value of Q_n, Q_{n+1} will be equal to 0.
3. For $S = 1$ and $R = 0$, regardless of the values of Q_n, Q_{n+1} will be equal to 1.
4. For $S = R = 1$, $Q_{n+1} = \bar{Q}_{n+1} = 0$, which is not allowed.

The logic-function table of the S–R latch is given in Fig. 4.2.1(a). Figure 4.2.1(b) shows a TTL circuit realization of a NOR S–R latch using two positive NOR gates ($\frac{1}{2}$SN7402).

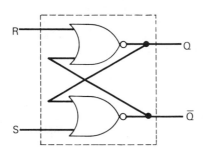

NOR S-R latch logic circuit

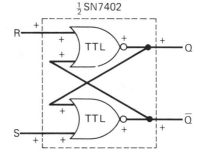

A circuit realization of a NOR S-R
latch using two positive NOR gates

Logic Function Table

Inputs		Outputs	
S	R	Q_{n+1}	Q_{n+1}
0	0	Q_n	$\bar{Q}_n$
0	1	0	1
1	0	1	0
1	1	0*	0*

*Not allowed

(a)

Voltage Function Table

Inputs		Outputs	
S	R	Q_{n+1}	$\bar{Q}_{n+1}$
L	L	No change	
L	H	L	H
H	L	H	L
H	H	L*	L*

*Not allowed

(b)

Fig. 4.2.1 NOR S-R latch.

The following important remarks about the S–R latch are in order.

1. *The S–R latch is a two-state ($Q = 0$ and $Q = 1$) sequential circuit.* From the logic-function table of Fig. 4.2.1(b), it is seen that the next state Q for $S = 0$ and $R = 0$ is equal to the present state of the circuit; thus the next state of this circuit is a function of its present inputs *and* its present state. A pictorial presentation of the

operational behavior of the S-R latch described by the logic-function table [Fig. 4.2.1(a)] is as follows:

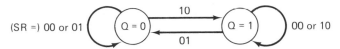

This diagram is called a *state transition diagram*. For example, if the present state of the circuit is $Q = 0$ and the present input is $SR = 10$ (i.e. $S = 1$ and $R = 0$), the next state of the circuit is $Q = 1$, as indicated by the transition line $\xrightarrow{10}$, drawn from state $Q = 0$ to state $Q = 1$.

2. *The S–R latch is an asynchronous sequential circuit.* Its state time depends solely upon the delays of the two NOR gates.

3. *The S–R latch is a memory element.*   It can be used to store a binary digit, 0 or 1. To store a binary digit 0 and 1, we simply apply $S = 0$ and $R = 1$, and $S = 1$ and $R = 0$, respectively, to the latch. This information remains in the latch as long as the inputs S and R are kept at 0 (i.e., at the low-voltage level, if the positive logic assignment is used) and the power supply to the device is kept on. Data in the memory are lost when the power is shut off.

The S–R latch can also be realized by cross-coupling two NAND gates as shown in Fig. 4.2.2(a). Comparing the function table of this circuit with that of Fig. 4.2.1(a), it is seen that if the S and R inputs of the circuit of Fig. 4.2.2(a) are inverted by two inverters, as shown in the circuit of Fig. 4.2.2(b), the function table of this circuit NAND S–R latch will be identical to that of the NOR S–R latch of Fig. 4.2.1(a) for the three allowable S–R inputs.

The S–R latch is limited to one mode of operation, asynchronous. But in most practical sequential circuit design, it is often desirable to have some form of clock input to the flip-flop so that the device may be operated simultaneously (synchronously) with all others in the system. It is this requirement that led to the development of the other types of flip-flops.

Edge-Triggered Flip-Flops   Most commonly used flip-flops are made in TTL. A TTL flip-flop has a single clock input. This single-input clock (CK) may be of the direct-coupled (d-c) or edge-triggered type or the double-rank or master–slave type, which subsequently described two types of flip-flops, the *edge-triggered flip-flop* and the *master–slave flip-flop*.

The d-c or edge-triggered clock is one that causes flip-flop operation at a particular voltage when either a positive (positive edge-trigger) or a negative (negative edge-trigger) transition occurs [see Fig. 4.2.3]. Either one or the other is recognized, but not both, for any chosen device. This type of clock enables the data inputs and transfers the data to the outputs simultaneously, resulting in a high-speed clocking technique that is relatively independent of clock rise and fall times. However, in TTL devices, noise immunity decreases as the rise and fall times increase beyond 150 ns.

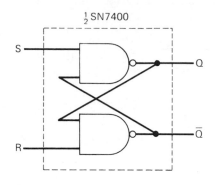

$\frac{1}{2}$ SN7400

Logic Function Table

Inputs		Outputs	
S	R	Q_{n+1}	$\overline{Q}_{n+1}$
0	0	1*	1*
0	1	1	0
1	0	0	1
1	1	Q_n	$\overline{Q}_n$

*Not allowed

(a)

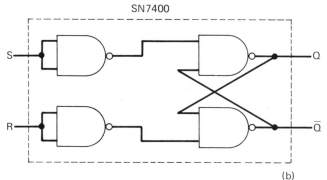

SN7400

Logic Function Table

Inputs		Outputs	
S	R	Q_{n+1}	$\overline{Q}_{n+1}$
0	0	Q_n	$\overline{Q}_n$
0	1	0	1
1	0	1	0
1	1	1*	1*

*Not allowed

(b)

Fig. 4.2.2 NAND S-R latch.

(a) Positive edge-trigger clock

(b) Negative edge-trigger clock

Fig. 4.2.3 Edge-trigger clock inputs.

The S–R edge-triggered flip-flop can be constructed from the NAND S–R latch by introducing a clock input (CK), as shown in Fig. 4.2.4(a), whose function table and symbolic representation are shown in the same figure, where Q_n and Q_{n+1} are the states (which are also outputs) of the flip-flop before and after clocking, respectively.

Besides the S–R edge-triggered flip-flop, there are three other types of commonly used flip-flops: J–K edge-triggered flip-flop, T-type edge-triggered flip-flop, and D-type edge-triggered flip-flop.

The J–K flip-flop has two data inputs, J and K, a single clock input, and two

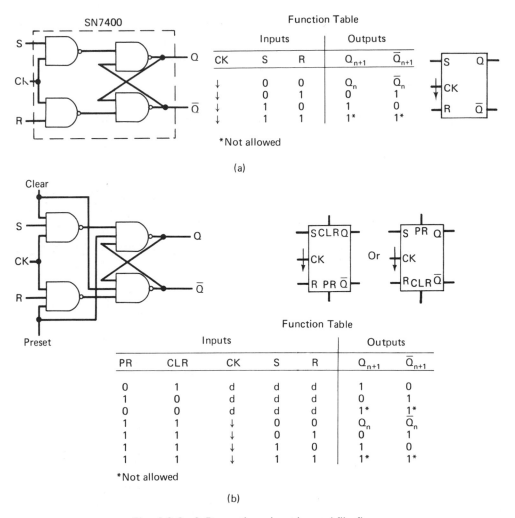

Fig. 4.2.4   S-R negative edge-triggered flip-flop.

outputs, Q and $\bar{Q}$. The operation of a J–K flip-flop is identical to that of a S–R flip-flop except that the former *allows* the input $J = K = 1$. When $J = K = 1$, the state Q will change state regardless of what the state Q was prior to clocking.

When the two inputs, J and K, are connected together, a T-type flip-flop is formed. This type of flip-flop has only one data input, T. When the input T is at a 0 level prior to a clock pulse, the Q output will not change with clocking. When the input T is at a 1 level, the Q output will be in the $\bar{Q}_n$ state after clocking (i.e., $Q_{n+1} = \bar{Q}_n$). This is called *toggling*, hence the name T flip-flop.

When the $S(J)$ input of an S–R flip-flop (J–K flip-flop) is connected to the inverted $R(K)$ input of the same flip-flop, a D-type flip-flop is formed. This type of flip-flop has only one data input, D. The operation of this type of flip-flop may be described as follows. When the D input is at a 0 level prior to a clock pulse, the Q output will be

0 after clocking. When the input D is at a 1 level, the Q output will be 1 after clocking. In other words, $Q_{n+1} = D_n$, where Q_{n+1} and D_n denote the values of the output Q and the input D at t_{n+1} (after clocking) and t_n (before clocking), respectively. For this type of flip-flop, the Q output is identical to the D input except with one pulse time *delay*; hence the name D flip-flop.

TABLE 4.2.1 Clocked J-K, D-type, and T-type Flip-flops Constructed from the Clocked S-R Flip-flop.

Circuit	Function table	Symbolic representation without CLEAR and PRESET	Symbolic representation with CLEAR and PRESET

All these flip-flops can be constructed from the S–R flip-flop. They are shown in Table 4.2.1. Their function tables and symbolic representations are also shown in the table. When the T input and the clock of a T flip-flop are tied together, the flip-flop circuit generates two clock signals of the same frequency as the input clock signal. This circuit is useful in sequential circuit design and can be constructed from an S–R flip-flop without using the two AND gates as shown in the last row of Table 4.2.1. It will be shown in Section 5.4 that a flip-flop of one type can always be obtained from a flip-flop of another type by introducing some appropriate input control logic. An example of illustrating the timing diagram of edge-triggered flip-flops is shown in Fig. 4.2.5.

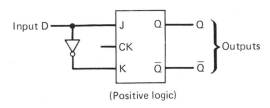

(Positive logic)

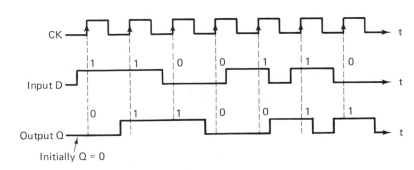

(a)   Timing for J-K positive edge-triggered flip-flop

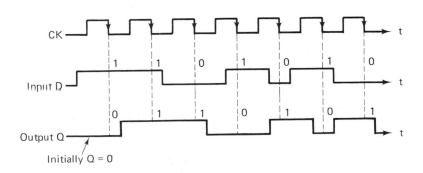

(b)   Timing for J-K negative edge-triggered flip-flop

Fig. 4.2.5 Example of a J-K edge-triggered flip-flop circuit.

Most commercially available flip-flops have two additional inputs: Clear (CLR) and Preset (PR). Their functions are as follows.

$$PR = 0, \quad Q = 1$$
$$CLR = 0, \quad Q = 0$$

PR	CLR	Q	$\overline{Q}$
0	1	1	0
1	0	0	1
0	0	1*	1*
1	1	No change	

*not allowed

When both PR and CLR are active (i.e., they are at the low voltage level for a positive-logic flip-flop, or both are at logic 0), $Q = \overline{Q} = 1$ which is not allowed for any flip-flops. These inputs are asynchronous and independent of the clock. While they are

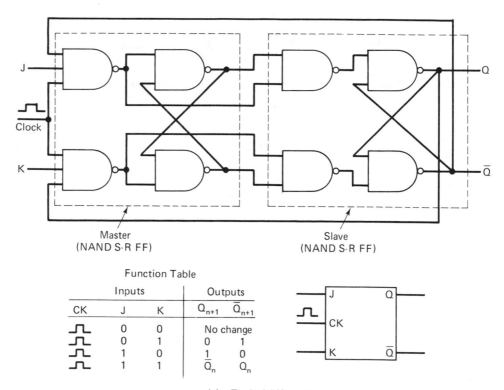

Master
(NAND S-R FF)

Slave
(NAND S-R FF)

Function Table

Inputs			Outputs	
CK	J	K	Q_{n+1}	$\overline{Q}_{n+1}$
⊓	0	0	No change	
⊓	0	1	0	1
⊓	1	0	1	0
⊓	1	1	$\overline{Q}_n$	Q_n

(a)   Typical J-K master-
slave flip-flop

Fig. 4.2.6

present, all other operations are inhibited. An S–R edge-triggered flip-flop with CLR and PR asynchronous inputs is shown in Fig. 4.2.4(b). Using this circuit as a building block, other types of flip-flops with CLR and PR can be constructed. The symbolic representations of the four types of flip-flops with CLR and PR are given in Table 4.2.1.

Master–Slave Flip-Flops   The circuit of the master–slave flip-flop is basically two latches connected serially. The first latch is called the *master*, and the second is termed the *slave*. For example, a J–K flip-flop is shown in Fig. 4.2.6(a). The J–K master–slave flip-flop with Clear and Preset asynchronous inputs is shown in Fig. 4.2.6(b). Normal action in clocking consists of four steps, as shown in Fig. 4.2.7(a). The important feature of this type of clocking is that the data input are never directly connected to the outputs at any time during clocking; this provides total isolation of

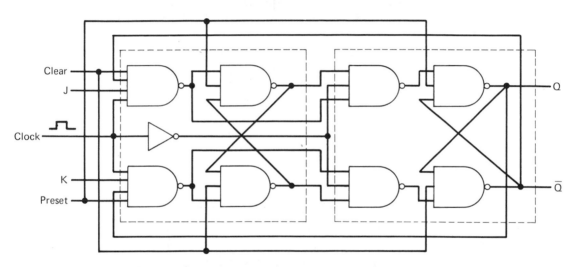

Function Table

	Inputs				Outputs	
PR	CLR	CK	J	K	Q_{n+1}	$\bar{Q}_{n+1}$
0	1	d	d	d	1	0
1	0	d	d	d	0	1
0	0	d	d	d	1*	1*
1	1	⊓	0	0	No change	
1	1	⊓	0	1	0	1
1	1	⊓	1	0	1	0
1	1	⊓	1	1	Q_n	Q_n

*Not allowed

(b)   J-K master-slave slip-flop
       with CLR and PR

Fig. 4.2.6   (continued)

outputs from data inputs. An example of a J–K master–slave flip-flop circuit and its input and output are shown in Fig. 4.2.7(b).

TTL/SSI Series 54/74 Flip-Flops and Their Applications   Table 4.2.2 shows some available TTL/SSI series 54/74 flip-flops. As stated earlier, the flip-flop is the most basic building block for the sequential system. For the reasons stated in Section 3.4, many frequently used sequential devices that are basically made up of flip-flops such as counters, shift registers, and memory devices are available in MSI chip form. These MSI chips will be discussed in detail in the next few sections, and therefore their realizations using flip-flops will not be discussed here. What we shall present here are two examples showing that many combinational logic design can be replaced by a

1. Isolate slave from master
2. Enable data inputs to master
3. Disable data inputs
4. Transfer data from master to slave

(a)   Master-slave clock

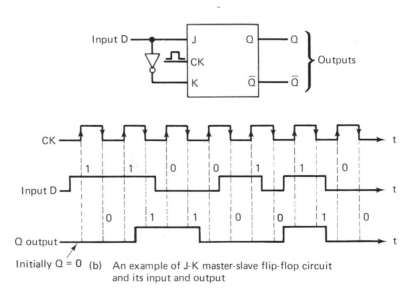

Initially Q = 0   (b)   An example of J-K master-slave flip-flop circuit
and its input and output

Fig. 4.2.7 J-K master-slave flip-flop.

TABLE 4.2.2   Some Available TTL/SSI Flip-Flops

Flip-Flop	Device
Edge-triggered J-K	54/7470
J-K master–slave	54/7472
Dual J-K master–slave	54/7473
Dual D-type edge-triggered	54/7474
Dual J-K master–slave, preset and clear	54/7476
J-K master–slave	54/74104
J-K master–slave	54/74105
Dual J-K master–slave	54/74107
Monostable nonretriggerable	54/74121
R-S master–slave	54/74L71
J-K master–slave	54/74L72
Dual J-K master–slave	54/74L73
Dual D-type edge-triggered	54/74L74
Dual J-K master–slave	54/74L78
J-K master–slave	54H/74H71
J-K master–slave	54H/74H72
Dual J-K master–slave	54H/74H73
Dual D-type edge-triggered	54H/74H74
Dual J-K master–slave with preset and clear	54H/74H76
Dual J-K master–slave	54H/74H78
J-K, negative edge-triggered	54H/74H101
J-K, negative edge-triggered	54H/74H102
Dual J-K, negative edge-triggered	54H/74H103
Dual J-K, negative edge-triggered	54H/74H106
Dual J-K, negative edge-triggered	54H/74H108

sequential logic design using flip-flops. The first example if a flip-flop circuit realization of a binary serial adder.

Example 4.2.1   Binary Serial Adder

The combinational circuit realization of a binary adder was presented in Section 2.2. The same device can also be realized by a sequential circuit using one flip-flop, as shown in Fig. 4.2.8(a). In this realization the two states "carry" and "no carry" are "remembered" by the state variable Q of the S–R flip-flop; that is, $Q = 0$ if there is no carry and $Q = 1$ if there is a carry. The "sum" output is achieved by a combinational circuit whose inputs are the augend, addend, and the state variable Q (and $\bar{Q}$). To illustrate the operation of this device, an example of adding 1011_2 and 0111_2 by this adder is given in Fig. 4.2.8(b).

Example 4.2.2   Parity Checker

The avoidance of error is of major interest in any data-handling system. One area where this is of special importance is in the transfer of information between the computer and its peripheral devices. It is common practice, for instance, to perform checks on items of data

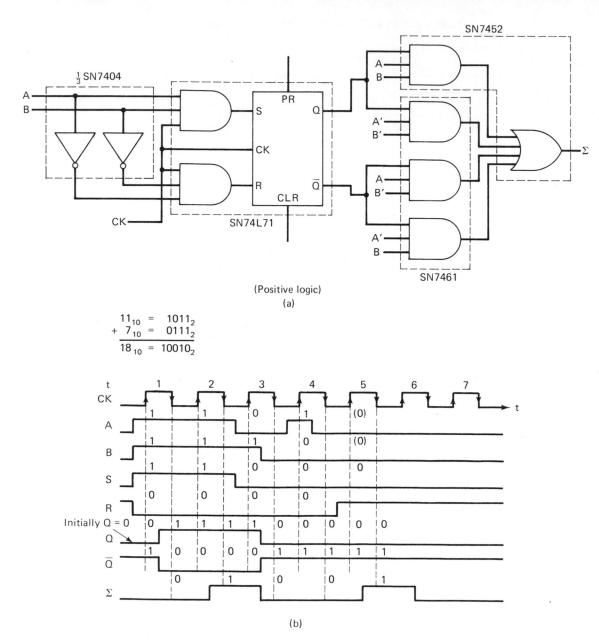

(Positive logic)

(a)

$$11_{10} = 1011_2$$
$$+ \ 7_{10} = 0111_2$$
$$18_{10} = 10010_2$$

(b)

Fig. 4.2.8 S-R flip-flop realization of binary serial adder.

read into the computer from a paper-tape reader or card reader. Similarly, information read from a magnetic tape unit or a magnetic drum is checked before it is accepted. A very popular method of error detection is to add an additional bit, called a *parity bit*, to each code combination. Odd or even parity may be used. If odd parity is required, the parity bit is made either

0 or 1, to ensure that the total number of 1's in any code combination is odd. If even parity is required, the total number of 1's is made even. The parity bits used to produce odd and even parities for the binary-coded decimal (BCD) numbers 0 to 7, for example, are shown in Table 4.2.3. The Boolean functions for checking odd and even parities are the classic example of

TABLE 4.2.3 Odd and Even Parities for the Binary- Coded Decimal Numbers 0 to 7

Odd Parity			Even Parity		
Parity Bit	BCD	Final Repre- sentation	Parity Bit	BCD	Final Repre- sentation
1	000	1000	0	000	0000
0	001	0001	1	001	1001
0	010	0010	1	010	1010
1	011	1011	0	011	0011
0	100	0100	1	100	1100
1	101	1101	0	101	0101
1	110	1110	0	110	0110
0	111	0111	1	111	1111

functions that are not efficiently realized in two levels. The function $f_0(x_1, x_2, x_3, x_4)$ [$f_e(x_1, x_2, x_3, x_4)$], which is 1 when an odd (even) number of the variable $x_1, x_2, x_3,$ or x_4 equals 1, is depicted in Fig. 4.2.9(a) [Fig. 4.2.9(b)]. An even-parity and an odd-parity check over any number of variables will appear as a checkerboard pattern on a Karnaugh map. Two combinational realizations of the function $f_0(x_1, x_2, x_3, x_4)$ are shown in Fig. 4.2.10.

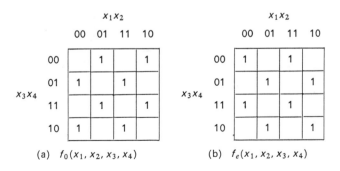

(a) $f_0(x_1, x_2, x_3, x_4)$ (b) $f_e(x_1, x_2, x_3, x_4)$

Fig. 4.2.9

Extending the argument to eight variables leads to the circuit of Fig. 4.2.11(a) (because the exclusive-or operation is associative). This is a combinational realization of the eight-variable odd-parity-check function. It requires 21 NAND gates and 14 inverters and eight lines, one for each bit.

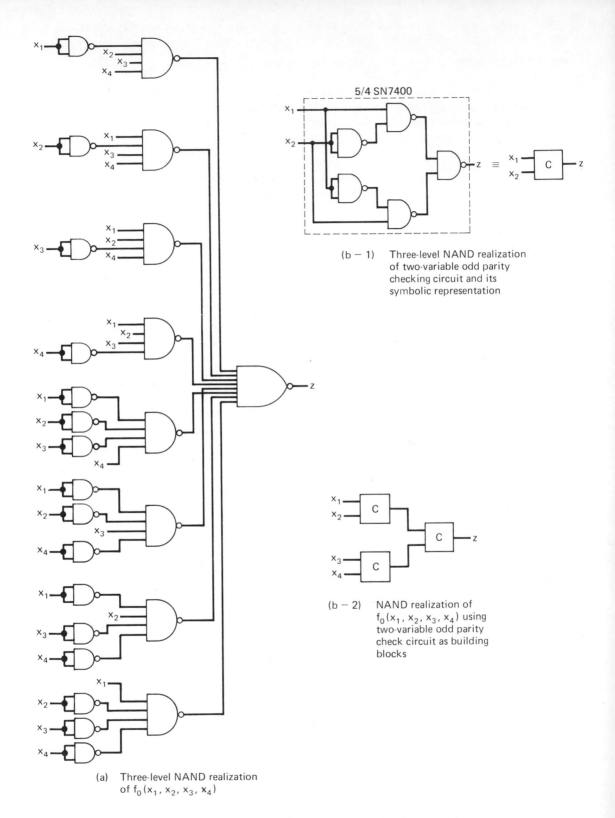

(b − 1) Three-level NAND realization
of two-variable odd parity
checking circuit and its
symbolic representation

(b − 2) NAND realization of
$f_0(x_1, x_2, x_3, x_4)$ using
two-variable odd parity
check circuit as building
blocks

(a) Three-level NAND realization
of $f_0(x_1, x_2, x_3, x_4)$

Fig. 4.2.10 NAND realization of $f_0(x_1, x_2, x_3, x_4)$.

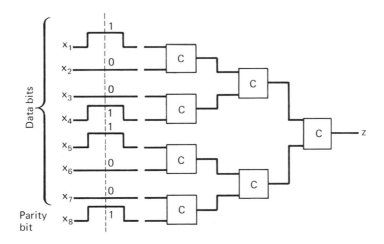

(a) Combinational circuit realization

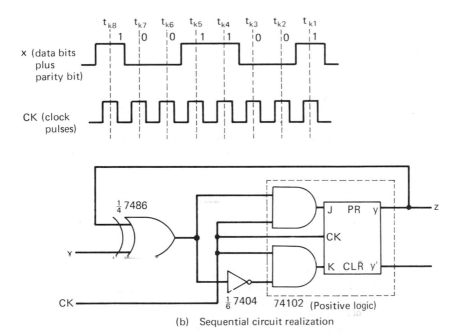

(b)   Sequential circuit realization

Fig. 4.2.11   Realization of eight-variable odd-parity check function.

Let us consider an alternative realization, namely, a sequential circuit realization. Based on the associativity property of the exclusive-or operation, we have

$$z = x_1 \oplus x_2 \oplus x_3 \oplus x_4 \oplus x_5 \oplus x_6 \oplus x_7 \oplus x_8$$
$$= ((((((x_1 \oplus x_2) \oplus x_3) \oplus x_4) \oplus x_5) \oplus x_6) \oplus x_7) \oplus x_8)$$

We may use a single line and send the eight bits in sequence, x_1 at time t_1, x_2 at time t_2, and so on. On a second line we send clock, or synchronizing, pulses to mark these times t_1, t_2, and so on. These two lines will provide the inputs to a circuit that is to put out a signal 1 if the number of i's in each group of eight bits is odd. Such a circuit using an EXCLUSIVE–OR gate, two AND gates, one inverter, and one S–R flip-flop is depicted in Fig. 4.2.11(b). Initially, the S–R flip-flop must be cleared to $z(t_{k1}) = 0$ before x_1 arrives, so the first equation is $x_1 \oplus z = x_1 \oplus 0 = x_1$. This is stored in the flip-flop when a clock pulse arrives at time t_{k1} [i.e., $z(t_{k2}) = x_1(t_{k1})$]. Then x_2 arrives and the exclusive-or forms $x_2 \oplus z = x_2 \oplus x_1$, which is stored in the flip-flop by the clock pulse at t_{k2}. Hence $z(t_{k3}) = x(t_{k2})$. This procedure continues until, after the clock pulse at t_{k8}, the output $z(t_{k9})$ will indicate whether there have been even or odd number of 1's in the eight bits. $z(t_{k9})$ is 1 if the number of 1's is odd, and 0 if the number of 1's is even. It should be noted that the circuit in Fig. 4.2.11(b) is a general n-bit odd-parity checker. It will provide an odd-parity check on whatever number of data bits have been received since the flip-flop was last cleared. The reader should have no difficulty in designing combinational as well as sequential even-parity checkers.

From the preceding two examples it is seen that the two sequential circuits are much simpler compared to their corresponding combinational circuits performing the same task typically at the expense of time. In practical digital communication systems and computers, sequential circuits are more frequently used than the combinational circuits. This is mainly because data always arrive sequentially at some level or another, usually for reasons of cost or hardware limitations rather than logical necessity. Table 4.2.4 shows three differences between a combinational circuit design and a sequential circuit design of a logic problem, such as the design of a parity checker just described. The sequential parity checker is slower than the combinational parity checker if we measure the speed of the combinational circuit from the time that all eight bits are available; but if the bits are going to arrive in sequence anyway, there is no difference

**TABLE 4.2.4 Comparison of a Combinational Circuit
Realization and a Sequential Circuit Realization**

Combinational Circuit Realization	Sequential Circuit Realization
Requires more hardware, thus more expensive	Requires less hardware, thus cheaper
Faster in speed†	Slower in speed†
Easier to design	Harder to design

†There is no difference if the bits arrive in sequence.

in speed. The sequential circuit, however, is simpler and therefore cheaper than the combinational circuit. But the former in general is harder to design compared to the latter. The design of sequential circuits will be discussed in Chapters 5 and 6.

4.3 Counters

Flip-flops programmed as counters are used in a wide variety of counting applications in scientific instruments, industrial controls, computers, and communication equipments, as well as in many other areas. The basic function of a counter is to "remember" how many clock pulses have been applied to the input; hence in the most basic sense, counters are memory systems. They are used for counting pulses, equipment operation sequencing, frequency division, and mathematical manipulation. Because of the wide range of applications for counters, as well as their large demand, manufacturers offer a wide selection of off-the-shelf MSI counter circuits, differing in complexity, functional versatility, speed, power, and cost.

Counters are classified into two types: the ripple or asynchronous counter and the synchronous counter. A *ripple counter* is an asynchronous sequential circuit, hence the name *asynchronous counter*. A *synchronous counter* is a synchronous sequential circuit. All flip-flops in a synchronous counter are under the control of the same clock pulse, whereas those in an asychronous counter are not. All counters are made up from a set of flip-flops in series. Although each individual flip-flop in a counter may be of edge-triggered or master–slave, the clock to the counter is of the edge-trigger type.

Asynchronous or Ripple Counters   To better visualize the general properties of this type of counter, we first present an example.

Example 4.3.1   A 4-Bit (or Four-Stage, or Modulo-16, or Divide-by-16) Binary Ripple Counter.

A simple 4-bit binary ripple counter that uses four J–K flip-flops is shown in Fig. 4.3.1. Suppose that the flip-flops are of negative edge-trigger type. Initially all flip-flops are in the logic 0 state ($Q_A = Q_B = Q_C = Q_D = 0$). A clock pulse is applied to the clock input of flip-flop A, causing Q_A to change from logic 0 to logic 1. At this moment, flip-flops B, C, and D do not change state (i.e., remain in the logic 0 state), since there is no negative-going edge of the clock pulse present at their clock input. If Q_A and Q_D denote the least significant bit (LSB) and most significant bit (MSB), respectively, after the first clock pulse is applied to the clock input to flip-flop Q_A, the counter will read

$$Q_D\ Q_C\ Q_B\ Q_A$$

$$0\quad 0\quad 0\quad 1$$

With the arrival of the second clock pulse to flip-flop A, Q_A goes from 1 to 0 (see the function

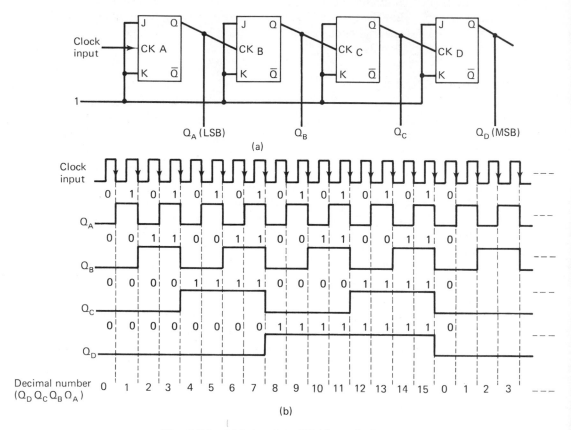

Fig. 4.3.1 4-bit (modulo-16) binary ripple counter.

table of the J–K flip-flop in Table 4.2.1). This change of state creates the negative-going pulse edge needed to trigger flip-flop B, and thus Q_B goes from 0 to 1. For the reason stated previously, at this moment flip-flops C and D still remain in the logic 0 state. Thus the counter will yield

$$
\begin{array}{cccc}
Q_D & Q_C & Q_B & Q_A \\
\hline
0 & 0 & 1 & 0
\end{array}
$$

It will count the input clock pulses in binary form as described above up to $Q_D = Q_C = Q_B = Q_A = 1$ (i.e., after 15 clock pulses). Clock pulse 16 causes all four flip-flops to go to 0—that is, the counter recycles! The state diagram of this counter is shown.

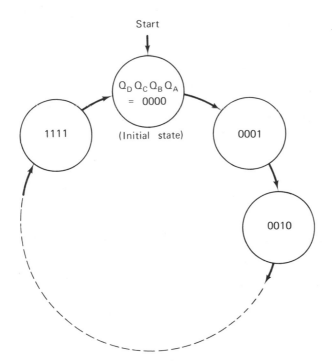

From the preceding example, it is seen that with a very minor modification, we can design counters that count finite N.

Design of Divide-by-N Ripple Counter   A general procedure for designing a divide-by-N ripple counter using J–K flip-flops with Preset is described as follows:

1.   Determine the number n of flip-flops required by the equation

$$n = \lceil \log_2 N \rceil$$

where the symbol $\lceil \log_2 N \rceil$ denotes the smallest integer that is greater than or equal to $\log_2 N$.

2.   Connect the n flip-flops as a ripple counter [see Fig. 4.3.1(a)].

3.   Find the binary representation of $N - 1$.

4.   Connect all flip-flop outputs that are 1 at the counter $N - 1$ as inputs to a NAND gate. Also feed the clock pulse to the NAND gate.

5.   Connect the NAND gate output to the Preset inputs of all the flip-flops for which $Q = 0$ at the count $N - 1$.

The counter resets in the following manner. At the positive-going edge of the Nth clock pulse all flip-flops are preset to the 1 state. On the trailing edge of the same clock pulse, all flip-flops count to the 0 state (i.e., the counter recycles). This procedure is illustrated by the following example.

Example 4.3.2   Decade (or Divide-by-10, or Modulo-10) Ripple Counter.

For $N = 10$, $n = \lceil \log_2 10 \rceil = \lceil 3.322 \rceil = 4$. Thus we use four J–K negative edge-triggered flip-flops and connect them as a ripple counter. Since $N = 9_{10} = 1001_2$, outputs Q_B and Q_C are connected to the NAND gate whose output is fed to the Preset's of the flip-flops. This decade ripple counter is shown in Fig. 4.3.2.

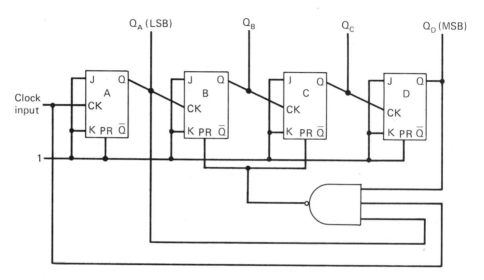

Fig. 4.3.2 Decade ripple counter.

Synchronous Counters   The synchronous counter eliminates the cumulative flip-flops delays seen in ripple counters. There are two methods of flip-flop control in synchronous counters, one with ripple carry and the other with parallel carry (or carry look-ahead). The latter is the faster of the two methods. But as the number of stages in a synchronous counter with parallel carry increases, the flip-flops must drive an ever-increasing number of NAND gates, and the number of inputs per control gate also increases.

Example 4.3.3   4-Bit Synchronous Counters

Two 4-bit synchronous counters using J–K positive edge-triggered flip-flops are shown in Fig. 4.3.3. Figure 4.3.3(a) shows a 4-bit synchronous counter with ripple carry, and Fig. 4.3.3(b) shows a 4-bit synchronous counter with parallel carry. It is easily seen that the repetition rates of these two counters are limited primarily by the delay of any one flip-flop plus the delay introduced by their corresponding control gating. Since the delay introduced by the control gating of the counter with parallel carry is shorter than that of the counter with

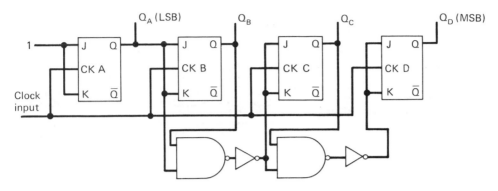

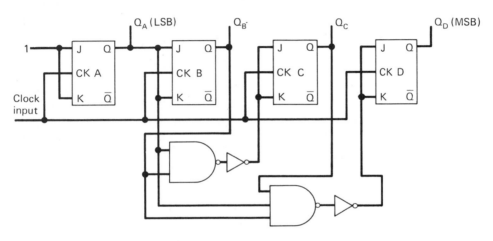

(a) 4-bit synchronous counters with ripple carry

(b) 4-bit synchronous counters with parallel carry

Fig. 4.3.3 4-bit synchronous counters.

ripple carry, the maximum clock frequency for the former is higher than that for the latter.

Just as the ripple counter can count any number that is not a power of 2, so can the synchronous counter.

Example 4.3.4 Decade Synchronous Counter

Figure 4.3.4 shows a decade synchronous counter using J–K flip-flops. The direct design of synchronous counters for any number base other than a power of 2 is more difficult than the direct design of a ripple counter. This will be discussed in Section 5.4.

From the above examples and discussion, it is interesting to formulate and compare the maximum clock frequencies for the three types of counters. They are indicated

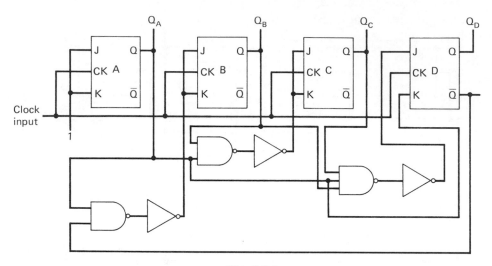

Fig. 4.3.4

in Table 4.3.1. For example, for a four-stage counter ($n = 4$), assuming that each flip-flop in the counter has a propagation delay (T_p) of 50 ns, decoding of any state (T_s) requires 100 ns, and propagation delay (T_g) from input to output of control gating (in this case, the delay of one NAND gate and one inverter) is 50 ns, the maximum clock frequencies for a ripple counter, a synchronous counter with ripple carry, and a synchronous counter with parallel carry are 3.67, 6.67, and 10 MHz, respectively.

TABLE 4.3.1   Formulas for Calculating Maximum Clock Frequency†

Type of Counter	Maximum Clock Frequency	Example: $N = 4$ $T_s = 100\ ns$ $T_p = 50\ ns$ $T_g = 50\ ns$
Ripple counter	$\frac{1}{f} \geq N(T_p) + T_s$	$\frac{1}{f} \geq 300$ ns or $f \leq 3.67$ MHz
Synchronous counter with ripple carry	$\frac{1}{f} \geq T_p + (N - 2)T_g$	$\frac{1}{f} \geq 150$ ns or $f \leq 6.67$ MHz
Synchronous counter with parallel carry	$\frac{1}{f} \geq T_p + T_g$	$\frac{1}{f} \geq 100$ ns or $f \leq 10$MHz

†N, number of flip-flop stages; T_p, propagation delay of one flip-flop; T_s, strobe time, width of decoded output pulse; T_g, propagation delay from input to input of control gating (in this case the delay of one NAND gate and one inverter).

Synchronous Up/Down Counters   The counters discussed so far are unidirectional counting up. Some applications require down counters, but often the design can be modified to use an up counter instead (e.g., by complementing the parallel input signals). In some cases, particularly in industrial applications, counters that count both up and down are needed. Figure 4.3.5 shows J–K flip-flop realizations of

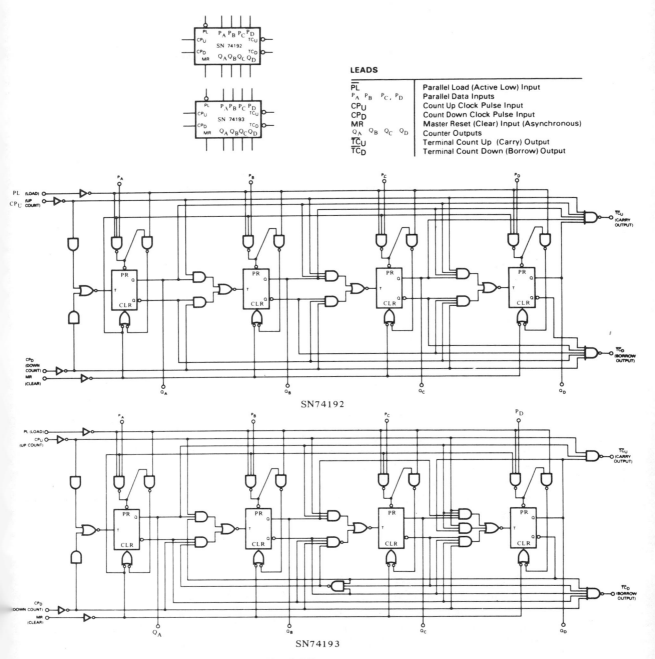

LEADS

$\overline{PL}$	Parallel Load (Active Low) Input
P_A P_B P_C, P_D	Parallel Data Inputs
CP_U	Count Up Clock Pulse Input
CP_D	Count Down Clock Pulse Input
MR	Master Reset (Clear) Input (Asynchronous)
Q_A Q_B Q_C Q_D	Counter Outputs
$\overline{TC}_U$	Terminal Count Up (Carry) Output
$\overline{TC}_D$	Terminal Count Down (Borrow) Output

SN74192

SN74193

Fig. 4.3.5

four-stage up/down binary counters, one with parallel carry and the other with ripple carry.

Some Available MSI/TTL Counters Table 4.3.2 shows some available TTL/ MSI counters. Two frequently used counters SN74192 up/down decade (8421) counter and SN74193 up/down 4-bit binary counter are shown in Fig. 4.3.6. Both devices are synchronous dual-clock up/down counters with asynchronous Parallel Load,

TABLE 4.3.2 Some Available TTL/MSI Counters

Function	Device
	Asynchronous counters (ripple clock)—negative-edge triggered
Decade	54/74L90, 54/74LS196, 54/7490, 54/7490A, 54/74176, 54/74196
4-bit binary	54/74L93, 54/74LS197, 54/74293, 54/7493A, 54/74177, 54/74197
Divide-by-12	54/7492A
	Synchronous counters—positive-edge triggered
Decade	54/74160, 54/74162
Decade up/down	54/74L192, 54/74190, 54/74LS190, 54/74192, 54/74LS192
4-bit binary	54/74L193, 54/74191, 54/74S191, 54/74193, 54/74LS193, 54/74161, 54/74163

asynchronous overriding Master Reset, and internal Terminal Count logic, which allows the counters to be easily cascaded without additional logic. The SN74192 and SN74193 can be used in many up/down counting applications, particularly when the initial count value must be loaded into the counter and multistage counting is required. The SN74192 and SN74193 can be Reset, Preset, and can count up and down. The operating modes of the counters are listed and are identical. The only difference is in their count sequences. Counting is synchronous with the outputs changing state after the Low-to-High transition of either the count-up clock (CP_U) or count-down clock (CP_D). The direction of the count is determined by the clock input, which is pulsed, while the other count input is High. Incorrect counting can occur if the count-up-clock and count-down-clock inputs are Low simultaneously. The counters respond to a clock pulse on either input by changing to the next appropriate state of the sequence. The SN74192 diagram shows the regular BCD (8421) sequence as well as the sequence of states when a code greater than 9 is preset into the counter. The SN74192 and SN74193 have an asynchronous parallel load capability that permits the counter to be preset. When the Parallel Load ($\overline{PL}$) and the Master Reset (MR) input are Low, information present on the parallel data inputs (P_0, P_1, P_2, and P_3) is loaded into the counter and appears on the outputs regardless of the conditions of the clock inputs. When the Parallel Load input goes High, this information is stored in the counter, and when the counter is clocked, it changes to the next appropriate state in the count sequence. The parallel inputs are inhibited when the Parallel Load is High and have no effect on the counter. A High on the asynchronous Master Reset (MR) input overrides both clocks and Parallel Load and clears the counter. Obviously, for predictable

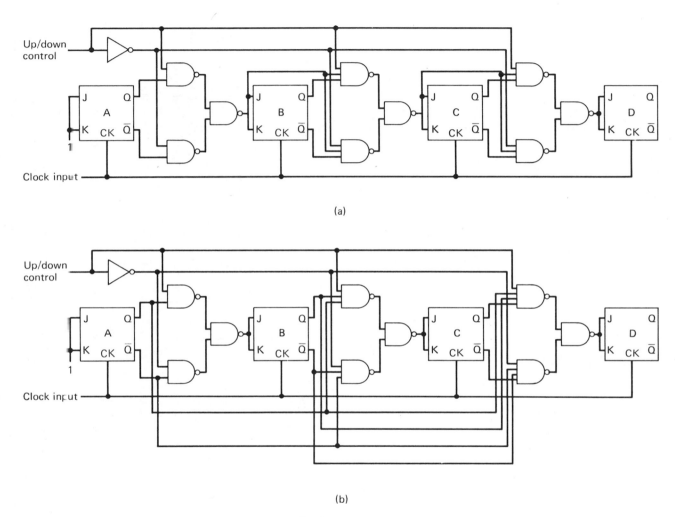

(a)

(b)

Fig. 4.3.6 Two MSI/TTL up/down counters.

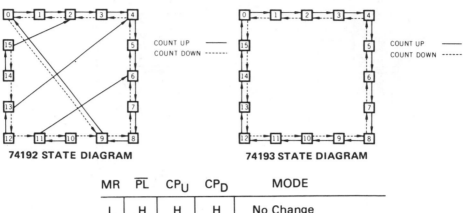

MR	$\overline{PL}$	CP_U	CP_D	MODE
L	H	H	H	No Change
L	H	CP	H	Count Up
L	H	H	CP	Count Down
L	L	X	X	Preset (Asynchronous)
H	L	X	X	Reset (Asynchronous)
H	H	X	X	Reset (Asynchronous)

Fig. 4.3.6 (continued)

operation, the Parallel Load and Master Reset must not be deactivated simultaneously. Both SN74192 and SN74193 have four master–slave flip-flops plus steering, Terminal Count decoding, and Preset logic. Each flip-flop is designed to toggle after each clock pulse. Counting occurs by steering clock pulses from either the up- or down-clock input to the appropriate flip-flops; and output changes are coincident, two gate delays after the rising clock edge. The steering logic in the SN74193 allows a particular flip-flop to receive an up-clock pulse when all preceding stages are 1 and to receive a down-clock pulse when all preceding stages are zero. The first flip-flop toggles if an up or down clock is received. The SN74192 incorporates slightly different steering logic to allow decade counting. Each flip-flop is a master–slave toggle flip-flop operating as follows. When the toggle clock input is Low, the slave is steady but the master is set to the opposite state of the slave. During the Low-to-High clock transition, the master is disabled so a later change in the slave outputs does not affect the master. Also, the information now in the master is transferred to the slave and appears at the output. When the transfer is completed, the master and slave are steady as long as the clock transition (the transfer path from master to slave) is inhibited, leaving the slave steady in its present state and allowing the master to be set to the opposite state of the slave. Asynchronous Preset and Clear inputs on each flip-flop allow the respective flip-flops to be set or cleared independent of the clock inputs. All the inputs are buffered and require only 1 unit load of drive; all outputs have drive capability of 10 unit loads. The input loading and drive capability of this device reduce the need for additional external buffers in digital systems.

The SN74192 and SN74193 have Terminal Count Up ($\overline{\text{TC}_U}$) and Terminal Count Down ($\overline{\text{TC}_D}$) outputs which allow multistage-ripple binary-decade-counter operations without additional logic. The Terminal Count Up output is Low while the up-clock input is Low and the counter is in its highest state (15 for the SN74193, 9 for the SN74192). Similarly, the Terminal Count Down output is Low while the down clock input is Low and the counter is in state zero. The logic equations for Terminal Counter are as follows: For SN74192,

$$\text{TC}_U = Q_0 \cdot \bar{Q}_1 \cdot \bar{Q}_2 \cdot Q_3 \cdot \overline{\text{CP}}_U$$
$$\text{TC}_D = \bar{Q}_0 \cdot \bar{Q}_1 \cdot \bar{Q}_2 \cdot \bar{Q}_3 \cdot \overline{\text{CP}}_D$$

and for SN74193,

$$\text{TC}_U = Q_0 \cdot Q_1 \cdot Q_2 \cdot Q_3 \cdot \overline{\text{CP}}_U$$
$$\text{TC}_D = \bar{Q}_0 \cdot \bar{Q}_1 \cdot \bar{Q}_2 \cdot \bar{Q}_3 \cdot \overline{\text{CP}}_D$$

Figure 4.3.7 shows counters that are cascaded by feeding the Terminal Count Up output to the up-clock input and the Terminal Count Down output to the down-clock input of the following (more significant) counter. Therefore, when an SN74193

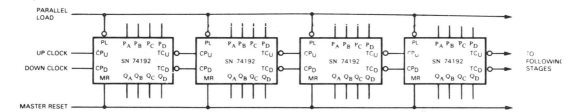

Fig. 4.3.7 Cascade counter.

counter is in state 15 and counting up or in state zero and counting down, a clock pulse will change the counter's state on the rising edge and simultaneously clock the following counter through the appropriate active Low terminal output. The operation of the SN74192 is the same, except when counting up, clocking occurs on state 9. The delay between the clock input and the Terminal Count output of each counter is two gate delays (typically 18 ns). Obviously, these delays are cumulative when cascading counters. When a counter is reset, the Terminal Count Down output goes Low if the down clock is Low and, conversely, if a counter is preset to its terminal count value, the Terminal Count Up output goes Low while the clock is Low.

Exercise 4.3

1. (a) Design a modulo-32 ripple counter using J–K flip-flops.
    (b) Design a modulo-60 ripple counter using J–K flip-flops with Preset.

2. Design a modulo-60 synchronous counter with (a) ripple carry and (b) parallel carry.

3. Assume that the propagation delay of one flip-flop, the strobe time, and the propagation delay from input to input of control gating are 50 ns, 100 ns, and 50 ns, respectively. Calculate the maximum clock frequencies for the modulo-60 counters obtained in problems 1 and 2.

4. Design a 12-hour digital clock with hour and minute displays. Assume that the input is a 110-volt 60 Hz AC constant voltage source.

5. Design an octal up/down counter using (a) an SN74192 and (b) an SN74193.

6. Design a counter that counts to 10. Reset by use of the configuration shown in Fig. 4.3.6 and verify its operation. The counter should be preset so that the counting sequence begins at 0000.

7. Design a cascade up/down counter that counts seven pulses. Reset by use of the configuration shown in Fig. 4.3.7. Check the circuit by verifying its operation.

8. A standard synchronous up/down counter, if clocked a sufficient number of times, will reset when the maximum counting sequence has been exceeded. The counting sequence may be modified, however, such that the counter will count to some number N and reset (N < maximum count). One method that may be used to construct an N-counter is shown in Fig. P4.3.8. The counter (SN74193) increments on each clock pulse

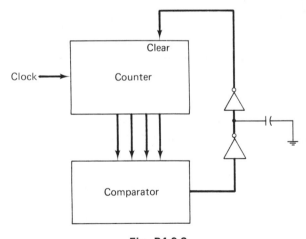

Fig. P4.3.8

and the output is compared (SN7485) to a number N. When the output of the counter is equal to N, the counter is reset. Design a counter that counts to 10 and resets.

9. A second method that may be used to realize an N-counter is shown in Fig. P4.3.9. The counter decrements on each clock pulse (starting from N). When the count reaches 0000, the next clock pulse causes a borrow which resets the counting sequence. Design a counter that counts seven pulses and resets.

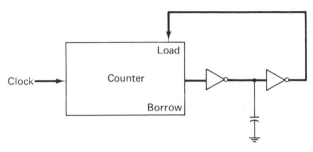

Fig. P4.3.9

4.4 Shift Registers

A group of cascaded flip-flops used to store related bits of information is known as a *register*. A register that is used to assemble and store information arriving from a serial source is called a *shift register*. Each flip-flop output of a shift register is connected to the input of the following flip-flop, and a common clock pulse is applied to all flip-flops, clocking them synchronously. Hence the shift register is a synchronous sequential circuits.

Registers are data-storage devices that are more sophisticated than latches. They use edge-triggered flip-flops. Shift registers can be classified into five classes:

1. Serial-in, serial-out shift registers.
2. Parallel-in, serial-out shift registers.
3. Serial-in, parallel-out shift registers.
4. Parallel-in, parallel-out shift registers.
5. Parallel-in, parallel-out bidirectional shift registers.

Table 4.4.1 shows some available MSI/TTL shift registers. An example of 8-bit shift register for each of the five classes is shown in Fig. 4.4.1.

TABLE 4.4.1 Some Available MSI/TTL Shift Registers

Function	No. of Bits	Device
Serial-in, serial-out	8	54/7491A, 54/74L91
Parallel-in, serial-out	4	54/7494
	8	54/74165, 54/74166
Serial-in, parallel-out	8	54/74164, 54/74L164
Parallel-in, parallel-out	4	54/74L95, 54/74L99, 54/74LS295, 54/74LS95A, 54/74LS195
		54/74178, 54/74179, 54/7495A, 54/74195, 54/74S195
	5	54/7496, 54/74L96
	8	54/74199
Parallel-in, parallel-out (bidirectional)	4	54/74194, 54/74S194, 54/74LS194
	8	54/74198

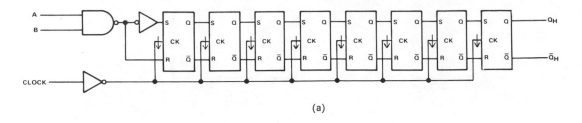

(a)

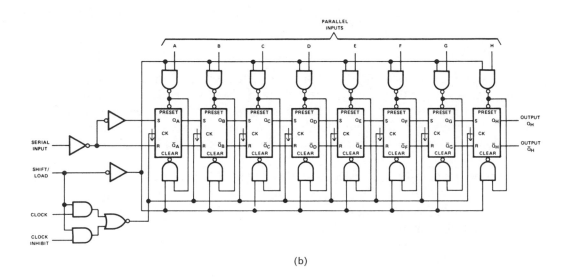

(b)

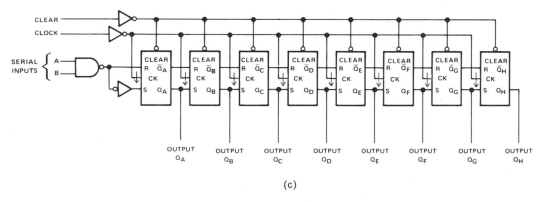

(c)

Fig. 4.4.1   Examples of MSI/TTL shift registers.

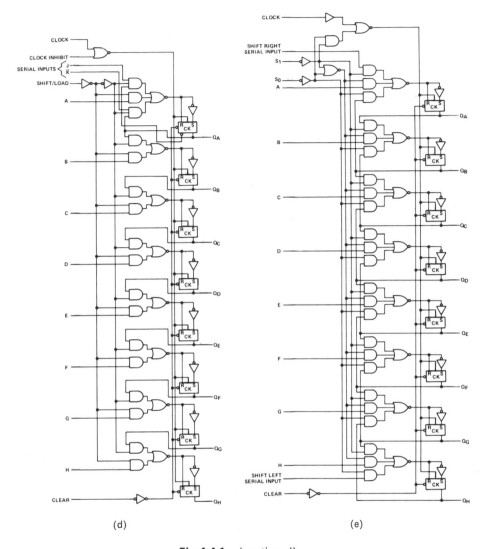

(d)

(e)

Fig.4.4.1 (continued)

Shift registers are used in a digital system for temporary information storage and data manipulation and transferring. In addition, they can be used in many counting circuits, including simple counters, variable modulo counters, up/down counters, and increment counters. Take the 4-bit parallel-in/parallel-out shift register SN74195, for example, whose block diagram and function table are shown in Fig. 4.4.2. A twisted ring counter constructed from an SN74195 and a NAND gate is depicted in Fig. 4.4.3.

Another approach to construct a counter is to use a shift register such as SN74195 with a feedback loop as shown in Fig. 4.4.4. Examination of the sequence indicated

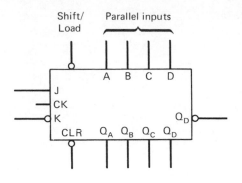

Shift/Load Parallel inputs

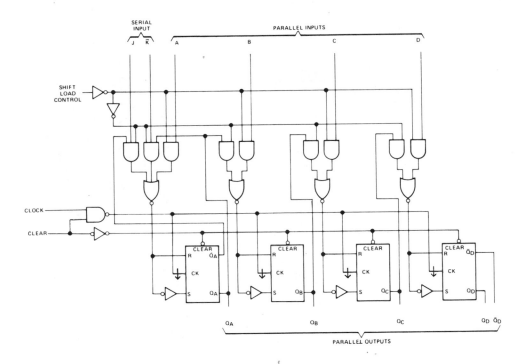

Fig. 4.4.2 4-bit parallel-in, parallel-out shift register.

FUNCTION TABLE

INPUTS									OUTPUTS				
CLEAR	SHIFT/ LOAD	CLOCK	SERIAL		PARALLEL				Q_A	Q_B	Q_C	Q_D	$\bar{Q}_D$
			J	$\bar{K}$	A	B	C	D					
L	X	X	X	X	X	X	X	X	L	L	L	L	H
H	L	↑	X	X	a	b	c	d	a	b	c	d	$\bar{d}$
H	H	L	X	X	X	X	X	X	Q_{A0}	Q_{B0}	Q_{C0}	Q_{D0}	$\bar{Q}_{D0}$
H	H	↑	L	H	X	X	X	X	Q_{A0}	Q_{A0}	Q_{Bn}	Q_{Cn}	$\bar{Q}_{Cn}$
H	H	↑	L	L	X	X	X	X	L	Q_{An}	Q_{Bn}	Q_{Cn}	$\bar{Q}_{Cn}$
H	H	↑	H	H	X	X	X	X	H	Q_{An}	Q_{Bn}	Q_{Cn}	$\bar{Q}_{Cn}$
H	H	↑	H	L	X	X	X	X	$\bar{Q}_{An}$	Q_{An}	Q_{Bn}	Q_{Cn}	$\bar{Q}_{Cn}$

H = high level (steady state)
L = low level (steady state)
X = irrelevant (any input, including transitions)
↑ = transition from low to high level
a, b, c, d = the level of steady-state input at A, B, C, or D, respectively
Q_{A0}, Q_{B0}, Q_{C0}, Q_{D0} = the level of Q_A, Q_B, Q_C, or Q_D, respectively, before the indicated steady-state input conditions were established
Q_{An}, Q_{Bn}, Q_{Cn} = the level of Q_A, Q_B, or Q_C, respectively, before the most-recent transition of the clock

204

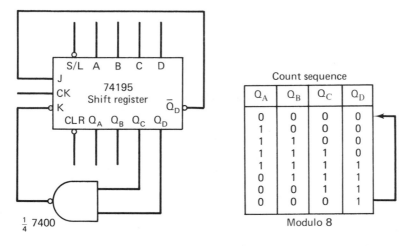

Count sequence

Q_A	Q_B	Q_C	Q_D
0	0	0	0
1	0	0	0
1	1	0	0
1	1	1	0
1	1	1	1
0	1	1	1
0	0	1	1
0	0	0	1

Modulo 8

Fig. 4.4.3 Twisted ring counter constructed from a feedback SN74195 shift register.

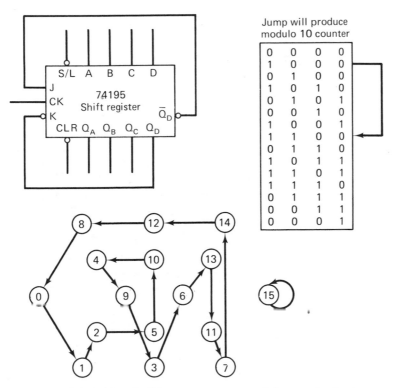

Jump will produce modulo 10 counter

0	0	0	0
1	0	0	0
0	1	0	0
1	0	1	0
0	1	0	1
0	0	1	0
1	0	0	1
1	1	0	0
0	1	1	0
1	0	1	1
1	1	0	1
1	1	1	0
0	1	1	1
0	0	1	1
0	0	0	1

Fig. 4.4.4 Modulo-15 counter constructed from a feedback SN74195 shift register.

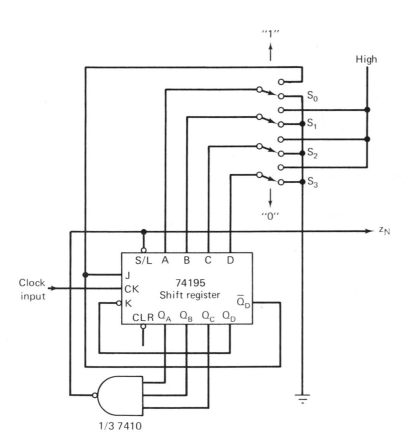

S_0	S_1	S_2	S_3	N
1	1	1	1	16
0	1	1	1	15
0	0	1	1	14
0	0	0	1	13
0	0	0	0	12
1	0	0	0	11
0	1	0	0	10
1	0	1	0	9
0	1	0	1	8
0	0	1	0	7
1	0	0	1	6
1	1	0	0	5
0	1	1	0	4
1	0	1	1	3
1	1	0	1	2
1	1	1	0	1

Fig. 4.4.5 Variable modulo counter constructed from a feedback.

shows that a 10-state counter results if a jump is caused from the 1000 state to the 1100 state. It is only necessary to inhibit the Reset of the first state to obtain this jump. Further examination of the sequence reveals that the inhibited function is simply the Q_C output. This approach creates exactly the same counter as obtained by the previous methods.

A variable modulo or divide-by-n counter can also be obtained by using an SN74195 shift register as depicted in Fig. 4.4.5. Again a simple feedback is used. This time a single gate produces the Parallel Load signal whenever the first three stages contain all ones. The parallel input combination to be loaded into the register is determined by the four switches. Note that this counter can divide by any integer up to and including 16.

Exercise 4.4

1. In Fig. P4.4.1 are shown the block diagrams of two ways to realize the binary fixed-point multiplication. Describe the operations of each of these two multipliers by a flow chart and tell their main differences.

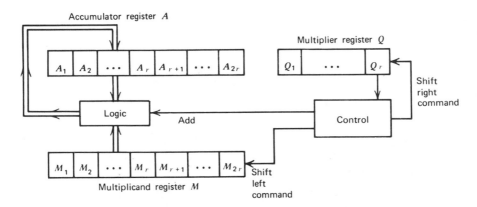

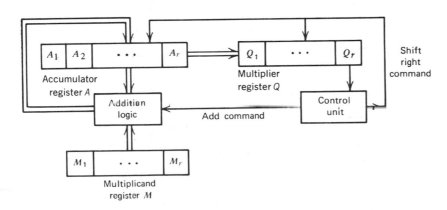

Fig. P4.4.1

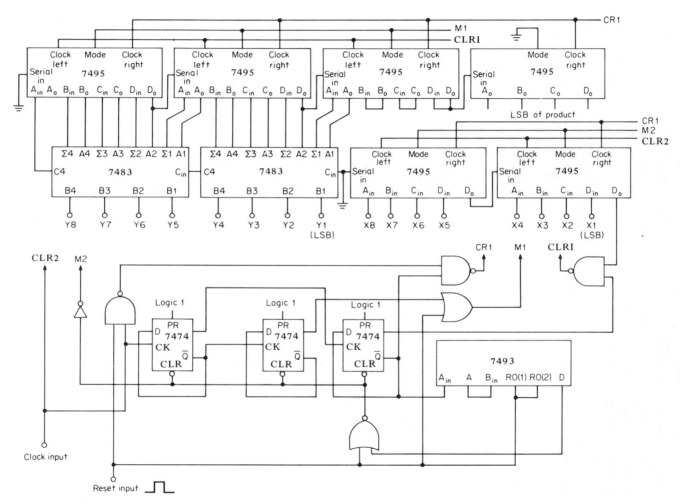

Fig. P4.4.2

2. The block diagram in Fig. P4.4.2 is a binary multiplier logic diagram. This multiplier is suitable for desk calculators. Show that it is equivalent to a pencil-and-paper method as shown below for two 4-bit numbers X and Y:

$$X = 1101$$
$$Y = 1011$$

$$
\begin{array}{r}
1101 \\
1101 \\
0000 \\
1101 \\
\hline
10001111
\end{array}
$$

Here the product is obtained by multiplying Y successively by the separate digits of X and summing the partial products. Shifting Y to the left by n bits is, of course, equivalent to multiplying by 2^n.

4.5 Programmable Logic Array

The simplest form of programmable logic array (PLA) contains the AND–OR ROM-pair matrices with clocked flip-flops in the feedback path, as shown in Fig. 4.5.1.

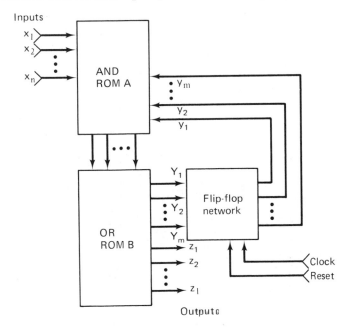

Note: $y_1, y_2 \ldots, y_m$ are present
state binary variables
$Y_1, Y_2 \ldots, Y_m$ are next
state binary variables

Fig. 4.5.1 Programmable logic array (PLA).

The ROM array described in Section 3.6 provides a matrix for design and production automation. A major advantage of the PLA results from the fact that a wide variety of sequential logic functions are economically obtained by designing for a single modification of the gate mask during circuit fabrication. Since ROM matrices may contain from 1,000 to 20,000 summing nodes, one can readily note that a great logical complexity is possible using the PLA circuit system. In the PLA system the feedback loop must be clocked carefully; otherwise, some undesirable transient phenomena† can develop if we feed the data back directly from the summing matrix into the product matrix. We shall therefore generally require that clocked flip-flops be connected into all feedback paths from the summing matrix (ROM B) to the product matrix (ROM A), as in Fig. 4.5.1. We can control the reset of the flip-flops to initialize the logic. The designer is free to choose the number of matrix points and flip-flops in his particular design consistent with whatever circuit design automaton techniques are used. A more detailed block diagram of the PLA is depicted in Fig. 4.5.2. Two design examples are given next to illustrate the design procedures for PLA implementation of sequential logic.

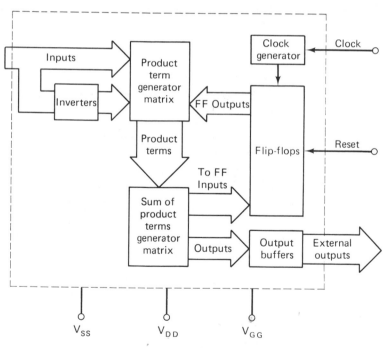

Fig. 4.5.2 Detailed block diagram of PLA.

†Such as race conditions and hazards, which will be discussed in the next chapter.

Example 4.5.1

Realize the circuit diagram for the binary serial adder in Fig. 4.2.8 using an SRFF PLA.

Solution: The PLA implementation of the circuit is shown in Fig. 4.5.3.

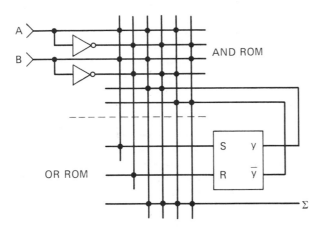

Fig. 4.5.3 Example 4.5.1.

Example 4.5.2

Implement the circuit diagram of the SN74195 4-bit parallel-in/parallel-out shift register in Fig. 4.4.2 using a DFF PLA.

Solution: The PLA implementation is shown in Fig. 4.5.4. Notice that in this realization, the SRFF's in the original circuit of Fig. 4.4.2 are replaced by DFF's (see Table 4.2.1 for an explanation).

Some PLA's are commerically available. For example, the TMC 2000 and TMC 2200 are groups of programmable logic arrays on a single MOS/LSI chip. In the TMS 2000, 17 external inputs and the 8 flip-flop outputs are combined by a product-term generator into 60 product terms. These are combined by a sum-of-product-terms generator into 16 lines for the 8 J and 8 K inputs to the 8 J–K master–slave flip-flops and into 18 external outputs. The flip-flop operation is controlled by a common reset input and a single clock. The block diagram of the TMS 2000 is shown in Fig. 4.5.5.

The TMS 2200 is similar to the TMS 2000. In the TMS 2200, 13 external inputs and the 10 flip-flop outputs are combined by a product-term generator into 72 product terms. These are then combined by a sum-of-product-terms generator into 20 lines for the 10 J and 10 K inputs to the 10 J–K master–slave flip-flops and into 10 external outputs. The flip-flop operation is also controlled by a common reset input and a single clock.

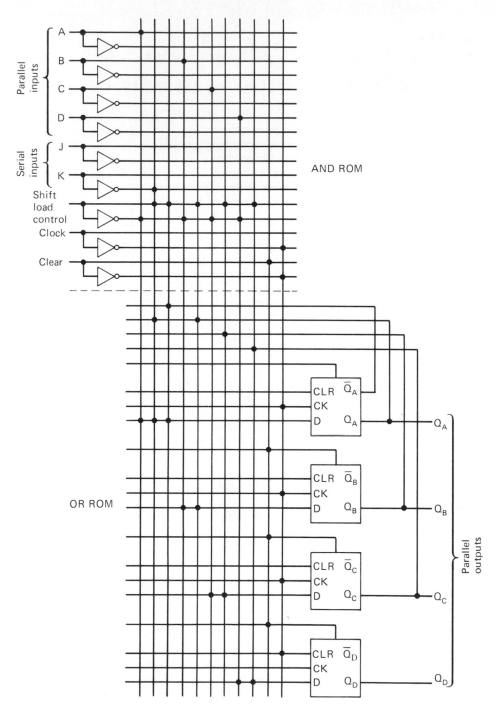

Fig. 4.5.4 PLA implementation of SN74195 4-bit parallel-in, parallel-out shift register.

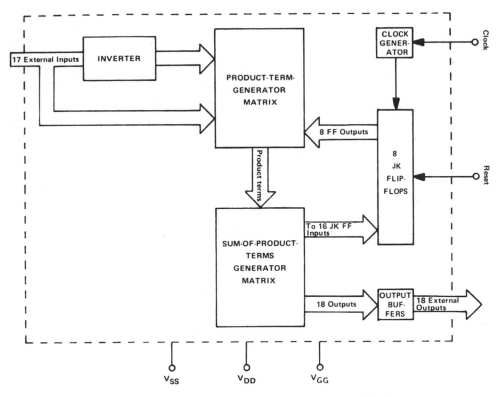

Fig. 4.5.5 Block diagram of programmable logic array TMS 2000.

It should be noted that any commercial PLA, such as the TMS 2000 and 2200 series, can be used as a ROM array (with all the flip-flops of the PLA unconnected). The programming of the PLA is similar to that of the ROM described in Section 3.5.

It should also be noted that in a situation where the type of flip-flop in a PLA does not match that in a circuit design which is to be implemented, one can always first convert the flip-flops in the circuit into flip-flops of the same type as the flip-flops in the PLA. A procedure for realizing a flip-flop of one type by a flip-flop of another type is presented in Section 5.4.

Modern digital-system implementation with MOS/LSI PLA's demands a more sophisticated technique than that offered by the gate-reduction method. To realize this improvement, we must first examine the overall logic of the system and break it down into functional blocks. We can then apply the PLA technique. The PLA represents an economical and efficient way to implement random logic using programmable techniques at the gate-oxide mask step (see Section A.4 of Appendix A). When a random logic circuit is implemented by conventional methods in MOS, a large part of the chip area is used for interconnection of cells. The PLA corrects this waste of silicon real estate.

Using the PLA technique, we can adjust our thinking toward the direction of building a hardware framework for an electronic system with a software program to drive it. This approach may be termed the "hardware framework driven by a software program" approach. Recent one-chip calculator and microprocessors are implemented using this approach. Figure 4.5.6 shows the block diagram of a typical MOS/LSI one-chip calculator in which ROM's, RAM's, and PLA's are used. The abbreviated terms ALU and DSR† in this diagram denote arithmetic logic unit and dynamic shift register, respectively. The symbols ⇨ and → denote parallel and serial interconnections. A photograph of a MOS/LSI silicon one-chip realization of the calculator design in Fig. 4.5.6 was given in Fig. 3.2.1.

Figure 4.5.7 shows a block diagram of the design of a modern digital system. From a given design specification, a system block diagram is first laid out. Logic equations for each part of the system will then be sought. Logic-circuit realizations of these

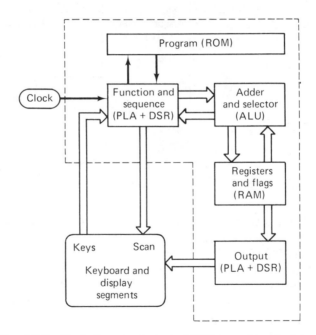

Fig. 4.5.6   Block diagram of a typical MOS one-chip calculator.

†Two circuit families of MOS registers—static and dynamic—are used in digital delay lines. The simplest circuit type is the dynamic register, which may contain as few as six MOS devices per master–slave delay cell. Dynamic registers depend upon charge storage for data retention. If the clock runs too slowly—or stops—in a dynamic register, the information stored is lost. The data-volatility problem, however, is offset by the economy and speed available with the dynamic circuitry. Static registers, in contrast, contain additional devices that form a bistable (static) latch within each register cell or delay element. The static register will not lose data when the clock is stopped. Static cells are generally larger and slower than dynamic cells.

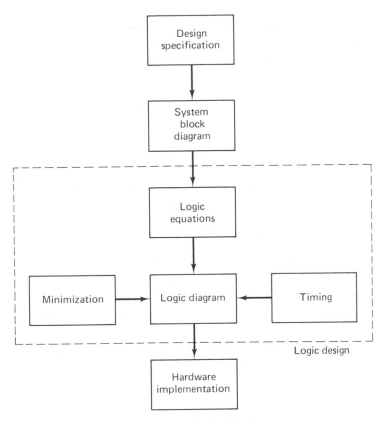

Fig. 4.5.7 Design of a digital system.

equations result in a logic diagram. In arriving at such a logic diagram, various important considerations, such as timing, component availability, cost, minimization, and so on, must be carefully considered. The final step is the hardware implementation of the logic diagram. Chapters 5 and 6 are devoted to the discussion of sequential logic design.

Exercise 4.5

1. (a) Implement the combinational realization of the eight-variable odd-parity checker in Fig. 4.2.10(a). Use a ROM array.
   (b) Implement the sequential realization of the eight-variable odd-parity checker in Fig. 4.2.11(b). Use a PLA.
2. Realize the sequential logic described in Fig. P4.5.2. Use a PLA.

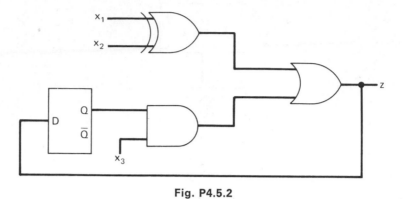

Fig. P4.5.2

3. Realize the five shift registers in Fig. 4.4.1. Use PLA's.

Bibliographical Remarks

References 1–7 of Chapter 3 are excellent references for commercially available flip-flops, counters, and shift registers and their applications. Following is a list of references on LSI/MOS PLA's.

References on LSI/MOS Memories

1. W. N. CARR, and J. P. MIZE, *MOS/LSI Design and Application*, McGraw-Hill, New York, 1972.

2. G. LUECKE, J. P. MIZE, and W. N. CARR, *Semiconductor Memory Design and Application*, McGraw-Hill, New York, 1973.

3. *MOS/LSI Standard Products Catalog*, Texas Instruments, Inc.

4. *McMOS Handbook*, Motorola Semiconductor Products, Inc.

5

Analysis and Synthesis of Sequential Circuits

In this chapter formal basic procedures for analyzing and synthesizing synchronous and asynchronous sequential circuits are presented. The synthesis procedure is useful in designing MSI and LSI sequential integrated circuits such as those discussed in Chapter 4. Moreover, it can be used to design any sequential circuits (that are not commercially available) using SSI integrated circuits.

Recall that in combinational circuit design (Section 1.8), the first step is to describe the given design problem by a logic truth table. In sequential circuit design, which will be introduced in this chapter, the first step is to describe the given design problem by a sequential machine. Two basic models of sequential machines, the Mealy machine and the Moore machine, and their conversions are introduced. The terms *sequential machine*, *finite-state machine*, *finite automaton*, and simply *machine* are synonyms. In essence, the synchronous and asynchronous sequential machines are mathematical models that describe the synchronous and asynchronous sequential circuits, respectively. Since a sequential machine is merely an abstract model, it may be used to describe the operational behavior of systems other than sequential circuits. Indeed, the term "machine" used here does not imply that a sequential machine has to be a real physical machine or a machine-like object. On the contrary, it does not even have to be tangible; any physical or abstract phenomenon may be called a sequential machine as long as it satisfies the axioms of this model.

The fundamental theory about the equivalence and minimization of a sequential machine is studied. Just like a switching function, which can be completely or incompletely specified, so can a sequential machine. Methods for minimizing both completely and incompletely specified sequential machines are presented.

The chapter is divided into two parts. The first part is devoted to the discussion of synchronous sequential machines and circuits, in which general procedures for analyzing and synthesizing synchronous flip-flop circuits are presented. The second part deals with asynchronous sequential machines and circuits. Two undesirable transient phenomena, races and hazards, are discussed in detail. Finally, design of race-free and hazard-free synchronous sequential circuits is given.

5.1 Basic Models of Sequential Machines

The theory of sequential machines is concerned with mathematical models of discrete, deterministic information-processing devices and systems, such as digital computers, digital control units, electronic circuits with synchronized delay elements, and so on. All these devices and systems have the following common properties, which are abstracted in the definition of sequential machine.

DEFINITION 5.1.1(a)

A sequential machine or *Mealy machine* is a system that can be characterized by a quintuple,

$$M = (\textstyle\sum, Q, Z, \delta, \lambda)$$

where $\sum$ = finite nonempty set of input symbols $\sigma_1, \sigma_2, \ldots, \sigma_l$
     Q = finite nonempty set of states, $q_1, q_2, \ldots, q_n$
     Z = finite nonempty set of output symbols, $z_1, z_2, \ldots, z_m$
     δ = next-state function, which maps $Q \times \sum \rightarrow Q$
     λ = output function which maps $Q \times \sum \rightarrow Z$

DEFINITION 5.1.1(b)

A sequential machine is said to be of the *Moore type* (*Moore machine*) if its output function is a function of its states only (i.e., $\lambda : Q \rightarrow Z$).

It will be shown later in this section that every Mealy machine can be converted into a Moore machine and vice versa.

Let $\sigma(t)$, $q(t)$, and $z(t)$ denote the input symbol, state, and output symbol at time t, where $t = 0, 1, 2, \ldots$ The next-state and output functions are expressed as

$$\delta(q(t), \sigma(t)) = q(t + 1) \tag{5.1.1}$$

and

$$\lambda(q(t), \sigma(t)) = z(t) \tag{5.1.2}$$

where $\sigma(t)$, $q(t)$, $q(t + 1)$, and $z(t)$ are the *present input*, *present state*, *next state*, and *present output* of the machine, respectively. An example of a sequential machine is given next.

Example 5.1.1

Consider a milk machine which sells milk that costs 15 cents per pint. The machine will accept nickles, dimes, and quarters, one at a time, and give the correct change. The coin-return control unit M_1, which computes the amount of change to be returned to the customer, can be modeled by a sequential machine as follows:

$$M_1 = (\textstyle\sum_1, Q_1, Z_1, \delta_1, \lambda_1) \tag{5.1.3}$$

where

$$\Sigma_1 = \{5\text{¢}, 10\text{¢}, 25\text{¢}\} \tag{5.1.4}$$

$$Q_1 = \{q_{0\text{¢}}, q_{5\text{¢}}, q_{10\text{¢}}\} \tag{5.1.5}$$

$$Z_1 = \{0\text{¢}, 5\text{¢}, 10\text{¢}, 15\text{¢}, 20\text{¢}\} \tag{5.1.6}$$

$$\left.\begin{array}{l} \delta_1(q_{0\text{¢}}, 5\text{¢}) = q_{5\text{¢}}, \ \delta_1(q_{5\text{¢}}, 5\text{¢}) = q_{10\text{¢}}, \ \delta_1(q_{10\text{¢}}, 5\text{¢}) = q_{0\text{¢}} \\ \delta_1(q_{0\text{¢}}, 10\text{¢}) = q_{10\text{¢}}, \ \delta_1(q_{5\text{¢}}, 10\text{¢}) = q_{0\text{¢}}, \ \delta_1(q_{10\text{¢}}, 10\text{¢}) = q_{0\text{¢}} \\ \delta_1(q_{0\text{¢}}, 25\text{¢}) = q_{0\text{¢}}, \ \delta_1(q_{5\text{¢}}, 25\text{¢}) = q_{0\text{¢}}, \ \delta_1(q_{10\text{¢}}, 25\text{¢}) = q_{0\text{¢}} \end{array}\right\} \tag{5.1.7}$$

and

$$\left.\begin{array}{l} \lambda_1(q_{0\text{¢}}, 5\text{¢}) = 0\text{¢}, \ \lambda_1(q_{5\text{¢}}, 5\text{¢}) = 0\text{¢}, \ \lambda_1(q_{10\text{¢}}, 5\text{¢}) = 0\text{¢} \\ \lambda_1(q_{0\text{¢}}, 10\text{¢}) = 0\text{¢}, \ \lambda_1(q_{5\text{¢}}, 10\text{¢}) = 0\text{¢}, \ \lambda_1(q_{10\text{¢}}, 10\text{¢}) = 5\text{¢} \\ \lambda_1(q_{0\text{¢}}, 25\text{¢}) = 10\text{¢}, \ \lambda_1(q_{5\text{¢}}, 25\text{¢}) = 15\text{¢}, \ \lambda_1(q_{10\text{¢}}, 25\text{¢}) = 20\text{¢} \end{array}\right\} \tag{5.1.8}$$

The terms $q_{0\text{¢}}$, $q_{5\text{¢}}$ and $q_{10\text{¢}}$ denote that the machine has received 0¢, 5¢, and 10¢, respectively. Since this system satisfies the mathematical model of a sequential machine described in Definition 5.1.1(a), it is a Mealy machine.

If, however, in addition to the states $q_{0\text{¢}}, q_{5\text{¢}}, q_{10\text{¢}}$, we add states $q_{15\text{¢}}, q_{20\text{¢}}, q_{25\text{¢}}, q_{30\text{¢}}$, and $q_{35\text{¢}}$ to denote that the machine has received 15¢, 20¢, 25¢, 30¢, and 35¢, respectively; that is,

$$Q_2 = \{q_{0\text{¢}}, q_{5\text{¢}}, q_{10\text{¢}}, q_{15\text{¢}}, q_{20\text{¢}}, q_{25\text{¢}}, q_{30\text{¢}}, q_{35\text{¢}}\} \tag{5.1.9}$$

and define

$$\delta_2(q_{x\text{¢}}, y\text{¢}) = \begin{cases} q_{(x+y)\text{¢}}, & \text{if } x = 0, 5, \text{ or } 10 \\ q_{y\text{¢}} & \text{if } x = 15, 20, 25, 30, \text{ or } 35 \end{cases} \tag{5.1.10}$$

$$\lambda_2(q_{x\text{¢}}) = \begin{cases} 0\text{¢} & \text{if } x = 0, 5, \text{ or } 10 \\ (x-15)\text{¢}, & \text{if } x = 15, 20, 25, 30, \text{ or } 35 \end{cases} \tag{5.1.11}$$

then the machine $M_2 = (\Sigma_2 = \Sigma_1, Q_2, Z_2 = Z_1, \delta_2, \lambda_2)$, satisfying Definition 5.1.1(b), is a Moore machine which is equivalent to machine M_1. Two machines with the same sets of input symbols and output symbols are equivalent if for *any* sequence of input symbols, they produce the identical sequence of output symbols. A detailed discussion of state and machine equivalences is given in the next section.

Three remarks about the models of sequential machine are in order.

1.   From the basic models of sequential machine, the next states and the sequence of output symbols of a machine for any sequence of input symbols $\tilde{x} = \sigma(0), \sigma(1), \ldots, \sigma(K)$ can be computed by

$$\delta(q(0), \sigma(0), \sigma(1), \ldots \sigma(K)) = \vartheta(\ldots \vartheta(\vartheta(\underbrace{\underbrace{\underbrace{q(0), \sigma(0)}_{= q(1)}, \sigma(1)), \ldots, \sigma(K))}_{= q(2)}}_{= q(K+1)} \tag{5.1.12}$$

$$\lambda(q(k), \sigma(k)) = z(k), k = 0, 1, \ldots, K \tag{5.1.13}$$

2. A sequential "machine" does not need to be physical, such as computers or computer-like machines; any discrete-time system, physical or abstract, as long as it can be described by the mathematical model of Definition 5.1.1(a) or 5.1.1(b), is a sequential machine.

3. The theory of sequential machines has a very widespread application, from linguistics to neurology, from computer science to political science, from the design of electronic logic circuits to the design of socioeconomic systems; in fact, it may be applied to practically every scientific field or discipline.

4. Our interest in studying the theory of sequential machines here is in its application to sequential logic design. Recall from Section 1.8 that the first step in combinational logic design is to describe the verbal description of a given design problem by a logic truth table. In this chapter we will see that in sequential logic design, the first step is to describe the verbal description of a given design problem by a sequential machine.

The Mealy machine and the Moore machine can be described by a transition table, shown in Table 5.1.1(a) and (b), respectively. For example, the machines of Exam-

TABLE 5.1.1 Transition-Table Description of Sequential Machine

(a) Transition-table description of the Mealy machine

Present input

σ \ q	σ_1	σ_2	...	σ_l
q_1	$\delta(q_1, \sigma_1), \lambda(q_1, \sigma_1)$	$\delta(q_1, \sigma_2), \lambda(q_1, \sigma_2)$	...	$\delta(q_1, \sigma_l), \lambda(q_1, \sigma_l)$
q_2	$\delta(q_2, \sigma_1), \lambda(q_2, \sigma_1)$	$\delta(q_2, \sigma_1), \lambda(q_2, \sigma_1)$	...	$\delta(q_2, \sigma_1), \lambda(q_2, \sigma_1)$
$\vdots$	$\vdots$	$\vdots$		$\vdots$
q_n	$\delta(q_n, \sigma_1), \lambda(q_n, \sigma_1)$	$\delta(q_n, \sigma_2), \lambda(q_n, \sigma_2)$	...	$\delta(q_n, \sigma_l), \lambda(q_n, \sigma_l)$

Present state

Next state, Present output

(b) Transition-table description of the Moore machine

Present input

σ \ q	σ_1	σ_2	...	σ_l	z
q_1	$\delta(q_1, \sigma_1)$	$\delta(q_1, \sigma_2)$	...	$\delta(q_1, \sigma_l)$	$\lambda(q_1)$
q_2	$\delta(q_2, \sigma_1)$	$\delta(q_2, \sigma_2)$	...	$\delta(q_2, \sigma_l)$	$\lambda(q_2)$
$\vdots$	$\vdots$	$\vdots$		$\vdots$	$\vdots$
q_n	$\delta(q_n, \sigma_1)$	$\delta(q_n, \sigma_2)$	...	$\delta(q_n, \sigma_l)$	$\lambda(q_n)$

Present state

Next state   Present output

ple 5.1.1 can be conveniently described by the transition tables of Table 5.1.2(a) and (b), and a 4-bit right-shift register and a modulo-16 (4-bit binary) counter can be described by a Mealy and Moore machine, respectively, which are shown in Table 5.1.2(c) and (d).

TABLE 5.1.2   Examples of Transition-Table Description
of Sequential Machine

(a)   Mealy machine M_1

q \ σ	5¢	10¢	25¢
$q_{0¢}$	$q_{5¢}$, 0¢	$q_{10¢}$, 0¢	$q_{0¢}$, 10¢
$q_{5¢}$	$q_{10¢}$, 0¢	$q_{0¢}$, 0¢	$q_{0¢}$, 15¢
$q_{10¢}$	$q_{0¢}$, 0¢	$q_{0¢}$, 5¢	$q_{0¢}$, 20¢

(b)   Moore machine M_2

q \ σ	5¢	10¢	25¢	z
$q_{0¢}$	$q_{5¢}$	$q_{10¢}$	$q_{25¢}$	0¢
$q_{5¢}$	$q_{10¢}$	$q_{15¢}$	$q_{30¢}$	0¢
$q_{10¢}$	$q_{15¢}$	$q_{20¢}$	$q_{35¢}$	0¢
$q_{15¢}$	$q_{5¢}$	$q_{10¢}$	$q_{25¢}$	0¢
$q_{20¢}$	$q_{5¢}$	$q_{10¢}$	$q_{25¢}$	5¢
$q_{25¢}$	$q_{5¢}$	$q_{10¢}$	$q_{25¢}$	10¢
$q_{30¢}$	$q_{5¢}$	$q_{10¢}$	$q_{25¢}$	15¢
$q_{35¢}$	$q_{5¢}$	$q_{10¢}$	$q_{25¢}$	20¢

(c)   Four-bit left-shift register (Mealy machine)

$y_1y_2y_3y_4$ \ x	0	1
0000	0000,0000	0001,0001
0001	0010,0010	0011,0011
0010	0100,0100	0101,0101
0011	0110,0110	0111,0111
0100	1000,1000	1001,1001
0101	1010,1010	1011,1011
0110	1100,1100	1101,1101
0111	1110,1110	1111,1111
1000	0000,0000	0001,0001
1001	0010,0010	0011,0011
1010	0100,0100	0101,0101
1011	0110,0110	0111,0111
1100	1000,1000	1001,1001
1101	1010,1010	1011,1011
1110	1100,1100	1101,1101
1111	1110,1110	1111,1111

TABLE 5.1.2   continued

(d)   Modulo-16 counter
(Moore machine)

$y_3 y_2 y_1 y_0$ \\ x	1	z
0000	0001	0
0001	0010	1
0010	0011	2
0011	0100	3
0100	0101	4
0101	0110	5
0110	0111	6
0111	1000	7
1000	1001	8
1001	1010	9
1010	1011	10
1011	1100	11
1100	1101	12
1101	1110	13
1110	1111	14
1111	0000	15

An alternative way of describing a sequential machine is to use the transition diagram. For example, the transition diagrams of machines M_1 and M_2 are given in Fig. 5.1.1(a) and (b), respectively.

It is quite easy to show that a Moore machine can always be converted into a Mealy (equivalent) machine. This can be done by omitting the rightmost column of the transition table of Table 5.1.1(b) and changing each next state $\delta(q_i, \sigma_j)$, $i = 1, 2, \ldots, n$ and $j = 1, 2, \ldots, l$, of the table by the next state and present output pair,

$$(\delta(q_i, \sigma_j), \lambda(\delta(q_i, \sigma_j)))$$

The resulting machine is a Mealy (equivalent) machine.

To show that a Mealy machine can always be converted into a Moore machine, we need the following definition.

DEFINITION 5.1.2

A state of a Mealy machine is *z-homogeneous* if the output associated with each transition line† incident to that state is the same; otherwise, it is *z-nonhomogeneous*. A Mealy machine is *z-homogeneous* if every state of it is *z*-homogeneous; otherwise, it is *z-nonhomogeneous*.

†A *transition line* is a line going from one state to another state in a transition diagram of a sequential machine.

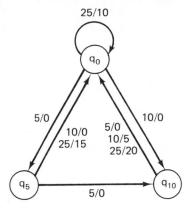

Note: The conventional input/output symbol is used. For example,

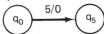

means that upon application of 5 cents (a nickel), the machine will go q_0 to q_5 with an output of 0 cents.

(a) Mealy machine M_1

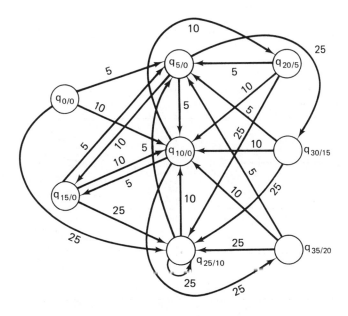

Note: $\left(\,q/z\,\right)$ means that $\lambda(q) = z$

(b) Moore machine M_2

Fig. 5.1.1 Transition diagrams of machines M_1 and M_2.

It is clear that a z-homogeneous Mealy machine is a Moore machine. Thus the question is how to convert a z-nonhomogeneous machine into a z-homogeneous one. To convert a z-nonhomogeneous state into a z-homogeneous state, four possible cases of transitions must be considered. They are:

1. Transition from a z-homogeneous state to a z-homogeneous state.
2. Transition from a z-homogeneous state to a z-nonhomogeneous state.
3. Transition from a z-nonhomogeneous state to a z-homogeneous state.
4. Transition from a z-nonhomogeneous state to a z-nonhomogeneous state.

The four basic types of conversions that will convert any z-nonhomogeneous Mealy machine into a z-homogeneous (Moore) machine are shown in Table 5.1.3. For example, by using these conversions M_1 can be converted into a z-homogeneous machine M_3, which is shown in Fig. 5.1.2(a). This z-homogeneous machine is now readily to be drawn in the type of Moore machine that is depicted in Fig. 5.1.2(b), in which states

TABLE 5.1.3 Transitions for Converting a z-nonhomogeneous Machine into a z-homogeneous Machine.

Case	z-nonhomogeneous Mealy machine	Equivalent z-homogeneous (Moore) machine
1	$H \xrightarrow[z]{x} H$	$H \xrightarrow[z]{x} H$
2	$H \xrightarrow[z]{x} N$	$H \xrightarrow[z]{x} N_z$
3	$N \xrightarrow[z]{x} H$	$N_{z_1}, N_{z_2}, \dots, N_{z_i}$ each $\xrightarrow[z]{x} H$
4	$N \xrightarrow[z]{x} N$	$N_{z_1}, N_{z_2}, \dots, N_{z_i}$ each $\xrightarrow[z]{x} N_z$

H: a z-homogeneous state
N: a z-nonhomogeneous state
$N_{z_1}, N_{z_2}, \dots, N_{z_i}$: equivalent z-homogeneous states of N which receive output $z_1, z_2, \dots z_i$, respectively

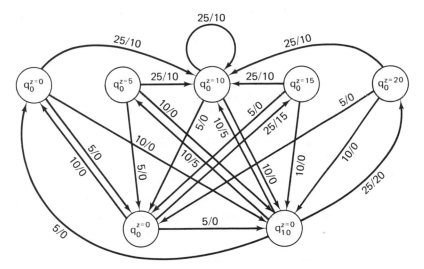

(a) z-homogeneous machine equivalent

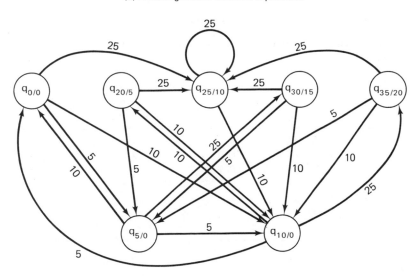

(b) Moore machine equivalent

Fig. 5.1.2 Moore machine equivalent M_3 of Mealy machine M_1.

q_0, q_{20}, q_{25}, q_{30}, and q_{35} represent states $q_0^{z=0}$, $q_0^{z=5}$, $q_0^{z=10}$, $q_0^{z=15}$, and $q_0^{z=20}$, respectively. The preceding discussion is summarized in the following theorem.

THEOREM 5.1.1

Every Mealy machine can be converted into a Moore equivalent machine, and vice versa.

A formal proof of this theorem will be given in the next section, when machine equivalence is discussed.

Exercise 5.1

1. (a) Give an example of a Moore machine.
    (b) Convert the sequential machine of (a) into a Mealy equivalent machine.
    (c) Give an example of a Mealy machine.
    (d) Convert the sequential machine of (c) into a Moore equivalent machine using the conversion table of Table 5.1.3.

2. Show that a combination lock can be considered as a sequential machine.

3. The input/output strings of machine M are made up of the symbols 0, 1, and π. M reverses any string of 0's and 1's that lie between two π's. For example,

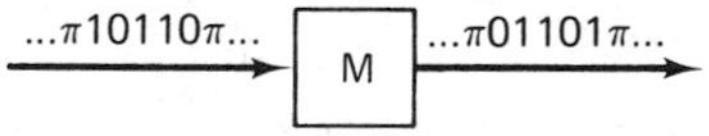

$$\xrightarrow{\ldots\pi10110\pi\ldots} \boxed{M} \xrightarrow{\ldots\pi01101\pi\ldots}$$

Fig. P5.1.3

Could M be a sequential machine? Justify your answer.

4. Determine which of the following Mealy machines are readily convertible to Moore machines.

q \ σ	a	b	c
A	B, 0	C, 1	A, 1
B	A, 1	B, 0	C, 0
C	C, 1	A, 0	B, 1

(a)

q \ σ	0	1
1	2, 1	3, 0
2	4, 1	2, 1
3	1, 0	3, 0
4	2, 1	4, 1

(b)

q \ σ	a	b	c
A	B, α	C, β	D, γ
B	A, α	A, β	A, γ
C	B, α	B, β	D, γ
D	B, α	C, β	C, γ
E	E, α	B, β	E, γ

(c)

q \ σ	a	b	c
1	5, γ	1, β	2, γ
2	6, β	3, α	4, α
3	3, α	2, γ	1, β
4	2, γ	4, α	2, γ
5	4, α	6, α	1, γ
6	1, β	3, α	6, β

(d)

Fig. P5.1.4

5. Fill up with proper states and outputs in the blanks of the transition table of Fig. P5.1.5, where $\Sigma = \{a, b, c, d\}$, $Q = \{A, B, C, D, E\}$, and $Z = \{\alpha, \beta, \gamma, \delta\}$ so that the machine is a Moore machine.

q \ σ	a	b	c	d
A	B, —	—, α	C, γ	—, α
B	—, δ	A, β	A, —	—, α
C	B, α	—, α	E, δ	—, α
D	D, —	E, —	—, —	D, α
E	—, γ	C, γ	—, —	A, β

Fig. P5.1.5

6. Show that the binary serial adder presented in Example 4.2.1 is a two-state sequential machine. (*Hint:* The two states of this machine are the no-carry and carry states.)

7. Show that the binary serial subtractor is a two-state sequential machine. (*Hint:* The two states of this machine are the no-borrow and borrow states.)

5.2 Equivalence and Minimization

In this section two important problems of sequential machine, equivalence and minimization, will be discussed. We begin with the discussion of the equivalence problem. Two equivalence relations, state equivalence and machine equivalence, will be considered. They are defined as follows:

DEFINITION 5.2.1

Let q_a and q_b be two states of machines M_a and M_b (M_a and M_b may be the same machine). States q_a and q_b are said to be *equivalent* if starting with q_a and q_b, for *any* sequence of input symbols applied to the two machines, the output sequences are identical. If q_a and q_b are not equivalent, we say they are *distinguishable*.

DEFINITION 5.2.2

Let M_a and M_b be two sequential machines. M_a and M_b are said to be *equivalent* if for every state of M_a there exists at least one equivalent state in M_b, and vice versa. Similarly, if M_a and M_b are not equivalent, we say they are *distinguishable*.

We shall first be concerned with state equivalence. In the above definition of state equivalence, instead of considering two machines, we could view these two machines as two submachines of a machine. In doing so, the discussion of state equivalence may be simplified considerably. This can be done as follows: Given two machines $M_a =$

$(\Sigma_a, Q_a, Z_a, \delta_a, \lambda_a)$ and $M_b = (\Sigma_b, Q_b, Z_b, \delta_b, \lambda_b)$, where $\Sigma_a = \Sigma_b = \Sigma, Z_a = Z_b = Z$ (i.e., M_a and M_b have the same set Σ of input symbols and the same set Z of output symbols), construct a machine $M = (\Sigma, Q, Z, \delta, \lambda)$ with

$$Q = Q_a \cup Q_b$$

$$\delta(q_i, \sigma_j) = \begin{cases} \delta_a(q_i, \sigma_j) & \text{for } q_i \in Q_a \\ \delta_b(q_i, \sigma_j) & \text{for } q_i \in Q_b \end{cases}$$

$$\lambda(q_i, \sigma_j) = \begin{cases} \lambda_a(q_i, \sigma_j) & \text{for } q_i \in Q_a \\ \lambda_b(q_i, \sigma_j) & \text{for } q_i \in Q_b \end{cases}$$

From now on, instead of considering the equivalence of two states in two machines M_a and M_b, we can simply consider the two states in the machine M defined above.

From the definition of state equivalence given in Definition 5.2.1, we find that we do not know how to design an experiment for determining the equivalence of two states of a sequential machine, because from Definition 5.2.1, it seems that we need to run an experiment with all possible sequences of input symbols (including sequences of infinite length), which is impossible. Moreover, the machine equivalence (Definition 5.2.2) is defined based on the state equivalence; therefore, we do not know how to design an experiment to determine whether two given machines are equivalent or not. In short, the definitions of state and machine equivalences given in Definitions 5.2.1 and 5.2.2 do not tell us how to determine the state equivalence of two states of a machine and the machine equivalence of two given machines, nor provide us any information as to whether such equivalences can always be determined.

In the following, several theorems will be developed. Based on them we can conclude that the equivalence of two states of a machine can *always* be determined by an experiment of *finite* length of input symbols. In fact, for an n-state sequential machine, the length of input sequences of an experiment needed for determining the state equivalence of any two states of the machine is *at most* $(n-1)$. Methods for determining the equivalence classes of states of a completely specified and incompletely specified machine,† thereby minimizing the machine, are presented below. The minimization of completely specified machines is presented first.

5.2.1 Minimization of Completely Specified Machines

The first method to be presented is the *k*-partition method. First we define

DEFINITION 5.2.3

Two states q_a and q_b in a machine M are said to be *k-equivalent* if q_a and q_b, when excited by an input sequence of k symbols, yield identical output sequences.

†A machine whose next-state function and/or output function are not completely specified is called an *incompletely specified machine*.

DEFINITION 5.2.4

The partition obtained by the k-equivalence relation is called the *k-partition*.

DEFINITION 5.2.5

The equivalence classes partitioned by the k-partition is called *k-equivalence classes*.

LEMMA 5.2.1

For any nontrivial machine, the 1-partition of the set Q of states has at least two nonempty 1-equivalence classes.

Proof: The proof is evident from the definition of nontrivial machine. ∎

LEMMA 5.2.2

For any n-state machine, the equivalence classes in Q obtained from 1-, 2-, . . . , $(n - 1)$-successive distinct partitions contain one and only one state.

Proof: From Lemma 5.2.1, a 1-partition of Q has at least two equivalence classes. If the 1-partition is not equal to the 2-partition, then the 2-partition must have at least three equivalence classes. By induction, the $(n - 1)$-partition must have n equivalence classes. Since the machine is assumed to have only n states, in each equivalence class of the $(n - 1)$-partition it can have at most one state; thus the n-partition must be equal to the $(n - 1)$-partition. Hence the lemma. ∎

From the lemma, we clearly see the following:

THEOREM 5.2.1

For any n-state machine, we can have at most $(n - 1)$ successive distinct partitions.

THEOREM 5.2.2

If the k-partition of the set Q of states is identical to the $(k + 1)$-partition of Q, then the states of the same equivalence class are equivalent.

Proof: Let q_a and q_b be two states in a k-equivalence class Q_1. Since the $(k + 1)$-partition is equal to the k-partition, Q_1 is also a $(k + 1)$-equivalence class. Now we apply one additional input symbol to q_a and q_b and assume that q_a and q_b go to two different $(k + 2)$-equivalence classes. If this is so, we could assume that the additional input symbol is applied to q_a and q_b of the k-equivalence class Q_1, since the k- and $(k + 1)$-equivalence classes are the same. Then q_a and q_b must be in two different $(k + 1)$-equivalence classes. This is a contradiction. ∎

COROLLARY 5.2.1

If two states q_a and q_b of an n-state machine M are $(n - 1)$-equivalent, then they are equivalent.

Proof: If q_a and q_b are $(n-1)$-equivalent, then, by definition of $(n-1)$-equivalence, they must be in the same $(n-1)$-equivalence class. This implies that q_a and q_b must be in the same $(n-2)$-equivalence class, since, by Lemma 5.2.2, the $(n-1)$-equivalence classes can have only one state if the 1-, 2-, . . . , $(n-1)$-partitions are all distinct. From our hypothesis we find that this is not the case. In other words, the partition of Q must terminate at the $(n-2)$-partition or before that, which means that the partition of Q by the $(n-1)$-partition must be identical to that resulting from the $(n-2)$-partition. By Theorem 5.2.2, q_a and q_b are equivalent. ∎

COROLLARY 5.2.2

For an *n*-state machine, the equivalence of any two states of the machine can *always* be determined by an experiment of input sequences of length $(n-1)$ input symbols.

ALGORITHM 5.2.1

All the classes of equivalent states of a completely specified *n*-state machine and its minimal form can be obtained by the following procedure:

1.   Find 1-equivalence classes, 2-equivalence classes, and so on, until the $(k+1)$-equivalence classes are identical to the *k*-equivalence classes, and then stop. The $(k+1)$-equivalence classes are equivalence classes of the set of states. There are at most $n-1$ distinct successive partitions of the states of the machine.
2.   Combine all the states in the same equivalence class into one state. If the given machine has *m* equivalence classes, then the minimized machine would have *m* states.

Let us illustrate this algorithm by the following example.

Example 5.2.1

Consider the Mealy machine M_4 in Table 5.2.1(a), which is obtained from M_2 by the method described in the previous section. For convenience, the symbols ¢'s have been omitted in this table.

1.   Find the 1-partition and 2-partition shown in Table 5.2.1(a) and (b), respectively. The second subscript (I, II, or III) of each next state in Table 5.2.1(b) denotes the 1-equivalence class to which the next state belongs. The 2-equivalence classes are formed from 1-equivalence classes by examining the second subscripts of the next states of each row. Two rows in the same 1-equivalence class having identical second subscripts indicate that the two 1-equivalent states are in fact 2-equivalent. All the 2-equivalence classes can be found by this way, and the *k*-equivalence classes for $k \geq 3$ can be obtained similarly. For this example, it is found that the 2-partition is identical to the 1-partition; hence the process stops. We find that there are three equivalence classes, denoted by E_1, E_2, and E_3: $E_1 = \{q_0, q_{15}, q_{20}, q_{25}, q_{30}, q_{35}\}$, $E_2 = \{q_5\}$, and $E_3 = \{q_{10}\}$.
2.   Combining the six states of E_1 into one state, denoted by q_0, the resultant machine is equivalent to M_1. This means that (1) M_4 is equivalent to M_1 and (2) M_1 is a minimized machine.

TABLE 5.2.1   Mealy machine M_4 of Example 5.2.1

(a)

σ \ q	5	10	25	1-partition
q_0	$q_5, 0$	$q_{10}, 0$	$q_{25}, 10$	I
q_5	$q_{10}, 0$	$q_{15}, 0$	$q_{30}, 15$	II
q_{10}	$q_{15}, 0$	$q_{20}, 5$	$q_{35}, 20$	III
q_{15}	$q_5, 0$	$q_{10}, 0$	$q_{25}, 10$	I
q_{20}	$q_5, 0$	$q_{10}, 0$	$q_{25}, 10$	I
q_{25}	$q_5, 0$	$q_{10}, 0$	$q_{25}, 10$	I
q_{30}	$q_5, 0$	$q_{10}, 0$	$q_{25}, 10$	I
q_{35}	$q_5, 0$	$q_{10}, 0$	$q_{25}, 10$	I

1-equivalence classes:
  Class I: $q_0, q_{15}, q_{20}, q_{25}, q_{30}, q_{35}$
  Class II: q_5
  Class III: q_{10}

(b)

1-Partition	σ \ q	5	10	25	2-Partition
I	q_0	$q_{5\text{II}}$	$q_{10\text{III}}$	$q_{25\text{I}}$	I
	q_{15}	$q_{5\text{II}}$	$q_{10\text{III}}$	$q_{25\text{I}}$	I
	q_{20}	$q_{5\text{II}}$	$q_{10\text{III}}$	$q_{25\text{I}}$	I
	q_{25}	$q_{5\text{II}}$	$q_{10\text{III}}$	$q_{25\text{I}}$	I
	q_{30}	$q_{5\text{II}}$	$q_{10\text{III}}$	$q_{25\text{I}}$	I
	q_{35}	$q_{5\text{II}}$	$q_{10\text{III}}$	$q_{25\text{I}}$	I
II	q_5	$q_{10\text{III}}$	$q_{15\text{I}}$	$q_{30\text{I}}$	II
III	q_{10}	$q_{15\text{I}}$	$q_{20\text{I}}$	$q_{35\text{I}}$	III

The 2-equivalence classes are identical to the 1-equivalence classes.

Example 5.2.2

As an example to show the minimization of a Moore machine, consider M_2. It is clear that states with different output are distinguishable states and therefore in different equivalence classes. Thus the states $q_{20¢}, q_{25¢}, q_{30¢},$ and $q_{35¢}$ are four distinct states which cannot be combined. The remaining four states, $q_{0¢}, q_{5¢}, q_{10¢},$ and $q_{15¢}$, which produce the same output, are therefore the candidates for minimization. Since the two rows of $q_{0¢}, q_{15¢}$ are identical, $q_{0¢}$ and $q_{15¢}$ are equivalent and thus can be combined. States $q_{5¢}$ and $q_{10¢}$, on the other hand, are distinct states. The minimized machine of M_2 should have seven states. Note that this minimized machine is the same as M_3 obtained from M_1 by using the transformation of Table 5.1.3 (see Fig. 5.1.2).

From Examples 5.2.1 and 5.2.2, it is seen that although a machine can be described either as Mealy type or Moore type, they require a different minimum number of states. For example, M_1 and the minimized M_2 both describing the same machine, but M_1 needs only three states to describe it, whereas the reduced M_2 requires a minimum of seven.

Proof of Theorem 5.1.1: The conversion from Moore machine to Mealy machine described in the previous section is evident. To show that every Moore machine obtained from a Mealy machine by using the conversions described in Table 5.1.3 is equivalent to the given Mealy machine, it is necessary to show that (1) all the new z-homogeneous states split from a z-nonhomogeneous state of the given machine are equivalent to each other, and (2) they are equivalent to the original z-nonhomogeneous state. Let $N_{z_1}, N_{z_2}, \ldots, N_{z_l}$ be l z-homogeneous states split from a z-nonhomogeneous state N. To prove that $N_{z_1}, N_{z_2}, \ldots, N_{z_l}$ are equivalent is to prove that they yield identical output sequences for any input sequence. This is readily seen from the way that the states $N_{z_1}, N_{z_2}, \ldots, N_{z_l}$ are constructed. Since $z_1, z_2, \ldots, z_l$ are the output symbols produced by the transitions incident to the state N, the transformations of Table 5.1.3 guarantee that the splitting of the state N into states $N_{z_1}, N_{z_2}, \ldots, N_{z_l}$ does not affect the other parts of the machine. The equivalence between state N and the states $N_{z_1}, N_{z_2}, \ldots, N_{z_l}$ is again immediately seen from the transformations of Table 5.1.3, because when the same input symbol is applied to them, the same output symbol will be observed. ∎

5.2.2 Minimization of Incompletely Specified Machines

In many practical design problems, the design specifications require only that a part of the transition table be specified; the rest is left blank or unspecified. Moreover, even for a given completely specified machine, the first step in realizing it using digital components (logic gates, flip-flops, etc.) is to code the states in binary codes and also the input and output symbols, if they are not binary. This generally results in an incompletely specified machine. Let us consider a logic design problem described by an incompletely specified machine M_5 whose transition table is given in Table 5.2.2(a), in which the d's stand for "unspecified" or "don't cares." If this machine is to be realized by a sequential circuit, its states must first be coded in binary codes. It is obvious that we need at least three state variables y_1, y_2, and y_3 to code the six states. Suppose that we arbitrarily choose an assignment for the states as follows:

State	$y_1 y_2 y_3$
1	000
2	001
3	010
4	011
5	100
6	101

The transition table in coded form resulting from this state assignment is shown in Table 5.2.2(b). Notice that there are six newly introduced unspecified next states and outputs. This indicates that even if M_5 were completely specified, after state assign-

TABLE 5.2.2 M_5

(a) Transition table

σ	0	1	z
q			
q_1	q_2	q_4	0
q_2	q_2	q_4	d_4
q_3	q_1	q_5	1
q_4	d_1	q_5	1
q_5	q_6	d_3	1
q_6	d_2	q_3	d_5

(b) Transition table in binary-coded form

σ	0	0	z
$y_1y_2y_3$			
000	001	011	0
001	001	011	d_4
010	000	100	1
011	d_1	100	1
100	101	d_3	1
101	d_2	010	d_5
110	d_6	d_7	d_{10}
111	d_8	d_9	d_{11}

newly introduced
unspecified entries

(c) Compatibility checking table

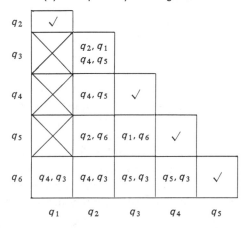

ment, the machine would still be an incompletely specified one. The problem of minimization of an incompletely specified machine is to maximize the utilization of the degrees of freedom offered by the unspecified next states and outputs (don't cares) in the transition table such that the minimal form of the machine would have the least number of states. To achieve this we can still use the k-partition method described in Section 5.2.1, but it will not be convenient. A more direct, less laborious method, known as the *compatibility method*, is described below. This method is applicable to the minimization of *both* completely specified and incompletely specified machines.

DEFINITION 5.2.6

An input sequence $\tilde{x}$ is an *applicable input sequence to a machine M in state q_i* if it will always lead to a specified next state after each input symbol of the sequence is applied. For example, the input sequences 011 and 0110100 are applicable sequences to M_5 in states q_3 and q_1, respectively, because

Input sequence: 0 1 1   0 1 1 0 1 0 0
State sequence: $q_3 q_1 q_4 q_5$ $q_1 q_2 q_4 q_5 q_6 q_3 q_1 q_2$
Output sequence: 1 0 1 1   0 d 1 1 d 1 0 d

But the sequence 101101 is not an applicable input sequence to M_5 in state q_2, because

Input sequence: 1 0 1 1 0 1
State sequence: $q_2 q_4 d$
Output sequence: d 1

DEFINITION 5.2.7

Two output sequences O_i and O_j are said to be *compatible* if and only if the corresponding symbols of the two sequences, whenever they are both specified, are identical. For example, let

$$O_1 = 1 \ 0 \ 1 \ 0 \ d \ 1 \ 1 \ 0 \ d \ 0$$
$$O_2 = 1 \ 0 \ d \ 0 \ d \ 1 \ 1 \ d \ d \ 0$$
$$O_3 = 1 \ d \ 0 \ d \ 0 \ 1 \ 1 \ 1 \ 1 \ 0$$

Then O_1 and O_2 are compatible, and O_2 and O_3 are also compatible, but O_1 and O_3 are not.

DEFINITION 5.2.8

Two states q_i and q_j of a machine M are said to be *compatible* if and only if, for *any* input sequence (applicable to both q_i and q_j) applied to M in q_i and q_j, the resulting two output sequences O_i and O_j are compatible. Two states q_i and q_j are said to be *incompatible* if and only if the two output sequences O_i and O_j are not compatible. Two states q_i and q_j are said to be *simply compatible* if (1) the outputs in any column, if both specified, are identical, and (2) the next states in any column, if both specified, are either identical or are q_i, q_j, or q_j, q_i. Two states q_i and q_j are said to be *simply incompatible* if the output symbols in at least one column of the rows of q_i and q_j of the transition table are distinct.

DEFINITION 5.2.9

A set of n states of a machine M is said to be *mutually compatible* if and only if every pair of states of the set is compatible. A set of n mutually compatible states is called a set of *maximal compatible* states if and only if it does not form a proper subset of another set of mutually compatible states.

ALGORITHM 5.2.2

The sets of maximal compatible states of an incompletely specified n-state machine M and the minimal form M_m of the machine can be obtained by the following procedure:

1.   Construct a compatibility checking table in a ladder shape, like the one in Table 5.2.2(c), and label the rows of the table $q_2, q_3, \ldots, q_n$ and the columns $q_1, q_2, \ldots q_{n-1}$. Examine each pair of states q_i and q_j and place

(a) a $\times$ in the (q_i, q_j) cell if q_i and q_j are simply incompatible.

(b) a $\sqrt{}$ in the (q_i, q_j) cell if q_i and q_j are simply compatible.

(c) the pairs of different next states and the pair of output symbols when they involve don't cares, in the (q_i, q_j) cell.

2. From the pairs of compatible states, form sets of $3, 4, \ldots$ mutually compatible states successively, by exhaustive search. Once all the sets of m mutually compatible states are formed, then the sets of $(m - 1)$ mutually compatible states, which are proper subsets of the m-state set, are deleted. When no more larger sets can be formed, the process is completed, and the undeleted sets are the sets of maximal compatible states.

3. Let the sets of maximal compatible states be $S_1, S_2, \ldots, S_l$ and the sets of states corresponding to the states of the minimal form of machine M be $S'_1, S'_2, \ldots, S'_l$. There are three requirements that have to be satisfied by $S'_1, S'_2, \ldots, S'_l$.

(a) *Minimality* l' must be minimal.

(b) *Completeness* The union of $S'_1, S'_2, \ldots$, and $S'_{l'}$ must contain all the states of M.

(c) *Closure* If we choose l' sets $S'_1, S'_2, \ldots, S'_{l'}$ and construct a machine M' with l' states $q'_1, q'_2, \ldots, q'_{l'}$ corresponding to $S'_1, S'_2, \ldots, S'_{l'}$, respectively, then the next states of $q'_1, q'_2, \ldots, q'_l$ must correspond to a q'_i or a set of q'_i, $i = 1, 2, \ldots, l'$. This requirement is the most stringent one, because sometimes while q' states corresponding to q' sets of maximal compatible states do not have the closure property, by changing some of the sets of maximal compatible states into sets of nonmaximal compatible states we can meet this requirement. Unfortunately, there does not exist any systematic method for doing this, and it is usually done by trial and error.

Let us illustrate this algorithm by an example.

Example 5.2.3

Consider the minimization of M_5. The compatibility checking table of this incompletely specified machine is given in Table 5.2.2(c). The successive formation of larger sets of compatible states for machine M_5 is as follows:

Two-state sets or pairs: ~~q_1, q_2~~ ~~q_1, q_6~~ ~~q_2, q_3~~ ~~q_2, q_4~~ ~~q_2, q_5~~ ~~q_2, q_6~~ ~~q_3, q_4~~ ~~q_3, q_5~~ ~~q_3, q_6~~ ~~q_4, q_5~~ ~~q_4, q_6~~ ~~q_5, q_6~~

Three-state sets: q_1, q_2, q_6 ~~q_2, q_3, q_4~~ ~~q_2, q_3, q_5~~ ~~q_2, q_4, q_5~~ ~~q_2, q_5, q_6~~ ~~q_2, q_4, q_6~~ ~~q_2, q_5, q_6~~ ~~q_3, q_4, q_5~~ ~~q_3, q_4, q_6~~ ~~q_3, q_5, q_6~~ ~~q_4, q_5, q_6~~

Four-state sets: ~~q_2, q_4, q_5, q_6~~ ~~q_3, q_4, q_5, q_6~~ ~~q_2, q_3, q_5, q_6~~ ~~q_2, q_3, q_4, q_5~~ ~~q_2, q_3, q_4, q_6~~

Five-state set: q_2, q_3, q_4, q_5, q_6

Thus the maximal compatible sets are

$$S_1 = \{q_1, q_2, q_6\}$$

$$S_2 = \{q_2, q_3, q_4, q_5, q_6\}$$

Let $S'_1 = S_1 = \{q_1, q_2, q_6\}$ and $S'_2 = \{q_3, q_4, q_5\}$. Since S'_2 is a subset of S_2, it is a set of compatible states. We find that this set, $\{S'_1, S'_2\}$, satisfies the requirements of minimality and completeness. Let q'_1 and q'_2 be the states corresponding to S'_1 and S'_2, respectively. It happens that this set of sets also meets the requirement of closure. The minimal form of M_5 is shown in Table 5.2.3.

TABLE 5.2.3 Minimal Form of M_5

q' \ σ	0	1	z
q'_1	q'_1	q'_2	0
q'_2	q'_1	q'_2	1

Two final important remarks are in order.

(1) The state equivalence and machine minimization of an incompletely specified machine only holds for all *applicable* input sequences to the machine, not all the input sequences.

(2) For a completely specified machine, the sets of maximal compatible states are sets of equivalent states of the machine; thus the compatibility method described in Algorithm 5.2.2 is a very efficient method for completely specified machine minimization.

Exercise 5.2

1. (a) Which of the following completely specified machines are minimal?
    (b) Minimize those which are not minimal.
    (c) Which of them are equivalent machines?

TABLE P5.2.1

(a) M_a

q \ σ	0	1
q_1	$q_1, 0$	$q_2, 1$
q_2	$q_2, 1$	$q_4, 0$
q_3	$q_3, 0$	$q_2, 0$
q_4	$q_2, 1$	$q_3, 1$

(b) M_b

q \ σ	0	1
q_1	$q_2, 0$	$q_5, 1$
q_2	$q_3, 1$	$q_4, 0$
q_3	$q_1, 1$	$q_4, 1$
q_4	$q_4, 1$	$q_1, 1$
q_5	$q_5, 0$	$q_3, 0$

(c) M_c

q \ σ	0	1
q_1	$q_3, 1$	$q_4, 1$
q_2	$q_2, 0$	$q_3, 1$
q_3	$q_3, 1$	$q_1, 0$
q_4	$q_4, 0$	$q_3, 0$
q_5	$q_5, 0$	$q_3, 0$
q_6	$q_6, 0$	$q_3, 1$

(d) M_d

q \ σ	0	1
q_1	$q_1, 0$	$q_7, 1$
q_2	$q_2, 0$	$q_4, 0$
q_3	$q_4, 1$	$q_5, 0$
q_4	$q_7, 1$	$q_5, 1$
q_5	$q_5, 0$	$q_7, 1$
q_6	$q_6, 0$	$q_4, 0$
q_7	$q_3, 0$	$q_6, 1$

2. Minimize the following incompletely specified machines.

TABLE P5.2.2

(a) M_e (b) M_f

σ \ q	0	1	z
q_1	q_3	d	1
q_2	d	q_3	d
q_3	d	q_2	d
q_4	q_4	d	0
q_5	q_5	q_4	d

σ \ q	0	1
q_1	$d, 0$	$q_2, 0$
q_2	$q_1, 1$	d, d
q_3	q_6, d	$q_1, 1$
q_4	$d, 1$	$q_5, 1$
q_5	$q_3, 0$	q_4, d
q_6	q_4, d	$q_3, 1$

3. Show that the sequential machines described by the transition tables in Table P5.2.3(a) and (b) can be minimized to a one-state machine (trivial machine) and a two-state machine, respectively.

TABLE P5.2.3

σ \ q	0	1
A	B, d_4	$D, 1$
B	B, d_5	$D, 1$
C	$A, 0$	$E, 1$
D	d_1, d_6	$E, 1$
E	F, d_7	d_3, d_9
F	d_2, d_8	$C, 1$

(a)

σ \ q	0	1
A	$B, 0$	$C, 0$
B	$A, 1$	D, d_3
C	$d_1, 1$	$A, 1$
D	C, d_4	$B, 0$

(b)

4. (a) Given a machine $M = (\Sigma, Z, Q, \delta, \lambda)$, where $|\Sigma| = 2$, $|Z| = 3$, and $|Q| = 12$. The symbol $|\ \ |$ denotes the number of elements of a set. Find an upper bound to the number of 1-equivalent classes.
   (b) Derive a general formula for computing this upper bound in terms of $|X|, |Z|$, and $|Q|$.

5. Let $M = (\Sigma, Q, Z, \delta, \lambda)$ be a sequential machine. Suppose there are two input symbols, two output symbols, and 10 states. The next-state function δ of M is known. However, the output function λ is an unknown.
   (a) How would you construct an input sequence x such that when x is applied to M (and the corresponding $\bar{z}$ recorded), λ can be determined?
   (b) Give an upper bound to the length of $\bar{x}$.

6. Consider the machine M whose input space and state space are

$$\Sigma = \{0, 1\}$$
$$Q = \{1, 2, \dots, 9\}$$

The four-equivalence partition P_4 is

$$P_4 = \{\{1\}, \{2, 3, 4, 5\}, \{6, 7\}, \{8\}, \{9\}\}$$

It is also known that when 0 is applied to states 4 and 5, the next states are 4 and 6, respectively. Determine three-equivalence partition P_3 of Q.

7. If for every pair of states q_i and q_j of a machine M, there exists an input sequence that takes M from q_i to q_j, then M is said to be *strongly connected*. Show that a minimal machine M' of M is strongly connected if M is strongly connected.

8. Let $M = (\Sigma, Z, Q, \delta, \lambda)$ be a sequential machine. Let Σ^* denote the set of all the input sequences and Z^* denote the set of all the output sequences. Define $\tilde{x}/\tilde{z}$ to be an *input/output sequence* of machine M if there is a state q in M such that

$$\tilde{z} = \lambda(q, \tilde{x})$$

where $\tilde{x} \in \Sigma^*$ and $\tilde{z} \in Z^*$. The (infinite) set of all input/output sequences of M is called the *input/output set* of M.

　　Consider the following statement: "If two machines have identical input/output sets, then they must be equivalent." Is this statement true or false? Prove your answer. How about the two machines that are strongly connected?

9. In addition to the transition table and transition diagram, we can use the transition matrix $[A]$ to describe an n-state machine.

Next states

$$[A] = \begin{array}{c} \text{Present} \\ \text{states} \end{array} \begin{array}{c} q_1 \\ q_2 \\ \cdot \\ \cdot \\ \cdot \\ q_n \end{array} \begin{array}{ccccc} q_1 & q_2 & \cdots & q_n \\ \left[\begin{array}{cccc} x_{11}/z_{11} & x_{12}/z_{12} & \cdots & x_{1n}/z_{1n} \\ x_{21}/z_{21} & x_{22}/z_{22} & \cdots & x_{2n}/z_{2n} \\ \cdot & \cdot & & \cdot \\ \cdot & \cdot & & \cdot \\ \cdot & \cdot & & \cdot \\ x_{n1}/z_{n1} & x_{n2}/z_{n2} & \cdots & x_{nn}/z_{nn} \end{array} \right] \end{array}$$

x_{ij} is the input that causes the state transition from state q_i to state q_j and z_{ij} is the corresponding output. Consider the machine M whose transition matrix is

$$[A] = \begin{array}{c} 1 \\ 2 \\ 3 \end{array} \begin{array}{ccc} 1 & 2 & 3 \\ \left[\begin{array}{ccc} 0/0 & 1/0 & 0 \\ 0/0 & 1/1 & 0 \\ 0 & 1/0 & 0/0 \end{array} \right] \end{array}$$

(a) Find the equivalent states of machine M.

(b) Show that if two rows r_i and r_j of the transition matrix of a sequential machine are identical, then the two states q_i and q_j are equivalent.

(c) A sequential machine is said to be *trivial* if its output function is a function of its input only. Show that a sequential machine is trivial if all the rows of its transition matrix are identical.

(d) Based on the result you obtained in (a), what more general statement can you make about the equivalence of two states of a machine from the observation of its transition matrix?

10. A sequential machine is *reversible* if the following condition is satisfied for all states of M: A state q_i is reachable from state q_j; then q_j is reachable from q_i.

   Suppose that M' is a minimal machine of machine M which is reversible.

   (a) Is M' reversible?
   (b) If not, construct a counterexample.
   (c) Conversely, if M' is reversible, is M reversible? If not, show a counterexample.

11. Show that if a sequential machine is strongly connected, then M is reversible but that the converse is not true.

12. The sequential machine M shown in black-box form in Fig. P5.2.12 has two input terminals (labeled A and B) and two output terminals (labeled C and D). An input or output appearing at a terminal may be either a 0 or a 1.

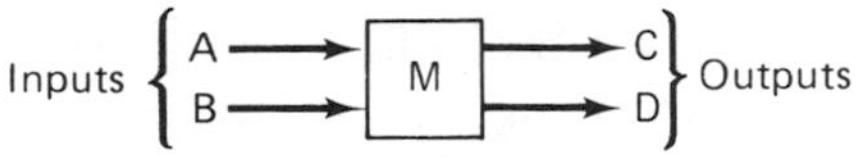

Fig. P5.2.12

Let

$$x_i(t) = \text{input symbol appearing at terminal } i \text{ at time } t$$
$$z_j(t) = \text{output symbol appearing at terminal } j \text{ at time } t$$

We are given the following incomplete equations that characterize the responses for the black box.

$$z_1(t) = x_1(t) \oplus \, ?$$
$$z_2(t) = (x_1(t) \oplus z_2(t - 2)) \oplus \, ?$$

It is not known which of the inputs (A or B) is number 1, which of the outputs (C or D) is number 1, and so on. The following table contains some information that is known about the black box.

TABLE P5.2.12

$q(t)$	$(x_A(t), x_B(t))$	$q(t+1)$				$(z_C(t), z_D(t))$			
		(0, 0)	(0, 1)	(1, 0)	(1, 1)	(0, 0)	(0, 1)	(1, 0)	(1, 1)
1		1				(0, 0)			(1, 0)
2						(0, 1)			
3		2							
4		4						(1, 0)	

(a) Give the complete equations for $z_1(t)$ and $z_2(t)$. Describe how you determined the unknown entries.

(b) Fill in the blank entries in Table P5.2.12.

5.3 General Procedure for Analyzing Synchronous Flip-Flop Circuits

In Chapter 4 several flip-flop circuits were presented and analyzed by a rather informal step-by-step trace method. In this section, a formal procedure for analyzing synchronous flip-flop circuits is presented that is based on the sequential-machine theory presented in the previous two sections. Before presenting the procedure, we want to show that:

1. A flip-flop satisfies the model of a sequential machine described in Section 5.1, and thus is a sequential machine.

2. The operation of a flip-flop can be described by a simple function called the *characteristic function* of the flip-flop.

In Section 4.2 the operation of a flip-flop was described by a logic function table. For example, the operations of an S–R flip-flop without and with Preset and Clear were described by logic function tables in Fig. 4.2.4(a) and (b), respectively. With Preset and Clear of the latter type set HIGH† (logic 1), these function tables can be described by a transition table, shown in Fig. 5.3.1; therefore, an S–R flip-flop is a

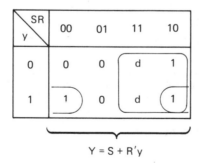

$$Y = S + R'y$$

Fig. 5.3.1 Transition table and characteristic function of S-R flip-flop.

two-state sequential machine. The transition table shown in Fig. 5.3.1 may be viewed as a regular three-variable Karnaugh map which can be used to minimize the next-state function Y. The function after minimized is

$$Y = S + R'y \qquad (5.3.1)$$

This equation is called the *characteristic function* of the S–R flip-flop.

†For illustration convenience, we shall assume that the flip-flops are positive TTL flip-flops, unless stated otherwise.

It is easy to show that other types of flip-flops, such as J–K flip-flops, D flip-flops, T flip-flops, are two-state sequential machines. As a matter of fact, every flip-flop is a two-state sequential machine. The characteristic functions of other types of flip-flops can be found in a way similar to that for the S–R flip-flop shown above. They are given in Table 5.3.1. These characteristic functions will be used to illustrate a general procedure for analyzing synchronous flip-flop circuits, described in Fig. 5.3.2. This procedure is best illustrated by examples.

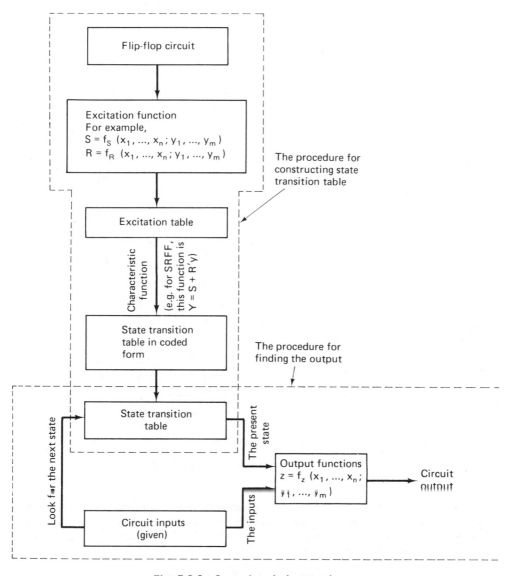

Fig. 5.3.2 General analysis procedure.

TABLE 5.3.1   **Characteristic Functions of Various Types of Flip-Flops**

Type of Flip-Flop	Inputs	Outputs	Characteristic Function	Subject to the Restrictions
SRFF	S, R	y, y'	$Y = S + R'y$	$S \cdot R = 1$
JKFF	J, K	y, y'	$Y = Jy' + K'y$	None
DFF	D	y, y'	$Y = D$	None
TFF	T	y, y'	$Y = T'y + Ty'$	None
			$= T \oplus y$	

Example 5.3.1

Find the transition table of the S–R flip-flop circuit (binary serial adder) of Fig. 4.2.8, which, for convenience, is repeated in Fig. 5.3.3.

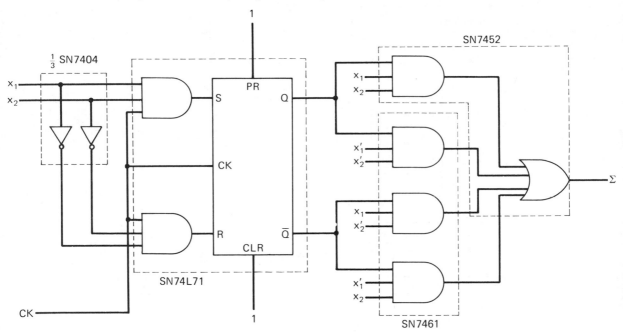

Fig. 5.3.3   Example 5.3.1.

The procedure for obtaining the state transition function and the output function of this circuit is as follows.

Step 1   Find the excitation functions in terms of the circuit inputs and the flip-flop outputs from the given circuit diagram. For example, the excitation functions of an S–R flip-flop are $S = f_S(x_1, \ldots, x_n; y_1, \ldots, y_m)$ and $R = f_R(x_1, \ldots, x_n; y_1, \ldots, y_m)$, and the excitation functions of the circuit of Fig. 5.3.3 are

$$S = x_1 x_2, \qquad R = x'_1 x'_2 \qquad (5.3.2)$$

Step 2   Obtain the excitation table. Display the functions S and R as shown in Table 5.3.2(a), which is known as the *excitation table*.

TABLE 5.3.2   Analysis of the Circuit of Fig. 5.3.3

(a)   Excitation Table

$x_1 x_2$ / y	00	01	11	10
0	01	00	10	00
1	01	00	10	00

$S\overset{\frown}{R}$

(b)   Binary Transition Table

$x_1 x_2$ / y	00	01	11	10
0	0, 0	0, 1	1, 0	0, 1
1	0, 1	1, 0	1, 1	1, 0

Y, z

(c)   Transition Table

$x_1 x_2$ / y	00	01	11	10
q_0	$q_0, 0$	$q_0, 1$	$q_1, 0$	$q_0, 1$
q_1	$q_0, 1$	$q_1, 0$	$q_1, 1$	$q_1, 0$

Next state, Present output

Step 3   Obtain the binary transition table from the excitation table using the characteristic function. The binary transition table of this example is shown in Table 5.3.2(b). The Y-component entities of this table are obtained by using the characteristic function of Eq. (5.3.1). The first-row and third-column entry, for example, is obtained by substituting the values of S, R, and y at that position into the characteristic function $Y = S + R'y = 1 + 0'0 = 1$. The rest are obtained similarly.

Step 4   The outputs of the circuit are determined by substituting the values of the circuit inputs and the next state into the output functions. For our example, the output function is

$$z = x_1 x_2 y + x_1' x_2' y + x_1 x_2' y' + x_1' x_2 y' \tag{5.3.3}$$

from which we obtain the z-component entities of the binary transition table. From Table 5.3.2(b), the circuit output to any input sequence can be uniquely determined. If the states

0 and 1 in Table 5.3.2(b) are used to represent the no-carry (q_0) and carry (q_1) states of the binary serial adder, the transition table of this machine is shown in Table 5.3.2(c) and the circuit of Fig. 5.3.3 is a realization of the binary serial adder.

Example 5.3.2

As another example, consider the TFF circuit of Fig. 5.3.4. The excitation functions of this circuit are

$$
\begin{aligned}
T_1 &= x \\
T_2 &= xy_1 \\
T_3 &= xy_1y_2 \\
T_4 &= xy_1y_2y_3
\end{aligned}
\qquad (5.3.4)
$$

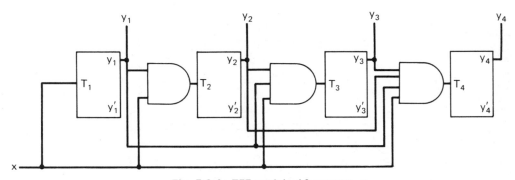

Fig. 5.3.4 TFF modulo-16 counter.

From these equations, the following excitation table can be constructed:

$y_4y_3y_2y_1$ \ x	0	1
0 0 0 0	0 0 0 0	0 0 0 1
0 0 0 1	0 0 0 0	0 0 1 1
0 0 1 0	0 0 0 0	0 0 0 1
0 0 1 1	0 0 0 0	0 1 1 1
0 1 0 0	0 0 0 0	0 0 0 1
0 1 0 1	0 0 0 0	0 0 1 1
0 1 1 0	0 0 0 0	0 0 0 1
0 1 1 1	0 0 0 0	1 1 1 1
1 0 0 0	0 0 0 0	0 0 0 1
1 0 0 1	0 0 0 0	0 0 1 1
1 0 1 0	0 0 0 0	0 0 0 1
1 0 1 1	0 0 0 0	0 1 1 1
1 1 0 0	0 0 0 0	0 0 0 1
1 1 0 1	0 0 0 0	0 0 1 1
1 1 1 0	0 0 0 0	0 0 0 1
1 1 1 1	0 0 0 0	1 1 1 1

$$T_4T_3T_2T_1$$

Applying the characteristic functions for the TFF's, $Y_i = T_i \oplus y_i$, $i = 1, 2, 3, 4$, to the excitation table, we obtain the transition table of the circuit:

$y_4 y_3 y_2 y_1$	x 0	1	Present output
0 0 0 0	0 0 0 0	0 0 0 1	0 0 0 0
0 0 0 1	0 0 0 1	0 0 1 0	0 0 0 1
0 0 1 0	0 0 1 0	0 0 1 1	0 0 1 0
0 0 1 1	0 0 1 1	0 1 0 0	0 0 1 1
0 1 0 0	0 1 0 0	0 1 0 1	0 1 0 0
0 1 0 1	0 1 0 1	0 1 1 0	0 1 0 1
0 1 1 0	0 1 1 0	0 1 1 1	0 1 1 0
0 1 1 1	0 1 1 1	1 0 0 0	0 1 1 1
1 0 0 0	1 0 0 0	1 0 0 1	1 0 0 0
1 0 0 1	1 0 0 1	1 0 1 0	1 0 0 1
1 0 1 0	1 0 1 0	1 0 1 1	1 0 1 0
1 0 1 1	1 0 1 1	1 1 0 0	1 0 1 1
1 1 0 0	1 1 0 0	1 1 0 1	1 1 0 0
1 1 0 1	1 1 0 1	1 1 1 0	1 1 0 1
1 1 1 0	1 1 1 0	1 1 1 1	1 1 1 0
1 1 1 1	1 1 1 1	0 0 0 0	1 1 1 1

$$Y_4 Y_3 Y_2 Y_1$$

Notice that this table describes the operation of a modulo-16 counter.

Exercise 5.3

All the circuits in this exercise are synchronous sequential circuits. If the clocks are not shown, they are implicit.

1. Analyze the parity-checking flip-flop circuit of Fig. 4.2.11.
2. Analyze the synchronous modulo-4 counting circuit of Fig. P5.3.2.
3. Show that the circuit shown in Fig. P5.3.3 is a decade synchronous counter.
4. Show that the circuits of Fig. P5.3.4 (a) and (b) are four-stage binary counters with ripple carry and parallel carry, respectively.
5. Prove that the sequential machine that describes the operation of the circuit of Fig. P5.3.5 is isomorphic† to the machine described in Table P5.3.5.

TABLE P5.3.5

q	x 0	1
1	2, 0	1, 1
2	4, 0	3, 0
3	2, 0	1, 0
4	4, 0	3, 0

†See the footnote on page 3.

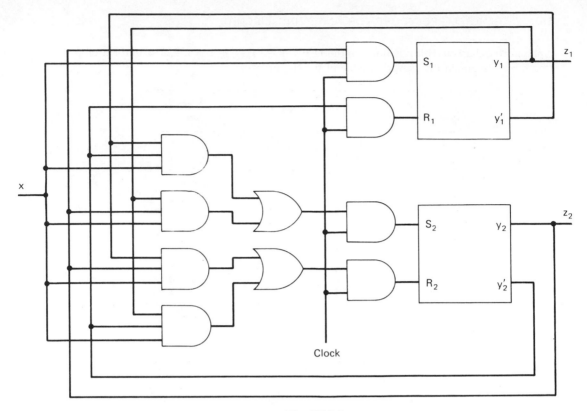

Fig. P5.3.2

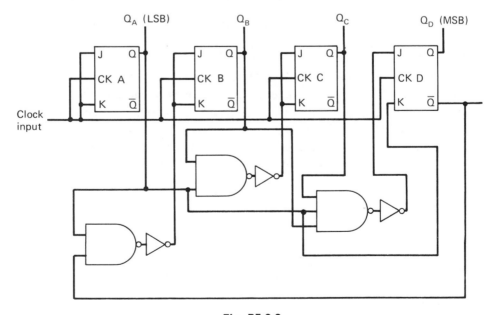

Fig. P5.3.3

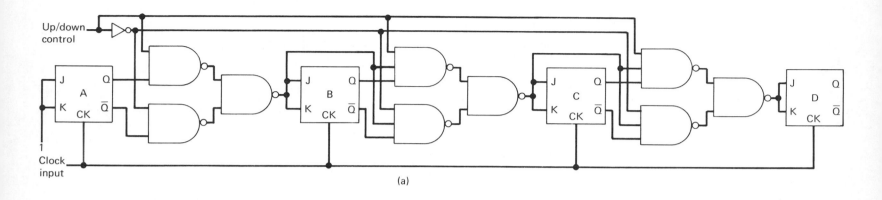

(a)

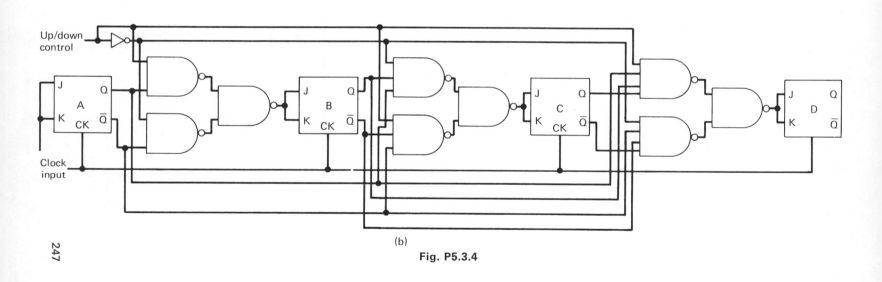

(b)

Fig. P5.3.4

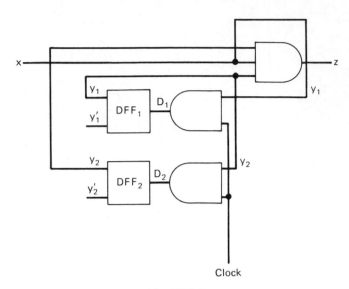

Fig. P5.3.5

6. Show that the synchronous DFF circuit of Fig. P5.3.6 can be modeled by a sequential machine that is isomorphic† to the machine whose transition table is described in Table P5.3.6.

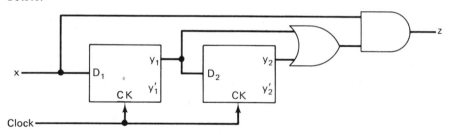

Fig. P5.3.6

TABLE P5.3.6

q \ σ	0	1
A	$A, 0$	$B, 0$
B	$A, 0$	$C, 1$
C	$B, 0$	$D, 1$
D	$B, 0$	$D, 1$

7. Consider the synchronous sequential circuit of Fig. P5.3.7.
   (a) Determine the state transition table in coded form.
   (b) Determine the state transition table.

†See the footnote on page 3.

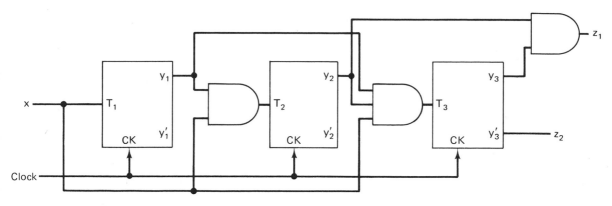

Fig. P5.3.7

(c) For an input signal:

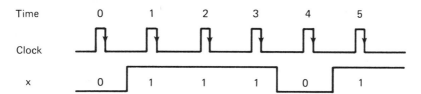

sketch the waveforms of the states and the outputs of the circuit for $t = 0, 1, 2, 3,$ 4, 5, and 6. Assume that positive logic is used throughout and the circuit is initially with all $y_i = 0$.

8. Analyze the synchronous J–K flip-flop circuit of Fig. P.5.3.8.

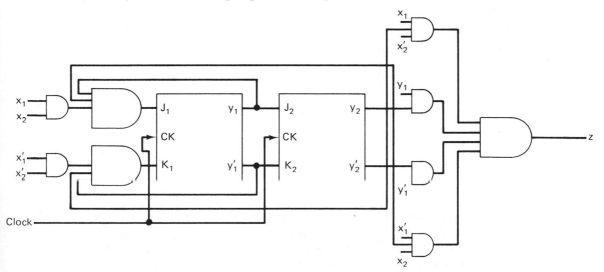

Fig. P5.3.8

9. (a) Show that the synchronous sequential circuit in Fig. P5.3.9 is a decade counter.
 (b) Implement this circuit using a PLA.

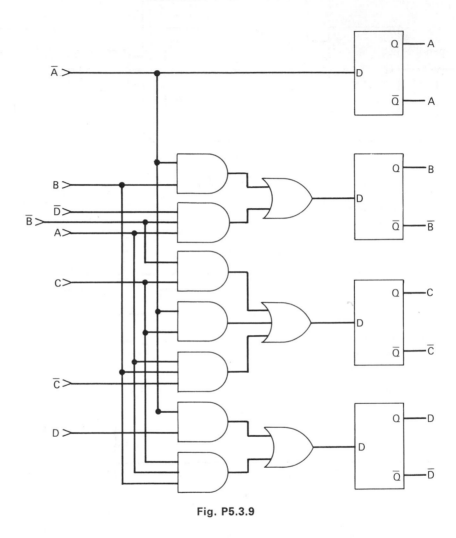

Fig. P5.3.9

5.4 General Procedure for Synthesizing Synchronous Flip-Flop Circuits

Having introduced the analysis of flip-flop circuits, we now turn to the synthesis problem: that is, given a machine specification (transition table or diagram), to find a flip-flop circuit that realizes it. The approach to be followed is to reverse the steps used in the analysis procedure described in the previous section. From Fig. 5.3.2 it is seen that the crucial step in analyzing a sequential circuit is that of going from an excitation table to the corresponding state transition table. This was achieved by use of the characteristic functions of flip-flops. In the synthesis of sequential circuits, this step in the reverse direction is still the crucial one. One method of achieving this is to make a truth table from the characteristic function, considering flip-flop inputs as dependent variables and y and Y as independent variables. This table is known as the *flip-flop application table*. Flip-flop application tables of various types of flip-flops are shown in Table 5.4.1.

A general procedure for synthesizing synchronous flip-flop circuits is described in Fig. 5.4.1. This procedure is illustrated by the following examples.

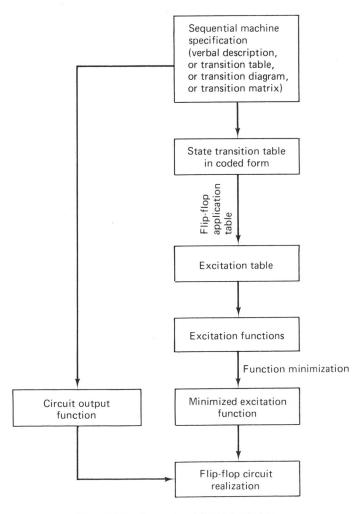

Fig. 5.4.1 General synthesis procedure.

TABLE 5.4.1 Flip-Flop Application Tables

y	Y	S	R
0	0	0	d
0	1	1	0
1	0	0	1
1	1	d	0

y	Y	J	K
0	0	0	d
0	1	1	d
1	0	d	1
1	1	d	0

y	Y	D
0	0	0
0	1	1
1	0	0
1	1	1

y	Y	T
0	0	0
0	1	1
1	0	1
1	1	0

(a) SRFF   (b) JKFF   (c) DFF   (d) TFF

Example 5.4.1

Find a synchronous-sequential-circuit realization of the binary serial adder whose state transition table and output function were described in Table 5.3.2(c) and Eq. (5.3.2), respectively.

Step 1   Find the state transition table in coded form. First we determine the number of logical variables. Let s, u, and v denote the minimum number of state, input, and output variables needed for coding n states, m input symbols, and p output symbols distinctly. Then

$$s = \lceil \log_2 n \rceil, \qquad u = \lceil \log_2 m \rceil, \qquad v = \lceil \log_2 p \rceil$$

The symbol $\lceil R \rceil$ denotes the smallest integer that is greater than or equal to the real number R. For example, since there are only two states in the machine, one state variable is sufficient to code them. If we code q_0 and q_1, respectively, by 0 and 1 and denote this variable by y, and denote the two input variables by x_1 and x_2, then the transition table of this machine in coded from will be as shown in Table 5.4.2(a).

TABLE 5.4.2 Synthesis of the Serial Binary Adder Using an S-R Flip-Flop

(a) Transition Table in Coded Form

y \ x_1x_2	00	01	11	10	00	01	11	10
0	0	0	1	0	0	1	0	1
1	0	1	1	1	1	0	1	0

$\underbrace{\qquad\qquad}_{Y}$ $\underbrace{\qquad\qquad}_{z}$

(b) Excitation Table

y \ x_1x_2	00	01	11	10
0	$0d$	$0d$	10	$0d$
1	01	$d0$	$d0$	$d0$

$\underbrace{\qquad\qquad}_{SR}$

(c) Karnaugh Map of the Excitation Function S

y \ x_1x_2	00	01	11	10
0	0	0	1	0
1	0	d	d	d

$$S = x_1 x_2$$

(d) Karnaugh Map of the Excitation Function R

y \ x_1x_2	00	01	11	10
0	d	d	0	d
1	1	0	0	0

$$R = x_1' x_2'$$

Step 2 Obtain the excitation table. Using the S–R flip-flop application table of Table 5.4.1, the excitation table of this example is shown in Table 5.4.2(b).

Step 3 Obtain the excitation functions. The excitation functions S and R presented by Karnaugh maps are shown in Table 5.4.2(c) and (d). After minimization, they are $S = x_1 x_2$ and $R = x_1' x_2'$.

Step 4 Realize the minimized excitation functions obtained in step 3 and the output function after minimization. The output function of the machine after minimization is $z = x_1 x_2 y + x_1 x_2' y' + x_1' x_2' y + x_1' x_2 y'$. The complete S–R flip-flop circuit realization of the binary serial adder was given in Fig. 5.3.3.

Example 5.4.2

Construct an SRFF, a JKFF, and a TFF from a DFF. The state transition tables of these flip-flops are as follows:

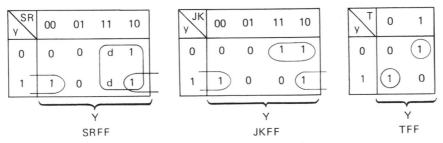

It is easy to show that these tables, with Y replaced by D, are the excitation tables for the realizations of these flip-flops using a DFF. Therefore, the corresponding minimized excitation functions are

$$D = S + R'y, \qquad D = Jy' + K'y, \qquad D = Ty' + T'y$$

The realizations are shown in Fig. 5.4.2.

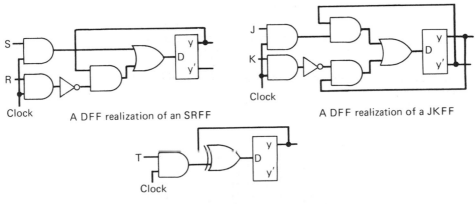

A DFF realization of an SRFF

A DFF realization of a JKFF

A DFF realization of a TFF

Fig. 5.4.2 Realization of an SRFF, a JKFF, and a TFF using a DFF.

Example 5.4.3

Design a synchronous J–K flip-flop circuit having one input and one output. The circuit outputs a 1 whenever three consecutive 1's are observed. Otherwise, the output will be 0.

Let q_0, q_1, and q_2 denote the states that observe no consecutive 1's, one consecutive 1, and two consecutive 1's, respectively. The transition table that describes the desirable operational behavior of this circuit is as follows:

q \\ x	0	1
q_0	$q_0, 0$	$q_1, 0$
q_1	$q_0, 0$	$q_2, 0$
q_2	$q_0, 0$	$q_2, 1$

A minimum of two state variables is needed to code these three states in binary code. Suppose that we choose the following codes:

q	$y_1 y_2$
q_0	00
q_1	01
q_2	11

The corresponding coded transition table is as follows;

$y_1 y_2$ \\ x	0	1
00	00, 0	01, 0
01	00, 0	11, 0
11	00, 0	11, 1
10	dd, d	dd, d

$$Y_1 Y_2, z$$

By applying the application table of the J–K flip-flop to this table, we obtain the following excitation table;

$y_1 y_2$ \\ x	0	1
00	$0d, 0d$	$0d, 1d$
01	$0d, d1$	$1d, d0$
11	$d1, d1$	$d0, d0$
10	dd, dd	dd, dd

$$J_1 K_1, J_2 K_2$$

From this table the minimized excitation functions can be obtained.

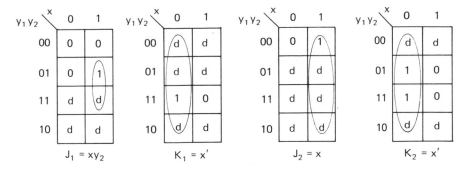

$y_1 y_2$ ╲ x	0	1
00	0	0
01	0	1
11	d	d
10	d	d

$J_1 = xy_2$

$y_1 y_2$ ╲ x	0	1
00	d	d
01	d	d
11	1	0
10	d	d

$K_1 = x'$

$y_1 y_2$ ╲ x	0	1
00	0	1
01	d	d
11	d	d
10	d	d

$J_2 = x$

$y_1 y_2$ ╲ x	0	1
00	d	d
01	1	0
11	1	0
10	d	d

$K_2 = x'$

From the transition table, we obtain the following minimized output function:

$y_1 y_2$ ╲ x	0	1
00	0	0
01	0	0
11	0	1
10	d	d

$z = xy_1$

Therefore, the desired circuit is

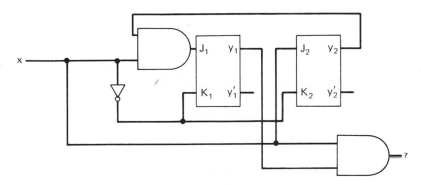

Example 5.4.4

Find a synchronous modulo-4 counter circuit which is described by

x q	0	1
0	0	1
1	1	2
2	2	3
3	3	0

Assume that J–K flip-flops are to be used. Note that the outputs of this (Moore) machine which have the same values as its states are not shown in the above table.

If we choose the following binary codes to code the four states:

q	$y_1 y_2$
0	00
1	01
2	10
3	11

we obtain the coded transition table as

x $y_1 y_2$	0	1
00	00	01
01	01	10
10	10	11
11	11	00

$$Y_1 Y_2$$

By applying the application table of a J–K flip-flop as given in Table 5.4.1(b) to the above binary-coded transition table, we obtain the excitation table as

x $y_1 y_2$	0	1
00	0d, 0d	0d, 1d
01	0d, d0	1d, d1
10	d0, 0d	d0, 1d
11	d0, d0	d1, d1

$$J_1 K_1, J_2 K_2$$

The excitation functions after minimization are

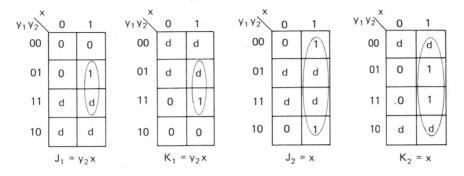

$$J_1 = y_2 x \qquad K_1 = y_2 x \qquad J_2 = x \qquad K_2 = x$$

The J–K flip-flop circuit realization of a modulo-4 counter is

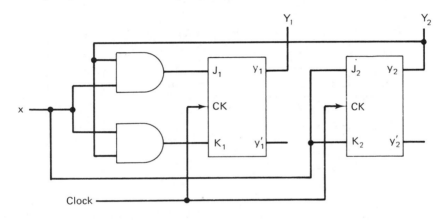

Example 5.4.5

As another example, realize a 3-bit serial-in/serial-out shift register whose transition table is as follows:

$y_1 y_2 y_3$ \\ x	0	1
000	000	100
001	000	100
010	001	101
011	001	101
100	010	110
101	010	110
110	011	111
111	011	111

$$Y_1 Y_2 Y_3$$

Again, the outputs of this (Moore) machine which have the same values as its states are not shown. Applying the synthesis procedure, we obtain the excitation table:

$y_1 y_2 y_3$ \ x	0	1
000	$0d, 0d, 0d$	$1d, 0d, 0d$
001	$0d, 0d, d1$	$1d, 0d, d1$
010	$0d, d1, 1d$	$1d, d1, 1d$
011	$0d, d1, d0$	$1d, d1, d0$
100	$d1, 1d, 0d$	$d0, 1d, 0d$
101	$d1, 1d, d1$	$d0, 1d, d1$
110	$d1, d0, 1d$	$d0, d0, 1d$
111	$d1, d0, d0$	$d0, d0, d0$

the excitation functions, $J_1 K_1, J_2 K_2,$ and $J_3 K_3$:

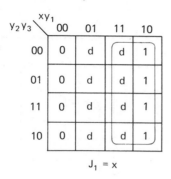

$J_1 = x$

$K_1 = x'$

$J_2 = y_1$

$K_2 = y_1'$

$J_3 = y_2$

$K_3 = y_2'$

and the J–K flip-flop circuit realization of a 3-bit shift register:

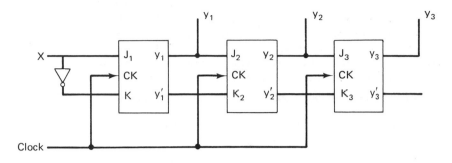

Exercise 5.4

1. Find a realization of an SRFF, a DFF, and a TFF from a JKFF.

2. (a) Realize the serial odd parity checker using (i) a TFF and (ii) a DFF.
   (b) Realize the serial even parity checker using (i) an SRFF, (ii) a JKFF, (iii) a TFF, and (iv) a DFF.

3. Synthesize the following synchronous counters using JKFF's:
   (a) Binary divide-by-3 counter.
   (b) Binary divide-by-5 counter.
   (c) Binary divide-by-8 counter.
   (d) Binary divide-by-12 counter.

4. (a) The characteristic function of a flip-flop is described by

$$Y = x_1 \oplus x_2 \oplus y$$

   Realize this flip-flop using (1) SRFF, (2) JKFF, (3) DFF, and (4) TFF.
   (b) Realize a modulo-4 synchronous counter using this type of flip-flop.
   (c) Realize a 3-bit serial-in/serial-out shift register by a flip-flop circuit using this type of flip-flop.

5. Design a synchronous controllable counter (using any type of flip-flops you like) which has the following characteristics:

Control Signals		Operation
c_1	c_2	
0	0	No change of state
0	1	Modulo-2 counter
1	0	Modulo-4 counter
1	1	Modulo-8 counter

6. (a) Give the transition table of a binary serial subtractor.
   (b) Realize the transition table obtained in (a) using a synchronous J–K flip-flop circuit.

7. Suppose that it is desired to design a digital system having two inputs, x_1 and x_2, and an output, z. It is assumed that the two inputs are restricted so that two pulses (two 1's) cannot occur simultaneously. The output z becomes 1 if, between two consecutive x_1 pulses, there have been *exactly* two x_2 pulses. Otherwise, the output remains at 0. Once the output becomes 1, it remains at 1 until an x_2 pulse is received. Then the output becomes 0. Find the transition table (with minimum number of states) that describes this machine and realize it with a minimum number of flip-flops. J–K flip-flops are preferred.

8. (a) Design a combinational circuit that will convert a two-digit 8421 BCD code into its binary equivalent.
    (b) Design a sequential 8421 BCD-to-binary converter.
    (c) Compare your results obtained in (a) and (b).

9. (a) Design a combinational circuit that will convert a four-bit binary number into its 8421 BCD equivalent.
    (b) Design a sequential binary-to-8421 BCD converter.
    (c) Compare your results obtained in (a) and (b).

5.5 Analysis of Asynchronous Sequential Circuits

After having discussed the synchronous sequential circuit, we now turn to the asynchronous sequential circuit. Although there are many forms that an asynchronous (sequential) circuit might take, the one shown in Fig. 5.5.1 is the most straightforward and widely used in the study of the theory of asynchronous circuits or machines. Externally, the inputs and outputs are represented by *levels rather than pulses*. Internally, it is characterized by the use of delay elements as memory devices. Symbolically, they are denoted by $\longrightarrow\boxed{\Delta}\longrightarrow$ where Δ denotes the amount of time delay between the input and the output. *It should be noted, however, that in practice, when the inherent delay of the combinational logic is large enough, the external delay elements may not be*

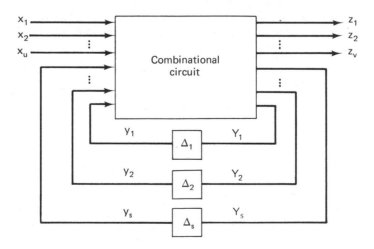

Fig. 5.5.1 Basic model of asynchronous sequential circuit.

necessary. But for clarity of presentation, we shall assume them to be present. The delay-type memory elements in Fig. 5.5.1 are characterized as follows:

$$y_i(t) = Y_i(t - \Delta_i), \qquad i = 1, 2, \ldots, s$$

Although the Δ_i's, in general, may not be equal, for illustration simplicity in this and the remaining sections, it is assumed that

1. $\Delta_1 = \Delta_2 = \ldots = \Delta_s = \Delta.$
2. The delay Δ is larger than any one may encounter in propagating a signal through the combinational circuit.

In Chapter 4 we have seen two examples of asynchronous sequential circuits. One was the S–R latch and the other was the ripple counter. As a matter of fact, these two examples represent typical examples of two types of asynchronous sequential circuits. The first one is the combinational circuit with loops, and the second is the flip-flop circuit whose flip-flops are not under the command of a single clock pulse. The following two examples show that both of these two types of asynchronous sequential circuits can be presented in the basic model form of Fig. 5.5.1.

Example 5.5.1

The NOR S–R latch can be presented in the basic model form, as shown in Fig. 5.5.2.

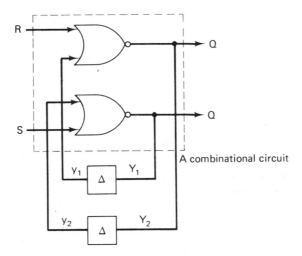

Fig. 5.5.2 Asynchronous NOR S-R latch presented in the basic model form.

Example 5.5.2

Using the result of Example 5.4.2, the 4-bit (modulo-16) binary ripple counter of Fig. 4.3.1 can be presented in the basic model form as shown in Fig. 5.5.3(b). Figure 5.5.3(a) shows that the DFF realization of a JKFF (see Fig. 5.4.3) can be represented by a combina-

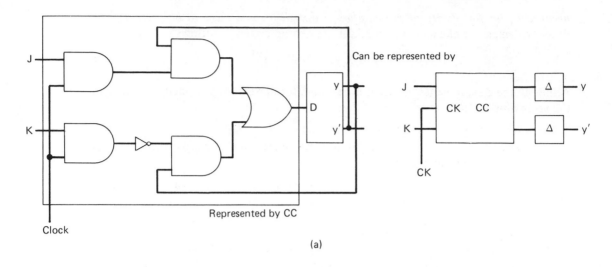

(a)

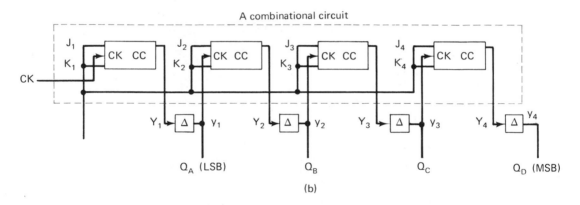

(b)

Fig. 5.5.3 Asynchronous 4-bit ripple counter presented in the basic model form.

tional circuit CC describing the logic part of the circuit, followed by a delay element to describe the delay part of the circuit. Using this representation of the J–K flip-flop, the asynchronous 4-bit ripple counter can be presented in the basic model form, which is depicted in Fig. 5.5.3(b).

In an asynchronous sequential circuit, an input symbol is implemented with a u-tuple, $(x_1, x_2, \ldots, x_u)$. Similarly, an output symbol is represented by a v-tuple $(z_1, z_2, \ldots, z_v)$. The present state and the next state of an asynchronous circuit are represented by s-tuples, $(y_1, y_2, \ldots, y_s)$ and $(Y_1, Y_2, \ldots, Y_s)$, respectively. At each operation there is a change of state, or a change of value of at least one of the state variables $\{y_i\}$. However, the state will not change further when $Y_i = y_i, i = 1, 2, \ldots,$

s. Then the machine will settle down indefinitely to equilibrium unless a further change of input symbol is made. The state in which a machine *stays* after the completion of a sequence of operations is called a *stable state*. Any state through which the machine passes temporarily during a sequence of computations is called an *unstable state* (or *transient state*). When the state variables assume their new values (i.e., the *y*'s become equal to the corresponding *Y*'s), the circuit enters its "next" stable state. Thus *a transition from one stable state to another occurs only in response to a change in the input variables*.

DEFINITION 5.5.1

A circuit is *stable internally* if none of the internal signals is unstable or changing.

DEFINITION 5.5.2

A sequential circuit is operating in the *fundamental mode* (asynchronous) if and only if the inputs are never changed unless the circuit is stable internally.

In practice, because of the stray delays and nonideal characteristics of electronic devices, it is impossible to have several input or state variables change at exactly the same instant. Thus to avoid possible undesirable operations, *only one input variable of an asynchronous sequential circuit is allowed to change at a time*. In other words, only *adjacent* input symbols follow each other in an input sequence.

DEFINITION 5.5.3

Two input symbols are *adjacent* if and only if they differ in value in one and only one input variable.

For example, for an asynchronous circuit with two inputs x_1 and x_2, the input symbols $x_1 x_2 = 00$ and $x_1 x_2 = 01$ are adjacent, but $x_1 x_2 = 10$ and $x_1 x_2 = 01$ are not. Thus input sequence 1 is an allowable input sequence for the circuit, but input sequence 2 is not:

Input sequence 1: $x_1 x_2 = 00$ 01   11   10   11   01   00   (allowable)

Input sequence 2: $x_1 x_2 = 00$ 10   01   00   11   01   00   (not allowable)

not adjacent   not adjacent

It is important to note that *every state of a synchronous circuit is a stable state, and only asynchronous circuits have unstable states*. The table that contains both stable and unstable states and describes the operational behavior of an asynchronous sequential circuit is called a *flow table*, and the state transition portion of it is called a *state flow table*. Apart from the unstable states, the behavior of an asynchronous circuit is similar to that of a synchronous circuit. If we consider the performance of a machine in stable states only (i.e., the transition from a stable state to a stable state and the outputs associated with transitions or with stable states), then an asynchronous circuit can be represented by a *transition table* or *diagram*, in the same way as a synchronous circuit.

The obtaining of the flow table of an asynchronous circuit is the same as that of the transition table of a synchronous circuit described in Section 5.3, except that the next states of an asynchronous circuit are classified into stable and unstable states. Consider the sequential circuit of Fig. 5.5.4, with the following assumptions:

1. No synchronizing clock is used.
2. Its inputs and outputs are level signals.
3. It gives an output only when it is in a stable state.
4. It is operating in the fundamental mode.

First consider the portion of the circuit within the dashed lines of Fig. 5.5.4. If the present state $y = 0$ and the input symbol applies to $x_1 = 1$ and $x_2 = 1$, from the function table we have $Y = 1$, expressed as

$$\delta(0, (1, 1)) = 1 \qquad (5.5.1)$$

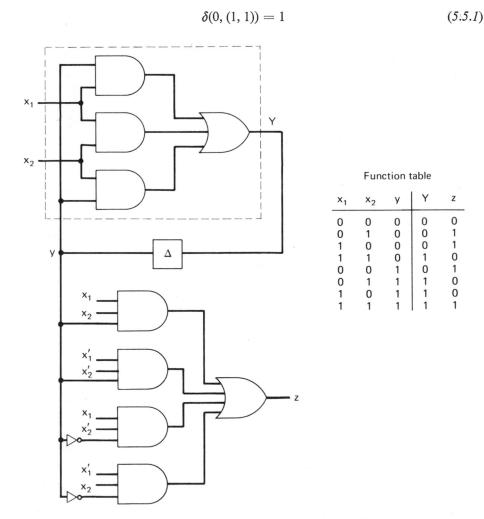

Function table

x_1	x_2	y	Y	z
0	0	0	0	0
0	1	0	0	1
1	0	0	0	1
1	1	0	1	0
0	0	1	0	1
0	1	1	1	0
1	0	1	1	0
1	1	1	1	1

Fig. 5.5.4 Sequential circuit operating in fundamental mode.

According to the preceding discussion, the next state, $Y = 1$, is an unstable state, because it is different from the value of y, which is 0. A second transition will take place *under the same input, $x_1 = 1$ and $x_2 = 1$,* immediately following the transition of Eq. (5.5.1), which is

$$\delta(1, (1, 1)) = 1 \qquad (5.5.2)$$

Note under the input symbol (1, 1) that the state changes from 0 to 1 and remains at 1, since now the values of y and Y are 1. It is customary to circle such a next state in the transition table, indicating that it is a stable state. When a machine lands at a stable state, it will stay in this stable state indefinitely, until a further change occurs in one of the input variables. All other state transitions can be determined in a similar way. The complete flow table is shown in Fig. 5.5.5(a). In this table, only the states in circles are stable states. Notice that although this flow table looks like the transition table of the binary serial adder shown in Table 5.3.2(c), they describe two totally different machines. Moreover, they have completely different operational behaviors. The circuit of Fig. 5.5.4 does not perform the addition operation at all. For example,

x / y	00	01	11	10
0	⓪,0	⓪,1	1,0	⓪,1
1	0,1	①,0	①,1	①,0

Y.z

(a) The flow table of the circuit of Fig. 5.5.4 when operating in fundamental or asynchronous mode in which only the states in circles are stable

t	0	1	2	3	4	5	6
x_1	1	1	1	0	0	0	1
x_2	0	1	1	1	0	0	0
y	⓪	⓪	1	①	①	0	⓪
Y	⓪	1	①	①	0	⓪	⓪
z	1	1		0	0		1

(b) The sequence of events of the response of the circuit of Fig. 5.5.4 when operating in fundamental mode with inputs $x_1 = 10011$ and $x_2 = 110$. The initial state of the circuit is assumed to be 0

Fig. 5.5.5 Example of asynchronous sequential circuit.

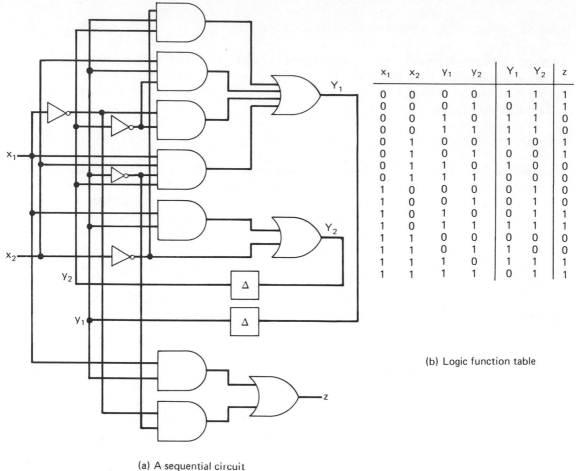

x_1	x_2	y_1	y_2	Y_1	Y_2	z
0	0	0	0	1	1	1
0	0	0	1	0	1	1
0	0	1	0	1	1	0
0	0	1	1	1	1	0
0	1	0	0	1	0	1
0	1	0	1	0	0	1
0	1	1	0	1	0	0
0	1	1	1	0	0	0
1	0	0	0	0	1	0
1	0	0	1	0	1	0
1	0	1	0	0	1	1
1	0	1	1	1	1	1
1	1	0	0	0	0	0
1	1	0	1	1	0	0
1	1	1	0	1	1	1
1	1	1	1	0	1	1

(b) Logic function table

(a) A sequential circuit operating in fundamental mode

$x_1 x_2$ / $y_1 y_2$	00	01	11	10
00	11**	10,1	⑩⓪,0	01,0
01	⓪①	00,1	10**,0	⓪①,0
11	⑪⑪	00*,0	01,1	⑪⑪,1
10	11	⑩⓪,0	11,1	01**,1

(c) Flow table

$x_1 x_2$ / $y_1 y_2$	00	01	11	10
0 0				
0 1				
1 1				
1 0				

A cycle

(d) Transition flow diagram (assuming no race conditions would occur, i.e., the changes in values of all the state variables occur simultaneously)

Fig. 5.5.6 Analysis of an asynchronous circuit.

the sequence of events of the response of the circuit of Fig. 5.5.4 when operating in fundamental mode with inputs $x_1 = 10011(19)$ and $x_2 = 110(6)$ is shown in Fig. 5.5.5(b). It is seen that the output sequence is $10011(19)$, which is not the sum of 10011 and 110.

In the example above, the circuit has only one state variable, which has made the analysis quite simple. Now let us consider an asynchronous circuit with more than one state variable. Consider the circuit of Fig. 5.5.6(a) operating in fundamental mode. From the function table of Fig. 5.5.6(b) we obtain the flow table of the circuit, which is depicted in Fig. 5.5.6(c). The next-state entries in the transition table can be replaced by arrows or direct paths. Let ○ denote a stable state and ● denote an unstable state. Then the transitions described in the flow table of Fig. 5.5.6(c) can be represented by the *transition flow diagram* of the circuit as shown in Fig. 5.5.6(d). For example, suppose that the present state of the circuit is $y_1 y_2 = 01$ and the input is $x_1 x_2 = 01$. Then the transition from state 01, upon the input 01, to other states of the circuit can be described by

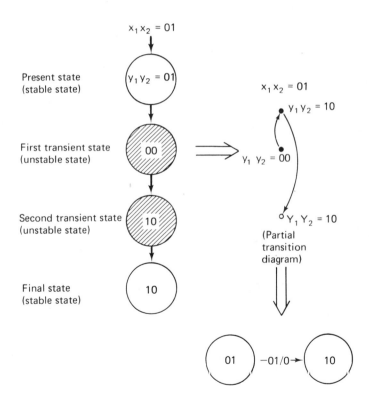

In a similar manner, we can construct all other such state transitions, which are shown in Fig. 5.5.6(d). Note that this transition flow diagram is constructed under the

assumption that the changes in values of all the state variables occur simultaneously. This, in practice, should never be expected to happen in asynchronous circuits. We shall discuss this subject in detail later, but first let us observe an undesirable transient phenomenon in the transition flow diagram. Under the input $x_1 x_2 = 11$ column, the three unstable 10, 01, and 11 states form a *cycle*, which indicates that if the present state is 01, 11, or 10 with an input 11, the circuit will circle around the three unstable stables 10, 01, and 11 under the input 11 column indefinitely, and the circuit will never reach a stable state. *When the transition flow diagram contains a cycle, the state transition diagram of the asynchronous circuit is undefined.*

There is another generally undesirable transient phenomenon in an asynchronous circuit called *race*.

DEFINITION 5.5.4

During a state transient, if there exist more than one unstable state variables (i.e., $y_i \neq Y_i$), we say the circuit is in a *race*. If the final stable state to which the circuit goes depends upon the *order* in which the state variables change, then we say the race is *critical*; otherwise, we say the race is *noncritical*.

There are four race conditions in the flow table of Fig. 5.5.6(c). Three of them are critical races and one is noncritical. The cirtical races and the noncritical race are indicated by a double and single asterisk (** and *), respectively. Suppose that the present state is $y_1 y_2 = 11$ and the input is $x_1 x_2 = 00$. The machine stays in the stable state ⑪ under $x_1 x_2 = 00$. Now if the value of the input 00 is changed to 01, the flow table of Fig. 5.5.6(c) indicates that the values for y_1 and y_2 should be changed. The changes are

$$
\begin{array}{ll}
\text{(1) the value of } y_1 \text{ from 1 to 0} & \\
\text{(2) the value of } y_2 \text{ from 1 to 0} & (5.5.3)
\end{array}
$$

Because of the way that the asynchronous circuit operates, one of the following three possible transitions can occur:

CASE 1

Changes 1 and 2 in Eq. (5.5.3) occur *simultaneously*, enabling the circuit to settle into state 00, as specified in the excitation table.

CASE 2

Change 1 occurs faster than change 2. In this case the immediate next state (a transient state) becomes state 01, which does not follow the specification of the problem.

CASE 3

Change 2 occurs faster than change 1. The immediate next state corresponding to this case is state 10, which again does not follow the specification of the problem.

The complete chains of state transitions for the three cases are given next.

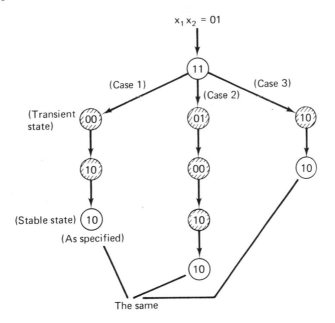

Notice that the final stable states for *all* the three cases are the same. This means that the final stable state to which the circuit goes does not depend upon the order in which the state variables change; hence this is a noncritical race and is marked by $*$. It is seen that a race of this type does not affect the correctness of the circuit performance.

The other three races, marked by $**$ in the excitation table, are critical races. Taking the one in the first row under the column $x_1 x_2 = 00$, for example, the chains of state transitions for all possible cases that may occur in the transition from $y_1 y_2 = 00$ to $y_1 y_2 = 11$ are given next.

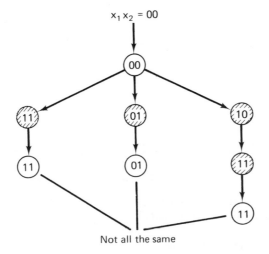

It is seen that the final stable states to which the circuit goes depends upon the order in which the state variables change, and thus this is a critical race. *When the transition flow diagram of an asynchronous circuit involves a critical race, the state transition diagram is again undefined.* The other two critical races involved in this circuit can be explained in a similar way (problem 1).

As a final remark, when either a cycle or a critical race exists in the transition flow diagram of an asynchronous circuit, the circuit operation becomes unpredictable; the circuit is therefore useless for any deterministic digital applications. There are two ways to eliminate critical races in a design:

1. *Insertion of controlled delays into the combinational circuitry.* Take the example above; the cirtical race involved in the transition from $y_1 y_2 = 00$ to 11 under the input $x_1 x_2 = 00$ can be eliminated by inserting an appropriate amount of delay into the combinational circuitry so that the change in value of y_1 from 0 to 1 will always be faster than that of y_2 from 0 to 1. This is because this race becomes noncritical and the correct result is always achieved.

2. *Elimination of critical races by an appropriate state assignment.* Methods of using this approach will be discussed in the next section. If no critical race is guaranteed by a state assignment, this assignment will also ensure that no cycles will exist. The operational behavior of the circuit obtained from such a state assignment, which is free of critical races and cycles, is completely predictable.

Exercise 5.5

1. Show the chains of state transitions for the two critical races under the columns $x_1 x_2 = 11$ and 10.

2. Suppose that the circuit of Fig. P5.5.2 is operating in the fundamental mode. Analyze the circuit by forming the
   (a) Flow table.
   (b) Transition flow diagram.
   (c) Transition diagram, if it exists.

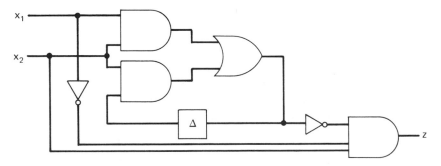

Fig. P5.5.2

3. Consider the asynchronous circuit of Fig. P5.5.3 operating in the fundamental mode.
   (a) Construct the flow table of the circuit.
   (b) Find all the race conditions in the flow table (if any), and determine which of them are noncritical and which are critical.

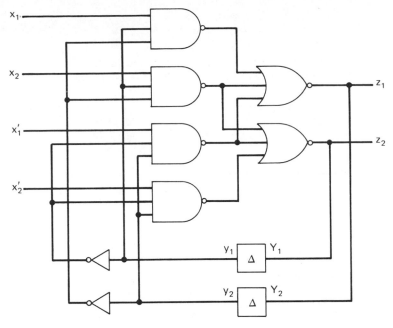

Fig. P5.5.3

4. Analyze the following multiple-output circuit which operates in the fundamental mode with the restriction that only one input variable can change at a time. Assume that the delays of the two flip-flops are the same. Determine whether there are any race conditions in this circuit. If so, are they critical?

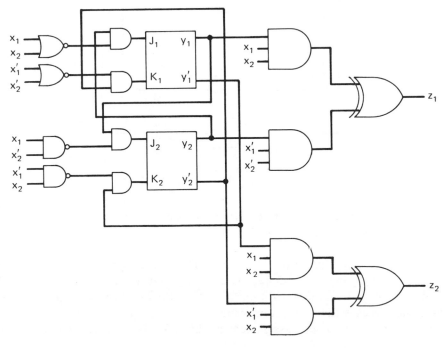

Fig. P5.5.4

5. (a) Show that there are two race conditions involved in the asynchronous circuit of Fig. P5.5.5 but that neither one is critical. Thus the circuit operational behaviors are completely predictable.
   (b) Find the transition diagram of the circuit.
   (c) Assume that the circuit is initially in state $y_1 y_2 = 00$. Determine the output sequence in response to the input sequence, $(01)(11)(10)(11)(01)(00)$.

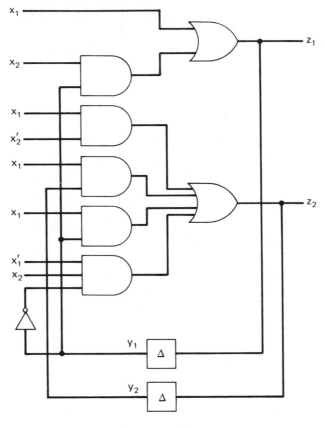

Fig. P5.5.5

5.6 State Assignments Without Critical Races for Asynchronous Circuits

The first step in realizing the state table of an asynchronous circuit is to find a state assignment (a binary code for each state) such that the resulting excitation table will not have any critical race conditions in it. Note that we do allow noncritical races in the table, because they do not affect the correctness of the circuit performance.

In this section two state-assignment methods are introduced. The first one is based on the state adjacency diagram. The state assignment obtained from this method employs the minimum number of state variables and is free of critical races. The second method is a general procedure that requires the use of n state variables (where n is the number of states) but guarantees a race-free assignment. These two methods are described below.

METHOD 1 Minimum-Variable State-Assignment Method Using the State Adjacency Diagram. A convenient way to describe the adjacency relation among the states of the state table of an asynchronous circuit is to use the *state adjacency diagram*. This diagram contains all the states in the state table. Two states are connected by a straight line only if one state is listed in the state table as the next state for the second state. For example, the state adjacency diagram of the state table of Fig. 5.6.1(a) is shown in Fig. 5.6.1(b). In this diagram, we see that each state has two adjacent states. For example, states B and C are adjacent states of state A. The machine has four states, and thus we need a minimum of two state variables to code them. With two state variables, any code can be adjacent to only two other codes. For example, the two adjacent codes to 01 are 00 and 11. If the number of adjacent states of each state of a machine does not exceed the number of state variables to be used, the existence of a race-free state assignment is then guaranteed. For our example, one such race-free state assignment for machine M_1 is

q	$y_2 y_1$
A	00
B	01
C	10
D	11

which is also displayed on the state adjacency diagram [Fig. 5.6.1(b)].

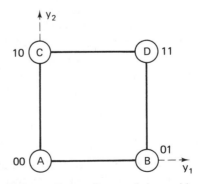

q \ $x_1 x_2$	00	01	11	10
A	(A)	B	C	(A)
B	A	D	(B)	(B)
C	A	D	(C)	D
D	(D)	(D)	A	(D)

(a) State flow table of machine M_1

(b) State adjacent diagram of the machine

Fig. 5.6.1

The preceding discussion may be summerized by the following theorem.

THEOREM 5.6.1

Let n be the number of states of a state flow table, $s = \lceil \log_2 n \rceil$ be the minimum number of state variables needed to code the n states, and m be the maximum number of adjacent states to a state in the state adjacency diagram. If $s \geq m$, the existence of a race-free minimum-variable state assignment is guaranteed.

It should be noted that Theorem 5.6.1 means that if $s < m$, a race-free minimum-variable state assignment is not guaranteed, but it does not mean that it never exists. An example of this is given in Fig. 5.6.2. From the state adjacency diagram of this

q \ $x_1 x_2$	00	01	11	10
A	(A)	B	C	D*
B	A	(B)	D	D
C	A	(C)	D	D
D	A*	C	(D)	(D)

(a) State flow table of machine M_2

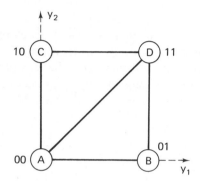

(b) State adjacent diagram of the machine

Fig. 5.6.2

machine, we see that both states A and D have three adjacent states. However, it is easy to see that the state assignment of Fig. 5.6.2(b) is a critical-race-free, minimum-variable state assignment for machine M_2, because the two races resulted from this assignment are noncritical and thus need not be eliminated. It is easy to show that if only one state is stable in a column containing a race condition, the race is not critical. The above is an example of this case.

Another situation for which a minimum-variable state assignment is possible should be mentioned here. Consider machine M_3 of Fig. 5.6.3(a), which is the same as machine M_2 except that the next state B of state A under the input $x_1 x_2 = 01$ is replaced by state D, which creates a critical race under the state assignment of Fig. 5.6.2(b). But this critical race can be eliminated by replacing this state D entry with state C, because.

   1. State C is the next state of state D under the $x_1 x_2 = 01$ column. Thus the final state will still be state C, as specified in machine M_3.

   2. State C is adjacent to state A.

   3. The machine operation is unchanged.

q \ $x_1 x_2$	00	01	11	10
A	(A)	D**	C	D*
B	A	(B)	D	D
C	A	(C)	D	D
D	A*	C	(D)	(D)

(a) Machine M_3

q \ $x_1 x_2$	00	01	11	10
A	(A)	C	C	D*
B	A	(B)	D	D
C	A	(C)	D	D
D	A*	C	(D)	(D)

(b) Machine M_3'

Fig. 5.6.3

This equivalent machine, denoted by machine M_3' [Fig. 5.6.3(b)] has the same state adjacent diagram as that of machine M_2, shown in Fig. 5.6.2(b). Again, even though $s < m$ for this machine, a critical-race-free, minimum-variable state assignment exists. The state assignment of Fig. 5.6.2(b) is such an assignment.

To summarize the preceding discussions, we have:

THEOREM 5.6.2

(a)   If only one state is stable in a column containing a race condition, the race is not critical and need not be eliminated.

(b)   Suppose that a transition from state A to state C ($A \longrightarrow C$) involves a critical race; if a multiple transition $A \longrightarrow B \longrightarrow C$ (in the same input column) can be found, then replacing this state C entry with state B will eliminate the critical race without changing the machine operation if state B can be made adjacent to state A.

METHOD 2   N-State-Variable State-Assignment Method. A general systematic procedure for obtaining a race-free state assingment can be obtained, which is described below. Let an asynchronous machine have a total of n stable and unstable states, $q_1, q_2, \ldots, q_n$. We use n state variables, $y_1, y_2, \ldots, y_n$.

Step 1   Assign each state an n-tuple $(e_1, e_2, \ldots, e_n)$, where e_i assumes the value of y_i, $i = 1, 2, \ldots, n$, which consists of only one 1:

State	Assigned n-Tuple
q_1	$(1, 0, \ldots, 0)$
q_2	$(0, 1, \ldots, 0)$
.	.
.	.
.	.
q_n	$(0, 0, \ldots, 1)$

Step 2 If there is a transition under an input symbol, x_α, from state q_i to state q_j, a transition state, q_{ij}, is created. State q_{ij} is assigned the n-tuple $(e_1, e_2, \ldots, e_n)$ with $e_k = 0$ if $k \neq i$ and $k \neq j$, and $e_k = 1$ if $k = i$ or $k = j$.

Step 3 The next-state function can be easily set as $\delta(q_{ij}, x_\alpha) = q_i$. For any input symbol, x_β, under which there is no transition from q_i to q_j, the next state, $\delta(q_{ij}, x_\beta)$, is undefined.

Step 4 For the output function for q_{ij}, we set $\lambda(q_{ij}, x_\alpha) = \lambda(q_i, x_\alpha)$ or $\lambda(q_{ij}, x_\alpha) = \lambda(q_j, x_\alpha)$; and $\lambda(q_{ij}, x_\beta)$ is unspecified.

The following example illustrates this procedure.

Example 5.6.1

Consider machine M_4, whose state table and state adjacent diagram are shown in Fig. 5.6.4(a) and (b), respectively. For our present purpose, the output function is not needed.

Step 1 Assign the following four-tuples to the four states:

State	Assigned Four-Tuple
1	0001
2	0010
3	0100
4	1000

Step 2 From the state table of machine M_4, we see that transitions exist in the pairs of states $(1, 2)$, $(1, 3)$, $(1, 4)$, $(2, 3)$, $(2, 4)$, and $(3, 4)$. We therefore create an intermediate state for each of them and assign them the appropriate codes, as follows:

Added New States	Assigned Codes
$(1, 2)$ or $(2, 1)$	0011
$(1, 3)$ or $(3, 1)$	0101
$(1, 4)$ or $(4, 1)$	1001
$(2, 3)$ or $(3, 2)$	0110
$(2, 4)$ or $(4, 2)$	1010
$(3, 4)$ or $(4, 3)$	1100

Step 3 The transition under input x_α from state q_i to state q_j now becomes

$$q_i \xrightarrow{x_\alpha} q_{ij} \xrightarrow{x_\alpha} q_j$$

In words, instead of going to state q_j directly, we specify that it goes to q_{ij} first, then to q_j. For example:

$$\text{state 1 (0001)} \xrightarrow{01} \text{state (1, 4) (1001)} \xrightarrow{01} \text{state 4 (1000)}$$

q \ $x_1 x_2$	00	01	11	10
1	①	4	3	①
2	1	②	4	②
3	4	2	③	1
4	④	④	④̄	2

(a) State flow table of machine M_4

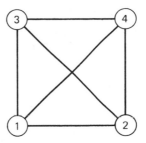

(b) State adjacency diagram

$y_1 y_2 y_3 y_4$ \ $x_1 x_2$	00	01	11	10
0001	0001	1001	0101	0001
0010	0011	0010	1010	0010
0100	1100	0110	0100	0101
1000	1000	1000	1000	1010
0011	0010	–	–	–
0101	–	–	0100	0001
1001	–	1000	–	–
0110	–	0010	–	–
1010	–	–	1000	0010
1100	1000	–	–	–

Original states: 0001, 0010, 0100, 1000

New transient states: 0011, 0101, 1001, 0110, 1010, 1100

(c) A race-free binary-coded state flow table
with six added intermediate states.

Fig. 5.6.4 Example to illustrate the procedure for obtaining a
race-free state assignment.

So in row 0001, under input 01, we fill it with state 1001. Because of the way in which we assign the codes to states 1, (1, 4), and 4 (they are adjacent codes), no race is ever possible. The entry of row 1001 under input 01 is filled with 1000 because in the original state table (row 1) there is a transition from state 1 to state 4 under input 01. All the remaining transitions can be constructed in a similar way. The complete excitation table is shown in Fig. 5.6.4(c). The entries marked with dashes are unspecified entries.

As a final remark, this assignment procedure ensures a race-free circuit but requires n state variables for a state machine, which is more than the minimum. Since by this method the transition between states of machines is always accomplished through an intermediate transient state, the operating speed of the resulting circuit is adversely affected. Nevertheless, this procedure is straightforward, easy to apply, and, more important, ensures a race-free state assignment for any asynchronous circuit.

Exercise 5.6

1. Give a state assignment without critical races to each of the following asynchronous machine using one of the methods described in this section. A state assignment using a minimum number of state variables is desirable whenever possible.

TABLE P5.6.1

q \ x	I_0	I_1	I_3	I_3
A	Ⓐ	B	Ⓐ	B
B	C	Ⓑ	Ⓑ	Ⓑ
C	Ⓒ	D	B	D
D	Ⓓ	Ⓓ	A	Ⓓ

(a)

q \ x	I_0	I_1	I_2	I_3
A	Ⓐ	C	D	Ⓐ
B	A	Ⓑ	C	Ⓑ
C	E	Ⓒ	Ⓒ	B
D	A	B	Ⓓ	A
E	Ⓔ	B	C	A

(b)

q \ x	I_0	I_1	I_2	I_3
A	Ⓐ	C	Ⓐ	B
B	A	Ⓑ	A	Ⓑ
C	Ⓒ	Ⓒ	E	D
D	C	B	Ⓓ	Ⓓ
E	Ⓔ	F	Ⓔ	D
F	E	Ⓕ	A	B

(c)

5.7 Hazard-Free Asynchronous Circuits

Having studied one transient phenomenon, race, and methods to eliminate it, we now study another transient phenomenon, *hazard*, which happens in both combinational and sequential circuits but which has serious effects only in asychronous sequential circuits. In asynchronous circuits, although only one input variable of an asynchronous sequential circuit is allowed to change at a time and only one state variable is to change at a time if a race-free state assignment [such as the one presented in Fig. 5.6.4(c)] is used, it is still possible to have some undesirable operating situations, because a variable has two forms, the uncomplemented form and the complemented form, which may not change at exactly the same time instant. Hazard is a transient phenomenon that happens during the change of such a single input variable.

The uncomplemented and complemented forms of a variable, say x_i and x_i', are usually obtained from two distinct lines. For instance, x_i and x_i' may be obtained from the two output terminals of a flip-flop, or the x_i' may be obtained from the x_i through an inverter, or vice versa. Thus when x_i changes state, say from 0 to 1, x_i' also changes state, from 1 to 0, but this change, which does not necessarily occur at exactly the same instant, may be at a slightly earlier or a slightly instant. If x_i' is obtained from x_i through an inverter, then the change of x_i' will be slightly later because of the delay in the inverter. Because of this slight difference in the time of change, the circuit may momentarily give some incorrect outputs. Since we consider the change of only one input variable at a time, the vertices representing the input combinations before and after the change are adjacent to each other. We define hazard as follows.

DEFINITION 5.7.1

Hazard is the transition between a pair of adjacent input states during which it is possible for a momentary incorrect output to occur.

In the following we shall first introduce and explain several types of hazards that occur in combinational circuits and how to avoid them by using examples.

DEFINITION 5.7.2

In a combinational circuit, if the outputs before and after change of the input are the same, then the hazard is called a *static hazard*.

There are two types of static hazard in a combinational circuit: static 1 hazard and static 0 hazard, which are defined below.

DEFINITION 5.7.3

In a combinational circuit, if the outputs before and after the change of input are both 1, with an incorrect output 0 in between (i.e., the output sequence is 1–0–1), then the hazard is called *a static 1 hazard*. If the outputs before and after the change of input are both 0, with an incorrect output 1 in between (i.e., the output sequence is 0–1–0), then the hazard is called a *static 0 hazard*.

First, let us study the static 1 hazard and a means to avoid it. Our discussion will be confined to the two-level AND–OR and two-level OR–AND realizations. The hazard situation in higher-level realizations is generally quite complicated and will not be discussed.

Example 5.7.1

Consider the realization of machine M_5, whose flow table is described in Fig. 5.7.1(a). The state adjacency diagram of this machine is shown in Fig. 5.7.1(b). It is seen that every state has two adjacent states. By Theorem 5.6.1, there exists a minimum-variable race-free state assignment for this machine. One way to obtain such a state assignment is to use the Gray code:

q	$y_1 y_2 y_3$
0	000
1'	001
1	011
2'	010
2	110
3'	111
3	101
0'	100

$$(5.7.1)$$

The corresponding excitation table is shown in Fig. 5.7.1(c), in which the four output symbols 0, 1, 2, and 3 are coded by $z_1 z_2 = 00, 01, 10$, and 11, respectively. The Y_1, Y_2, Y_3, z_1, and z_2 in minimum sum-of-products form are

$$Y_1 = xy_1 + y_1 y_3 + x'y_2 y_3' \qquad (5.7.2)$$

$$Y_2 = xy_2 + y_2 y_3' + x'y_1' y_3 \qquad (5.7.3)$$

$$Y_3 = x'y_3 + xy_1 y_2 + xy_1' y_2' \qquad (5.7.4)$$

$$z_1 = y_1 y_3 + y_2 y_3' \qquad (5.7.5)$$

and

$$z_2 = y_3 \qquad (5.7.6)$$

Since the output functions z_1 and z_2 do not involve the input variable x, it is impossible to have any hazard; thus we do not need to consider them.

Let us first consider Y_1. The following sequence of changes of the values of the variables may occur in the two-level AND–OR realizations of Y_1. Let the input combination before time t_1 be $xy_1 y_2 y_3 = 1110$. The circuit response of Y_1 of Eq. (5.7.3) to this input at time $t < t_1$ is

$$Y_1 = 1 \cdot 1 + 1 \cdot 0 + 0 \cdot 1 \cdot 1 = 1 \qquad (5.7.7)$$

Now suppose that x changes from 1 to 0 at t_1. Note that x' is considered as a distinct variable which is obtained from x through an inverter. Suppose that x' changes from 0 to 1 at a slightly later time t_2. During the time interval $t_1 < t < t_2$, variable values are $y_1 = 1$, $y_2 = 1$, $y_3 = 0$, and *both x and x' are equal to 0*. The Y_1 during this time interval is computed to be

$$Y_1 = 0 \cdot 1 + 1 \cdot 0 + 0 \cdot 1 \cdot 1 = 0 \qquad (5.7.8)$$

Finally, consider the circuit response of Y_1 to the input combination after t_2, which is 0011. The responses are found to be

$$Y_1 = 0 \cdot 1 + 1 \cdot 0 + 1 \cdot 1 \cdot 1 = 1 \qquad (5.7.9)$$

From Eqs. (5.7.7)–(5.7.9), the following sequence of changes of the values of the variables is observed.

	$t < t_1$	$t_1 < t < t_2$	$t > t_2$
x	1	0	0
x'	0	0	1
y_1	1	1	1
y_2	1	1	1
y_3	0	0	0
Y_1	1	0	1

q \ x	0	1
0	⓪,0	1′,–
1	①,1	2′,–
2	②,2	3′,–
3	③,3	0′,–
1′	1,–	①′,1
2′	2,–	②′,2
3′	3,–	③′,3
0′	0,–	⓪′,0

(a)

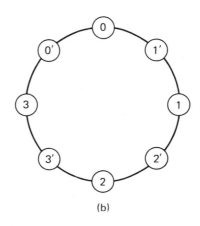

(b)

$y_1 y_2 y_3$ \ x	0	1
000	⟨000⟩,00	001,dd
011	⟨011⟩,01	010,dd
110	⟨110⟩,10	111,dd
101	⟨101⟩,11	100,dd
001	011,dd	⟨001⟩,01
010	110,dd	⟨010⟩,10
111	101,dd	⟨111⟩,11
100	000,dd	⟨100⟩,00

$Y_1 Y_2 Y_3, z_1 z_2$

(c)

Fig. 5.7.1 (a) Flow table of machine M_5; (b) state adjacent diagram; (c) binary-coded flow table; (d) minimized next-state functions and output functions.

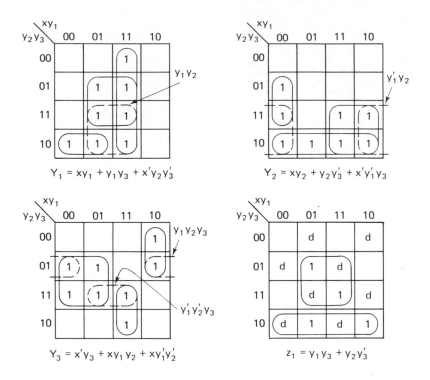

(d)

Fig. 5.7.1   (continued)

This indicates that the two-level AND–OR realization of Y_1 has static 1 hazard, because they produce an incorrect output 0 during the transient from $x110$ to $x'110$. Similarly, we can show that the two-level AND–OR realizations of Y_2 and Y_3 have the following static 1 hazards:

	$t < t_1$	$t_1 < t < t_2$	$t > t_2$
x	1	0	0
x'	0	0	1
y_1	0	0	0
y_2	1	1	1
y_3	1	1	1
Y_2	1	0	1

This static-1 hazard occurs during the transient from $x011$ to $x'011$.

	$t < t_1$	$t_1 < t < t_2$	$t > t_2$
x	1	0	0
x'	0	0	1
y_1	0	0	0
y_2	0	0	0
y_3	1	1	1
Y_3	1	0	1

This static-1 hazard occurs during the transient from $x001$ to $x'001$.

	$t < t_1$	$t_1 < t < t_2$	$t > t_1$
x	1	0	0
x'	0	0	1
y_1	1	1	1
y_2	1	1	1
y_3	1	1	1
Y_3	1	0	1

This static-1 hazard occurs during the transient from $x111$ to $x'111$.

These static 1 hazards can be seen from the Karnaugh maps of the functions Y_1, Y_2, and Y_3 shown in Fig. 5.7.1(d). A static 1 hazard will occur in the two-level AND–OR realization of a next-state function Y if:

1. The function Y has two product terms π_1 and π_2, π_1 containing an x and π_2 containing an x'.

2. There exists an input combination $y_1 = \sigma_1, y_2 = \sigma_2, \ldots, y_s = \sigma_s$, where the σ_i's are 0 and 1, such that $\pi_1 = 1$ when $xy_1y_2 \ldots y_s = 1\sigma_1\sigma_2 \ldots \sigma_s$ and $\pi_2 = 1$ when $x'y_1y_2 \ldots y_s = 0\sigma_1\sigma_2 \ldots \sigma_s$. This implies that the function $Y = 1$ when $xy_1y_2 \ldots y_s = 1\sigma_1\sigma_2 \ldots \sigma_s$ and $0\sigma_1\sigma_2 \ldots \sigma_s$.

3. The function $Y = 0$ when x and x' are considered as two independent variables and $x = 0$, $x' = 0$, $y_1 = \sigma_1$, $y_2 = \sigma_2, \ldots, y_s = \sigma_s$.

For example, the two terms xy_1 and $x'y_2y_3'$ of Y_1 satisfy the above three conditions, since

(a) for $x = 1$, $y_1 = 1$, $y_2 = 1$, and $y_3 = 0$, $xy_1 = 1$, which implies that $Y = 1$.
(b) For $x = 0$, $y_1 = 1$, $y_2 = 1$, and $y_3 = 0$, $x'y_2y_3' = 1$, which implies that $Y = 1$
(c) For $x = 0$, $x' = 0$, $y_1 = 1$, $y_2 = 1$, and $y_3 = 0$, the function $Y = 0$.

Every static hazard in the two-level AND–OR and two-level OR–AND realizations can be avoided by using a sort of redundancy. For example, the static 1 hazard of Y_1 can be avoided by adding the product term $y_1y_2y_3'$ to Y_1 "to bridge the gap" between the terms xy_1 and $x'y_2y_3'$, as indicated in the following Karnaugh map.

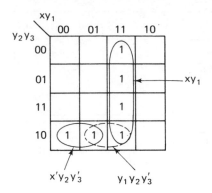

Note that the inclusion of the term $y_1 y_2 y_3'$ in Y_1 does not change the function. This term can be realized by an AND gate whose output will be 1 before t_1, between t_1 and t_2, and after t_2, because $y_1 y_2 y_3'$ is independent of x. Hence the static 1 hazard is avoided. It is worth noting that in searching for a term to avoid a static 1 hazard, we should select the one with the fewest number of literals. For example, to avoid the above static 1 hazard, we should use the term $y_1 y_2$ instead of $y_1 y_2 y_3'$, since the inclusion of the term $y_1 y_2$ in Y_1 will also not change the function and makes $Y_1 = 1$ for $x = 0$, $x' = 0$, $y_1 = 1$, $y_2 = 1$, and $y_3 = 0$ [see Fig. 5.7.1(d)]. Referring to Fig. 5.7.1(d), the three static-1-hazard-free next-state functions are

$$Y_1 = xy_1 + y_1 y_3 + x' y_2 y_3 \mid + y_1 y_2 \mid \tag{5.7.10}$$

$$Y_2 = xy_2 + y_2 y_3' + x' y_1' y_3 \mid + y_1' y_2 \mid \tag{5.7.11}$$

$$Y_3 = x' y_3 + xy_1 y_2 + xy_1' y_2' \mid + y_1 y_2 y_3 + y_1' y_2' y_3 \mid \tag{5.7.12}$$

The terms in dashed frames are the terms added for avoiding static 1 hazards. The complete race-free and static-1-hazard-free realization of machine M_5 is shown in Fig. 5.7.2.

To explain the static 0 hazard in a combinational circuit, let us consider the complement of one of the functions Y_1, Y_2, and Y_3, say Y_1 of Eq. (5.7.2).

$$Y_1' = (x_1' + y_1')(y_1' + y_3')(x + y_2' + y_3) \tag{5.7.13}$$

It is easy to show by duality that the two-level OR–AND realization of Eq. (5.7.13) has a static 0 hazard during the transition from 1110 to 0110, since the following transient phenomenon would occur:

Function	$t < t_1$	$t_1 < t < t_2$	$t > t_2$
Y_1'	0	1	0

We see that a static 1 hazard in a two-level AND–OR realization of a function f corresponds to a static 0 hazard in a two-level OR–AND realization of the complement of

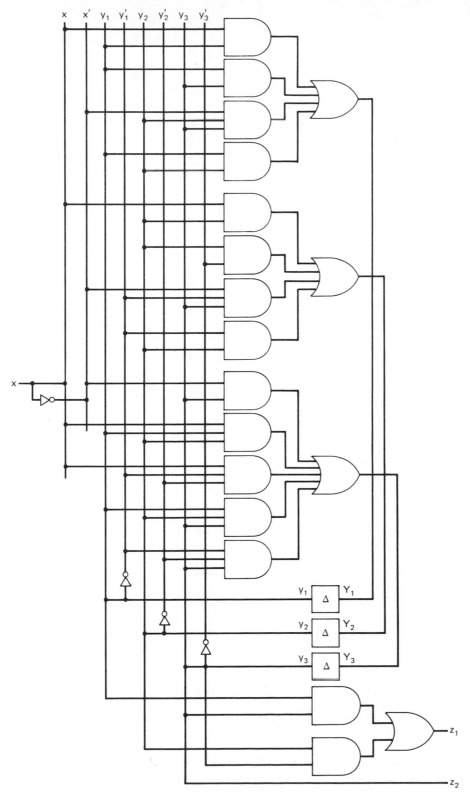

Fig. 5.7.2 Race-free hazard-free realization of machine M_5.

f. Therefore, similar techniques, described above for avoiding static 1 hazards, can be applied to avoiding static 0 hazards. From the Karnaugh map of Y'_1 [see Fig. 5.7.3(a)], it is seen that if we add the term $(y'_1 + y'_2)$, realized by a first-level OR gate, to the realization of Y'_1, the resulting circuit will be static-0-hazard-free; the output of this added OR gate will be 0, for the time before t_1, between t_1 and t_2, and after t_2, since $(y'_1 + y'_2)$ is independent of x. Figure 5.7.3(b) shows a static-0-hazard-free realization of Y'_1 of

$$Y'_1 = (x' + y'_1)(y'_1 + y'_3)(x + y'_2 + y_3)(y'_1 + y'_2) \qquad (5.7.14)$$

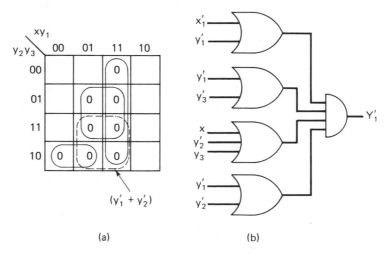

(a)   (b)

Fig. 5.7.3 (a) Karnaugh map of Y'_1; (b) static-0-hazard-free realization of Y'_1.

LEMMA 5.7.1

No static 0 (static 1) hazard can happen in a two-level AND–OR (OR–AND) realization of a switching function *f*.

Proof: To show that no static 0 hazard can happen in a two-level AND–OR realization of a switching function is to show that no adjacent input–state input (e.g., $y_1 y_2 y_3 x = 1110$ and 0110 in the above example) can produce an output sequence 0–1–0. To be able to output a 0 before t_1 and after t_2, the cells corresponding to the two adjacent input combinations in the Karnaugh map must be of value 0. In the sum-of-products Karnaugh map of *f*, it is known that for an input combination, $f = 1$ if and only if *f* contains the minterm corresponding to the input combination (i.e., the cell corresponding to this input combination is a 1-cell). Hence no static 0 hazard can happen in a two-level AND–OR realization of *f*. That no static 1 hazard can happen in a two-level OR–AND realization of *f* can be proved similarly. ∎

From this lemma we have

THEOREM 5.7.1

If a two-level AND–OR (OR–AND) realization of a switching function *f* is static-1 (static-0)-hazard-free, then it is free of all static hazards.

According to Theorem 5.7.1, we find that the realizations of Y_1, Y_2, and Y_3 in Fig. 5.7.2 and the realization of Y'_1 in Fig. 5.7.3(b) are free of all static hazards.

Besides static 1 and static 0 hazards, there is another type of hazard, dynamic hazard.

DEFINITION 5.7.4

In a combinational circuit, if the outputs before and after the change of input are different, and the output changes three times instead of once and passes through an additional temporary sequence of 01 or 10 in going to the final output (i.e., the output sequence is either 1–0–1–0 or 0–1–0–1), the hazard so created is called a *dynamic hazard*.

As mentioned above, our main interest is to avoid hazard in the class of two-level AND–OR (OR–AND) circuits. It can be shown that dynamic hazards do not happen in two-level AND–OR and two-level OR–AND circuits. The proof is left to the reader (problem 1).

After having discussed the hazards in combinational circuits, we shall now study a type of hazard in sequential circuits. It is called *essential hazard*. This type of hazard happens when the changes of one form of an input variable x_i, say the complemented form, is slower than the change of the state variable resulting from a change in the uncomplemented form of the input variable. This can best be explained through a simple example. Consider machine M_6, whose state flow table is given in Fig. 5.7.4(a). The next-state functions are

$$Y_1 = y_1 x + y_2 x' \qquad (5.7.15)$$

$$Y_2 = y'_1 x + y_2 x' \qquad (5.7.16)$$

To avoid static hazards, redundant terms are added to the expressions of Y_1 and Y_2.

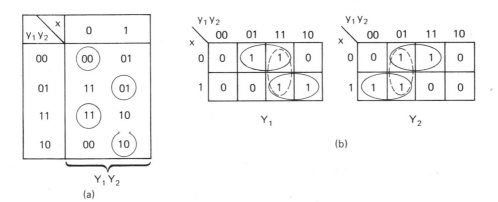

(a)

(b)

Fig. 5.7.4 (a) State flow table of machine M_6; (b) Karnaugh maps of Y_1 and Y_2.

With reference to the Karnaugh maps of Y_1 and Y_2 in Fig. 5.7.4(b), it is found that the static-hazard-free next-state functions are

$$Y_1 = y_1 x + y_2 x' + y_1 y_2 \qquad (5.7.17)$$
$$Y_2 = y_1' x + y_2 x' + y_1' y_2 \qquad (5.7.18)$$

Assume that initially $x = 0$, $y_1 = 0$, and $y_2 = 0$. Thus $Y_1 = 0$, $Y_2 = 0$ and the circuit is in stable state 00. Now suppose that x changes from 0 to 1 and *at the same time* or *time interval* $t_1 < t < t_2$, where $t = t_2 - t_1$ is very small. The circuit will land at stable state 01. This may be described by the following table:

	$t < t_1$	$t_1 < t < t_2$	$t > t_2$
x	0	1	1
x'	1	0	0
y_1	0	0	0
y_2	0	0	1
Y_1	0	0	0
Y_2	0	1	1
$Y_1 Y_2$	00	01	01

But x changes from 0 to 1 at time t_1 and x' changes from 1 to 0 at a later time, t_2, after some computation of Y_1 and Y_2 has been completed. Here x and x' are considered to be two distinct variables. The sequence of changes of the values of the variables are as follows:

	$t < t_1$	$t_1 < t < t_2$	$t = t_2$	$t > t_2$
x	0	1	1	1	1
x'	1	1	1	0	0
y_1	0	0	0	1	1
y_2	0	0	1	1	0
Y_1	0	0	1	1	1
Y_2	0	1	1	0	0
$Y_1 Y_2$	00	01 11	10	10

From the analysis we see that an essential hazard can happen in the machine of Fig. 5.7.4(a), and this essential hazard has caused an incorrect transition of the asynchronous circuit. Therefore, essential hazards cannot be tolerated. Fortunately, essential hazards can be avoided by inserting delay units in the feedback paths from the next-state variables, thus delaying the change of state variables until the change of input variables has been completed.

Exercise 5.7

1. Prove that dynamic hazards do not happen in two-level AND–OR and two-level OR–AND circuits.

2. Realize each of the following switching functions with a (static) hazard-free two-level AND–OR or OR–AND circuit:

   (a) $f(x_1, x_2, x_3, x_4) = x_1 x_2 + x_1' x_3 x_4$

   (b) $f(x_1, x_2, x_3, x_4) = x_1' x_3 + x_2 x_4 + x_1 x_3' x_4'$

   (c) $f(x_1, \ldots, x_5) = x_2 x_3' x_4' + x_3' x_4' x_5 + x_1 x_2 x_4 x_5 + x_2 x_3 x_4 x_5$
   $+ x_1 x_2' x_4' x_5 + x_1' x_2' x_3' x_5'$

   (d) $f(x_1, x_2, x_3, x_4) = (x_2' + x_3 + x_4)(x_1' + x_4)(x_1 + x_3' + x_4')(x_2 + x_3 + x_4')$

   (e) $f(x_1, x_2, x_3, x_4) = (x_1' + x_4)(x_2 + x_3 + x_4')(x_3' + x_4')$

   (f) $f(x_1, \ldots, x_6) = (x_3 + x_4 + x_5' + x_6')(x_2' + x_3 + x_6')(x_1' + x_2 + x_4')$
   $\cdot (x_1 + x_3' + x_4' + x_5')(x_3' + x_4 + x_6)$

3. Realize each of the following sets of next-state functions of an asynchronous machine with a hazard-free circuit:

   (a) $Y_1 = y_1 y_3 + x y_1 + x' y_2 y_3'$
   $Y_2 = x y_2 + y_2 y_3 + x' y_1' y_3$
   $Y_3 = x' y_3 + x y_1 y_2 + x y_1' y_2'$

   (b) $Y_1 = y_1 y_3 + x y_1 + x' y_2 y_3' + y_1 y_2$
   $Y_2 = x y_2 + y_2 y_3' + x' y_1' y_3 + y_1' y_2$
   $Y_3 = x' y_3 + x y_1 y_2 + x y_1' y_2' + y_1' y_2' y_3 + y_1 y_2 y_3$

4. Show that the following asynchronous machines have essential hazards.

TABLE P5.7.4

$y_1 y_2$ \ x	0	1
00	00	01
01	11	01
10	00	10
11	11	10

(a)

$y_1 y_2 y_3$ \ x	0	1
010	110	010
110	110	111
111	101	111
101	101	111

(b)

5.8 Synthesis of Asynchronous Sequential Circuits

After having presented the analysis of asynchronous sequential circuits, we now turn to the reverse problem, the synthesis problem. A list of assumptions that are made when the basic model of Fig. 5.5.1 is used for the synthesis of asynchronous circuits is as follows:

A1. The combinational logic contains only bounded stray delays.

A2. Input changes are restricted to the change of a single binary input.

A3.   The combinational logic is hazard-free.

A4.   Input changes occur only when the circuit is in a stable state (fundamental mode operation).

A5.   The delays in the feedback loops are pure delays and are sufficiently large to allow the combinational logic to stabilize from an input change before that input change causes a change of state.

A6.   For any input combination, the circuit becomes stable in some state (i.e., $Y_i = y_i$, for $i = 1, \ldots, s$).

A7.   For each stable state and each allowable input change, the circuit does not have any critical races among the next-state variables.

A8.   The circuit is assumed to respond to input changes only; repetitions of the same input cannot be allowed.

The synthesis procedure for asynchronous circuits is described as follows:

Step 1   Find the transition table from a verbal statement of the desired circuit performance. This is the least systematic step in the synthesis procedure, which has been discussed in detail and illustrated by several examples in Section 5.1.

Step 2   Find the primitive flow table from the transition table obtained in step 1. A primitive flow table is a transition table in which only one stable state appears in each row and the outputs are specified for stable states only.

Step 3   Minimize the primitive flow table. Since a primitive flow table is a transition table with unspecified outputs, we can apply the method described in Section 5.2.2 to it to obtain a minimized flow table. It should be noted that whenever a circled entry and an uncircled entry are to be combined, the resulting entry is circled, since the corresponding state must be stable.

Step 4   Assign state variables to the rows of the minimized primitive flow table such that no critical races and cycles will occur.

Step 5   Assign the input and output variables to code the input symbols and output symbols of the machine, respectively.

Step 6   Derive the simplest hazard-free next-state and output functions from the excitation and output tables.

Step 7   Construct the circuit.

This procedure is illustrated by the following example.

Example 5.8.1

Design an asynchronous circuit with one-level input x. The circuit is to serve as a modulo-4 counter, which counts the number of times the value of x changes from 0 to 1 and ignores the value of x changing from 1 to 0.

Step 1 The transition table of a modulo-4 counter is given in Table 5.8.1(a). Note that all the next states in this table are stable states.

TABLE 5.8.1 Transition and Flow Tables of Example 5.8.1

q \ x	0	1
0	0, 0	1, 1
1	1, 1	2, 2
2	2, 2	3, 3
3	3, 3	0, 0

(a) Transition table

q \ x	0	1
0	(0)0,	
1	(1)1	
2	(2)2	
3	(3)3	
1'		(1) 1
2'		(2) 2
3'		(3) 3
0'		(0) 0

(b) Partial primitive flow table

q \ x	0	1
0	(0)0	1', —
1	(1)1	2', —
2	(2)2	3', —
3	(3)3	0', —
1'	1, —	(1) 1
2'	2, —	(2) 2
3'	3, —	(3) 3
0'	0, —	(0) 3

(c) Primitive flow table, which is also the minimized flow table

Step 2 Since a primitive flow table allows only one stable state in each row and we have eight next states in the transition table which are stable states, we need at least eight rows to construct the primitive flow table of this circuit. Expand the transition table as follows: Leave the four next states and outputs under the input column $x = 0$ as they are and circle the states to indicate that they are stable states. Move the four next states and outputs under the input column $x = 1$ down to rows 5–8 and circle the states. This diagram is shown in Table 5.8.1(b). Next we want to fill in the four entries on the right upper corner and the four entries on the lower left corner. From the transition table of Table 5.8.1(a), we find that the four next states on the right upper corner should be states 1', 2', 3', and 0', and the four next states on the left lower corner should be states 1, 2, 3, and 0. Since they are all unstable states, no output specification is necessary. The unspecified outputs are indicated by the dashes. The complete primitive flow table is shown in Table 5.8.1(c).

Step 3 To minimize the primitive flow table of Table 5.8.1(c), we apply the method described in Section 5.2.2. It is easy to show that no reduction in the number of states is possible, regardless of how the unspecified outputs are assigned. Hence the primitive flow table of this example is also the minimized primitive flow table.

By now the reader must have noticed that this machine is the same as machine M_6 described in Fig. 5.7.1(a). The remaining synthesis steps (steps 4–7) of realizing the primitive flow table of Table 5.8.1(c) were given in Example 5.7.1. A race-free, hazard-free realization of this machine (modulo-4 counter) was given in Fig. 5.7.2.

Exercise 5.8

1. The primitive flow table of an asynchronous circuit is shown in Table P5.8.1.
   (a) Minimize the flow table.

TABLE P5.8.1

q \ x_1x_2	00	01	11	10	z
q_1	q_1	—	q_7	—	0
q_2	q_2	q_8	—	—	0
q_3	q_7	q_3	—	q_4	0
q_4	—	q_8	q_6	q_4	1
q_5	—	q_8	q_6	q_5	1
q_6	q_2	—	q_6	q_5	1
q_7	q_7	—	q_6	q_4	0
q_8	q_1	q_8	—	q_5	0

(b) Give a state assignment and find the functions of the next-state and output variables.

(c) Realize this machine with an asynchronous circuit. Use delay memory devices.

2. (a) An asynchronous circuit has two input variables, x_1 and x_2, and one output, z. z is 1 if x_1 changes from 0 to 1, and 0 if x_2 changes from 0 to 1. Find a primitive flow table of this circuit.

(b) Find the minimum-state flow table of this machine.

(c) Give a state assignment and find the next-state and output functions.

(d) Realize the machine with an asynchronous circuit using J–K flip-flops.

3. An asynchronous circuit is to be designed with two input variables, x_1 and x_2, and two corresponding output variables, z_1 and z_2. z_i, $i = 1, 2$, is 1 if and only if x_i changes from 0 to 1. The output z_1 (z_2), once it becomes 1, will remain at 1 until x_2 (x_1) changes from 1 to 0.

(a) Construct a primitive flow table and simplify it, if possible.

(b) Give a state assignment and find the next-state and output functions.

(c) Find a TFF asynchronous circuit that realizes this machine.

4. A J–K flip-flop asynchronous circuit is to be designed which has two input variables x_1 and x_2, and a single output variable z. z is 1 if and only if the same input variable changes two or more times consecutively. It is required that this circuit possesses the following desired features: (1) race-free and (2) hazard-free.

Bibliographical Remarks

    The basic sequential machine theory can be found in any standard textbook on switching circuit theory and logic design, such as references 1–4. A good treatment on the analysis and synthesis of sequential circuits is given in reference 2. For general references on the theory of asynchronous circuits or machines, references 1 (Chapters 9 and 10), 2 (Chapters 5 and 6), 3 (Chapter 10), 4 (Chapter 22), and 5 are recommended. References 7, 8, 13, 17, and 19 are original papers on the analysis and synthesis of asynchronous circuits. Analysis and synthesis of asynchronous sequential circuits using edge-triggered flip-flops can be found in reference 23. The problem of state assignment for asynchronous machine is treated extensively in references 10, 16, 24, and 26. A considerable amount of research work is still being carried out concerning the state-assignment problem. The analysis of various types of hazards in asynchronous circuits and the design of hazard-free sequential circuits can be found in references 6, 9–12, 14, 15, 18, and 25. The fault diagnosis of asynchronous machines and the design of fault-tolerant asynchronous machines are reported in references 20, 21, and 22.

References

Books

1. D. D. Givone, *Introduction to Switching Circuit Theory*, McGraw-Hill, New York, 1970.

2. E. J. McCluskey, *Introduction to the Theory of Switching Circuits*, McGraw-Hill, New York, 1965.

3. C. L. Sheng, *Introduction to Switching Logic*, Intext, Scronton, Pa., 1972.

4. H. C. Torng, *Switching Circuits Theory and Logic Design*, Addison-Wesley, Reading, Mass., 1972.

5. S. H. Unger, *Asynchronous Sequential Switching Circuits*, Wiley–Interscience, New York, 1970.

Papers

6. D. B. Armstrong, A. D. Friedman, and P. R. Menon, "Realization of Asynchronous Sequential Circuits Without Inserted Delay Elements," *IEEE Trans. Computers*, Vol. C-17, Feb. 1968, pp. 129–134.

7. D. D. Aufenkamp and F. E. Hohn, "Analysis of Sequential Machines," *IRE Trans. Electron. Computers*, Vol. EC-6, No. 4, 1957, pp. 276–285.

8. D. D. Aufenkamp, "Analysis of Sequential Machines II," *IRE Trans. Electron. Computers*, Vol. EC-7, No. 4, 1958, pp. 299–306.

9. E. B. Eichelberger, "Hazard Detection in Combinational and Sequential Switching Circuits," *IBM J. Res. Develop.*, Vol. 9, Mar. 1965, pp. 90–99.

10. A. D. Friedman, R. L. Graham, and J. D. Ullman, "Universal Single Transition Time Asynchronous State Assignments," *IEEE Trans. Computers*, Vol. C 18, June 1969, pp. 541–547.

11. J. Hlavicka, "Essential Hazard Correction Without the Use of Delay Elements," *IEEE Trans. Computers*, Vol. C-19, Mar. 1970, pp. 232–238.

12. A. B. Howe and C. L. Coates, "Logic Hazards in Threshold Networks," *IEEE Trans. Computers*, Vol. C-17, Mar. 1968, pp. 238–251.

13. D. A. Huffman, "The Synthesis of Sequential Circuits," *J. Franklin Inst.*, Vol. 257, Pt. 1, Mar. 1954, pp. 161–190; Pt. 2, Apr. 1954, pp. 275–303. Also available in *Sequential Machines: Selected Papers*, E. F. Moore (ed.), Addison Wesley, Reading, Mass., 1964, pp. 3–62.

14. D. A. Huffman, "Design and Use of Hazard Free Switching Networks," *J. ACM*, Vol. 4, Jan. 1957, pp. 47–62.

15. S. B. Lerner, "Hazard Correction in Asynchronous Sequential Circuits," *IEEE Trans. Electron. Computers*, Vol. EC-14, Apr. 1965, pp. 265–267.

16. C. N. Liu, "A State Variable Assignment Method for Asynchronous Sequential Machines," *J. ACM*, Vol. 10, Apr. 1963, pp. 209–216.

17. G. K. Maki, J. H. Tracey, and R. J. Smith II, "Generation of Design Equations in Asynchronous Sequential Circuits," *IEEE Trans. Computers*, Vol. C-18, May 1969, pp. 467–472.

18. W. S. Meisel and R. S. Kashef, "Hazards in Asynchronous Sequential Circuits," *IEEE Trans. Computers*, Vol. C-18, Aug. 1969, pp. 752–759.

19. D. E. Muller and W. S. Bartky, "A Theory of Asynchronous Circuits," *Proc. Int. Symp. Theory of Switching, Part II*, Harvard University, Apr. 1959, pp. 51–58.

20. W. W. Patterson and G. A. Metze, "A Fault-Tolerant Asynchronous Machines," *Proc. 1972 Int. Symp. Fault-Tolerant Computing*, June 1972, pp. 176–181.

21. D. K. Pradhan and S. M. Reddy, "Fault-Tolerant Asynchronous Networks," *IEEE Trans. Computers*, Vol. C-22, No. 7, 1973, pp. 662–668.

22. S. Seshu and D. N. Freeman, "The Diagnosis of Asynchronous Sequential Switching Systems," *IRE Trans. Electron. Computers*, Vol. EC-11, August 1962, pp. 459–465.

23. J. R. Smith, Jr. and C. H. Roth, Jr., "Analysis and Synthesis of Asynchronous Sequential Networks Using Edge-Sensitive Flip-Flops," *IEEE Trans. Computers*, Vol. C-20, Aug. 1971, pp. 847–855.

24. J. H. Tracey, "Internal State Assignment for Asynchronous Sequential Machines," *IEEE Trans. Electron. Computers*, Vol. EC 15, Aug. 1966, pp. 551–560.

25. S. H. Unger, "Hazards and Delays in Asynchronous Sequential Switching Circuits," *IRE Trans. Circuit Theory*, Vol. CT-6, Mar. 1959, pp. 12–25.

26. S. H. Unger, "A Row Assignment for Delay-Free Realizations of Flow-Tables Without Essential Hazards," *IEEE Trans. Computers*, Vol. C-17, Feb. 1968, pp. 146–151.

6

Design of Sequential Circuits
Using Sequential Machine
Flow Charts

A combinational system may be expressed in several forms, including truth tables, Boolean expressions, and Karnaugh maps, as we have seen in previous chapters. Similarly, a sequential machine may be described by several methods, such as transition tables and transition diagrams, as in Chapter 5. This chapter presents another representation, called a *sequential machine flow chart* (or *flow-chart machine* or *algorithmic state machine chart*), which lends itself to the realization of even very complex design algorithms. The design of digital circuits and systems using the sequential machine flow chart has found great acceptance among practicing engineers in the industry, yet is so far little known in academic circles.

The most familiar type of flow chart to us is the algorithmic or operational flow chart, and the most familiar realization of it is the software realization (i.e., a computer program). In this chapter a hardware realization of an algorithmic flow chart via a sequential machine flow chart is presented. As mentioned in the previous chapters, in a digital circuit design, it is impossible to develop a formal procedure for going from a word statement to a transition table; however, obtaining a sequential machine description from a verbal description of a design problem can be made easier through use of the sequential machine flow chart. This is because we are familiar with the ordinary (algorithmic) flow chart, and the conversion from an ordinary flow chart to a sequential machine flow chart is straightfoward.

Once a sequential-machine flow-chart description of a logic design problem is formed, the next-state functions and the output functions of the sequential machine that describe the design can be obtained readily, and hence the circuit realization. The block diagram shown shows the software and hardware implementations of an algorithmic flow chart that describes a given design specification.

This chapter is divided into six sections. The first four are indicated in the block diagram. Section 6.5 discusses the problem of state assignment which was studied in Section 5.6 for the purpose of using it to eliminate races in the design of asynchronous. The purpose of studying the state assignment problem here, however, is to minimize the circuit complexity of a logic design. Two examples, a design of an accumulator

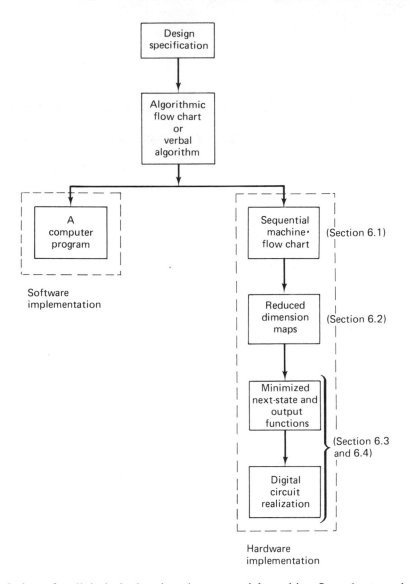

and a design of a digital clock using the sequential-machine flow-chart method are presented in Section 6.6.

The design method introduced in this chapter is completely general and is applicable to the design of any synchronous and asynchronous sequential circuits.

6.1 Sequential Machine Flow Chart

The *sequential machine flow chart*, or *SM chart*, is a state diagram that describes the overall behavior of a sequential machine. Before introducing it, a system of mnemonics for defining input and output terminals and some semantics for defining the meaning associated with each terminal must first be described. The terminals that form the inputs and outputs of the sequential machine are given names, called *mnemonics*, consisting of letters and numbers, which are used as memory aids in recalling the functions of operations associated with the logic levels on the terminals. These names are short, making them easy to manipulate in the design process. Although the choice of a naming system is personal, the system chosen is recommended. This system fulfills four objectives, which are to provide:

1. A single name for each common logic line.
2. A consistent logic level throughout the machine.
3. An indication of the terminal type (input, output).
4. A means of identifying the logic interpretation.

The basic mnemonic is a group of three or four letters which is usually formed from the first letters of the words describing the operation or function of the related terminal. For example, an output terminal might be called OUT. An initial letter H, L, Y, or N is used to signify the logic level and the type (input or output) of terminal. The prefixes H (high) and L (low) are used to signify the HIGH and LOW signal levels of an output, respectively; whereas the prefixes Y (yes) and N (no) are used to signify the logic levels 1 and 0 of an input, respectively, if the test of the input is true. For example, HOUT2 signifies that the output OUT2 occurs on a 1 (HIGH) signal and YX means that the logic level is 1 if X is true. Another prefix, I, is useful in the interpretation of the output instruction terminals. I stands for IMMEDIATE function, which means a task that is completed in the present state time. H or L without a prefix indicates a delay function, which means a task that is completed in the next state time. For example, IHOUT1 signifies that the output OUT1 occurs on a 1 (HIGH) signal in the present state time. Table 6.1.1 summarizes the interpretation of the basic prefix letters.

TABLE 6.1.1   Prefix Letters Used in SM Chart

Type of Logic Terminal	Mnemonic Initial Letter	Meaning for Logic Terminal Equal to	
		1	0
Output or instruction	H	Operation performed	Inactive
	L	Inactive	Operation performed
Input or qualifiers	Y	Statement is true	Statement is false
	N	Statement is false	Statement is true
Immediate	I	Precedes either H or L to designate an immediate function	

Figure 6.1.1 shows the model of sequential machine and the letters used for denoting the inputs, outputs, next-state function, output function, and state when the

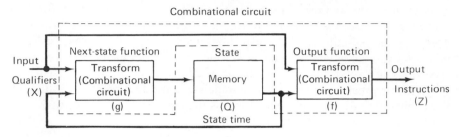

Fig. 6.1.1 More detailed sequential machine model.

SM chart design technique is employed. Note that this model is equivalent to the one given in Fig. 4.1.1(b), except the latter has combined the realizations of the next-state function and output function into one. In this model the letters X (Qualifiers) and Z (Instructions) are used for representing the inputs and outputs of the machine, the letter Q is used to represent the state, and the letters g and f are used to denote the next-state function and output function, respectively. In addition, letters $A, B, C, \ldots$ are

TABLE 6.1.2 Symbols Used in Algorithmic Flow Chart

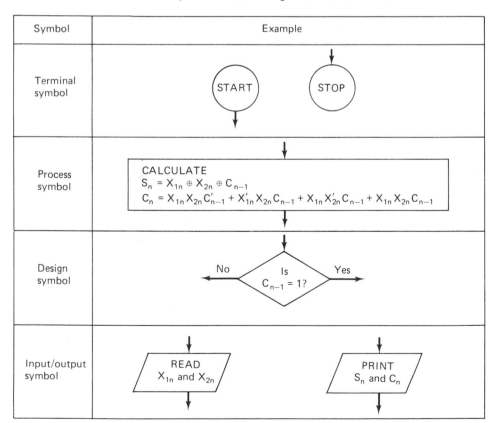

used to denote Boolean variables and lowercase letters with a circle are used to denote states of a machine. F and G are frequently used to denote Boolean functions. The don't cares in this chapter will be represented by a dash instead of using d, to distinguish them from states.

Now let us review the (ordinary) algorithmic or operational flow chart. The standard symbols used in algorithmic flow charts are shown in Table 6.1.2. A simple algorithmic flow chart describing the operation of a serial binary adder is shown in Fig. 6.1.2. The basic building blocks of an SM chart are different, however, from

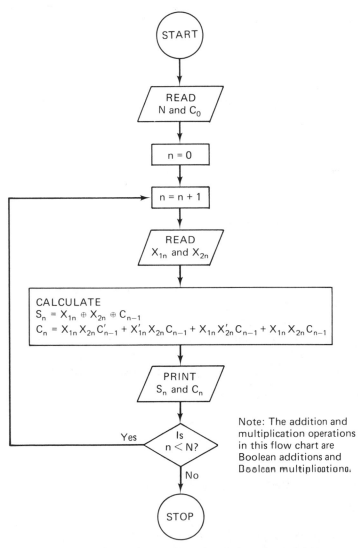

START

READ
N and C_0

$n = 0$

$n = n + 1$

READ
X_{1n} and X_{2n}

CALCULATE
$S_n = X_{1n} \oplus X_{2n} \oplus C_{n-1}$
$C_n = X_{1n} X_{2n} C'_{n-1} + X'_{1n} X_{2n} C_{n-1} + X_{1n} X'_{2n} C_{n-1} + X_{1n} X_{2n} C_{n-1}$

PRINT
S_n and C_n

Is
$n < N$?

Yes

No

STOP

Note: The addition and multiplication operations in this flow chart are Boolean additions and Boolean multiplications.

Fig. 6.1.2 Algorithmic flow chart description of a serial binary adder.

TABLE 6.1.3 Basic Building Blocks of an SM Chart

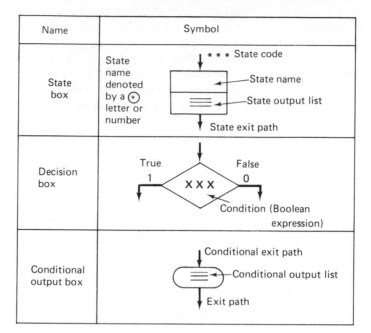

those of the algorithmic flow chart. An SM chart contains three basic elements: the state box, the decision box, and the conditional output box, which are shown in Table 6.1.3.

State Box   The *state box* is used to represent a single state of the sequential machine. The letter or number of the state is encircled on the left or right of the state box, and the binary code for the state is written along the upper edge of the box. If there are state outputs of the state, they are listed in the state box. If the output is to become true immediately, an I precedes the output name. Delayed outputs have no prefix. The state box has a single exit path, which may lead to decision boxes, conditional output boxes, or other state boxes.

Decision Box   The *decision box* describes the inputs or qualifiers to the sequential machine. Each decision box has two exit paths, one of which is taken when the Boolean expression in the box is true and the other which is taken when the condition is false. Arrows are used to indicate a conditional exit path. It should be noted that the exit paths do not indicate any time dependence. Only state boxes represent functions of time.

Conditional Output Box   The *conditional output box* is used to describe outputs that are dependent on one or more inputs as well as the state of the machine. The

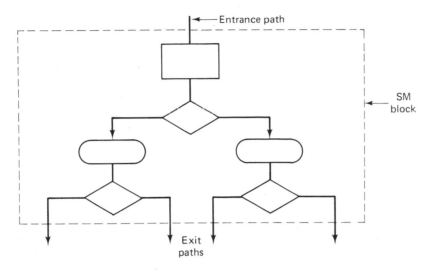

Fig. 6.1.3   SM block.

conditional outputs are listed in the box with immediate and delay operations permitted. The input to the box must be a conditional exit path, and there must be only a single exit path.

A state box and a network of decision boxes and conditional output boxes are enclosed within an SM block. An SM block is denoted by a dashed line, as illustrated in Fig. 6.1.3. There is one entrance to a block and any number of exit paths, depending on the network of decision boxes within the block. Each exit path must connect to a state, and each possible path from one state to the next is called a *link path*. Therefore, each exit path is a link path. It is noted that in practice the dashed line is often omitted as a shortcut, but the blocks are still evident since they consist of a state box and all the conditional boxes between that state and the next.

As an example of showing the derivation of an SM chart from an algorithmic flow chart, let us consider the flow chart of Fig. 6.1.2. Replacing the complex Boolean operations in this flow chart by a set of more basic operations, we obtain first the flow chart of Fig. 6.1.4(a) and then the flow chart of Fig. 6.1.4(b). From the latter it is observed that there are two states in this flow chart, the $C_n = 0$ state and the $C_n = 1$ state. By redrawing the flow chart of Fig. 6.1.4(b) using SM chart symbols, we obtain the SM chart description of the serial adder of Fig. 6.1.5(a). Just as the algorithmic-flow-chart representation of a mathematical or logical computation is not unique, neither is the SM chart representation of a sequential machine. For example, the two SM charts of Fig. 6.1.5(a) and (b) are SM chart representations of the serial adder; the qualifiers of the former are Boolean variables, whereas those of the latter are Boolean functions. As a final remark, *the operations indicated by the SM chart occur simultaneously, not sequentially as in an algorithmic flow chart.* This is an important point.

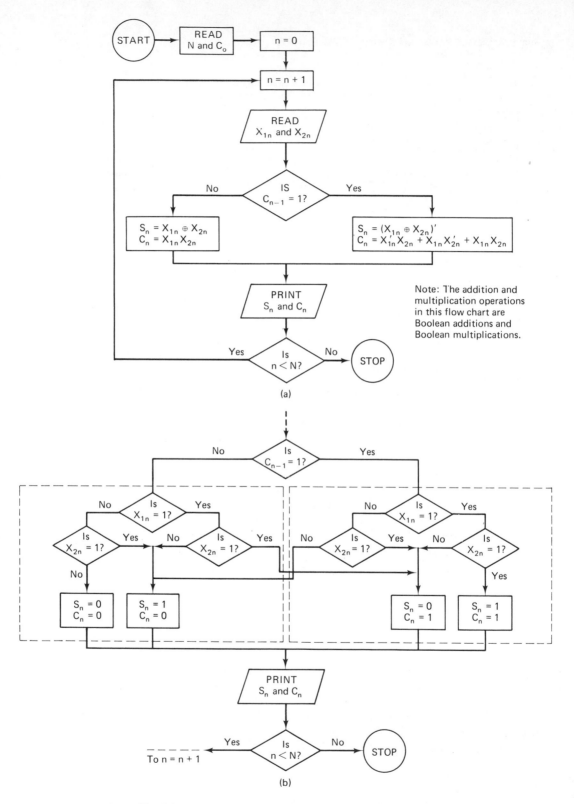

Note: The addition and multiplication operations in this flow chart are Boolean additions and Boolean multiplications.

Fig. 6.1.4 Two algorithmic flow charts of serial sequential adder.

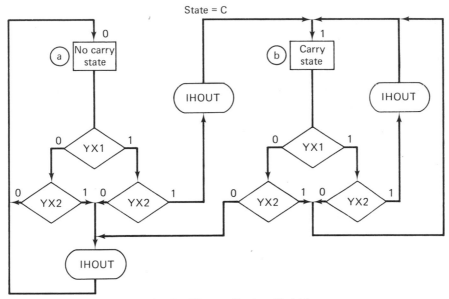

(a)   Qualifiers are Boolean Variables

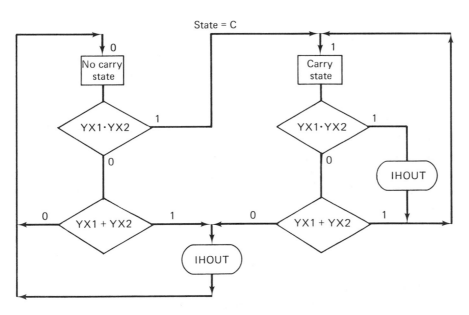

(b)   Qualifiers are Boolean Functions

Fig. 6.1.5 Two SM chart representations of serial sequential adder.

6.2 Reading Reduced-Dimension Maps

As we have seen, the Karnaugh map is an invaluable tool in studying switching functions. We have used map entries of 0, 1, and don't care, but now let us consider including variable terms as entries on our maps. Previously, all variables were used as map variables, with the dimension of the map equal to the number of variables. However, we can reduce the dimension of our maps by entering map variables as entries on a *reduced-dimension map* (RDM). These entries are called *map-entered variables*.

Consider the function $F = A \cdot D + B \cdot C$ and its four-variable map shown here. Using D as a map-entered variable, D can be eliminated as a map variable, thereby reducing the map to a three-variable map.

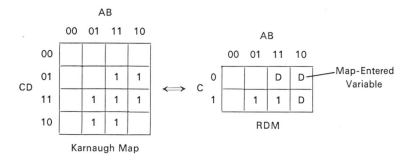

Karnaugh Map

Entries on the RDM are made after examining the locations on the original map where the map-entered variable D is true and where false, with other map variables held constant. The location of the entry on the RDM is the map-variable combination that was held constant while examining the value of the function when the map-entered variable is equal to 0 and when the map-entered variable is equal to 1.

When the function is 0 in both locations, a 0 is entered on the RDM. When the function is 1 in both locations, a 1 is entered on the RDM. If the function is 1 when the map-entered variable is 1 and 0 when the map-entered variable is 0, then the map-entered variable is entered on the RDM. For the case where the function is 0 when the map-entered variable is 1 and the function is 1 when the map-entered variable is 0, the complement of the map-entered variable is placed on the RDM. Table 6.2.1 lists RDM entries for the cases that might be encountered in generating RDM's.

Example 6.2.1

The function represented on the five-variable Karnaugh map in Fig. 6.2.1(a) may also be represented on a four-variable RDM as shown in Fig. 6.2.1(b). Figure 6.2.1(c) is an RDM containing map-entered variables E and F. Using Table 6.2.1, it can be reduced to the three-variable RDM shown in Fig. 6.2.1(d).

As seen from Fig. 6.2.1(d), an RDM can be reduced further. Further reductions can reduce a map until only a single equation remains. For our purposes, however, it

TABLE 6.2.1 Rules for Eliminating a Map Variable of a Karnaugh Map or a RDM

Value of Function for Map-Entered-Variable Location		RDM Entry†
$X = 0$	$X = 1$	
0	0	0
0	1	X
1	0	$\overline{X}$
1	1	1
0	F	$X \cdot F$
1	F	$\overline{X} + F$
F	0	$\overline{X} \cdot F$
F	1	$X + F$
F	$\overline{F}$	$X \oplus F$
$\overline{F}$	F	$\overline{X \oplus F}$
F	F	F
F	G	$\overline{X}F + XG$
F	$F + G$	$F + XG$
$F + G$	F	$F + \overline{X}G$
F	$F \cdot G$	$F(\overline{X} + G)$
$F \cdot G$	F	$F(X + G)$
$F + G$	$F + H$	$F + \overline{X}G + XH$
$F \cdot G$	$F \cdot H$	$F(\overline{X}G + XH)$
$F + G$	$F \cdot H$	$FH + \overline{X}(F + G)$
$F \cdot G$	$F + H$	$FG + X(F + H)$
0	—	$(0, X)^*$
1	—	$(1, \overline{X})^*$
—	0	$(0, \overline{X})^*$
—	1	$(1, X)^*$
—	F	$(X \cdot F, \overline{X} + F)^*$
F	—	$(\overline{X} \cdot F, X + F)^*$

†$(a, b)^*$ indicates that the RDM may be either a or b, whichever most simplfies the RDM.

is more important to be able to read a map than to reduce one, so the remainder of this section will be devoted to techniques for reading reduced-dimension maps.

A reduced-dimension map is read in three steps.

Step 1: All map-entered variables are set to zero and the terms from the 1's and don't-care entries are read from the map.

Step 2: The terms containing the map-entered variables are obtained by considering 1's as don't-care entries. The terms from the encirclements are ANDed with the expression within the circle.

Step 3: The function is then read by combining the results of steps 1 and 2. The following examples illustrate the map-reading process.

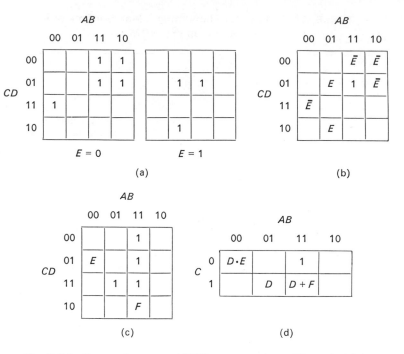

Fig. 6.2.1　Karnaugh maps and RDM representations of Example 6.2.1.

Example 6.2.2

The function represented on the RDM of Fig. 6.2.2 can be read by combining the adjacent 1 terms, yielding a term A, then by treating the 1's as don't cares and combining the C

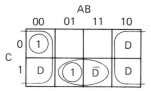

Fig. 6.2.2　RDM of Example 6.2.2.

with a 1, yielding a term $\bar{B}$, which is ANDed with C to give $\bar{B}\cdot C$, and finally by ORing the results of the first two steps to obtain $F = A + \bar{B}\cdot C$. The process is:

Step 1　A

Step 2　$\bar{B}\cdot C$

Step 3　$F = A + \bar{B}\cdot C$

Example 6.2.3

The function represented on the RDM of Fig. 6.2.3 may be read by:

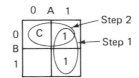

Fig. 6.2.3 RDM of Example 6.2.3.

Step 1 $\bar{A}\bar{B}\bar{C} + \bar{A}BC$

Step 2 $\bar{B}D + BC\bar{D}$

Step 3 $F = \bar{A}\bar{B}\bar{C} + \bar{A}BC + \bar{B}D + BC\bar{D}$

Don't-care entries may be utilized in both steps 1 and 2, as demonstrated in the following example.

Example 6.2.4

The don't-care entries* of Fig. 6.2.4 are used in step 1 to yield B and in step 2 to obtain $\bar{C}D + AC\bar{D}$.

Step 1 B

Step 2 $\bar{C}D + AC\bar{D}$

Step 3 $F = B + \bar{C}D + AC\bar{D}$

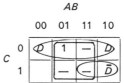

Fig. 6.2.4 RDM of Example 6.2.4.

Example 6.2.5

The don't-care entries in Fig. 6.2.5 are combined with the 1 entries, and AB with the map-entered variable E and the Boolean function $E \cdot F$.

Step 1 $\bar{B}D$

Step 2 $\bar{A}DE + ABC(E \cdot F)$

Step 3 $F = \bar{B}D + \bar{A}DE + ABCEF$

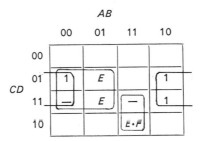

Fig. 6.2.5 RDM of Example 6.2.5.

*Recall that don't-care entries in the Karnaugh map of this chapter are denoted by dashes instead of d or X, to avoid possible confusion with state designators or state variables.

When adjacent locations contain common terms, some simplifications may be made. As a rule, such simplifications are found by manipulating the Boolean expressions for the individual terms from the RDM. When the adjacent locations contain common OR terms, the common terms may be encircled to obtain one term, with the other OR terms treated separately. This is illustrated by the following example.

Example 6.2.6

A minimized expression for the function represented on the RDM of Fig. 6.2.6 is $F = BE + B\bar{D}F$.

Proof:
$$F = B\bar{D}(E + F) + BDE$$
$$= BE(\bar{D} + D) + B\bar{D}F$$
$$= BE + B\bar{D}F \quad \blacksquare$$

CD \ AB	00	01	11	10
00		E + F	F + E	
01		E	E	
11		E	E	
10		E + F	F + E	

Fig. 6.2.6 RDM of Example 6.2.6.

The following examples present several possible combinations found when reading RDM's. Several other combinations are left as exercises.

Example 6.2.7

A minimized expression for the function represented on the RDM of Fig. 6.2.7 is $F = AC(B + D)$.

Proof:
$$F = A\bar{B}CD + ABC$$
$$= AC(\bar{B}D + B)$$
$$= AC(B + D) \quad \blacksquare$$

B \ A	0	1
0		C·D
1		C

Fig. 6.2.7 RDM of Example 6.2.7.

Example 6.2.8

A minimized expression for the function represented on the RDM of Fig. 6.2.8 is $F = C(\bar{A} + D)$.

Proof:
$$F = \bar{A}C + ACD$$
$$= C(\bar{A} + AD)$$
$$= C(\bar{A} + D) \quad \blacksquare$$

B \ A	0	1
0	C	C·D
1	C	C·D

Fig. 6.2.8 RDM of Example 6.2.8.

Example 6.2.9

A minimized expression for the function represented on the RDM of Fig. 6.2.9 is $F = C(\bar{A} + \bar{B} + E) + A\bar{B}D$.

Proof:
$$\begin{aligned} F &= \bar{A}C + A\bar{B}(C + D) + ABCE \\ &= \bar{A}C + A\bar{B}C + A\bar{B}D + ABCE \\ &= C(\bar{A} + A\bar{B} + ABE) + A\bar{B}D \\ &= C(\bar{A} + A(\bar{B} + BE)) + A\bar{B}D \\ &= C(\bar{A} + \bar{B} + E) + A\bar{B}D \quad \blacksquare \end{aligned}$$

B \ A	0	1
0	C	C + D
1	C	C·E

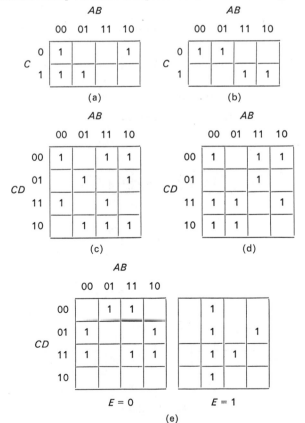

Fig. 6.2.9 RDM of Example 6.2.9.

As we have seen, map-entered variables are useful in representing functions of many variables on reduced-dimension maps with only a few variables. We shall see that map-entered variables are a natural consequence of designing sequential circuits with SM charts. The next three sections describe how to obtain and use RDM's in the design of sequential circuits.

Exercise 6.2

1. Reduce the number of map variables to represent the functions in Fig. P6.2.1. Use RDM's.

(a)

C \ AB	00	01	11	10
0	1			1
1	1	1		

(b)

C \ AB	00	01	11	10
0	1	1		
1			1	1

(c)

CD \ AB	00	01	11	10
00	1		1	1
01		1		1
11	1		1	
10		1	1	1

(d)

CD \ AB	00	01	11	10
00	1		1	1
01			1	
11	1	1		1
10	1	1		

(e)

E = 0

CD \ AB	00	01	11	10
00		1	1	
01	1			1
11	1		1	1
10				

E = 1

CD \ AB	00	01	11	10
00		1		
01		1		1
11		1	1	
10		1		

Fig. P6.2.1

2. Represent the functions on the maps in Fig. P6.2.2 on RDM's having one less map variable than the map shown.

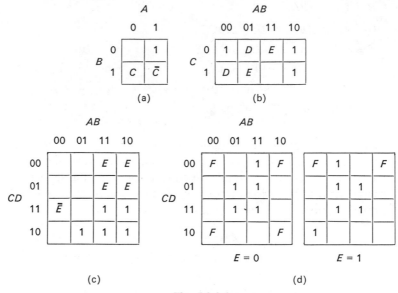

(a)   (b)

(c)   (d)

Fig. P6.2.2

3. Obtain expressions for the functions represented on the maps in problem 2.

4. Obtain expressions for the functions represented on the RDM's obtained in problem 2. Compare your results with those obtained in problem 3.

5. Obtain expressions for the functions represented on the maps in Fig. P6.2.5.

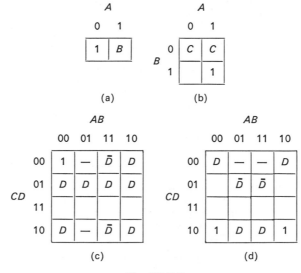

(a)   (b)

(c)   (d)

Fig. P6.2.5

6. Obtain expressions for the functions represented on the maps in Fig. P6.2.6.

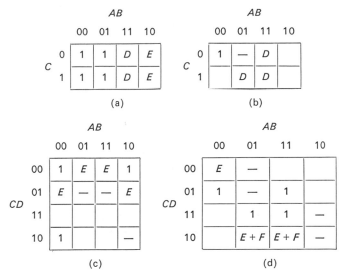

(a)   (b)

(c)   (d)

Fig. P6.2.6

7. Obtain expressions for the functions represented on the maps in Fig. P6.2.7.

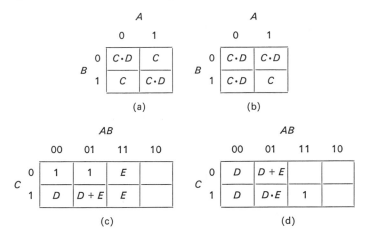

(a)   (b)

(c)   (d)

Fig. P6.2.7

6.3 Output-Function Synthesis

Output-function maps may be obtained directly from SM charts, as will be seen in this section. They are made from the state-assignment map, which has the state variables $(A, B, \ldots)$ as map variables and has SM chart state labels $(a, b, \ldots)$ as entries for the particular state-variable combinations.

The entries on the output-function maps are found by examining the paths from the states for which the output is true. The possible map entries are 0, 1, don't care, and qualifiers.

1. A 1 or 0 is entered in a state location on the map when from that state the output is always true or false, respectively.

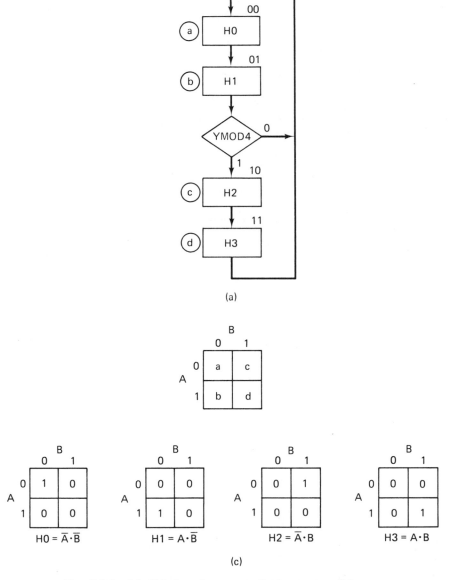

Fig. 6.3.1 (a) SM chart for a controllable counter; (b) state assignment map; (c) output function maps and output functions.

2.   A don't-care condition is entered in all unassigned state locations.

3.   A qualifier (input or Boolean expression) or its complement is entered on the map for a conditional output, depending on whether the qualifier is true or false, respectively.

The qualifiers are ANDed and entered in the state location on the map when more than one decision box is encountered.

If several paths lead to a true output, the conditions of all paths are ORed together for the map entry. This procedure is best described by the use of an example. Consider the controllable counter of Fig. 6.3.1(a), which functions as a module-2 or module-4 counter, depending on the qualifier YMOD4. There are four output maps with the entries as shown in Fig. 6.3.1(c) for the state assignment Fig. 6.3.1(b). Since H0 is true only for state a, a 1 is entered in location a on the H0 output map and 0's are entered in all other locations. The other maps for H1, H2, and H3 are obtained similarly.

The machine of Fig. 6.3.2 contains state outputs as well as conditional outputs. Also, a don't-care entry is made on the maps since state assignment $\bar{A}B$ is unassigned. The output functions are obtained from the maps using the steps for reading RDM's.

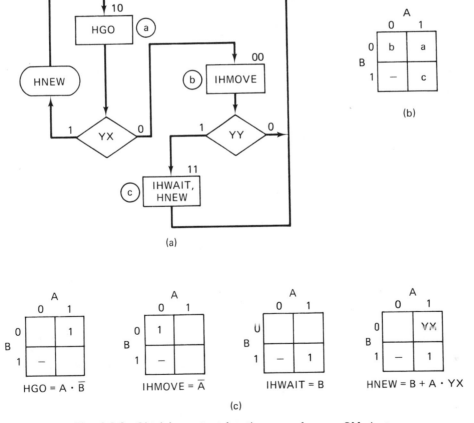

Fig. 6.3.2 Obtaining output-function maps from an SM chart.

The output of a sequential machine may be coded so that fewer output lines are needed. For example, five outputs could be coded such that only three output lines are required. This means that fewer output maps will be needed when obtaining the output-function equations. The modulo-6 counter of Fig. 6.3.3(a) uses coding to reduce

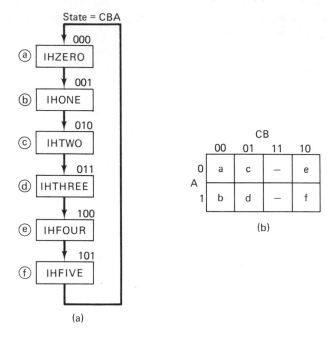

(a)

(b)

Output	Output variables		
	z_1	z_2	z_3
Zero	0	0	0
One	0	0	1
Two	0	1	0
Three	0	1	1
Four	1	0	0
Five	1	0	1

(c)

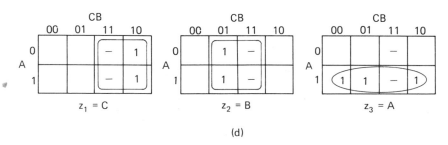

(d)

Fig. 6.3.3 Modulo-6 counter with coded outputs.

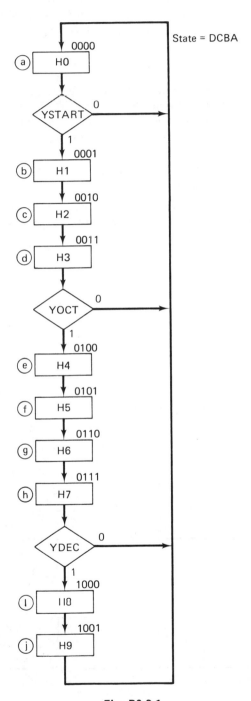

Fig. P6.3.1

the number of output lines from six to three. For output coding as shown in Fig. 6.3.3(c), the output maps for Z_1, Z_2, and Z_3 in Fig. 6.3.3(d) could be obtained using the coding table of Fig. 6.3.3(c) and the state-assignment map of Fig. 6.3.3(b).

Exercise 6.3

1. The sequential machine shown in the SM chart of Fig. P6.3.1 is a controllable counter with the operation controlled by inputs YSTART, YOCT, and YDEC. The machine functions as a modulo-4 counter, a modulo-8 counter, or a modulo-10 counter, as determined by the inputs. Obtain expressions for the output functions for the controllable counter.

2. Use coding to obtain the output functions of the controllable counter of problem 1 using a minimum number of output lines.

3. A sequential machine is to serve as a sequence detector which has a serial input and has output HZERO or HONE when a sequence 110 or 111 is detected, respectively. The

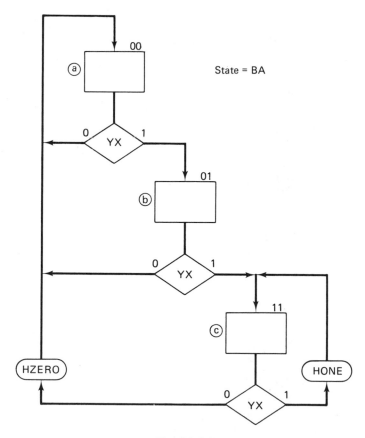

Fig. P6.3.3

machine has no output for other sequences. Obtain the output functions for the sequence detector as shown in the SM chart of Fig. P6.3.3.

6.4 Next-State-Function Synthesis

Having synthesized the output functions, the next task is to realize the next-state functions from the SM chart. The next-state function $g(Q, X)$ consists of partial next-state functions $g_A(Q, X)$, $g_B(Q, X)$, ... A partial next-state function is the next-state function for a single state variable. The partial next-state function is obtained from the next-state map, which can be obtained from the SM chart as from the output map described in the previous section. Whereas the output-function-map entries were for conditions for which the outputs were true, *the next-state-map entries are for conditions for which the partial next-state variable is 1*. The partial next-state functions and their constructions will be made clear in the following example.

Example 6.4.1 Consider the SM chart of Fig. 6.4.1. The state assignment of this machine is shown in Fig. 6.4.2(a), and the partial next-state functions can be obtained from the partial next-state RDM's which are shown in Fig. 6.4.2(b). The construction of the partial next-state RDM's needs some explanation which is following.

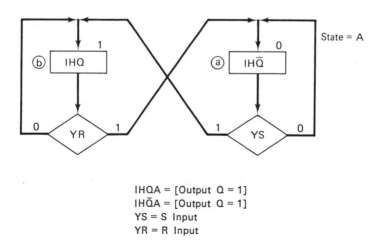

IHQA = [Output Q = 1]
IHQ̄A = [Output Q = 1]
YS = S Input
YR = R Input

Fig. 6.4.1 An example for illustrating the construction of partial next-state functions from an SM chart.

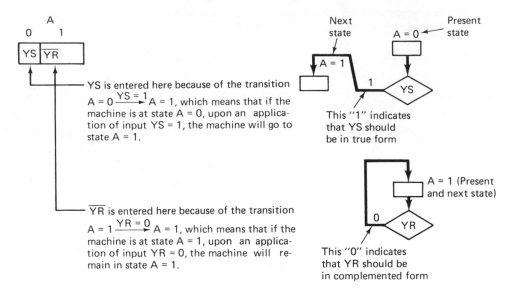

The explanation for the partial next-state $\bar{A}$ RDM is similar.

6.4.1 Realization Using Gates

The partial next-state functions can be realized by the use of gates. The method for realization will be presented through the following example.

Example 6.4.2 Design of Master-Slave Flip-Flops

Figure 6.4.3 presents an SM chart for the master-slave DFF. The output-function maps are obtained as before and are presented in Fig. 6.4.4. The next-state function is realized by deriving partial next-state functions for each state variable. However, the functions are found using map-entered variable maps obtained directly from the SM chart. The map for a single state variable is obtained by entering the conditions for all link paths that lead to a next state of 1 for that state variable. This is illustrated in Fig. 6.4.5.

The partial next-state map for state variable A was found by observing all link paths for

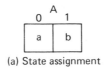

(a) State assignment

(b) Partial next-state functions obtained from RDM

Fig. 6.4.2 State assignment and partial next-state function.

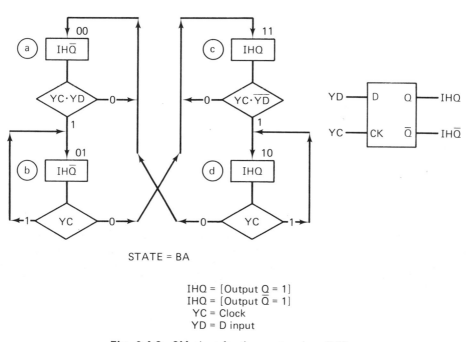

STATE = BA

IHQ = [Output Q = 1]
IHQ̄ = [Output Q̄ = 1]
YC = Clock
YD = D input

Fig. 6.4.3 SM chart for the master-slave DFF.

which the next state of A was 1. From state a, the next state A will be 1 if the qualifier $YC \cdot YD$ is true, so $YC \cdot YD$ is entered in the state a location. Both exit paths of the decision box in the SM block for state b lead to a next state of 1 for A, so a 1 is entered in location b. For state c, the qualifier $YC \cdot \overline{YD}$ must be false for the next-state variable A to be a 1.

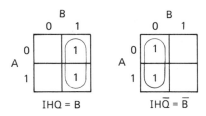

Fig. 6.4.4   Output maps for the DFF.

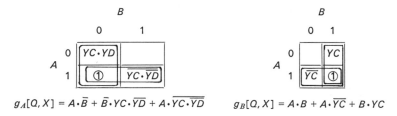

$$g_A[Q, X] = A \cdot \overline{B} + \overline{B} \cdot YC \cdot \overline{YD} + A \cdot \overline{YC \cdot YD} \qquad g_B[Q, X] = A \cdot B + A \cdot \overline{YC} + B \cdot YC$$

Fig. 6.4.5   Partial next-state functions for the DFF.

Therefore, $\overline{YC \cdot \overline{YD}}$ is entered in location c. The map is completed with a 0 in the state d location, since no link path leads from d to a next state where A is 1. The partial next-state map for B is obtained similarly, and the partial next-state function is read from the map. The next-state functions may be manipulated using Boolean algebra into a form realizable from the gates available. Suppose NAND gates are available; we then express the partial next-state functions A and B as:

$$A = \overline{(A \cdot B \cdot \overline{YC \cdot YD}) \cdot (\overline{B} \cdot \overline{YC \cdot YD})}$$

$$B = \overline{\overline{B} \cdot (\overline{A \cdot YC}) \cdot (\overline{A \cdot \overline{YC}})}$$

A realization of the master-slave DFF using NAND gates is shown in Fig. 6.4.6.

The methods for realizing other types of master-slave flip-flops using the SM chart are

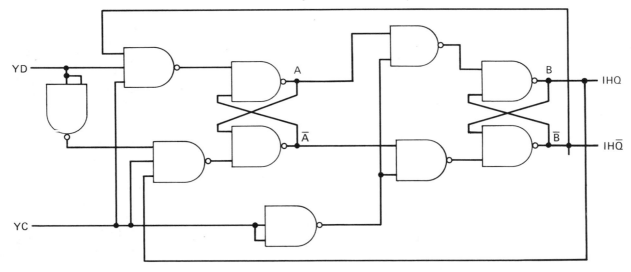

Fig. 6.4.6 NAND-gate realization of DFF.

similar. For example, the SM charts of the master-slave TFF and the master-slave JKFF are shown in Figs. 6.4.7 and 6.4.8, respectively. Their realizations are left to the reader as an exercise.

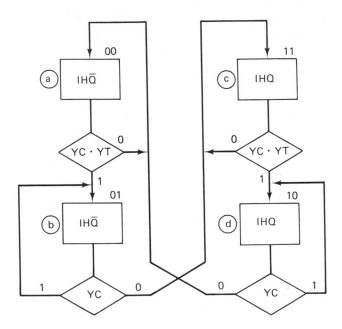

Fig. 6.4.7 SM chart for TFF.

6.4.2 Realization Using Flip-Flops

Now that flip-flops have been realized, let us use flip-flops as components in designing logic systems. This can be done by using application tables to convert the next-state functions into flip-flop input or excitation equations.

Flip-flop application tables of unconditional transitions ($y = 0 \rightarrow Y = 0$, $y = 0 \rightarrow Y = 1$, $y = 1 \rightarrow Y = 0$, and $y = 1 \rightarrow Y = 1$) were introduced in Chapter 5. Table 6.4.1 presents the application tables for the S–R, D, T, and J–K flip-flops. There are two additions to the application table of Chapter 5 which are used to determine flip-flop inputs for conditional transitions ($y = 0 \rightarrow Y = 0$ or 1 and $y = 1 \rightarrow Y = 0$ or 1) on the SM chart.

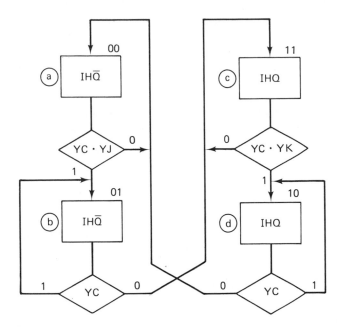

Fig. 6.4.8 SM chart for JKFF.

TABLE 6.4.1 Application Tables for S–R, D, T, and J–K Flip-Flops

Present State	Next State	FF Input Map Entries					
y	Y	R	S	D	T	J	K
0 ⟶ 0		—	0	0	0	0	—
0 ⟶ 1		0	1	1	1	1	—
1 ⟶ 0		1	0	0	1	—	1
1 ⟶ 1		0	—	1	0	—	0
Conditional 0 ⟶ 0 or 0 ⟶ 1		∩	Cond. for 0 ⟶ 1	Cond. for 0 ⟶ 1	Cond. for 0 ⟶ 1	Cond. for 0 ⟶ 1	—
Conditional 1 ⟶ 0 or 1 ⟶ 1		Cond. for 1 ⟶ 0	0	Cond. for 1 ⟶ 1	Cond. for 1 ⟶ 0	—	Cond. for 1 ⟶ 0

Unconditional transitions

Conditional transitions

Flip-flop input maps may be generated directly from the SM chart. First, any unused locations on the state-assignment map are filled as don't cares. Next, the remaining locations are filled by observing the transition for each state variable on the SM chart and then consulting the application table to find the input for that transition. The input equations may then be obtained for the state-variable flip-flops.

Example 6.4.3

An SM chart for a JKFF is given in Fig. 6.4.8. Suppose it is desired to realize it using SRFF and NOR gates. The output functions are found from the output maps:

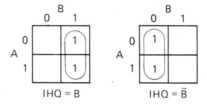

Next, observing the transitions on the SM chart, the flip-flop input maps are obtained:

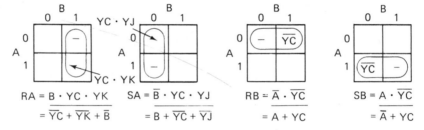

The SRFF-NOR realization of the JKFF is shown in Fig. 6.4.9.

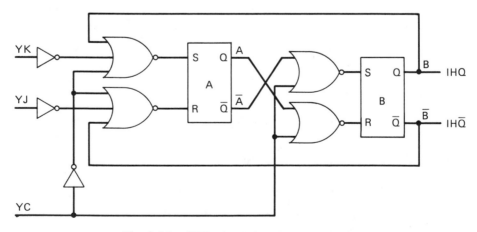

Fig. 6.4.9 JKFF using NOR gates and SRFF.

1. The SM chart in Fig. P6.4.1 describes an edge-triggered J–K flip-flop which permits the inputs to change even while the clock is high, as long as the inputs are stable during the 1-to-0 transition of the clock. Realize this JKFF using NAND gates.

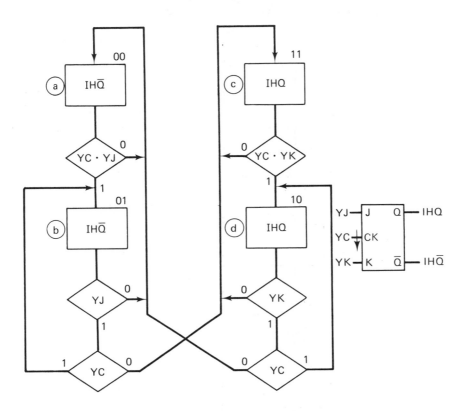

Fig. P6.4.1

2. Synthesize a gated latch that allows inputs to pass to the output only when the gate is 1. An SM chart of a gated latch is shown in Fig. P6.4.2. Use NAND gates to realize the gated latch.

3. Realize the sequential machine of problem 1 in Exercise 6.3 using JKFF and NAND gates.

4. Realize a sequential adder using JKFF and NAND gates. Two possible SM charts for the adder were presented in Fig. 6.1.6.

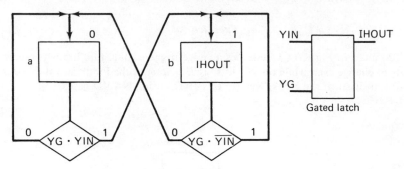

Fig. P6.4.2

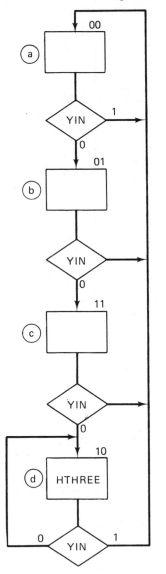

Fig. P6.4.5

5. A sequential machine is to examine an input sequence and to output HTHREE if three or more consecutive 0's follow the last 1 input. Use JKFF's and NAND gates to realize this sequential machine, which could be represented by the SM chart in Fig. P6.4.5.

6.5 State Assignment

Each state indicated on an SM chart is assigned a state name and a binary code. The set of codes given to the states of a sequential machine is called the *state assignment*. The choice of a state assignment affects the complexity of gating to realize the next-state function. The assignments of the input variables also affect gating complexity. The problem then is to find a state assignment and an input variable assignment that minimize the complexity of gating to realize the sequential circuit. Finding the optimum assignment is often an extremely difficult task, possible requiring a computer search of possible assignments. Two approaches, the minimum state locus and reduced dependency methods are presented in this section yield an optimum or near-optimum solution by attempting to minimize the probable contribution by each state to the complexity of next-state-function gating.

6.5.1 Minimum State Locus

The first approach considered is that of the *minimum state locus*. The premise of this method is that the state assignment resulting in the minimum state locus will have the simplest next-state-function gating. The state locus is the sum of the bit distances of all the transitions between states of a sequential machine. The bit distance between two states is the number of state variables involved in the transition between the states. On an SM chart, the state locus is the sum of the bit distances of all the link paths on the chart. The state locus does not generate a state assignment, but rather is a criterion on which to judge the probable contribution to gating complexity of a given state assignment. The minimum-state-locus approach is best applied to complex link structures with conditional branching and asymetric structures. On simpler problems, the minimum state locus is not a reliable criterion, as shown in the following example.

Example 6.5.1

Two state assignments for a modulo-6 counter whose SM chart was given in Fig. 6.3.3(a) are presented, one using a Gray-code assignment and the other a binary-code assignment.

Gray Code	Binary Code
0 0 0	0 0 0
0 0 1	0 0 1
0 1 1	0 1 0
0 1 0	0 1 1
1 1 0	1 0 0
1 0 0 state locus	1 0 1 state locus
= 6	= 10

The minimized partial next-state functions for the two state assignments may be obtained as follows.

Gray-Code Assignment

Present State CBA	Next State CBA
0 0 0	0 0 1
0 0 1	0 1 1
0 1 1	0 1 0
0 1 0	1 1 0
1 1 0	1 0 0
1 0 0	0 0 0

Binary-Code Assignment

Present State CBA	Next State CBA
0 0 0	0 0 1
0 0 1	0 1 0
0 1 0	0 1 1
0 1 1	1 0 0
1 0 0	1 0 1
1 0 1	0 0 0

Next-state functions:

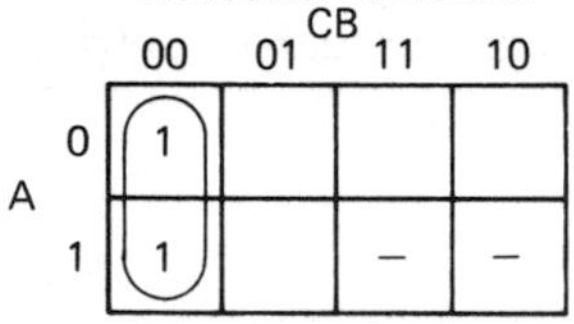

$$g_A(Q) = \bar{C}\bar{B}$$

Next-state functions:

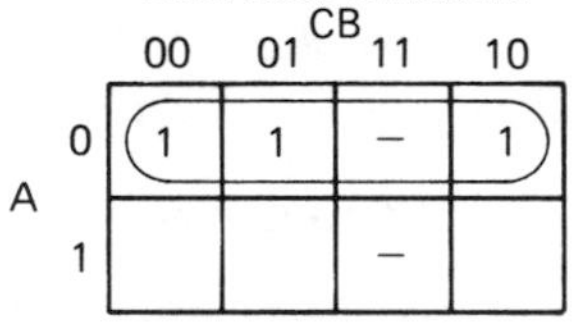

$$g_A(Q) = \bar{A}$$

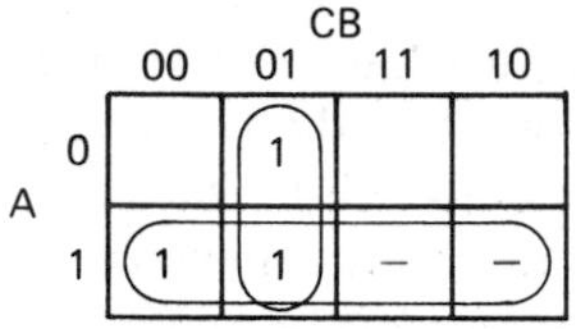

$$g_B(Q) = A + \bar{C}B$$

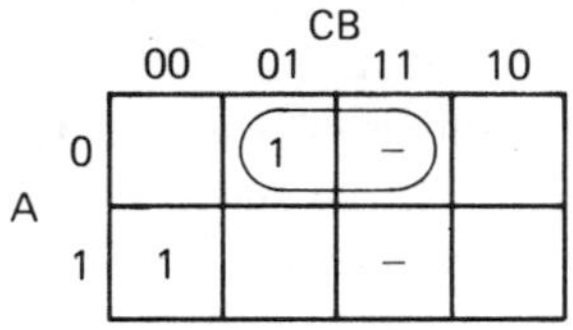

$$g_B(Q) = \bar{C}\bar{B}A + B\bar{A}$$

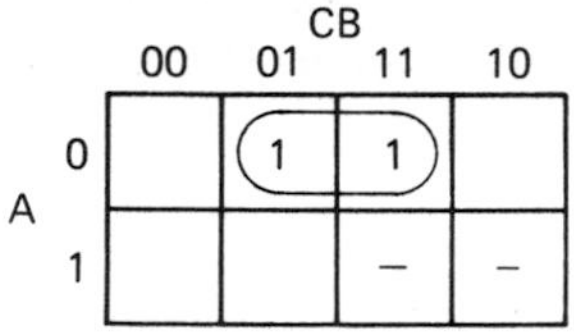

$$g_C(Q) = B\bar{A}$$

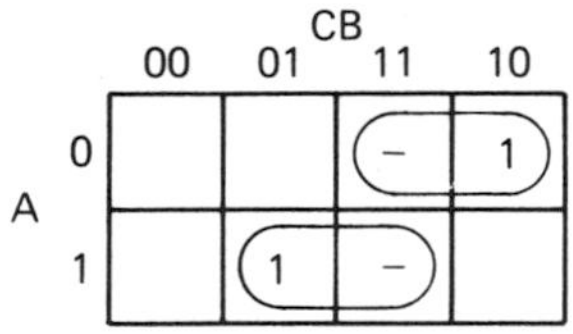

$$g_C(Q) = C\bar{A} + BA$$

To summarize, we have

Gray-Code Assignment	Binary-Code Assignment
$g_A[Q] = \bar{B}\bar{C}$	$g_A[Q] = \bar{A}$
$g_B[Q] = A + B\bar{C}$	$g_B[Q] = A\bar{B}\bar{C} + \bar{A}B$
$g_C[Q] = \bar{A}B$	$g_C[Q] = AB + \bar{A}C$

The Gray-code assignment with a smaller state locus yields simpler equations when gates are used, but when synthesized using JKFF, the following result shows that the binary-code assignment yields simpler gating:

Gray-Code Assignment

Present State	Next State	Excitation Functions of J–K Flip-Flops					
		JKFF C		JKFF B		JKFF A	
CBA	CBA	JC	KC	JB	KB	JA	KA
0 0 0	0 0 1	0	–	0	–	1	–
0 0 1	0 1 1	0	–	1	–	–	0
0 1 1	0 1 0	0	–	–	0	–	1
0 1 0	1 1 0	1	–	–	0	0	–
1 1 0	1 0 0	–	0	–	1	0	–
1 0 0	0 0 0	–	1	0	–	0	–

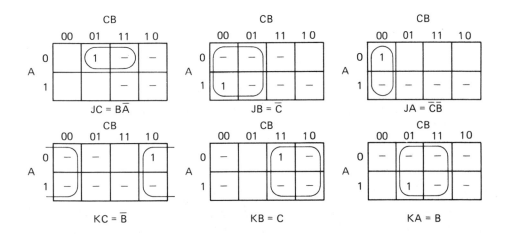

$JC = B\overline{A}$ $JB = \overline{C}$ $JA = \overline{C}B$

$KC = \overline{B}$ $KB = C$ $KA = B$

Binary-Code Assignment

Present State	Next State	Excitation Functions of J–K Flip-Flops					
		JKFF C		JKFF B		JKFF A	
CBA	CBA	JC	KC	JB	KB	JA	KA
0 0 0	0 0 1	0	–	0	–	1	–
0 0 1	0 1 0	0	–	1	–	–	1
0 1 0	0 1 1	0	–	–	0	1	–
0 1 1	1 0 0	1	–	–	1	–	1
1 0 0	1 0 1	–	0	0	–	1	–
1 0 1	0 0 0	–	1	0	–	–	1

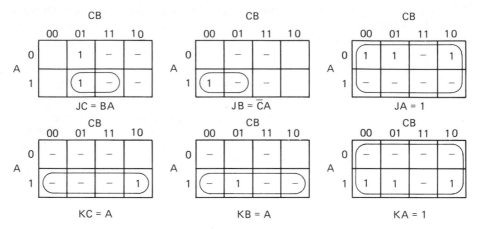

To summarize, we have

	Gray-Code Assignment		Binary-Code Assignment	
$JA = \bar{B}\bar{C}$	$KA = B$		$JA = 1$	$KA = 1$
$JB = \bar{C}$	$KB = C$		$JB = A\bar{C}$	$KB = A$
$JC = \bar{A}B$	$KC = \bar{B}$		$JC = AB$	$KC = A$

In the binary-code assignment, state variable A is made independent of the machine state, which simplifies gating. The simple structure and the J–K-like symmetry of the count cause the state-locus approach to be misleading, but the concept may be applied with good results to more complicated sequential machines.

6.5.2   Reduced Dependency

The minimum-state-locus approach considered the probable contributions due only to the state variables, but now consider a method that also includes the inputs. Such a technique uses the concept of reduced dependency, which attempts to minimize the probable contribution from a given state.

Consider the SM chart with state assignments as in Fig. 6.5.1(a) and (b). The partial next-state functions for each assignment show that for the second assignment, state variable A always becomes 1 in state transitions from state a. The state assignment of Fig. 6.5.1(b) is called a *reduced-dependency assignment* and saves two gate inputs over the minimum-state-locus assignment of Fig. 6.5.1(a).

When making a reduced-dependency assignment, a particular state of the sequential machine is chosen as the initial state for which all the next states are examined. On the SM chart, this means selecting a state and considering all the exit paths of the SM block for that state. Usually, the initial state is chosen because its SM block has the greatest number of exit paths, so the maximum number of states is considered in the assignment process.

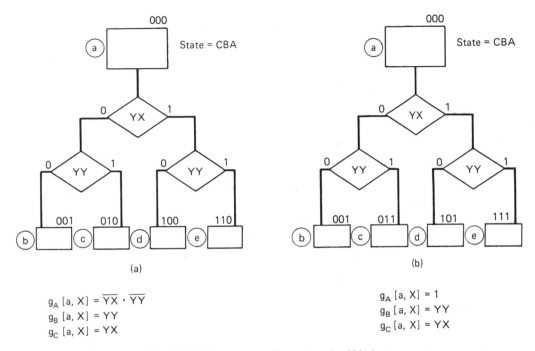

(a)

$$g_A [a, X] = \overline{YX} \cdot \overline{YY}$$
$$g_B [a, X] = YY$$
$$g_C [a, X] = YX$$

(b)

$$g_A [a, X] = 1$$
$$g_B [a, X] = YY$$
$$g_C [a, X] = YX$$

Fig. 6.5.1 Two state assignments of a SM chart.

Let us first consider an initial state a that has conditional transitions to b and c. A minimum-state-locus assignment would result in the change vectors shown in Fig. 6.5.2. The change vector a to c could also be considered the resultant vector of a vector from a to b, then from b to c, as in Fig. 6.5.3. The reduced-dependency change vector from a to c consists of two single-variable change vectors. The change vector from a to b takes part in either next-state transition, but in the transition from a to c, the change vector from b to c also appears. The common change vector from a to b is what allows the reduced dependency in the next-state-function gating.

Figure 6.5.4 shows the change vectors for the SM chart of Fig. 6.5.1(b). The

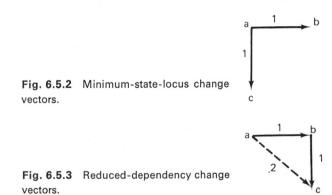

Fig. 6.5.2 Minimum-state-locus change vectors.

Fig. 6.5.3 Reduced-dependency change vectors.

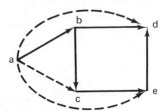

Fig. 6.5.4 Reduced-dependency change vectors.

dashed lines are the independent transitions; the solid lines indicate the reduced-dependency change vectors.

Finding common change vectors is aided by using a dependency map which has as its map variables all the qualifiers in the SM block of the initial state. The entries on the map are the next states that occur for each particular input combination. Thus the dependency map shows the relation between inputs and next states and can be used to find equations for the change vectors. Example 6.5.2 uses a dependency map to obtain a reduced-dependency assignment.

Examples 6.5.2

Find a reduced-dependency assignment for the transitions shown in the SM chart of Fig. 6.5.1.

The dependency map will have YX and YY as map variables, as shown below. When YX is false, the next state is b or c, depending on the value of YY. Similarly, if YX is true, the next state is d or e, depending on YY. This relation is indicated on the dependency map of Fig. 6.5.5.

An encirclement on the dependency map indicates a change vector. When more than one element is contained with an encirclement, the elements within the encirclement have a common change vector. In Example 6.5.2 the entire map was enclosed,

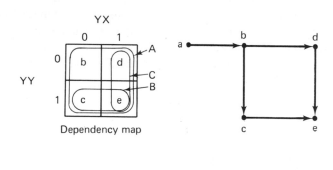

Dependency map

	C B A
a	0 0 0
b	0 0 1
c	0 1 1
d	1 0 1
e	1 1 1

$g_A [a, X] = 1$

$g_B [a, X] = YY$

$g_C [a, X] = YX$

Fig. 6.5.5 Example 6.5.2.

so there was one change vector common to all transitions. This common change vector
is from *a* to *b*. The other two encirclements indicate common change vectors in
transitions to *c* or *e* and to *d* or *e*. The diagram of reduced-dependency change
vectors shows that change vector *b* to *c* is common to transitions *b* to *c* and
b to *e*. Similarly, *b* to *d* is common to *b* to *d* and *b* to *e*.

The partial next-state function for transitions from the initial state may be read
from the encirclements on the map. Each grouping or change vector is assigned a
state variable. When making state assignments the entries enclosed by a change vector
are given one condition of the state variable, and entries outside the enclosure are given
the complemented condition of the state variable. Figure 6.5.5 shows how a state
assignment is made from the dependency map. The partial next-state function for a
particular state variable is read from the map considering the true condition of the
variable as 1 entries.

The most desirable reduced-dependency assignment is one made from a simple
nested enclosure. An example of such an assingment is seen in Example 6.5.3.

Example 6.5.3

A reduced-dependency assignment for the SM chart of Fig. 6.5.6(a) is shown in Fig.
6.5.6(b).

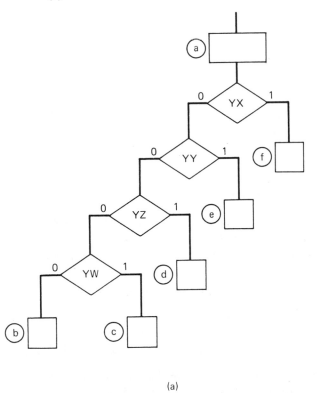

YX, YY

YZ, YW

	00	01	11	10
00	b	e	f	f
01	c	e	f	f
11	d	e	f	f
10	d	e	f	f

Dependency map

A reduced dependency assignment

State	E	D	C	B	A
a	0	0	0	0	0
b	0	0	0	0	1
c	0	0	0	1	1
d	0	0	1	1	1
e	0	1	1	1	1
f	1	1	1	1	1

(a)

(b)

Fig. 6.5.6 Example 6.5.3.

When the next state is the same as the initial state, a 0 is entered on the map for that combination. This is illustrated in the following example.

Example 6.5.4

Consider the SM chart of Fig. 6.5.7(a). It is seen that the next state of the initial state upon an input $YX = 1$ is itself. Thus a 0 is entered on the map for $YX = 1$, as shown in Fig. 6.5.7(b). A reduced-state assignment for this sequential machine is given in Fig. 6.5.7(b).

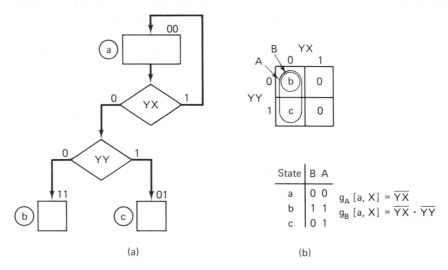

State	B	A
a	0	0
b	1	1
c	0	1

$$g_A [a, X] = \overline{YX}$$
$$g_B [a, X] = \overline{YX} \cdot \overline{YY}$$

(a)   (b)

Fig. 6.5.7 Example 6.5.4.

In summary, a state assignment using single-variable transition vectors, which are described by simple nested encirclements in the dependency map, is the most desirable reduced-dependency assignment for the minimum probable contribution to the next state.

Exercise 6.5

1. Make a state assignment for a decade (0–9) counter which has the least probable contribution to the next-state function gating when JKFF's are used in the realization.

2. Synthesize a decade counter using TFF and NAND gates. Use a state assignment that simplifies next-state function gating.

3. Synthesize a decade counter using DFF and NOR gates after making a reduced-dependency state assignment for the counter.

6.6 Design Examples

In this section two logic design examples to illustrate the SM chart method are presented. One is a design of an accumulator and the other is a design of a digital clock.

Example 6.6.1

Design an accumulator which is to operate as described below.

1. On the first clock pulse, the input X is loaded into register U and the overflow bit is cleared.

2. On the second clock, the contents of register U are loaded into register V.

3. On the third clock pulse, another input is loaded into register U.

4. The fourth clock pulse causes $U + V$ to be loaded into register V and the overflow bit to be set if an overflow occurred.

The design may be broken into several parts, including the control unit, the registers and associated logic circuitries, and the logic circuit, which presents U or $U + V$ to register V.

An SM chart for the control unit is shown in Fig. 6.6.1.

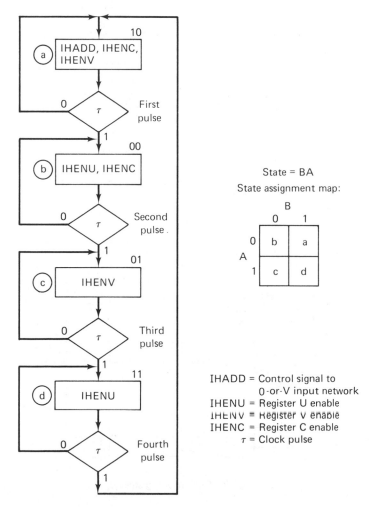

State = BA

State assignment map:

	B	
	0	1
A 0	b	a
A 1	c	d

IHADD = Control signal to
            0-or-V input network
IHENU = Register U enable
IHENV = Register V enable
IHENC = Register C enable
τ = Clock pulse

Fig. 6.6.1 SM chart for control unit.

Registers U, V, and C have similar logic design, so designing one will determine the logic circuit needed for each of the registers. An SM chart for register U is presented in Fig. 6.6.2 along with the design of the register and logic circuit.

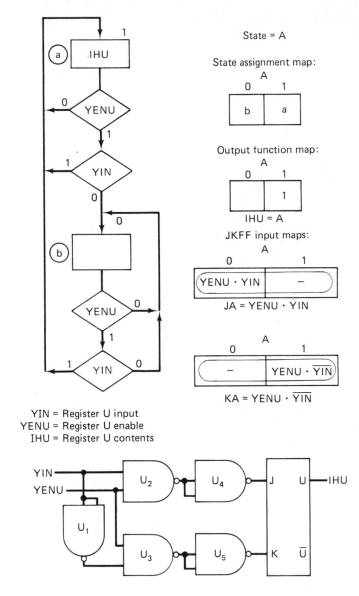

YIN = Register U input
YENU = Register U enable
IHU = Register U contents

Fig. 6.6.2 Design of register U.

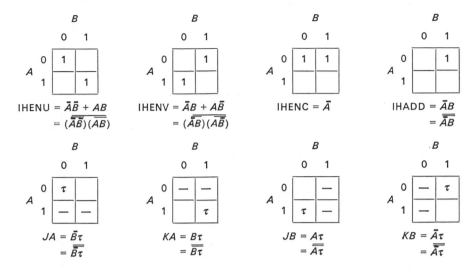

IHENU = $\bar{A}\bar{B}$ + AB
     = $(\overline{\bar{A}B})(\overline{A\bar{B}})$

IHENV = $\bar{A}B$ + $A\bar{B}$
     = $(\overline{\bar{A}B})(\overline{A\bar{B}})$

IHENC = $\bar{A}$

IHADD = $\bar{A}B$
     = $\overline{A\bar{B}}$

JA = $\bar{B}\tau$
   = $\overline{B\tau}$

KA = $B\tau$
   = $\overline{\bar{B}\tau}$

JB = $A\tau$
   = $\overline{\bar{A}\tau}$

KB = $\bar{A}\tau$
   = $\overline{A\tau}$

U may be loaded into register V by adding $U + 0$ and loading the sum into register V. Then the circuit needed to load either U or $U + V$ into V may consist of an adder which always has input U and which has either 0 or V as the other input. We must design a logic circuit that will present 0 when $YADD$ is 0 and which presents V when $YADD$ is 1. This combinational circuit might have the SM chart of Fig. 6.6.3.

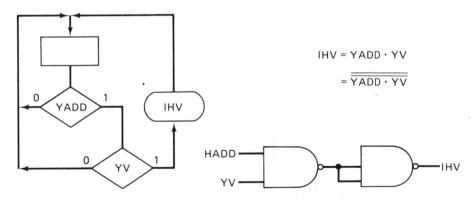

IHV = YADD · YV

    = $\overline{\overline{YADD \cdot YV}}$

Fig. 6.6.3 0-or-V input circuit.

Then the accumulator may be realized by connecting the subsystems as shown in Fig. 6.6.4. Figure 6.6.4 shows the accumulator for only one bit, but a word of any number of bits may be realized by duplicating the logic shown.

Example 6.6.2

Design the control module for a digital clock whose operational flow chart is described in Fig. 6.6.5.

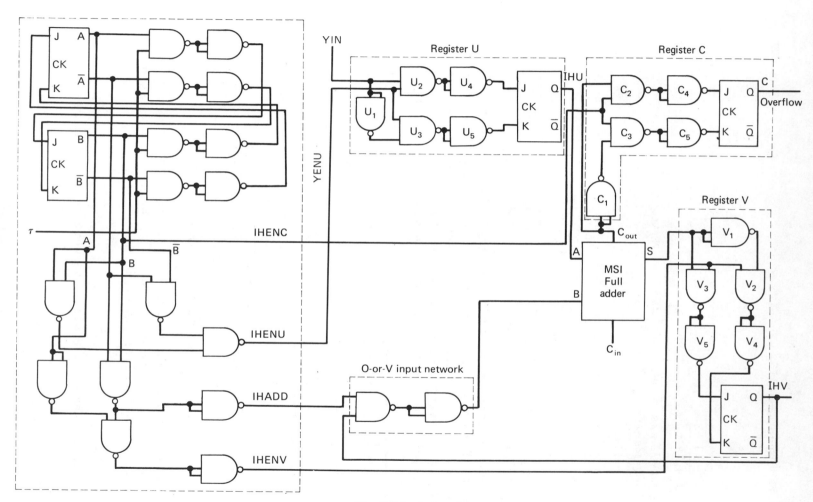

Fig. 6.6.4 Accumulator.

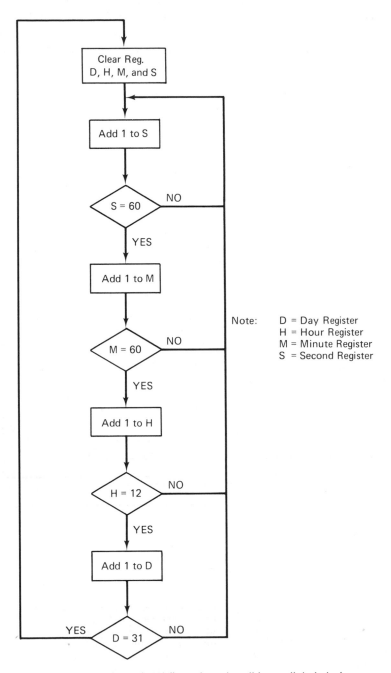

Fig. 6.6.5 Operational flow chart describing a digital clock.

Figure 6.6.6 shows a block diagram definition for the digital clock. The definitions of the qualifiers and output instructions of this digital clock machine are indicated in the figure. The SM chart of the machine is shown in Fig. 6.6.7. It is seen that five states denoted by *a*, *b*, *c*, *d*, and *e* are used to describe this machine; and thus a minimum of three variables is necessary to code the five states. Suppose the following state assignment is used.

State	CBA
a	0 0 0
b	0 0 1
c	0 1 1
d	0 1 0
e	1 1 0

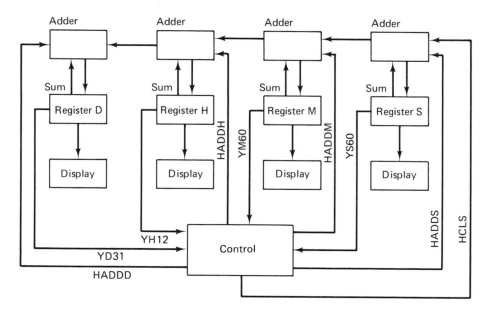

Qualifiers: YS60 = Does S register = 60?
                YM60 = Does M register = 60?
                YH12 = Does H register = 12?
                YD31 = Does D register = 31?

Output Instructions: HCLS   =   Set the sum registers D, H, M, and S to 0
                      HADDS =   Add 1 to the content of register S
                      HADDM =   Add 1 to the content of register M
                      HADDH =   Add 1 to the content of register H
                      HADDD =   Add 1 to the content of register D

Fig. 6.6.6 Block diagram definition for the digital clock machine.

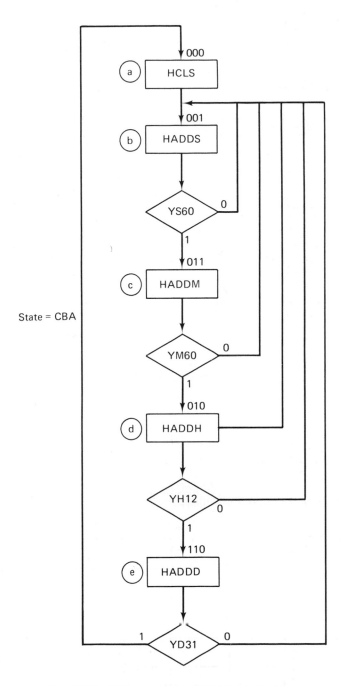

Fig. 6.6.7   SM chart of the digital clock machine.

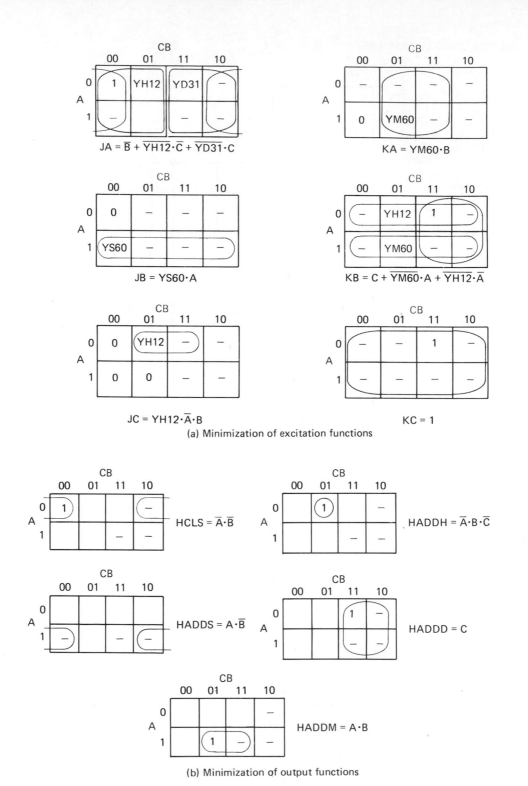

(a) Minimization of excitation functions

(b) Minimization of output functions

Fig. 6.6.8

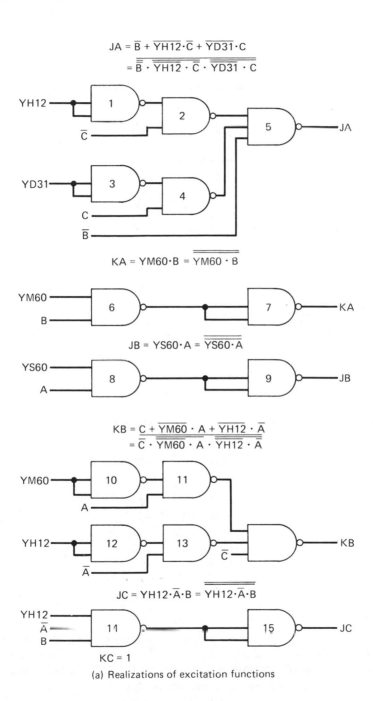

$$JA = \bar{B} + \overline{YH12 \cdot \bar{C}} + \overline{YD31 \cdot C}$$
$$= \overline{\bar{B} \cdot \overline{YH12 \cdot \bar{C}} \cdot \overline{YD31 \cdot C}}$$

$$KA = YM60 \cdot B = \overline{\overline{YM60 \cdot B}}$$

$$JB = YS60 \cdot A = \overline{\overline{YS60 \cdot A}}$$

$$KB = \underline{C + \overline{YM60 \cdot A} + \overline{YH12 \cdot \bar{A}}}$$
$$= \overline{\bar{C} \cdot \overline{YM60 \cdot A} \cdot \overline{YH12 \cdot \bar{A}}}$$

$$JC = YH12 \cdot \bar{A} \cdot B = \overline{\overline{YH12 \cdot \bar{A} \cdot B}}$$

$$KC = 1$$

(a) Realizations of excitation functions

Fig. 6.6.9

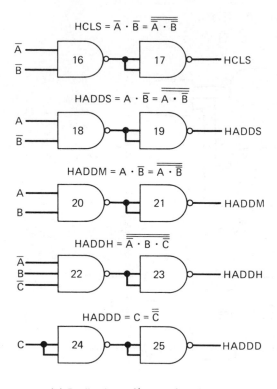

(b) Realizations of output functions   **Fig. 6.6.9.**   (continued)

The above state assignment may be expressed in a map form as

$$
\begin{array}{c}
 & CB \\
 & \begin{array}{cccc} 00 & 01 & 11 & 10 \end{array} \\
A\ \begin{array}{c} 0 \\ 1 \end{array} &
\begin{array}{|c|c|c|c|}
\hline
a & d & e & - \\
\hline
b & c & - & - \\
\hline
\end{array}
\end{array}
$$

Suppose it is desired to realize this machine using JKFF whose application table was shown in Table 6.4.1. Observing the transition on the SM chart, we can obtain the excitation functions for the three JKFF's to be used for realizing this machine directly from the SM chart. They are shown in Fig. 6.6.8(a). Figure 6.6.8(b) shows the output functions which can also be obtained directly from the SM chart. The realizations of the excitation functions and output functions after minimization are shown in Fig. 6.6.9. Finally, the complete logic circuit realization of the digital clock machine (control module) is given in Fig. 6.6.10.

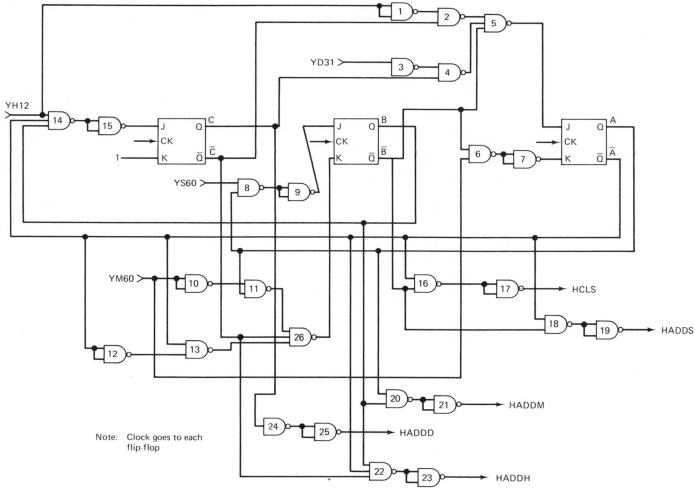

Fig. 6.6.10

Note: Clock goes to each
flip-flop

Bibliographical Remarks

The general concept of the SM chart was published in references 1–3. The first formal report on the SM chart was given in reference 4. Many symbols and notations of the SM chart used in the this chapter follow those in reference 5, which is also an excellent reference for this subject.

References

1. C. BOHM and G. JACOPINI, "Flow Diagrams, Turing Machines and Languages with Only Two Formation Rules," *J. ACM*, Vol. 9, No. 5, 1964, pp. 366–371.

2. R. NARASHMAN, "Programming Languages and Computers: A Unified Metatheory," in *Advances in Computers*, (F. L. Alt and M. Rubinolf, eds.), Vol. 8, 1967, pp. 189–224.

3. D. M. KAPLAN, *The Formal Theoretic Analysis of Strong Equivalence for Elemental Programs*, Thesis, Stanford University, Tech. Rept. CS-101, 1968.

4. DINES BJØRNER, *Flowchart Machines*, IBM Research Laboratory, San Jose, Calif., RJ-685 (No. 13346), Apr. 7, 1970.

5. CHRISTOPHER R. CLARE, *Designing Logic Systems Using State Machines*, McGraw-Hill, New York, 1973.

7

Fault Detection and Location in Combinational Circuits

In recent years, the development of integrated-circuit technology has accelerated rapidly; MSI and LSI techniques promise to make today's functional-level devices tomorrow's (even today's) basic components. Accordingly, digital systems are built with more and more complexity; the fault testing and diagnosis of digital circuits becomes an important and indispensible part of the manufacturing process. The problem of fault testing and diagnosis consists of the following two subproblems:

1. The fault-detection problem.
2. The fault-location problem.

The first problem is the determination of whether or not any of a prescribed list of faults, which may include all possible single faults and multiple faults, has occurred. The second problem, in turn, consists of the following subproblems:

2(a) Determination of the location of the fault that has occurred.

2(b) Determination of the location of the fault within a module (package or subnetwork) in which it has occurred.

A number of basic analytic and heuristic methods, including the fault-table method, the path-sensitizing and equivalent-normal-form (ENF) method, the Karnaugh map and tabular method, the ENF–Karnaugh map method, the Boolean difference method, and the SPOOF method are presented. The fault-table method is the most classic approach to the problem. It is completely general and always yields a minimum set of diagnostic tests. However, it suffers from the fact that it requires a very large fault table to be constructed. To overcome the problem of not requiring the construction of a very large fault table, the concept of path sensitizing is introduced. The path-sensitizing method requires no fault table, basically only work for the class of tree-like circuits and some special non-tree-like circuits. A heuristic, systematic procedure derived from the concept of path sensitizing, known as the equivalent-normal-form (ENF) method, is then introduced. Although this method has eliminated the two imperfections that the fault table and path-sensitizing methods have, it introduces an unattractive new feature, the requirement of the cumbersome computation of a "score function" for every literal in the ENF and the complemented ENF of the circuit. Finally, a ENF–Karnaugh map method is presented which eliminates the above three shortcomings. This method is a combination of the ENF method and the Karnaugh map and tabular method. Both the ENF method and the ENF–Karnaugh map method do not guarantee minimal experiments, nor is there a guarantee that a set of sensitized paths can be found for every circuit. Finally, the Boolean difference method and the SPOOF methos are two convenient general methods for deriving tests for detecting any single and/or multiple faults in any part of the circuit without using any fault tables or maps.

Based on these methods, computer programs can be written to generate complete fault-detection test sets for checking combinational circuits, such as those combinational TTL/MSI and MOS/LSI ROM IC chips. Two such computer programs DALG-II (D-Algorithm Version II) and TEST-DETECT are available.

7.1 Fault Detection and Fault Location:
Classical Methods

After a digital circuit is designed and built, it is always desirable to know whether the circuit is constructed without any faults. If it is properly constructed and in use, it may be disabled by almost any internal failure. The process of applying tests and determining whether a digital circuit is fault-free or not is generally known as *fault detection*. If a known relationship exists between the various possible faults and the deviations of output patterns, it is possible to identify the failures and to classify them (at least) within the smallest possible set of components. This process is termed *fault location*. *Fault diagnosis* is generally referred to as the composite of both the above processes.

The diagnostic techniques of digital systems have progressed from the state of testing the operation codes of the machine instruction set in the early days to the current hardware-assisted software aids for speedy determination and identification of faulty elements. In the coming LSI (large-scale-integration) era, the problem of diagnosis is shifted from locating the faulty discrete component to the identification of the replaceable module (chip or flat pack) in which the faulty component is resident.

Testing strategies generally depend on the system design and the architecture. Generally, the diagnostic engineer partitions the system into levels of hierarchy (systems, functional subsystems, modules, chips, etc.) and uses these levels in an organized fashion to isolate faults. Test points and the like provide mutual isolations among the partitions at various levels and help to reduce the size and time of testing. Here we shall only be concerned with the fault detection and location problem at the circuit level.

A fault-detection test set is said to be *minimal* if it is a minimal set that can detect the existence of each of the faults under consideration. A fault-location test set is said to be *minimal* if it is a minimal set that can locate each of the faults under consideration to the smallest identifiable part of the logic circuit. To construct a minimal fault-detection test set or minimal fault-location test set for combinational logic circuits, it is often necessary to have the complete test set of every fault under consideration; where a test of a fault is defined as the application of an input pattern to the primary input terminals and the observation of the corresponding outputs appearing at the primary output terminals so that the existence of the fault can be detected, the complete test set of a fault is defined as the set that contains all the tests of the fault. When it is not feasible to generate the complete test set of every fault under consideration because of a limitation on the amount of storage space, a near-optimal fault-detection test set or a near-optimal fault-diagnosis test set should be constructed. In this case the more tests that we can find, the more nearly optimal the fault-detection and fault-location test sets that can be constructed. The derivations of optimal fault-detection tests and optimal fault-location tests will be discussed.

Even when we are concerned only with the fault detection and location problem at the circuit level, without making certain realistic assumptions about the types of circuits and faults, the problem is still too broad to be discussed. The following are the assumptions made about the types of circuits and faults under investigation.

1. *Assumptions about the type of digital circuit to be considered*
   (a) It is assumed that the *digital* circuits under study are combinational circuits.
   (b) The circuits are assumed to be composed of exclusively AND, OR, NOT, NAND, and NOR gates. It should be pointed out that almost all the practical digital circuits in use are made up of these five types of gates. Furthermore, the methods derived for obtaining tests for this class of circuits are general enough to be applied to circuits consisting of gates other than these five types, such as the XOR (EXCLUSIVE–OR) gate, with only some slight modifications.
   (c) It is assumed that the response delays of all the gate elements are the same.
2. *Assumptions about the type of fault to be considered*
   (a) The faults considered here are assumed to be fixed or permanent or nontransient faults, by which we mean that without having them fixed or repaired, the fault will be permanently there.
   (b) Most of the faults occurred in currently used circuits, such as resistor–transistor logic circuits (RTL), diode–transistor logic circuits (DTL), and transistor–transistor logic circuits (TTL), are those which cause a wire to be (or appear logically to be) stuck-at-zero (s-a-0) or stuck-at-one (s-a-1). Restricting our consideration to just a class of faults is technically justified, since most circuit failures exhibit symptomatically identical effects, and this class of faults covers the faults in circuits with discrete components as well as integrated circuits. A multiple fault is defined as the simultaneous occurrence of any possible combination of s-a-0 and s-a-1 faults. In this chapter, both single and multiple fault detections will be considered.

Note that there are some faults in a circuit which are undetectable. For example, the s-a-0 fault on line α and s-a-1 fault on line β of the circuit of Fig. 7.1.1(a), denoted by α_0 and β_1, respectively, are undetectable because the output functions z^{α_0} and z^{β_1} of the faulty versions of the circuit are identical to the output function z of the normal circuit. Thus we define

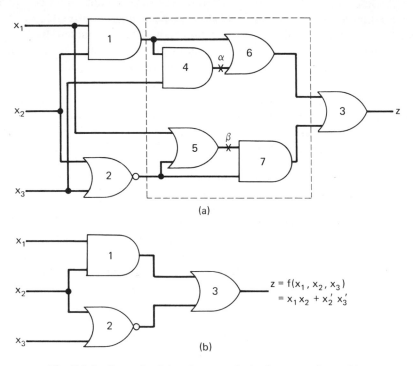

Fig. 7.1.1   Example of showing some faults that are undetectable.

DEFINITION 7.1.1

A fault of a combinational circuit is said to be *detectable* if there exists a test by which we can judge whether or not the circuit has such a fault; otherwise, we call the fault *undetectable*.

DEFINITION 7.1.2

A combinational circuit is said to be *irredundant* if any logic fault that occurred at any part of the circuit will cause a change in the switching function that the fault-free circuit realizes.

THEOREM 7.1.1

All s-a-1 and s-a-0 faults in a circuit are detectable if and only if the circuit is irredundant and if and only if the function that the circuit realizes is a minimized function.

The proof of this theorem is quite obvious and is omitted.

Consider the output function z of the circuit in Fig. 7.1.1(a):

$$z = x_1 x_2 + \overline{\left| (x_1 x_2)x_3 + (x_1 + x_2' x_3') \right|} (x_2' x_3') \qquad (7.1.1)$$

Note that the terms $(x_1 x_2)x_3$ and $x_1 + x_2' x_3'$ in the dashed frame can be deleted from the equation without changing the function. The deletion of these two terms corresponds to the removal of the four gates in the dashed frame in Fig. 7.1.1(a). It should also be noted that the function of Eq. (7.1.1) after the terms in the frame are removed is minimal, and the circuit of Fig. 7.1.1(a) after the four gates in the frame are removed (see Fig. 7.1.1(b)) is irredundant. Hence every s-a-0 and s-a-1 fault in the irredundant circuit of Fig. 7.1.1(b), wherever it occurs, will be detectable.

DEFINITION 7.1.3

A test set of a circuit is said to be *complete* if it detects every fault of the circuit under consideration. A *minimal complete* test set of a circuit is a complete test set that contains a minimum number of tests.

It is obvious that the set of all possible combinations of the values of the inputs to a combinational circuit constitute a complete test set for the circuit; that is to say that, by checking the entire truth table of a combinational circuit we can always determine whether the circuit is faulty. This method requires that every row of the truth table be tested. For a circuit with a large number of inputs n, the number of rows of the truth table of the circuit which is equal to 2^n is large. Now the question is: Can a subset of the set of the 2^n rows constitute a complete test set for the circuit? The answer is yes. This can be demonstrated by the following simple example.

Consider the NAND gate in Fig. 7.1.2. Let A and B denote the two inputs and C denote the output. The two input wires and the output wire are denoted by α, β, and γ, respectively. Suppose that we want to test a fault s-a-0 on wire α. The test that

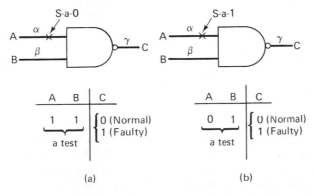

Fig. 7.1.2 Examples of illustrating the derivation of a test.

can detect this fault can be found as follows. First we find that the input B must be 1 because otherwise the output C will be 1 regardless of the value of the input A. Then we find that to test the fault s-a-0 on wire α, the input A must also be 1. Therefore the test is $A = 1$ and $B = 1$ which can be represented by a decimal number 3. By applying this test to the NAND gate and observing its output, we can determine whether the gate has the s-a-0 fault. The gate does not have the fault s-a-0 on wire α if the output is 0; otherwise, it is faulty. This is shown in Fig. 7.1.2(a). The test for detecting a s-a-1 fault on wire α which is $A = 0$ and $B = 1$ can be derived in a similar way. The normal and faulty outputs of this test are 1 and 0, respectively. This is shown in Fig. 7.1.2(b).

Now let us consider the problem of having a complete "check-up" for all possible logic faults that may occur in this gate. There are two possible faults (s-a-0 fault and s-a-1 fault) on each of the three wires which gives a total of six; namely, wire α s-a-0, wire α s-a-1, wire β s-a-0, wire β s-a-1, wire γ s-a-0, and wire γ s-a-1.

It is easy to show that the test $A = 1$ and $B = 1$ can detect the following logic faults:

$$\text{wire } \alpha \text{ s-a-0}$$

$$\text{wire } \beta \text{ s-a-0}$$

$$\text{wire } \gamma \text{ s-a-1,}$$

The test $A = 0$ and $B = 1$ can detect the following logic faults:

$$\text{wire } \alpha \text{ s-a-1}$$

$$\text{wire } \gamma \text{ s-a-0,}$$

and the test $A = 1$ and $B = 0$ can detect the following logic fault:

$$\text{wire } \beta \text{ s-a-1.}$$

The following table summarizes the above discussions.

Test	A	B	C	C
This test is redundant → 0	0	0	1 ⎫	0 ⎫
1	0	1	1 ⎬ Normal	or 0 ⎬ Faculty
2	1	0	1 ⎭ output	or 0 ⎭ output
3	1	1	0 ⎭	or 1 ⎭

It is seen that test 0 is redundant, but all the other three tests are necessary. Hence the test set $\{1, 2, 3\}$ is a minimal complete test set for a NAND gate, which shows a 25% savings of testing time over the test set using the complete truth table. The percentage of the rows of a truth table that are necessary for a complete testing of a combinational circuit decreases as the number of inputs to the circuit increases. The main objective of this chapter is to present methods for deriving a minimal complete test set (experiment) for a given combinational circuit.

The most classic method for constructing a set of tests to detect and locate a prescribed list of permanent logic faults, single or multiple, in a combinational circuit is the fault-table method. Basically, a test set (experiment) derived by this method is obtained by comparing the truth tables of the normal and faulty circuits. Two types of testing experiment will be presented. One is *fixed-scheduled*, which means that the choice of test schedules is completely independent of the outcome of the individual tests in the sequence, and the other is *adaptive (sequential)-scheduled*, which means that the choice of test schedules depends upon the outcome of the individual tests in the sequence. Derivations of fixed-scheduled and adaptive-scheduled experiments for fault detection and fault location will be discussed. In the sequel it will be shown that the length of a minimal adaptive fault-detection test set is the same as that of a minimal fixed-scheduled fault-detection test set. Hence there is no advantage in using an adaptive fault-detection experiment. But for fault location and fault location-to-within-modules, the possible reductions in test-schedule length are quite substantial when adaptive experiments are used.

Fixed-Scheduled Fault Detection   The procedure of obtaining a minimal fixed-scheduled fault-detection test set (experiment) using a fault table consists of the following three steps:

Step 1   Construction of the fault table   If $x_1, x_2, \ldots, x_n$ are the input variables to a single output circuit whose fault-free (correct) output is $z = z(x_1, \ldots, x_n)$ and $z^{\alpha_1}, z^{\alpha_2}, \ldots, z^{\alpha_l}$ are the erroneous outputs, each corresponding to one of the possible faults $\alpha_1, \alpha_2, \ldots, \alpha_l$, a multiple-output table of the combinations may be obtained:

Row Number r	x_1	x_2	...	x_n	z	z^{α_1}	z^{α_2}	...	z^{α_l}
0	0	0	...	0	0	1	0	...	0
1	0	0	...	1	1	1	0	...	1
.			.		.	.	.		.
.			.		.	.	.		.
$2^n - 1$	1	1	...	1	0	0	0	...	1

This is called a *fault table*, F. The complete test set for any α_i is the set of input combinations $x^j = (x_1^j, x_2^j, \ldots, x_n^j)$ such that

$$z(x^j) \oplus z^{\alpha_i}(x^j) = 1 \qquad \text{for all } 1 \leq i \leq l \qquad (7.1.2)$$

DEFINITION 7.1.4

Two faults are said to be *indistinguishable* if the truth tables of the output functions of the circuits with these two faults are completely identical. In other words, we cannot find a test such that the two faults can be distinguished based on the information of the output. If, on the other hand, there exists such a test, by applying it, the output values of the circuit having the two faults are different; we say that these two faults are *distinguishable*.

The multiple-output table above may be simplified somewhat as follows:

1.   Delete the columns that correspond to undetectable faults (see Definition 7.1.1); that is, delete the columns z^{α_i} in the table that are identical to z.
2.   Combine the columns that correspond to indistinguishable faults; that is, combine the columns z^{α_i} and z^{α_j} if $z^{\alpha_i} = z^{\alpha_j}$.

The reduced fault table F^* still contains the complete information about the complete test sets for detecting all the detectable faults. In fact, the complete test set for detecting each detectable fault α_i may be obtained by taking exclusive–or operations on column z^{α_i} with column z as in Eq. (7.1.2).

Step 2   Construction of the fault-detection table   The fault-detection table or fault-detection matrix G_D is constructed as follows: Each column $i, i = 1, 2, \ldots, l$ of G_D presents a distinguishable fault under detection and the entries of the column are obtained from the reduced-fault table by $z(x^j) \oplus z^{\alpha_i}(x^j)$. The rows of G_D are the input combinations of the union of all the complete test sets for detecting the faults. If there are m distinguishable faults and the union of all the complete test sets contain p tests, the size of G_D is $p \times m$. An entry in row j and column i is 1 if row j is a test for fault i, and is 0 if row j is not a test for fault i. Since there is no need for displaying both the 0's and 1's entries in the table G_D, it is customary to display the 1's only and to leave all the 0 entries blank.

Step 3   Determination of a minimal test set   The first step toward obtaining a minimal test set from a fault-detection table is simplification of the table by deleting certain superfluous rows and columns following the two rules:

RULE 1

Delete any row whose 1's all fall in the same columns as the 1's in some other row. That is, delete any row that is covered by, or is the same as, some other row.

RULE 2

Delete any column that has 1's in all the rows in which another column has 1's. That is, delete any column that covers, or is the same as, some other column.

These steps may be applied in any order until neither is applicable. The reduced table G_D^* has distinct rows and distinct columns; also, no row covers another row, and no column covers another column.

Let us observe the following analogies:

	Test-Set Minimization	Switching-Function Minimization
Rows	⟷ Tests	⟷ Implicants
Columns	⟷ Faults	⟷ Minterms
Table	⟷ Fault table	⟷ Implicant table
Reduced table	⟷ Reduced fault table	⟷ Prime-implicant table

Therefore, the problem of choosing a minimal set of tests that can detect all the faults under detection may be transformed into the familiar problem of minimizing a switching function, which has been solved in Sections 1.4–1.7.

Fixed-Scheduled Fault Location The difference between the fault-detection problem and the fault-location problem is that the former is concerned only with determining whether a circuit has a detectable fault, whereas the latter, in addition to the determination of whether the circuit is faulty or not, demands determination of the location of the fault that has occurred. There are two types of approaches to the latter: the fixed-scheduled experiment and the adaptive-scheduled experiment. The method for deriving a shortest fixed-scheduled fault-location experiment using the fault table is quite similar to the one for fault detection just described.

The fault-table method for constructing a shortest fixed-scheduled experiment for fault location consists of three steps:

Step 1 Construction of the fault table.

Step 2 Construction of the fault-location table. The fault-location table G_L is constructed from the reduced fault table F^*. A column of G_L is formed by taking the exclusive-or operation on a pair of columns of F^*. If we let f_0 be the output of an n-input single-output combinational circuit with no detectable faults, and $f_1, f_2, \ldots, f_m$ be the outputs of the circuit with m faults, respectively, then the fault-location table has 2^n rows denoted by $0, 1, \ldots, 2^n - 1$, and $m(m + 1)/2$ columns, each of which is obtained by $f_i \oplus f_j$, $0 \leq i < j \leq m$. If it is only required to locate the faults within the limits of the modules of the circuit, the table G_M for fault location-to-within-modules is constructed from F in the same way as G_L, except for one column for each pair of columns of F^* that belong to different modules. It should be noted that G_D is always contained in G_L and G_M.

Step 3 Find a minimal cover of G_L or G_M.

The fault-table method for fault detection and fault location is illustrated by the following example.

Example 7.1.1

Consider the circuit of Fig. 7.1.3. The output function of this circuit is $z = f(x_1, x_2, x_3)$ $= x_1x_2 + x_2x_3$. The fault table for detection and location of all the single faults of the circuit is shown in Table 7.1.1. In this table, f_0 denotes the output function of the faultless version of the circuit, and f_{ij} represents the output function of the circuit with its ith line stuck at 0 ($j = 0$) or 1 ($j = 1$). It is observed that all the faults are detectable, and four groups of faults $\{f_{10}, f_{20}, f_{50}\}, \{f_{11}, f_{41}\}, \{f_{30}, f_{40}, f_{60}\}$, and $\{f_{51}, f_{61}\}$ in this table are indistinguishable faults. We combine them, choose one function from each group, and delete the rest of them from the

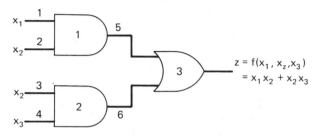

Fig. 7.1.3 Example to illustrate fault location using the fault table.

TABLE 7.1.1   Fault Table for Detection and Location of all Single Faults of the Circuit of Fig. 7.1.2

r	x_1 x_2 x_3	f_0	f_{10}	f_{11}	f_{20}	f_{21}	f_{30}	f_{31}	f_{40}	f_{41}	f_{50}	f_{51}	f_{60}	f_{61}
0	0 0 0	0	0	0	0	0	0	0	0	0	0	1	0	1
1	0 0 1	0	0	0	0	0	0	1	0	0	0	1	0	1
2	0 1 0	0	0	1	0	0	0	0	0	1	0	1	0	1
3	0 1 1	1	1	1	1	1	0	1	0	1	1	1	0	1
4	1 0 0	0	0	0	0	1	0	0	0	0	0	1	0	1
5	1 0 1	0	0	0	0	1	0	1	0	0	0	1	0	1
6	1 1 0	1	0	1	0	1	1	1	1	1	0	1	1	1
7	1 1 1	1	1	1	1	1	1	1	1	1	1	1	1	1

| | | f_0 | f_1 | f_2 | | f_3 | f_4 | f_5 | f_6 | | | | |

table. Name the six distinct output functions which represent six distinguishable faults to be $f_1, \ldots, f_6$, as indicated at the bottom the table. From the functions $f_0, \ldots, f_6$, we form the fault-location table G_L, which is shown in Table 7.1.2. The column label kl represents the functions f_k and f_l, from which the column is constructed ($f_k \oplus f_l$), where $0 \le k < l \le m$, and m denotes the total number of distinguishable faults. So there are $C_2^m = m(m-1)/2$ columns in G_L. Note that it is not necessary to show all the 0 entries of this table, and that the subtable G_D of this table, composed of the first six columns, is, in fact, the fault-detection table needed to detect the above-designated faults. From this subtable it is seen that tests 2, 3, and 6, which cover columns 01, 02, 04, and 06, are essential, and columns 03 and 05 can be covered by test 5. Thus the minimal complete test set for detection of all single faults

TABLE 7.1.2 Fault-Location Table G_L of Example 7.1.1, Which Includes the Fault-Detection Table as a Subtable

Test \ Fault	01	02	03	04	05	06	12	13	14	15	16	23	24	25	26	34	35	36	45	46	56
0						1			1					1		1				1	1
1				1	1					1	1		1	1		1	1	1	1		
*2	(1)					1	1				1	1	1	1				1		1	1
*3				(1)				1						1			1		1	1	
4			1			1	1				1	1		1	1	1				1	1
5			1		1	1	1		1	1	1		1	1	1				1	1	
*6	(1)						1	1	1	1	1										
7																					

Fault-detection table G_D

of this circuit is $\{2, 3, 5, 6\}$. Note that in this example we only need four out of a total of eight tests to completely check all the single faults of the circuit, which shows a 50% savings of testing time over the test time using the complete truth table. A close examination of the faults that are covered by each of these four tests shows that

Test	x_1	x_2	x_3	
3	0	1	1	} test all the s-a-0 faults
6	1	1	0	
2	0	1	0	} test all the s-a-1 faults
5	1	0	1	

Now let us examine the fault-location table, which is the entire table of Table 7.1.2. It is observed that tests 2, 3, and 6 are essential tests which cover all the columns except the four columns 03, 05, 26, and 35. With these columns and the nonessential rows (tests), we construct the fault-location table:

Test \ Fault	03	05	26	35
0			1	
1		1	1	1
4	1		1	1
5	1		1	

Test 0 is dominated by test 1 and column 26 dominates all the other three columns, so they can be removed from the table. The remaining table shows that any choice of two tests from the three tests 1, 4, and 5, plus the essential tests, will cover all the columns of G_L (i.e., will locate all six distinguishable faults). Hence the minimal fault-location experiment for this circuit should consist of tests 2, 3, and 6, plus any two of the three tests 1, 4, and 5, as shown:

	Test	x_1	x_2	x_3	Response of Faulty Circuits						
					f_0	f_1	f_2	f_3	f_4	f_5	f_6
	2	0	1	0	0	0	1	0	0	0	1
	3	0	1	1	1	1	1	1	0	1	1
	6	1	1	0	1	0	1	1	1	1	1
plus { or {	1	0	0	1	0	0	0	0	0	1	1
	4	1	0	0	0	0	0	1	0	0	1
	1	0	0	1	0	0	0	0	0	1	1
	5	1	0	1	0	0	0	1	0	1	1
or {	4	1	0	0	0	0	0	1	0	0	1
	5	1	0	1	0	0	0	1	0	1	1

Now let us examine these tests and the faults that they detect and locate. This is illustrated in Table 7.1.3. It is seen that the test set containing only three tests, 2, 3, and 6, cannot detect f_3 and f_5 because they are identical to f_0. A test set consisting of these three sets plus one additional test may be able to detect all the single faults. But this additional test must be test 5 and not test 1 or test 4, because test set {2, 3, 6, 1} cannot detect f_3, and test set {2, 3, 6, 4} cannot detect f_5. As a consequence, a test set consisting of five tests, tests 2, 3, and 6, plus two of the three tests 5, 1, and 4, will detect and locate all the six distinguishable single faults. The test set {2, 3, 6, 5, 1 or 4} and the responses of the circuit under no fault and a single (distinguishable) fault condition are shown in Table 7.1.3.

TABLE 7.1.3 Fault-Detection and Fault-Location Tests and the Responses of the Circuit of Fig. 7.1.3 with No Fault and a Single Distinguishable Fault

Test \ Fault	f_0	f_1	f_2	f_3	f_4	f_5	f_6
2	0	0	1	0	0	0	1
3	1	1	1	1	0	1	1
6	1	0	1	1	1	1	1
(Test set {2, 3, 6} cannot detect f_3 and f_5)							
5	0	0	0	1	0	1	1
[Test set {2, 3, 6, 5} can detect all the faults but cannot locate (distinguish) faults between f_3 and f_5]							
1	0	0	0	0	0	1	1
or 4	0	0	0	1	0	0	1
(All the faults are detected and located)							

Now suppose that instead of locating each individual fault, we only want to know which of the three gates of the circuit is faulty. By examining the definition of faults f_{ij} and distinguishable faults $f_1, \ldots, f_6$, we find that the following relation exists between the individual faults and the three gates:

Faults		Faulty Gate
f_1, f_3	indicate	gate 1
f_4, f_5	indicate	gate 2
f_2	indicates	gate 1 or gate 2
f_6	indicates	gate 3

The third row in this table demands an explanation. Fault f_2 represents two indistinguishable faults, f_{11} and f_{41}, which indicate a fault in gate 1 and gate 2, respectively. In other words, fault f_2 may come either from f_{11}, which indicates that gate 1 is faulty, or from f_{41}, which indicates that gate 2 is faulty. Thus the presence of f_2 may indicate either gate 1 or gate 2 being faulty (but not gate 3, of course). The G_M for this case is the same as G_L of Table 7.1.2, except that certain columns representing pairs of faults (i.e., 13 and 45) within the same module (gate) need not be included. It can easily be shown that the same test set for locating each individual single fault is required for detecting and locating a faulty gate of the circuit.

Suppose that the circuit is constructed by two modules, module 1, containing the two AND gates, and module 2, containing the OR gate. In this case the relation between the individual single fault and the faulty modules indicated by them is

Faults		Faulty Modules
f_1, f_2, f_3, f_4, f_5	indicate	module 1 (gates 1 and 2)
f_6	indicates	module 2 (gate 3)

By deleting columns 12, 13, 14, 15, 23, $\ldots$, 45, from the G_L of Table 7.1.2 and finding the minimal cover for it, we find that the fault-detection test set $\{2, 3, 6, 5\}$ obtained previously will locate any single fault within the two modules.

Let N_D, N_L, and N_M be the number of test inputs in the minimal experiments for fault detection, fault location, and fault location-to-within-modules. It should be remarked that for detecting or locating a set of faults of a circuit, the following relation always holds:

$$N_D \leq N_M \leq N_L$$

For example, in the example above, $N_D = 4$ and $N_L = 5$. $N_M = 5$ if each gate is considered as a module, and $N_M = 4$ if the two AND gates and the OR gate are considered as two modules of the circuit.

We have been discussing fixed-scheduled fault detection and location. In the following we will discuss the same problems but the test schedule will be adaptive. It will be shown that the use of an adaptive test schedule will be of no benefit in fault detection, but in fault location and fault location-to-within-modules, the possible reductions in test-schedule length are substantial. A heuristic procedure is given which is easy to carry out and reasonably effective, although it does not necessarily lead to a test whose length is absolutely minimal.

Adaptive-Scheduled Fault Detection and Location It may be observed that the choice of test schedules derived above is completely independent of the outcome of the individual tests in the sequence; moreover, the length of the schedule is independent of the order in which the tests are performed. It is quite conceivable, however, that after the first test input of a schedule has been applied and the output noted, the residual test schedule, which is minimal with respect to a 0 output, is not the same as that which is minimal with respect to a 1 output. Similarly, after two test inputs have been applied, the four partial test schedules that should follow may be all different in content and length, and so on for successive test inputs. We consider here the economies that can be achieved by choosing each test input to be applied to the circuit on the basis of the outcomes of all previous tests in the schedule. A convenient way to present such a sequence of tests and their outcomes is to use a diagnosing tree, which is defined as follows:

DEFINITION 7.1.5

A *diagnosing tree* is a directed graph whose nodes are tests. The outgoing branches from a node represent the different outcomes of the particular test.

It is easy to show that adaptive-scheduled fault detection requires the same test length as that of fixed-scheduled fault detection. Take the fault detection of the circuit of Fig. 7.1.3, for example. A diagnosing tree of the fixed-scheduled fault-detection experiment {2, 3, 5, 6} for the circuit may be constructed from the fault table (Table 7.1.1), as shown in Fig. 7.1.4(a). From this diagnosing tree, we see that the circuit is fault-free if and only if the output sequence to the sequence (2, 3, 6, 5) is (0, 1, 1, 0), as indicated by the path in dark lines. Thus this tree can be simplified to the one shown in Fig. 7.1.4(b). Applying tests 2, 3, 5, 6 in any order will not shorten the length of the experiment. Another diagnosing tree for test set {2, 3, 5, 6} is shown in Fig. 7.1.4(c), which still requires all four tests. Since the adaptive-scheduled fault detection has no advantage over the fixed-scheduled fault detection, we can completely ignore it.

Unlike fault detection, adaptive-scheduled fault location in general yields a shorter experiment. For example, the fault location of the circuit of Fig. 7.1.3, using the fixed-scheduled experiment, requires a minimal length of 5 but needs only 4 when the adaptive-scheduled experiment is used. The diagnosing trees of the fixed-scheduled fault-location experiment {5, 4, 6, 3, 2} and an adaptive-scheduled fault-location experiment for this circuit are shown in Fig. 7.1.5(a) and (b), respectively.

Now the question is: How to find an adaptive schedule of tests such that the number of the test levels is minimal? First, several interesting observations are in order.

1. Since our goal has now been changed from finding a minimal test set to finding an (adaptive) test schedule with a minimal number of levels, in constructing such

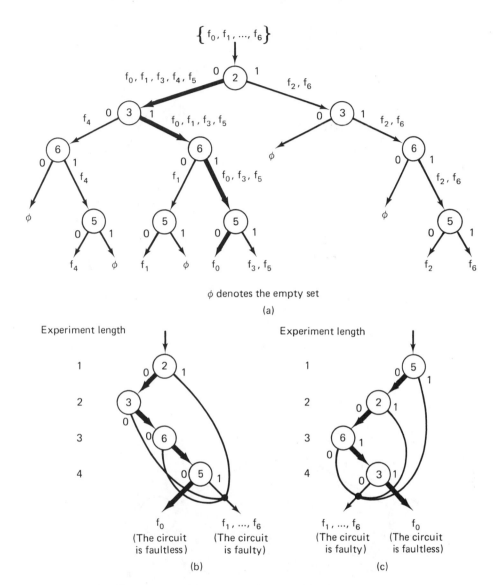

ϕ denotes the empty set

(a)

Fig. 7.1.4   (a) Diagnosing tree for fault detection of the circuit of Fig. 7.1.3; (b) simplified diagnosing tree; (c) another diagnosing tree for the same test set [2, 3, 5, 6]

an experiment, tests chosen must not be confined to any particular given set of tests (e.g., any of the minimal fixed-schedule fault-location test sets). They must be chosen from the rows of the fault table.

2.   The construction of an adaptive schedule of tests for fault location with a minimal number of levels needs at least N_L tests.

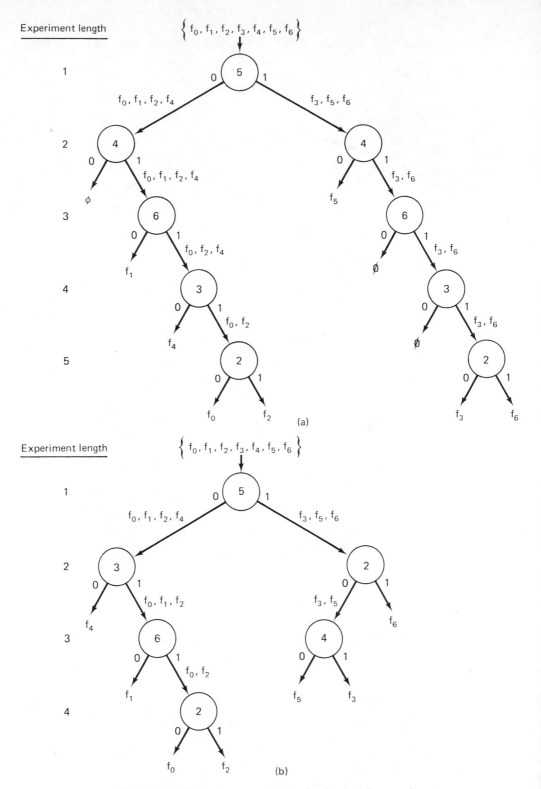

Fig. 7.1.5 Diagnosing tree of the fixed-schedule experiment $\{5, 4, 6, 3, 2\}$.

3. In constructing this experiment, at each step the test that will distinguish between the largest number of faults not already distinguished should be chosen.

4. One way to select the appropriate row (test) having the desired property described in observation 3 at each step of the procedure is to try all possible remaining rows (tests). For even a small fault table, however, the table number of possible graph labelings that must be tried to determine the minimal number of levels is astronomical. This approach is therefore impractical.

Let w_{i0} and w_{i1} denote the numbers of 0's and 1's in row i, respectively. A simple heuristic method for finding a nearly minimal adaptive-scheduled fault-location experiment is: Select row i, which maximizes the number of $(0, 1)$ pairs between digits in that row, that is, which maximizes the exprsssion

$$R_i = w_{i0}w_{i1} \qquad (7.1.3)$$

This number is optimized if the row that has the most nearly equal distribution of 0's and 1's, [i.e., the row (or one of the subset of rows) for which $|w_{i0} - w_{i1}|$ is minimal] is selected. The use of this criterion appears to work very well for many problems. This method is illustrated by the following example.

Example 7.1.1 (Continued)

Consider the construction of an adaptive-scheduled fault-location experiment for the circuit of Fig. 7.1.3 whose fault table (Table 7.1.1) may be presented in matrix form as

$$F^* =
\begin{array}{c}
 \\
\begin{array}{ccccccc} f_0 & f_1 & f_2 & f_3 & f_4 & f_5 & f_6 \end{array} \\
\left[\begin{array}{ccccccc}
0 & 0 & 0 & 0 & 0 & 0 & 1 \\
0 & 0 & 0 & 0 & 0 & 1 & 1 \\
0 & 0 & 1 & 0 & 0 & 0 & 1 \\
1 & 1 & 1 & 1 & 0 & 1 & 1 \\
0 & 0 & 0 & 1 & 0 & 0 & 1 \\
0 & 0 & 0 & 1 & 0 & 1 & 1 \\
1 & 0 & 1 & 1 & 1 & 1 & 1 \\
1 & 1 & 1 & 1 & 1 & 1 & 1
\end{array}\right]
\begin{array}{c}
0 \\ 1 \\ 2 \\ 3 \\ 4 \\ 5 \\ 6 \\ 7
\end{array}
\end{array}
\qquad (7.1.4)$$

For this matrix, row 5 should be chosen first, since it has 4 0's and 3 1's and $|w_{50} - w_{51}| = 1$, and all the other rows of this matrix whose $|w_{i0} - w_{i1}|$ are greater than 1. Selection of row 5 yields the two submatrices

$$F_0^*(5) =
\begin{array}{c}
\begin{array}{cccc} f_0 & f_1 & f_2 & f_4 \end{array} \\
\left[\begin{array}{cccc}
0 & 0 & 0 & 0 \\
0 & 0 & 0 & 0 \\
0 & 0 & 1 & 0 \\
1 & 1 & 1 & 0 \\
0 & 0 & 0 & 0 \\
1 & 0 & 1 & 1 \\
1 & 1 & 1 & 1
\end{array}\right]
\begin{array}{c}
0 \\ 1 \\ 2 \\ 3, \\ 4 \\ 6 \\ 7
\end{array}
\end{array}
\qquad
F_1^*(5) =
\begin{array}{c}
\begin{array}{ccc} f_3 & f_5 & f_6 \end{array} \\
\left[\begin{array}{ccc}
0 & 0 & 1 \\
0 & 1 & 1 \\
0 & 0 & 1 \\
1 & 1 & 1 \\
1 & 0 & 1 \\
1 & 1 & 1 \\
1 & 1 & 1
\end{array}\right]
\begin{array}{c}
0 \\ 1 \\ 2 \\ 3 \\ 4 \\ 6 \\ 7
\end{array}
\end{array}
\qquad (7.1.5)$$

The selection of subsequent rows for the two submatrices should be considered separately. For $F_0^*(5)$, there are three rows, rows 2, 3, and 6, whose $|w_{i0} - w_{i1}| = 2$, and those of the rest of the rows are 4. Thus any of the three rows may be chosen. Say that we choose row 3. Now consider $F_1^*(5)$. It is obvious that any rows of this matrix may be chosen except rows 3, 6, and 7. Suppose that we choose row 2. Then the two submatrices are further partitioned into four smaller submatrices:

$$
F_{00}^*(5, 3) = \begin{bmatrix} 0 \\ 0 \\ 0 \\ 0 \\ 1 \\ 1 \end{bmatrix} \begin{matrix} f_4 \\ \hline 0 \\ 1 \\ 2 \\ 4 \\ 6 \\ 7 \end{matrix}, \qquad
F_{01}^*(5, 3) = \begin{bmatrix} 0 & 0 & 0 \\ 0 & 0 & 0 \\ 0 & 0 & 1 \\ 0 & 0 & 0 \\ 1 & 1 & 1 \\ 1 & 1 & 1 \end{bmatrix} \begin{matrix} f_0\ f_1\ f_2 \\ \hline 0 \\ 1 \\ 2 \\ 4 \\ 6 \\ 7 \end{matrix}
$$

$$
F_{10}^*(5, 2) = \begin{bmatrix} 0 & 0 \\ 0 & 1 \\ 1 & 1 \\ 1 & 0 \\ 1 & 1 \\ 1 & 1 \end{bmatrix} \begin{matrix} f_3\ f_5 \\ \hline 0 \\ 1 \\ 3 \\ 4 \\ 6 \\ 7 \end{matrix}, \qquad
F_{11}^*(5, 2) = \begin{bmatrix} 1 \\ 1 \\ 1 \\ 1 \\ 1 \\ 1 \end{bmatrix} \begin{matrix} f_6 \\ \hline 0 \\ 1 \\ 3 \\ 4 \\ 6 \\ 7 \end{matrix}
$$

$$(7.1.6)$$

Now we have two submatrices, $F_{00}^*(5, 3)$ and $F_{11}^*(5, 2)$, which have only one column. This means that the faults denoted by f_4 and f_6 are already located by the sequences of tests (5, 3) and (5, 2), respectively. Continue this process until all the submatrices form a single-column matrix. One such adaptive experiment having the minimal number of levels obtained by this procedure is shown in Fig. 7.1.5(b).

The method for deriving the shortest test sequence to locate any individual faults just described applies with little change to fault location-to-within-modules. We now modify the criterion of row acceptance to count not all $(0, 1)$ pairs, but only those pairs in each row in which the 0 and the 1 fall in different module classes. This quantity is most easily calculated by subtracting the sum of the number of $(0, 1)$ pairs that fall entirely within the individual classes from the total number of $(0, 1)$ pairs:

$$R_i = w_{i0}w_{i1} - \sum_{j=1}^{p} w_{ij0}w_{ij1} \qquad (7.1.7)$$

where w_{ij0} and w_{ij1} are the number of 0's and 1's, respectively, in the jth module class in row i. The row to be selected is the one with the largest row count R_i.

Example 7.1.1 (Continued)

As an example for adaptive-scheduled fault location-to-within-modules, suppose that in the above F^* matrix, faults f_1, f_2, and f_4 are associated with a single module, as are faults

$f_3, f_5,$ and f_6. The row counts on the eight rows of F^* are

$$0: \quad 6 - 0 - 2 = 4$$
$$1: \quad 10 - 0 - 2 = 8$$
$$2: \quad 10 - 2 - 2 = 6$$
$$3: \quad 6 - 2 - 0 = 4$$
$$4: \quad 10 - 0 - 2 = 8$$
$$5: \quad 12$$
$$6: \quad 6 - 2 - 0 = 4$$
$$7: \quad 0$$

Clearly, row 5 should be selected first, and the same two submatrices $F_0^*(5)$ and $F_1^*(5)$ that appeared in Eq. (7.1.5) arise here. $F_0^*(5)$ falls entirely within (and in fact covers exactly) module class ($f_3, f_5,$ and f_6); therefore, it need not be reduced further. For the rows of $F_0^*(5)$, the row-count values are

$$0: \quad 0$$
$$1: \quad 0$$
$$2: \quad 3 - 2 = 1$$
$$3: \quad 3 - 2 = 1$$
$$4: \quad 0$$
$$6: \quad 3 - 2 = 1$$
$$7: \quad 0$$

Any of 2, 3, or 6 should be chosen. Selecting 2, only the $F_{00}^*(5, 2)$ need be further reduced, and any of the nonconstant rows 3 or 6 can be used. Figure 7.1.6(b) shows the resulting diagnosing tree. The diagnosing tree for fixed-scheduled fault location-to-within-modules for the same modules is shown in Fig. 7.1.6(a). It is seen that for locating faults $f_3, f_5,$ and f_6 within the module, one test (test 5) is sufficient if the adaptive-scheduled experiment is used.

From the above discussion, the following bounds can readily be established.

$$1 \leq l_D = N_D \leq m$$
$$1 + \lceil \log_2 m \rceil \leq l_L \leq N_L \leq m$$
$$1 + \lceil \log_2 p \rceil \leq l_M \leq N_M \leq m$$

where m and p denote the number of distinguishable faults under investigation and the number of module classes, respectively; l denotes the number of levels in adaptive test schedules. For most problems, of course, we can expect l_L to be much smaller than N_L.

The heuristic adaptive-scheduled fault-location method provides a good but not necessarily optimal solution. The procedure is quite straightforward and easy to apply.

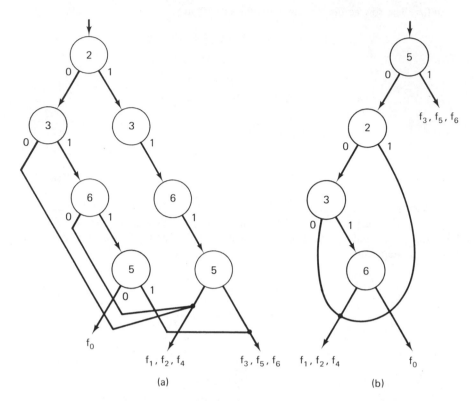

Fig. 7.1.6 Fault location-to-within-modules of the circuit of Fig. 7.1.3. (a) Diagnosing tree of the fixed-scheduled experiment {2, 3, 6, 5}. (b) Diagnosing tree of the adaptive-scheduled experiment {5, 2, 3, 6}.

The drawback of this method is that it requires a large amount of computer storage space to store the fault table. When the size of the fault table exceeds the amount available, we modify the above method to the *hybrid method*, which is described below.

Let n be the number of primary inputs of a circuit and m be the number of single faults of the circuit under consideration. Then the size of the fault table will be $2^n(m + 1)$. Suppose that the available memory of a computer can take only s columns of the $m + 1$ columns at a time. The hybrid method consists of the following steps:

1.   Enter the memory with s arbitrary columns of the fault table and obtain an adaptive-scheduled experiment for it by the heuristic method described above. Record this experiment and then clean up the memory.

2.   Enter the rest columns into the computer, s columns at a time, and apply the obtained adaptive-scheduled experiment to each of these $2_n \times s$ submatrices. Note that the application of the adaptive-scheduled experiment to these sets of faults is fixed-scheduled.

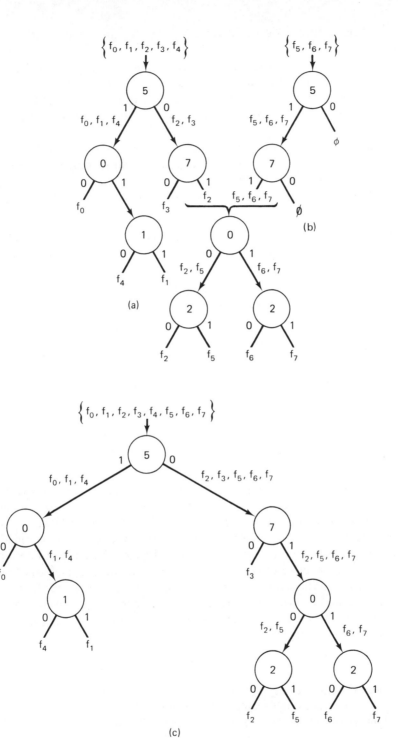

Fig. 7.1.7 Example to illustrate the hybrid method.

3. Superimpose all the diagnosing trees obtained in steps 1 and 2 such that the faults associated with a branch of the resulting diagnosing tree is the union of the faults associated with the corresponding branches of its component trees.

4. Terminate the branches that have one fault. For those which have more than one fault:

(a) If the number of faults exceeds s, repeat steps 1–3 for these faults.

(b) If the number of faults is not greater than s, apply the heuristic method to them and obtain an adaptive testing schedule as in step 1.

5. Repeat steps 1–4 until every terminal branch of this tree has only one fault associated with it.

The following example illustrates the principal idea of this method. For convenience, let us use the fault matrix of Example 7.1.1, which is repeated below.

$$F = \begin{array}{c} \begin{array}{cccccccc} f_0 & f_1 & f_2 & f_3 & f_4 & f_5 & f_6 & f_7 \end{array} \\ \left[\begin{array}{cccccccc} 0 & 1 & 0 & 1 & 1 & 0 & 1 & 1 \\ 1 & 1 & 0 & 0 & 0 & 0 & 1 & 0 \\ 0 & 1 & 0 & 1 & 0 & 1 & 0 & 1 \\ 0 & 0 & 0 & 1 & 0 & 1 & 1 & 1 \\ 1 & 0 & 1 & 1 & 1 & 1 & 1 & 1 \\ 1 & 1 & 0 & 0 & 1 & 0 & 1 & 1 \\ 0 & 0 & 0 & 0 & 1 & 1 & 1 & 1 \\ 0 & 0 & 0 & 1 & 1 & 1 & 0 & 0 \end{array}\right] \begin{array}{c} 0 \\ 1 \\ 2 \\ 3 \\ 4 \\ 5 \\ 6 \\ 7 \end{array} \end{array}$$

Suppose that the available storage space of a computer can only store five columns of this matrix ($s = 5$) and that we have arbitrarily chosen the first five columns. From the heuristic method, we obtain a minimum-level adaptive fault-location experiment, which is shown in Fig. 7.1.7(a). Application of this experiment to the remaining columns (since it is less than five) gives the tree in Fig. 7.1.7(b). Superimposing the two trees, we find that in the resulting tree, all the terminal branches have one fault, except that one of the branches, which comes from node (test) 7, has four faults, which is, however, less than five. We therefore reenter these four columns and apply the heuristic method to this set of faults. The final result is shown in Fig. 7.1.7(c).

Exercise 7.1

1. Consider the circuit of Fig. P7.1.1.
   (a) Find three undetectable faults in the circuit.
   (b) Find three pairs of indistinguishable faults in the circuit.

2. Find a minimal complete test set for detecting all distinguishable single faults in the irredundant circuit of Fig. P7.1.2 by the fault-table method.

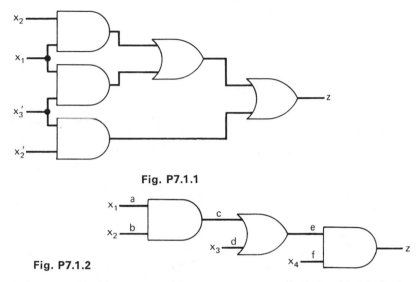

Fig. P7.1.1

Fig. P7.1.2

3. Find a fixed-scheduled fault-location experiment to locate all the distinguishable faults in the circuit of Fig. P7.1.2.

4. Derive an adaptive-scheduled fault-location experiment for the circuit of Fig. P7.1.2 and compare it with the result obtained in problem 3.

5. (a) Show the output sequences to the tests of the experiments of Fig. 7.1.4(a) and (b) and indicate the faults they locate.
   (b) Show the output sequences to the tests of the experiments of Fig. 7.1.5(a) and (b) and indicate the faults they locate within the modules.

6. For the circuit of Fig. 7.1.1(b):
   (a) Derive a minimal complete fixed-scheduled fault-detection experiment.
   (b) Derive a minimal complete fixed-scheduled fault-location experiment.
   (c) Derive a minimal complete adaptive-scheduled fault-location experiment.
   (d) Draw the diagnosing trees of the experiments obtained in (a), (b), and (c).
   (e) Show the output sequences to the tests of these experiments and indicate the faults they detect and locate.

7. Consider a fault table F of some three-input circuit which is presented in matrix form as follows:

$$
F = \begin{array}{cccccccccc}
 & f_0 & f_1 & f_2 & f_3 & f_4 & f_5 & f_6 & f_7 & \\
 & \begin{bmatrix} 0 \\ 1 \\ 0 \\ 0 \\ 1 \\ 1 \\ 0 \\ 0 \end{bmatrix} & \begin{matrix} 1 \\ 1 \\ 1 \\ 0 \\ 0 \\ 1 \\ 0 \\ 0 \end{matrix} & \begin{matrix} 0 \\ 0 \\ 0 \\ 0 \\ 1 \\ 0 \\ 0 \\ 0 \end{matrix} & \begin{matrix} 1 \\ 0 \\ 1 \\ 1 \\ 1 \\ 0 \\ 0 \\ 1 \end{matrix} & \begin{matrix} 1 \\ 0 \\ 0 \\ 0 \\ 1 \\ 1 \\ 1 \\ 1 \end{matrix} & \begin{matrix} 0 \\ 0 \\ 1 \\ 1 \\ 1 \\ 0 \\ 1 \\ 1 \end{matrix} & \begin{matrix} 1 \\ 1 \\ 0 \\ 1 \\ 1 \\ 0 \\ 1 \\ 0 \end{matrix} & \begin{bmatrix} 1 \\ 0 \\ 1 \\ 1 \\ 1 \\ 0 \\ 1 \\ 0 \end{bmatrix} & \begin{matrix} 0 \\ 1 \\ 2 \\ 3 \\ 4 \\ 5 \\ 6 \\ 7 \end{matrix}
\end{array}
$$

Show that a complete fixed-scheduled fault-location experiment of this circuit requires four tests, whereas a complete adaptive-scheduled fault-location experiment requires only three levels.

7.2 Path-Sensitizing Method

The fault-table method presented in the previous section requires construction of the fault table, which in general is not practical. For example, if a circuit has, say, 20 inputs and if there are 30 lines within the circuit, then there will be over 30,000,000 entries in the associated fault table. In this section we shall show that a fault-detection test may be found by examining the paths of transmission from the location of an assumed fault to one of its primary outputs. This is the principal idea behind the path-sensitizing method.

DEFINITION 7.2.1

In a logic circuit, a *primary output* is a line whose signal output is accessible to the exterior of the circuit, and a *primary input* is a line that is not fed by any other line in the circuit.

DEFINITION 7.2.2

A *transmission path*, or simply path of a combinational circuit, is a connected, directed graph containing no loops from a primary input or internal line to one of its primary outputs.

DEFINITION 7.2.3

A path is said to be *sensitized* if the inputs to the gates along this path are assigned values so as to propagate any change on the faulty line along the chosen path to the output terminal of the path.

For example, consider the irredundant circuit of Fig. 7.1.1(b), which, for convenience, is repeated in Fig. 7.2.1. Each line of the circuit is labeled by a letter, $a, b,$

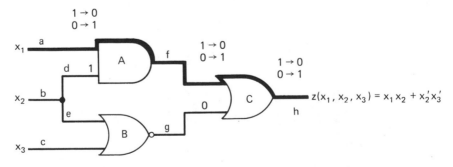

Fig. 7.2.1 Circuit to illustrate the concept of sensitized path.

$c, \ldots$ For convenience, we use the subscripts 0 and 1 to denote the s-a-0 and s-a-1 faults, respectively. Suppose that we want to detect the fault a_0 (line a s-a-0). The connected directed graph afh is a path containing the faulty line a. It is interesting to observe that if we assign a value 0 to the input g of the C gate (an OR gate) and a value 1 to the input d of A gate (an AND gate), we then see that

Value Change on Line a	Corresponding Change in Value on Line f	Corresponding Change in Value on Line h
$0 \longrightarrow 1$	$0 \longrightarrow 1$	$0 \longrightarrow 1$
$1 \longrightarrow 0$	$1 \longrightarrow 0$	$1 \longrightarrow 0$

The actual changes in values along path afh upon occurrence of the fault at line a are indicated in Fig. 7.2.1. Path afh is thus said to be *sensitized*. From the value of line $g = 0$ it is required that c or e or both equal to 1. If we choose to make $e = 1$ and $c = 0$ or 1 (don't care), then the values of d and e are 1 and thus are *consistent*; hence a test exists for detecting the fault a_0 using this method. Moreover, we observe that

Path	Input Variable Values Required to Sensitize the Path	Value of the Input Variable of the Path	Response
afh	$x_2 = 1,$ $x_3 = 0$ or 1 (don't care)	$x_1 = 1$	1, if the circuit is faultless 0, if the circuit has fault a_0
		$x_1 = 0$	0, if the circuit is faultless 1, if the circuit has fault a_1

Furthermore, it is apparent that a test which detects fault a_0 also detects faults f_0 and h_0 and that a test which detects fault a_1 also detects faults f_1 and h_1. Thus the above two tests together would detect *all* the s-a-0 and s-a-1 faults on this path.

From the above simple example, a systematic procedure for detecting a single fault by path sensitizing may be described as follows:

Step 1 Choose a path from the faulty line to one of the primary outputs.

Step 2 Assign the faulty line a value 0 (1) if the fault is a s-a-0 (s-a-0) fault.

Step 3 Along the chosen path, except the lines of the path, assign a value 0 to each input to the OR and NOR gates in the path and a value 1 to each input to the AND and NAND gates in the path.

Step 4 Trace back from gates along the sensitized path toward the circuit inputs. If a consistent input combination (a test) exists, the procedure is terminated. If, on the other hand, a contradiction is encountered, choose another path which starts at the faulty line, and repeat the above procedure.

To appreciate the advantage of this procedure over the fault-table method, an example of deriving a complete test set using the path-sensitizing method is given below.

Example 7.2.1

Consider the same circuit of Example 7.1.1 that was used to illustrate the fault-table method. By applying the above procedure to sensitize the four paths of the circuit, *afh*, *bdfh*, *begh*, and *cgh*, the required input-variable values, the assigned value of the input variable of the paths, and the faults detected by the tests are shown in Table 7.2.1. It is seen that the faults listed in the last column of this table include all the single faults of the circuit. From this table it is an easy matter to obtain a complete test set, which is $\{0, 2, 5, 7\}$. This set is one of the four minimal complete test sets obtained in Example 7.1.1, but the effort as well as the information storage space required here for obtaining it is much less compared to that required by the fault-table method.

**TABLE 7.2.1   Construction of the Complete Test Set
of the Circuit of Fig. 7.2.1 by Sensitizing Its Four Paths**

Path Number	Path	Input-Variable Values Required by Sensitizing the Path	Assigned Value of the Input Variable of the Path	Test	Faults Detected by the Test
1	*afh*	$x_2 = 1$ and $x_3 = 0$ or 1	$x_1 = 1$	6 or 7	a_0, f_0, h_0
			$x_2 = 0$	2 or 3	a_1, f_1, h_1
2	*bdfh*	$x_1 = 1$ and $x_3 = 0$ or 1	$x_2 = 1$	6 or 7	b_0, d_0, f_0, h_0
		$x_1 = 1$ and $x_3 = 1$	$x_2 = 0$	5	b_1, d_1, f_1, h_1
3	*begh*	$x_1 = 0$ and $x_3 = 0$	$x_2 = 1$	2	b_0, e_0, g_1, h_1
		$x_1 = 0$ or 1 and $x_3 = 0$	$x_2 = 0$	0 or 4	b_1, e_1, g_0, h_0
4	*cgh*	$x_1 = 0$ or 1 and $x_2 = 0$	$x_3 = 1$	1 or 5	c_0, g_1, h_1
			$x_3 = 0$	0 or 4	c_1, g_0, h_0

An important question arises: Will this method always yield a test for detecting a fault? The answer is no. However, it always yields a test for a class of combinational circuits, known as tree-like circuits, which is defined as follows.

DEFINITION 7.2.4

A *tree-like circuit* is defined as a circuit in which (1) each input is an independent input line to the circuit and (2) the fan-out† of every gate is 1.

The circuits of Fig. 7.2.2(a) and (b) are examples of tree-like circuits. The reason that the path-sensitizing method always yields a test for detecting any single fault in a tree-like circuit is because that a consistent input combination always exists for any

†The definition of the fan-out of a gate was given on page 126.

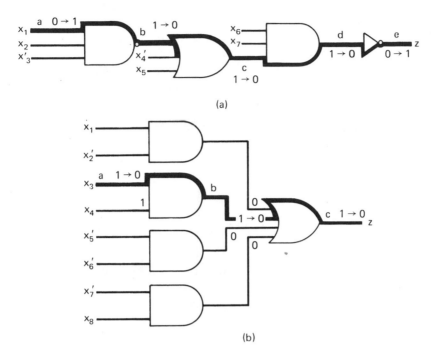

(a)

(b)

Fig. 7.2.2   Two tree-like circuits.

sensitized path in the circuit. For example, the faults a_1, b_0, c_0, d_0, and e_1 of the circuit of Fig. 7.2.2(a) can be detected by the test $x_1 = 0, x_2 = 1, x_3' = 1, x_4' = 0, x_5 = 0, x_6 = 1, x_7 = 1$, and the faults a_0, b_0, and c_0 of the circuit of Fig. 7.2.2(b) can be detected by the test $x_1 = x_2' = x_5' = x_6' = x_7' = x_8 = 0$ and $x_3 = x_4 = 1$.

From the above discussions, we have:

THEOREM 7.2.1

For any tree-like circuit, a complete test set can always be found by the path-sensitizing method.

Proof: Since every path of a tree-like circuit is sensitizable, we can find a complete test set for detecting any fault for every path. The union of the complete test sets for all the paths of the circuit is a complete test set for detecting any fault in the circuit. The determination of a minimal complete test set from this test set is similar to step 3 of the fault-table method for fault detection. ∎

For general combinational circuits, a contradiction of line value may occur at the gates, which are points of reconvergence for two or more fan-out paths from some preceding gate. For example, if we want to test whether the output of the NOR gate B of the circuit in Fig. 7.2.3(a) is s-a-0. In order to detect the s-a-0 fault, we require $h = 1$, which in turn requires $c = d = 0$. There are two paths from line h to the outputs z_1 and z_2, paths hjl and hkm. Let us first consider the path hjl. According to step 3, line g should be set to logic value 0, which requires $a = b = 1$. But both lines b and c connect directly to the input x_2, and they should have the same value. This leads to a contradiction. Hence the path hjl is not sensitizable, and consequently the fault line h s-a-0 cannot be detected at z_1. However, it can be easily shown that this fault can be detected at z_2 by sensitizing the path along hkm, as shown in Fig. 7.2.3(b), and that the test is $x_2 = x_3 = x_4 = 0$ and x_1 may be either 0 or 1.

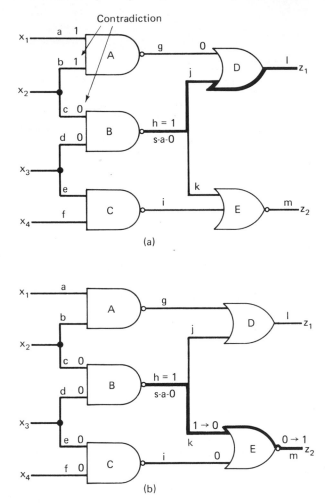

Fig. 7.2.3 Circuit of Example 7.2.2. (a) unsensitizable path hjl; (b) sensitizable path hkm.

It should be pointed out that there are cases where none of the individual paths of a circuit is sensitizable, but if we sensitize a group of them simultaneously, they become sensitizable. For example, none of the individual paths of the circuit of Fig. 7.2.4(a) from x_1 to z for detecting the fault line a s-a-0 is sensitizable, but if we sensitize *all the three paths* from x_1 to z as shown in Fig. 7.2.4(b) simultaneously, they become sensitizable.

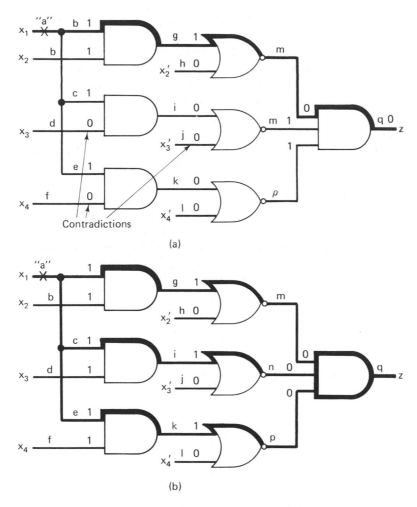

Fig. 7.2.4 Example of sensitizing several paths simultaneously.

This example shows that in searching out the sensitizable path for a particular fault, all possible combinations of single paths sensitized simultaneously must be included. We therefore add one additional step to the above procedure to make it complete.

Step 5 Apply steps 1–4 to all the single paths. If none of them is sensitizable, apply the procedure to all possible pairs of paths, then to all possible groups of three paths, and so on, until all possible combinations of the single paths are tried.

Referring to the simultaneously sensitized paths *agmq*, *acinq*, and *aekpq* of the circuit of Fig. 7.2.4(b), it is easily seen that the test 1111 would detect a set of s-a-0 and s-a-1 faults on the lines along the paths, and the test 0111 would detect a set of s-a-0 and s-a-1 faults on the lines along the paths. Since the two tests detect the two sets of faults on the lines along the paths, and one is the complementary set of faults to the faults of the other, the two tests together would detect *all* s-a-1 and s-a-0 faults on these three paths. This suggests that if tests are chosen *which detect all the faults on the circuit inputs*, all faults in the circuit will be detected, *provided* only that the set of tests chosen sensitizes a set of paths containing all connections in the circuit. This, in fact, is guaranteed to happen if there is no fan-out in the circuit, because in this case there is only one path from each circuit input to the outputs; therefore, the set of paths originating at circuit inputs will contain all connections. The path-sensitizing method is therefore very attractive from the point of view of not requiring the construction of the fault-table and is useful for the fault detection of tree-like circuits. But for general non-tree-like circuits, the process of exhausting all possible single paths, then all possible pairs of paths, then all possible groups of three paths, and so on, involves quite a lot of searching and computation, even when done by a computer. When fan-out exists, there occurs the additional problem of sensitizing a set of paths which, in fact, contain all connections. It would be desirable if such tests could be found by direct inspection of the circuit. Unfortunately no such direct technique has been discovered. However, in the sequel, it will be shown that if the circuit is first reduced to its equivalent normal form (ENF), a heuristic, systematic procedure can then be applied to find these "good" tests. This method is described in the next section.

Exercise 7.2

1. Find a minimal complete test set for the irredundant circuits in Fig. P7.2.1.

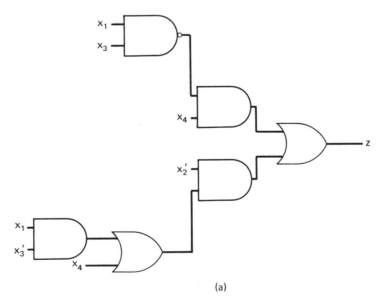

(a)

Fig. P7.2.1

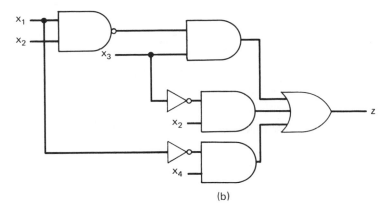

(b)

Fig. P7.2.1 (continued)

2. Show that the three paths indicated in the circuit of Fig. P7.2.2 cannot be sensitized individually but can be sensitized simultaneously.

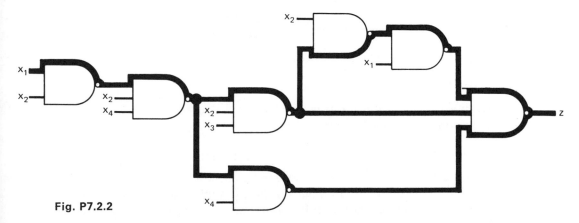

Fig. P7.2.2

7.3 Equivalent-Normal-Form Method

The equivalent normal form of a combinational circuit is defined below.

DEFINITION 7.3.1

The *equivalent normal form* (ENF) of a circuit is obtained by expressing the output of each gate as a sum-of-products expression of its inputs and preserving the identity of each gate by a suitable subscript.

Example 7.3.1

The ENF of the circuit of Fig. 7.2.1 is

$$h = (f + g)_h = (ad)_{fh} + (e'c')_{gh}$$
$$= (x_1)_{afh}(x_2)_{bdfh} + (x_2')_{begh}(x_3')_{cgh}$$

$$(7.3.1)$$

Example 7.3.2

As another example, consider the circuit of Fig. 7.3.1. The ENF for the circuit is

$$h = (x_1)_{acfh}(x_1')_{acgh}(x_2')_{bcgh}(x_3)_{dfh} + (x_1')_{acgh}(x_2')_{bcgh}$$
$$\cdot (x_2)_{bcfh}(x_3)_{dfh} + (x_1)_{acfh}(x_3)_{dfh}(x_4')_{egh} + (x_2)_{bcfh} \qquad (7.3.2.a)$$
$$\cdot (x_3)_{dfh}(x_4')_{egh}$$
$$= (x_1)_{acfh}(x_3)_{dfh}(x_4')_{egh} + (x_2)_{bcfh}(x_3)_{dfh}(x_4')_{egh} \qquad (7.3.2)$$

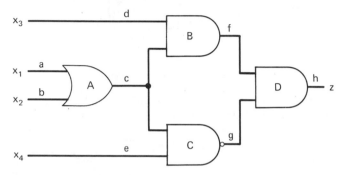

Fig. 7.3.1　Circuit of Example 7.3.2.

DEFINITION 7.3.2

Each subscripted input variable in an ENF is called a *literal*. An appearance of a literal in a term is also called a *literal*.

The ENF of a combinational circuit possesses the following characteristics:

1.　The ENF is a sum-of-products expression, hence the name "normal form."

2.　Each literal consists of an input variable, possibly primed, subscripted by a sequence of gate numbers. The variable and its subscript sequence identifies a path from the corresponding circuit input to the output. For example, the literal $(x_1)_{afh}$ in the first term of Eq. (7.3.1) identifies the path from input x_1 through wires a, f, and h to the output of the circuit.

3.　From the method of constructing the ENF, it follows that before any term of the ENF being discarded for the reason such as described in the next characterstic, there is at least one literal appearing in the ENF for every possible path from each input to the outputs.

4.　Any term in the ENF may be discarded if it contains a variable and its complement. For example, the first two terms of Eq. (7.3.2a) may be discarded.

5.　A literal may make several appearances in an ENF. For example, the literals $(x_3)_{dfh}$ and $(x_4')_{egh}$ appear in both terms of Eq. (7.3.2).

6.　The ENF may contain redundant terms and literals even though the original circuit is irredundant. An example that illustrates this characteristic will be given later [see Eq. (7.3.3)].

7.   If the path associated with a literal in the ENF contains an odd number of inversions (an inversion is produced by each NAND, NOR, and NOT element in the path), the priming on the literal will be opposite that on the corresponding input in the original circuit. If the number of inversions on the path is even, the priming on the literal will be the same as on the corresponding input in the original circuit. For example, the literals $(x_1)_{afh}$ and $(x_3')_{cgh}$ of the ENF of Eq. (7.3.1) are associated with the path *afh* with an even number (zero) of inversions and the path *cgh* with an odd number (one) of inversions, respectively. The two literals correspond to the input x_1 on gate *A* and the input x_3 on gate *B*.

8.   Every ENF corresponds to a hypothetical two-level AND–OR equivalent circuit. For example, the ENF of the circuit of Fig. 7.2.1 corresponds to the hypothetical two-level equivalent circuits of Fig. 7.3.2.

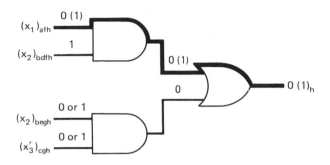

(a) Hypothetical equivalent two-level AND-OR circuit of the circuit in Fig. 7.2.1 derived from the ENF.

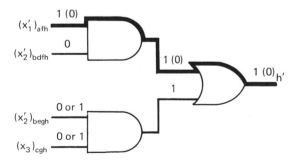

(b) Hypothetical equivalent two-level OR-AND circuit of the complement of the circuit in Fig. 7.2.1 derived from the complemented ENF.

Fig. 7.3.2 Hypothetical equivalent circuits of Example 7.3.1.

According to the rules of path sensitizing and the eighth characteristic described above, to test a particular literal for s-a-1, that literal must be assigned value 0, while the remaining literals in the term are assigned value 1. Also, at least one literal in each remaining term must be 0, in order that all remaining inputs to the OR gate be 0. By similar reasoning, a test that assigns value 1 to all literals of a particular term and assigns at least one 0 to literal in each remaining term tests all literals in that particular term for s-a-0. These rules will be referred to as the *literal sensitizing rules*.

THEOREM 7.3.1

A test devised for a literal appearance in the ENF sensitizes the path in the original circuit associated with that literal appearance.

Proof: This theorem may be proved by mathematical induction. The complete proof is quite lengthy and will not be included here. It can be found in reference 7. ∎

THEOREM 7.3.2

If a set of literals $\{L\}$ is selected whose paths contain every connection in the circuit, and if a set of tests $\{T\}$ can be found which tests at least one appearance of each of these literals for s-a-1 and s-a-0, then the set of tests detects every s-a-1 and s-a-0 fault in the circuit.

Since this theorem follows sufficiently directly from Theorem 7.3.1, the proof is omitted.

Example 7.3.1   (Continued)

The four literals $(x_1)_{afh'}$, $(x_2)_{bdfh'}$, $(x_2')_{begh'}$, and $(x_3')_{cgh}$ identifying the four paths in the circuit of Fig. 7.2.1 can be tests for s-a-0 and s-a-1, which are shown in Table 7.3.1. Since the paths associated with these four literals containing all the connections of the circuit and the set of tests of the rightmost column of Table 7.3.1 tests each of the four literals for s-a-0 and s-a-1, by Theorem 7.3.2, this set of tests detects every s-a-0 and s-a-1 fault in the circuit. From this table it is seen that tests 2 and 5 are essential, and these two tests plus two other tests, test 0 or 4 and test 6 or 7, constitute a minimal complete test set for detecting every

TABLE 7.3.1   Tests for Detecting the Four Literals of the ENF of Eq. (7.3.1) for s-a-0 and s-a-1

ENF	$h = (x_1)_{afh}(x_2)_{bdfh} + (x_2')_{begh}(x_3')_{cgh}$				Tests
Test $(x_1)_{afh}$ for s-a-0	1	1	0	d	6 or 7
Test $(x_1)_{afh}$ for s-a-1	0	1	0	d	2 or 3
Test $(x_2)_{bdfh}$ for s-a-0	1	1	0	d	6 or 7
Test $(x_2)_{bdfh}$ for s-a-1	1	0	1	0	5
Test $(x_2')_{begh}$ for s-a-0	d	0	1	1	0 or 4
Test $(x_2')_{begh}$ for s-a-1	0	1	0	1	2
Test $(x_3')_{cgh}$ for s-a-0	d	0	1	1	0 or 4
Test $(x_3')_{cgh}$ for s-a-1	d	0	1	0	1 or 5

single s-a-0 and s-a-1 fault. It is left to the reader to show that the same sets of tests can be obtained by using the fault-table method, but the equivalent-normal-form method requires much less effort compared to the fault-table method.

DEFINITION 7.3.3

The *ENF of a multiple-output* circuit is defined to be the set of ENF's for the individual outputs of the circuit. These individual ENF's will be referred to as *sub-ENF's*.

From the above discussions, the following theorems are seen immediately.

THEOREM 7.3.3

A s-a-0 (s-a-1) test for a particular literal in the ENF of a single-output circuit (or sub-ENF of a multiple-output circuit) is a s-a-1 (s-a-0) test for the corresponding literal in the complemented ENF (or sub-ENF), denoted by ENF'.

Example 7.3.1 (Continued)

For example, the ENF' of the circuit of Fig. 7.2.1 is

$$h' = (x_1')_{afh} + (x_2')_{bdfh})((x_2)_{begh} + (x_3)_{cgh} \qquad (7.3.3)$$

whose hypothetical two-level OR–AND equivalent circuit is shown in Fig. 7.3.2(b). The s-a-0 and s-a-1 tests for literal $(x_1)_{afh}$ in the ENF are the s-a-1 and s-a-0 tests for the corresponding literal $(x_1')_{afh}$ in the ENF', as shown in Fig. 7.3.3. The complete tests for detecting the four literals of the ENF' of Eq. (7.3.3) for s-a-0 and s-a-1 are given in Table 7.3.2.

For brevity, replace the subscript sequences in the ENF of Eq. (7.3.1) and in the complemented ENF of Eq. (7.3.3) as follows: $afh \rightarrow 1$, $bdfh \rightarrow 2$, $begh \rightarrow 3$, and $cgh \rightarrow 4$. Denote the four paths afh, $bdfh$, $begh$, and cgh by P_1, P_2, P_3, and P_4, respectively.

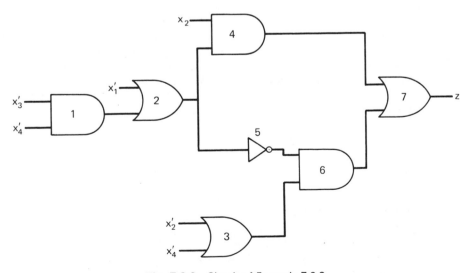

Fig. 7.3.3 Circuit of Example 7.3.2.

**TABLE 7.3.2 Tests for Detecting the Four Literals of the
ENF′ of Eq. (7.3.3) for s-a-0 and s-a-1**

ENF′	$h' = ((x'_1)_{afh} + (x_2)_{bdfh})((x_2)_{begh} + (x_3)_{cgh})$				Tests
Test $(x'_1)_{afh}$ for s-a-0	1	1	0	d	2 or 3
Test $(x'_1)_{afh}$ for s-a-1	0	1	0	d	6 or 7
Test $(x'_2)_{bdfh}$ for s-a-0	1	1	0	d	5
Test $(x'_2)_{bdfh}$ for s-a-1	1	0	1	0	6 or 7
Test $(x_2)_{begh}$ for s-a-0	d	0	1	1	2
Test $(x_2)_{begh}$ for s-a-1	0	1	0	1	0 or 4
Test $(x_3)_{cgh}$ for s-a-0	d	0	1	1	1 or 5
Test $(x_3)_{cgh}$ for s-a-1	d	0	1	0	0 or 4

From Tables 7.3.1 and 7.3.2, we obtain the following relation:

$$T(P_{10}) \longleftrightarrow T((x_1)_{afh,0}) = \{11d\} = \{6, 7\} = T((x'_1)_{afh,1}) \longleftrightarrow T(P'_{11})$$

$$T(P_{20}) \longleftrightarrow T((x_2)_{bdfh,0}) = \{11d\} = \{6, 7\} = T((x'_2)_{bdfh,1}) \longleftrightarrow T(P'_{21})$$

$$T(P_{30}) \longleftrightarrow T((x'_2)_{begh,0}) = \{d00\} = \{0, 4\} = T((x_2)_{begh,1}) \longleftrightarrow T(P'_{31})$$

$$T(P_{40}) \longleftrightarrow T((x'_3)_{cgh,0}) = \{d00\} = \{0, 4\} = T((x_3)_{cgh,1}) \longleftrightarrow T(P'_{41})$$

$$T(P_{11}) \longleftrightarrow T((x_1)_{afh,1}) = \{01d\} = \{2, 3\} = T((x'_1)_{afh,0}) \longleftrightarrow T(P'_{10})$$

$$T(P_{21}) \longleftrightarrow T((x_2)_{bdfh,1}) = \{101\} = \{5\} = T((x'_2)_{bdfh,0}) \longleftrightarrow T(P'_{20})$$

$$T(P_{31}) \longleftrightarrow T((x'_2)_{begh,1}) = \{010\} = \{2\} = T((x_2)_{begh,0}) \longleftrightarrow T(P'_{30})$$

$$T(P_{41}) \longleftrightarrow T((x'_3)_{cgh,1}) = \{d01\} = \{1, 5\} = T((x_3)_{cgh,0}) \longleftrightarrow T(P'_{40})$$

$T(\cdot)$ denotes the test set for testing $\cdot$. The second subscript of each path denotes the type of corresponding literal fault that the set of tests is designed to test. The prime denotes that the tests are obtained from the complemented ENF. This result indicates that s-a-0 (s-a-1) tests can be obtained for literals in an ENF by generating the complemented ENF and finding s-a-1 (s-a-0) tests for the corresponding literals in the latter. Consequently, s-a-0 (s-a-1) tests for paths P_k can be obtained by finding s-a-1 (s-a-0) tests for P'_k. This property is useful when we write a computer program to compute the tests. We need only to write a program for calculating either s-a-0 or s-a-1 tests for the literals of the ENF and repeat using the same program to its complement to obtain the complete test set.

It should be pointed out that although the path-sensitizing method and the equivalent-normal-form method are both based on the concept of path sensitizing, the latter has several advantages over the former:

1. The ENF method is an analytical method that provides a vehicle for systematically finding the most desirable tests, those which each detect many faults in the circuit. The derivation of a near-minimal set of fault detection tests using the ENF method will be discussed later in this section.

2. Single paths that are not sensitizable are usually easily seen from the ENF. For instance, when both a variable x and its complement x' are contained in the same

term of an ENF, it indicates that the paths represented by the corresponding two literals of x and x' cannot be sensitized. Thus these literals cannot be tested for either s-a-0 or s-a-1 fault. The reason for this is that due to reconvergent fan-out, no matter how we try to sensitize (i.e., to sensitize) one of the two paths or both paths simultaneously and what faults we sensitize, the s-a-0 or s-a-1 fault, a contradiction will always result. Another example is that when two or more literals pertaining to the same input variable are contained in the same term, the s-a-1 test for testing either of the literals individually is impossible.

3. The way to derive tests using the ENF is very simple.

Note that Theorem 7.3.2 does not guarantee finding a set $\{T\}$ which tests every literal in $\{L\}$ for s-a-1 and s-a-0. However, Armstrong (reference 7) has conjectured that such a set of tests can be found. More precisely, it is believed that the following result holds; there exist at least one set of literals $\{L\}$ and an associated set of tests $\{T\}$ that tests an appearance of every literal in $\{L\}$ for s-a-1 and s-a-0 and also detects every fault in the net. As yet this result has not been proved; however, efforts to find a counterexample have been unsuccessful.

Several important remarks about the equivalent-normal-form method should be made here.

1. The path-diagnostic test set for a path of a circuit obtained from the ENF or its complement should not be expected to be complete (problem 3).

2. Although Theorem 7.3.3 states that a s-a-0 (s-a-1) test for a particular literal in the ENF of a single-output circuit (or sub-ENF of a multioutput circuit) is a s-a-1 (s-a-0) test for the corresponding literal in the complemented ENF (or sub-ENF), not every test obtainable from one of two functions (the ENF and its complement) is obtainable from the other.

3. If what Armstrong has conjectured is true, certain terms in the ENF (or its complement) may be omitted, since they cannot be used for testing any literal [e.g., the first two terms in Eq. (7.3.2a)]. If terms that contain complementary variables are discarded, the reduced ENF so obtained may not contain literals corresponding to some connections in the circuit. It has not been shown that the faults on these connections will be detected by tests derived for other literals in the reduced ENF or its complement, but no counterexample has been found.

In the following, a heuristic, systematic procedure for finding most desirable tests, namely, those which each detect many faults, will be presented. This procedure is derived based on the following two main heuristic rules employed in constructing s-a-1 tests:

RULE 1

A term in the ENF exhibiting the fewest number of variables is chosen.

RULE 2

Then that literal in the term chosen by rule 1 such that the prime of the variable appearing in this literal has the most number of appearances in the remaining terms of the ENF is chosen. Value 0 is then assigned to this literal and value 1 to the remaining literals in this term.

The first heuristic rule assures that when the selected term has these values assigned, as many variables as possible remain unassigned, thus giving maximum flexibility in assigning values in the remaining terms. The second heuristic rule results in many 1's being assigned in all terms, a desirable goal since the ideal condition is to have just a single 0 assigned in each term and the remaining literals assigned 1's. Together rules 1 and 2 result in constructing a test that simultaneously tests a literal in each of many terms for s-a-1.

In practice, the order of testing literals is determined by assigning a numerical score to each literal appearance. At any stage of formation of a test, variable values are assigned so as to test the as-yet untested literal appearance of highest score, provided no other appearance of this literal has already been tested. Otherwise, this literal appearance is passed over. The scoring function assigns appropriate weights to rules 1 and 2, which is

$$(Sc_1)_k = \left(1 - \frac{V_j}{V}\right) + \frac{\lambda_k}{L} \qquad (7.3.4)$$

$$\underset{\substack{\text{(provides for}\\\text{rule 1)}}}{\uparrow} \qquad \underset{\substack{\text{(provides for}\\\text{rule 2)}}}{\uparrow}$$

where $(Sc_1)_k \equiv$ s-a-1 score for the kth literal in the ENF

$V_j \equiv$ number of variables exhibited in the jth term of the ENF, where this term contains the kth literal

$V \equiv$ total number of variables

$L \equiv$ total number of literal appearances in the ENF

$\lambda_k \equiv \begin{cases} Ai', \text{ if the } k\text{th literal is unprimed} \\ Ai, \text{ if the } k\text{th literal is primed} \end{cases}$ †

$Ai \equiv$ number of unprimed appearances of the ith variable in the ENF

$Ai' \equiv$ number of primed appearances of the ith variable in the ENF

The construction of s-a-0 tests requires somewhat different heuristics from the two used for s-a-1 tests, thus requiring a new scoring function. Theorem 7.3.3 states that s-a-0 tests can be obtained for literals in an ENF by generating the complemented ENF and finding s-a-1 tests for the corresponding literals in the latter. Therefore, both the ENF and the complemented ENF will be employed and only s-a-1 tests will be constructed, using the s-a-1 scoring function for both ENF's. During construction of tests, the next test to be constructed will start with the as-yet-untested literal appearance of highest score in either the ENF or the complemented ENF. The remainder of the construction of this test will be confined to the particular ENF in which it started.

†It is assumed that the kth literal represents an appearance of the ith variable.

To illustrate this procedure, the following example was given by Armstrong (reference 7).

Example 7.3.3

Consider the irredundant circuit of Fig. 7.3.3 whose ENF and ENF′ are

$$z = (x_1')_{247}(x_2)_{47} + (x_2)_{47}(x_3')_{1247}(x_4)_{1247} + (x_1)_{2567}(x_3)_{12567}(x_2')_{367}$$
$$+ (x_1)_{2567}(x_3)_{12567}(x_4')_{367} + (x_1)_{2567}(x_4')_{12567}(x_2')_{367} \tag{7.3.5}$$
$$+ (x_1)_{2567}(x_4')_{12567}(x_4')_{367}$$

and

$$z' = (x_1')_{2567}(x_2')_{47} + (x_1)_{247}(x_1')_{2567}(x_3)_{1247} + (x_1)_{247}(x_1')_{2567}(x_4')_{1247}$$
$$+ (x_2')_{47}(x_3')_{12567}(x_4)_{12567} + (x_1)_{247}(x_3)_{1247}(x_3')_{12567}(x_4)_{12567} \tag{7.3.6}$$
$$+ (x_1)_{247}(x_3')_{12567}(x_4')_{1247}(x_4)_{12567} + (x_2')_{47}(x_2)_{367}(x_4')_{367}$$
$$+ (x_1)_{247}(x_2)_{367}(x_3)_{1247}(x_4)_{367} + (x_1)_{247}(x_2)_{367}(x_4')_{1247}(x_4)_{367}$$

Note that in the above expressions, instead of line labels, gate labels are used to describe the subscript sequence of a literal. We purposely choose to do so to show this alternative. Again for brevity, replace the subscript sequences in the ENF [Eq. (7.3.5)] and in the complemented ENF [Eq. (7.3.6)] as follows: $247 \rightarrow \alpha$, $47 \rightarrow \beta$, $1247 \rightarrow \gamma$, $2567 \rightarrow \delta$, $367 \rightarrow \epsilon$, and $12567 \rightarrow \lambda$. Two tables are constructed, Table 7.3.3(a) for the ENF and Table 7.3.3(b) for the complemented ENF. Row 1 in each table exhibits the relevant ENF. Only five of the nine terms comprising the complemented ENF appear in Table 7.3.3(b). The remaining four terms were eliminated because none of their literals are testable for s-a-1.

To illustrate the use of the scoring function, the score for a particular literal in the ENF of Table 7.3.3(a) is computed in the following table. First, the values of Ai and Ai' for each variable in the ENF are tabulated:

Variable	Ai	Ai′
x_1	4	1
x_2	2	2
x_3	2	1
x_4	1	4

For example, the first entry in column Ai indicates that there are four appearances of variable x_1 in the ENF. These appearances are in the literal $x_{1\delta}$, which occurs in each of the last four terms. In this example the number of variables V is 4. The total number of literal appearances L in the ENF is 17.

To compute the score of the literal $x_{1\alpha}'$ in the first term of the ENF in Table 7.3.3(a), observe that the number of variables exhibited in this term is 2, and λ_k for

TABLE 7.3.3

(a) Tests for ENF

ENF	1 $x_{1\alpha'}$	4 $x_{2\beta}$	√ $x_{2\beta}$	× $x_{3\gamma'}$	1 $x_{4\gamma}$	× $x_{1\delta}$	√ $x_{2\epsilon'}$	× $x_{4\lambda}$	8 $x_{1\delta}$	√ $x_{2\epsilon'}$	1 $x_{3\lambda}$	8 $x_{1\delta}$	√ $x_{2\epsilon'}$	√ $x_{3\lambda}$	1 $x_{4\epsilon'}$	4 $x_{1\delta}$	× $x_{4\lambda'}$	× $x_{4\epsilon}$
Sc1	200	168	100	100	132	84	100	84	84	100	84	84	84	84	152	—	—	—
Ordering of literals																		
Test 1	1	2	5	6	4	9	7	10	11	8	12	13	14	15	3			
Test 4	0	1	1	0	1	1	0	1	0	1	1	1	0	1	—			
Test 8	0	0	0	1	0	1	1	0	1	0	1	0	1	0	—			

(b) Tests for Complemented ENF

Complemented ENF	3 $x_{1\delta'}$	2 $x_{2\beta'}$	√ $x_{2\beta'}$	3 $x_{3\lambda'}$	7 $x_{4\lambda}$	√ $x_{2\beta'}$	3 $x_{2\epsilon}$	× $x_{4\epsilon}$	2 $x_{1\alpha}$	√ $x_{2\epsilon}$	5 $x_{3\gamma}$	√ $x_{4\epsilon}$	× $x_{1\alpha}$	× $x_{2\epsilon}$	× $x_{4\gamma'}$	× $x_{4\epsilon}$
Sc1	170	187	119	85	85	187	187	—	17	51	17	17	—	—	136	85
Ordering of literals																
Test 2	4	1	6	7	8	2	3	1	11	10	12	13	15	3	5	9
Test 3	1	0	0	0	1	0	1	1	0	1	0	1	0	1	0	1
Test 5	0	1	1	1	1	1	0	1	1	1	1	1	1	1	0	0
Test 6	0	0	0	0	0	0	1	0	1	1	0	0	1	1	1	1
Test 7	0	1	1	0	0	0	0	0	1	0	0	0	1	1	0	0

this literal is 4. Therefore, the score for this literal is

$$Sc_1 = \left(1 - \frac{2}{4}\right) + \frac{4}{17} = \frac{50}{68} = \frac{200}{272}$$

All literal scores in Table 7.3.3(a) and (b) have been brought to the common denominator 272, and this denominator is omitted throughout.

The third row gives an ordering of the literal appearances within the particular table according to their scores. The ordering is arbitrary in case of tie scores. Henceforth, any literal appearance is referred to by the number assigned in this row. The remaining rows contain the binary values assigned to each literal by s-a-1 tests generated by the algorithm. Derivation of the first two tests is now described.

First, the literal appearance of highest score in either table is selected. This is literal 1 of Table 7.3.3(a); therefore, the first test will be constructed entirely within Table 7.3.3(a). Literal 1 is assigned value 0, and the remaining literal in this term is assigned value 1. This assigns value 1 to variable x_1 and x_2. Next, assign all other appearances of these variables their appropriate values in Table 7.3.3(a). Now check to see if the values of additional variables are forced to prevent the test from aborting. A test aborts if the variable assignment produces one of more terms of all 1's. Variable x_4 falls in this category. It must be assigned 1, because if it is assigned 0, all literals in the last term will be 1. It is now apparent that variable x_3 must be assigned 1; otherwise, all literals in the second term will be 1, again aborting the test. The test is now complete, because all variables have been assigned.

Above those literal appearances in Table 7.3.3(a) which were tested by this test, the number of the test is written. Four distinct literals are seen to have been tested. A check mark is also placed above all other appearances in the ENF of these literals, since only one appearance of each need be tested.

To start the second test, the highest-scoring literal appearance in Table 7.3.3(a) or (b), which does not represent an already-tested literal, is chosen. In this case it is literal 1 of Table 7.3.3(b), so the second test is constructed entirely in Table 7.3.3(b). Literal 1 is assigned value 0, and the other literal in this term is assigned 1. This assigns values 0 and 1 to variables x_1 and x_2, respectively. Next, assign all other appearances of these variables their appropriate values in Table 7.3.3(b). It is now observed that no further variables need be assigned at this time solely for the purpose of avoiding aborting the test. Therefore, the as-yet-untested literal appearance of highest score, which is literal 2 appearing in the third term, is sought. Since another appearance of this literal has already been tested by this test, it is passed over. Further search indicates that only one other literal can be tested by the current test, literal 11 in the fourth term. To test it, variables x_3 and x_4 must both be assigned value 1, completing the construction of this test. The literals tested are $x_{1\alpha}$ and $x'_{2\beta}$. If this test is applied to the ENF, it will be seen that it tests literals $x'_{1\alpha}$ and $x_{2\beta}$ in the first term of the latter for s-a-0, as expected.

Proceeding as previously stated, eight tests, which test every distinct, testable literal in both the ENF and complemented ENF for s-a-1, were constructed. [Literals having a $\times$ above them in Table 7.3.3(a) and (b) are untestable.] It is found, by constructing and reducing the complete fault table for this circuit, that six of these eight tests form a minimal fault-detection set. Two such minimal test sets are $\{\underline{1}, \underline{3}, 4, 5, \underline{6}, \underline{7}\}$ and $\{\underline{1}, 2, \underline{3}, \underline{6}, \underline{7}, 8\}$. The underlined tests are essential; they must be included since they each detect faults not detected by any other test.

Exercise 7.3

1. Show that the test set described in Table P7.3.1 obtained from Eq. (7.3.2) will detect all the detectable single faults along the four paths of the circuit of Fig. 7.3.1.

TABLE P7.3.1 Test Sets for Detecting the s-a-0 and s-a-1 Faults along the Paths Represented by the Literals in Eq. (7.3.2)

Literal	Path Represented by the Literal	Test Sets That Detect s-a-0 and s-a-1 Faults on the Path
$(x_1)_{acfh}$	acfh	$\{1010, 0010\}$
$(x_2)_{bcfh}$	bcfh	$\{0110, 0010\}$
$(x_3)_{dfh}$	dfh	$\{1010, 1d00\}$ or $\{0110, d100\}$
$(x_4')_{egh}$	egh	$\{1010, 1d11\}$ or $\{0110, d111\}$

2. (a) Find the ENF of the circuit of Fig. 7.2.4.
   (b) Show that the test set $\{3, 5, 6, 7, 15\}$ obtained from this ENF will detect all the detectable single faults in the circuit.

3. (a) Show that the path-diagnostic test for the path abcd of the circuit in Fig. P7.3.3 obtained from the ENF is $\{1010, 1110\}$.
   (b) Show that there are other tests which can be used to detect faults occurring along the path abcd.

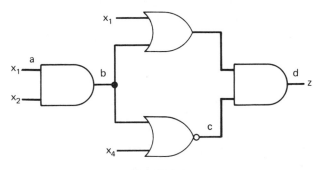

Fig. P7.3.3

4. Using the scoring function, find a near-minimal complete test set for the circuit in Fig. P7.3.4 using the ENF method. Note that only the faults on the labeled lines are considered.

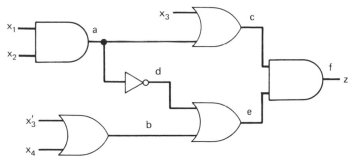

Fig. P7.3.4

7.4 Two-Level-Circuit Fault Detection

So far we have discussed methods of construction of a complete fault-detection test set for a combinational circuit using the two basic approaches: (1) examining each "individual fault" (the fault-table method) and (2) examining each "path" (the path-sensitizing method and the ENF method). In this section, a third approach to the problem is introduced. Instead of examining each individual fault or each path, it is proposed to examine each gate of the circuit. A very simple and direct method for constructing a minimal complete fault-detection test set for any two-level AND–OR (OR–AND, NAND–OR, etc.) irredundant circuit using this approach is presented. It utilizes certain basic properties of the prime implicants of a minimized function when displayed on the Karnaugh map. By using these properties, not only s-a-0 and s-a-1 tests for each gate can be directly "read out" from the map, but the information about the effect of the variable perturbation on the output (i.e., whether the output changes from a 1 to a zero or a zero to a 1) is also provided. In practice, it is necessary to know what the output should be for a specific test so that the result of the applied test can be monitored at the circuit output.

This method may be considered to have two versions: a graphical and a tabular. The graphical version will first be presented which uses the Karnaugh map, hence is convenient to apply to circuits with a small number of input variables, say no more than six, but preferably no more than four. Then, just as what was done in the minimization of switching functions (Sections 1.4–1.7), this Karnaugh map version is extended to a tabular method in a similar manner as that of Quine–McCluskey (Section 1.5). It uses exactly the same principles but without maps, allows the circuit to have any finite number of input variables, and hence is particularly attractive from a machine-computation point of view. First, we present the graphical version of this method to two-level AND–OR circuits with input variables no greater than four.

Karnaugh Map Method Consider the irredundant two-level AND–OR circuit shown in Fig. 7.4.1. Since it is assumed to be an irredundant circuit, the algebraic

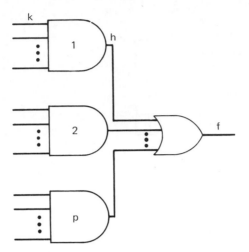

Fig. 7.4.1 Gereral two-level AND-OR circuit.

expression describing the circuit is an irredundant sum of prime implicants. In fact, each AND gate realizes one prime implicant,

$$F = f(X) = \Pi_1(X) + \Pi_2(X) + \ldots + \Pi_p(X) \qquad (7.4.1)$$

where $X = (x_1, x_2, \ldots, x_n)$ and $\Pi_i(X)$, $i = 1, 2, \ldots, p$, are p prime implicants. The subscripts $1, 2, \ldots, p$ denote p AND gates. In analyzing the possible faults on the internal wires connecting the AND gates with the OR gate, it is evident that each such fault has an equivalent fault (or faults) on the inputs of the circuit (i.e., the inputs of the AND gates). For example, an s-a-0 fault on wire h is equivalent to an s-a-0 fault on one or more of the inputs of gate 1. Consequently, an experiment designed to detect all faults on the external inputs of the two-level AND–OR circuit will detect all the faults within the circuit as well.

To detect an s-a-0 fault on wire k at the input to gate 1, it is necessary to assign the value 1 to the variable associated with this input and apply 1's to all the remaining inputs of gate 1. The yet-unassigned variables must be assigned values in such a way that the outputs of all the other AND gates will be 0. Then, if the circuit output is 1, no s-a-0 fault exist on wire k. However, if the output is 0, the gate is faulty, because one (or more) of its inputs or its output is s-a-0. Clearly, such a set of values assigned to the variables will detect s-a-0 faults at *any* input of gate 1. Consequently, it constitutes an s-a-0 test for gate 1.

An s-a-1 fault in a gate input is detected in a similar manner. The input in question is assigned the value 0, while all other inputs to that AND gate are assigned 1's. The remaining variables are again assigned values in such a way that the output of each AND gate will be 0. Then, if the circuit output is 0, no s-a-1 fault exists on the wire being tested. But if the output is 1, the circuit has an s-a-1 fault. The identification of the particular faulty wire, however, is not always possible without additional tests, since any of the AND gate outputs may be s-a-1, not necessarily the AND gate in question. This point will be discussed further.

An s-a-0 fault in an input of an AND gate causes the output of this gate to be s-a-0, regardless of the values of the remaining variables. *Such a fault in effect elimina-tes one prime implicant from the function realized by the circuit.* Thus, to check whether a given prime implicant has completely "vanished," it is sufficient to test one minterm which is covered by that prime implicant and by no other prime implicant, and to verify that this minterm is indeed realized by the circuit. (Clearly, if the prime implicant is not redundant, such a minterm always exists.) The requirement that the minterm chosen for testing must be one that is not covered by more than one prime implicant is essen-tial, since a minterm covered by two (or more) prime implicants is realized by two (or more) AND gates, and the failure of one gate will not be detected if at least one of the remaining gates is operating correctly. From a path-sensitizing point of view, this is to say that each path from a primary input to the circuit output must be sensitized with a test condition that does not simultaneously sensitize another path. *A single unique minterm from a prime implicant will sufficiently describe an s-a-0 test for the AND gate that realizes the prime implicant.* If the output of the block is s-a-0 when the test is applied, the output of the circuit will immediately go to zero since, in effect, that prime implicant has now been eliminated from the circuit. The same is also true if any of the input wires (literals) to the AND gate should fail to a zero. Thus a complete set of tests for s-a-0 faults for a two-level AND–OR circuit consists of p tests, correspond-ing to the p prime implicants in f. This set of tests can be easily determined from the Karnaugh map of the function. For example, consider

$$f(x_1, x_2, x_3, x_4) = x_2x_4 + x_3'x_4 + x_1x_2x_3 \qquad (7.4.2)$$

whose two-level AND–OR realization is shown in Fig. 7.4.2(a) and whose Karnaugh-map representation is shown in Fig. 7.4.2(b). The three subcubes describing the three prime implicants that are realized by the AND gates are indicated in the map.

DEFINITION 7.4.1

A *true subcube* is a subcube in the map describing a prime implicant for which the function $f = 1$.

For example, the three true subcubes of the function of Eq. (7.4.2) are (1, 5, 9, 13), (5, 7, 13, 15), and (14, 15). To test these three true subcubes for s-a-0 faults we select three minterms, one for each subcube, such that each minterm is contained in *only one* true subcube. If more than one minterm is exclusively contained in a true subcube, any one of them may be chosen to test the s-a-0 fault of the AND gate that realizes this true subcube (prime implicant). From Fig. 7.4.2(b) we see that the minimal com-plete test set T_0 for detecting all s-a-0 faults is therefore $T_0 = \{7, 1 \text{ or } 9, 14\}$. It should be noted that since the circuit is assumed as having no redundancy (i.e., the function that it realizes is minimal), any prime implicant of the function has at least one minterm that is not contained in any other true subcubes. Therefore, *such an s-a-0 test set always exists.*

Now, consider the effect on the circuit output of an s-a-1 fault in one of the AND-gate inputs. The output of the faulty AND gate will be independent of the variable

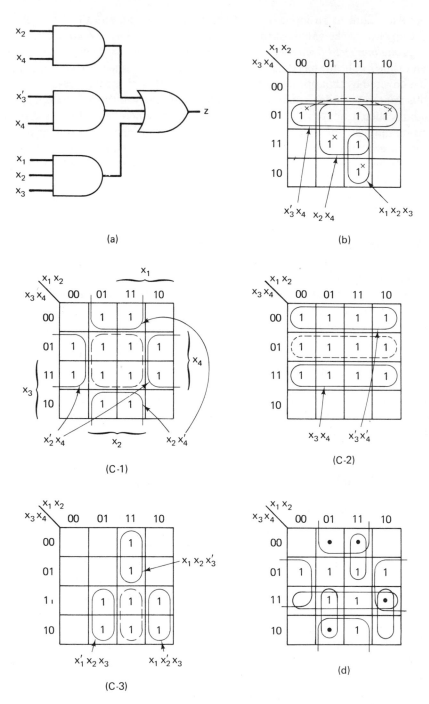

Fig. 7.4.2 Two-level AND-OR circuit for illustrating the true subcubes and adjacent subcubes associated with a circuit.

associated with the s-a-1 input. Consequently, the gate will realize a product that covers the original prime implicant and is independent of one of its variables.

DEFINITION 7.4.1(a)

Two product terms of the same variables are said to be *adjacent* if they differ in the value of just one variable.

For example, the adjacent product terms of the prime implicants of the function Eq. (7.4.2) are:

Prime Implicants	Adjacent Product Terms of the Prime Implicant
$x_2 x_4$	$x'_2 x_4$ and $x_2 x'_4$
$x'_3 x_4$	$x_3 x_4$ and $x'_3 x'_4$
$x_1 x_2 x_3$	$x'_1 x_2 x_3$, $x_1 x'_2 x_3$, and $x_1 x_2 x'_3$

Suppose that input x_2 in gate 1 is s-a-1; the output of that gate will depend only on input x_4. The output of gate 1 under the fault s-a-1 on the x_2 input can be represented by the expression

$$1 \cdot x_4 = (x_2 + x'_2)x_4 = x_2 x_4 + \underbrace{x'_2 x_4}_{\text{an additional term}}$$

This indicates that an s-a-1 fault on the x_2 input is equivalent to that an additional term $x'_2 x_4$, that is adjacent to $x_2 x_4$, is realized by gate 1. Since any of the inputs to gate 1 may be s-a-1, a test must be made to verify that none of the subcubes adjacent to $x_2 x_4$ (i.e., $x'_2 x_4$ and $x_2 x'_4$) is realized by gate 1. Similarly, to test whether gates 2 and 3 have any s-a-1 faults, we must verify that they do not realize any (additional) product terms adjacent to the prime implicants that they realize.

DEFINITION 7.4.1(b)

Two subcubes are said to be *adjacent* if their corresponding product terms are functions of the same variables and differ in the value of just one variable.

For example, the adjacent product terms of the three prime implicants are shown in Fig. 7.4.2(c). Figure 7.4.2(d) shows all the adjacent subcubes to the three prime implicants of the function $f = x_2 x_4 + x'_3 x_4 + x_1 x_2 x_3$. In this map, only the 1's of the true subcubes of the function are shown. This map offers a convenient way to select the cells (tests) to test all the adjacent subcubes. A set of s-a-1 tests should be selected with the following properties:

1. Each adjacent subcube must be tested in at least one of its cells.
2. The value of f in the cell selected for testing the adjacent subcube must be zero.
3. It is allowed (and generally desirable) to select cells common to a number of adjacent subcubes.

The reason that each adjacent subcube must be tested in at least one of its cells is that every input of every AND gate must be checked to see whether it is s-a-1, and the reason for having the third property is simply for minimizing the number of tests. The need for the second property demands an explanation. Note that an s-a-1 fault is detected by a 1 at the circuit output. If some other AND gate is set to 1, the circuit output will be 1, regardless of the fault at the gate being tested. For example, if we choose input combination 13 (i.e., 1101) to detect the s-a-1 of input x_3 of gate 3 [this is equivalent to saying that we choose an input combination to test whether gate 3, in addition to realizing the prime implicant $x_1 x_2 x_3$, also realizes its adjacent product term $x_1 x_2 x_3'$ represented by the adjacent subcube (12, 13) shown in Fig. 7.4.2(c–3), for which the circuit output will be 1], then there is no way to tell whether input x_3 is s-a-1. If, on the other hand, input combination 12 is selected, then the s-a-1 fault on input x_3 of gate 3 will be detected, since the circuit is 0 if the fault is not present and is 1 if the fault is present. It is obvious that *if the circuit is irredundant, it is always possible to find in each subcube adjacent to a true subcube at least one cell for which the value of f is zero.*

The problem now is to select a set of s-a-1 tests such that each adjacent subcube is checked at least once and the total number of tests is minimal.

DEFINITION 7.4.2

An input combination is said to be an *essential* test if it is the only input combination that can be used to check an adjacent subcube.

For example, the combination 6 and 12 are essential, since they are the only tests that can be used to test the subcubes (6, 7) and (12, 13), respectively; whereas the rest of the subcubes can be tested by at least two input combinations. For example, the subcube (1, 3, 9, 11) can be tested by either combination 3 or 11. Thus combinations 3 and 11 are not essential. It is clear that input combinations 6 and 12 must be selected in our s-a-1 test set. Referring to Fig. 7.4.2, it is seen that combinations 6 and 12 test not only subcubes (6, 7) and (12, 13), but also two other subcubes, (4, 6, 12, 14) and (0, 4, 8, 12). The subcubes left to be tested are (1, 3, 9, 11), (3, 7, 11, 15), and (10, 11). Based on the consideration that the total number of tests be as small possible, combination 11 should be chosen, since it is contained in all three subcubes. The set of tests composed of combinations 6, 11, and 12 constitute the s-a-1 test set T_1 (i.e., $T_1 = \{6, 11, 12\}$). Hence the minimal complete test set T is $T = T_0 \cup T_1 = \{7, 1$ or $9, 14, 6, 11, 12\}$, which will detect any number of s-a-0 and s-a-1 faults within the circuit.

Several final remarks are made here.

1. *The upper bound of the length of the experiment.* Suppose a two-level AND–OR irredundant circuit that realizes a function having p prime implicants. Let $q_1, q_2, \ldots, q_k$ be the numbers of the literals of the p prime implicants. A minimal complete test set can always be found by the method described in this section, and the upper bound of the length of the experiment $l_{u.b.}$ using the method described above is

$$l_{u.b.} = p + \sum_{i=1}^{p} q_i$$

2. An s-a-0 fault is detected by a 0 at the circuit output (i.e., for detecting any s-a-0 fault in a two-level AND–OR circuit by a test obtained by the method described above, the circuit output is 1 if the s-a-0 fault under detection is not present and is 0 if the fault is present), and an s-a-1 fault is detected by a 1 at the circuit output (i.e., for detecting any s-a-1 fault by a test obtained by the method, the circuit output is 0 if the s-a-1 fault under detection is not present and is 1 if the fault is present).

3. Although this method was demonstrated for the case of AND–OR logic, it can be easily extended to OR–AND logic.

Tabular Method   The procedure just outlined for the determination of the set of tests in a fault-detection experiment suffers from the limitation inherent in any operation with a map. It is very useful for circuits with a small number of variables, and becomes complicated when the number of variables increases. To overcome this difficulty, the tabular method can be used, which is described as follows:

Step 1   Determination of a minimal complete s-a-0 test set T_0. From the given input implicants, find all the combinations for which each AND gate output is 1. Then find those combinations that are *not* common to two or more gates. Let $S_1, S_2, \ldots, S_p$ be the p sets of input combinations for which the p prime implicants or the p AND-gate outputs are 1. The s-a-0 test set T_{0j} for AND gate j can be found by the equation

$$T_{0j} = S_j - \sum_{\substack{k=1 \\ k \neq j}}^{p} (S_j \cap S_k) \qquad \text{for } j = 1, 2, \ldots, p \qquad (7.4.3)$$

Note that $T_{0j} \neq \emptyset$ for all j, where $\emptyset$ denotes the empty set. Choose a combination from each set T_{0j} that will form a minimal complete s-a-0 test set.

Step 2   Determination of a minimal complete s-a-1 test set. For each prime implicant m_i, find all its adjacent product terms $m_{i1}^*, m_{i2}^*, \ldots, m_{iq}^*$, where q is the number of literals in the prime implicant m_i^*. Then for each m_{ij}^*, find the set R_{ij} of all the combinations for which m_{ij}^* is 1. The complete s-a-1 test set for the jth input of ith AND gate

$$T_{1ij} = R_{ij} - \sum_{k=1}^{p} S_k \qquad \text{for } i = 1, 2, \ldots, p; j = 1, 2, \ldots, q_i \qquad (7.4.4)$$

From the test sets T_{1ij}, for $i = 1, 2, \ldots, p$, and $j = 1, 2, \ldots, q_i$, we can form an s-a-1 fault-detection table. A minimal cover T_1 can be obtained from it using the standard procedure described in the previous section.

Step 3   The minimal complete s-a-0 and s-a-1 test set T is the union of T_0 and T_1.

To illustrate this method, consider the construction of a minimal complete test set of the function of Eq. (7.4.2) using the tabular method.

TABLE 7.4.1

Prime Implicant	Binary-Number Representation	S_j	s-a-0 Test	Input at Which an s-a-0 Fault Will Be Detected
$x_2 x_4$	$d1d1$	$S_1 = \{5, 7, 13, 15\}$	$T_{01} = S_1 - \sum_{k=1}^{3} S_1 \cap S_k = \{7\}$	At any input of gate 1
$x_3' x_4$	$dd01$	$S_2 = \{1, 5, 9, 13\}$	$T_{02} = S_2 - \sum_{k=1,3} S_2 \cap S_k = \{1, 9\}$	Any input of gate 2
$x_1 x_2 x_3$	$111d$	$S_3 = \{14, 15\}$	$T_{03} = S_3 - \sum_{k=1}^{2} S_3 \cap S_k = \{14\}$	Any input of gate 3

TABLE 7.4.2

Prime Implicant	Adjacent Product Terms	Binary-Number Representation	R_{ij}	s-a-1 Test	Input at Which an s-a-1 Fault Will Be Detected
$x_2 x_4$	$x_2' x_4$	$d0d1$	$R_{11} = \{1, 3, 9, 11\}$	$T_{111} = \{3, 11\}$	Input x_2 of gate 1
	$x_2 x_4'$	$d1d0$	$R_{12} = \{4, 6, 12, 14\}$	$T_{112} = \{4, 6, 12\}$	Input x_4 of gate 1
$x_3' x_4$	$x_3' x_4'$	$dd00$	$R_{21} = \{3, 7, 11, 15\}$	$T_{121} = \{3, 11\}$	Input x_3' of gate 2
	$x_3' x_4'$	$dd00$	$R_{22} = \{0, 4, 8, 12\}$	$T_{122} = \{0, 4, 8, 12\}$	Input x_4 of gate 2
$x_1 x_2 x_3$	$x_1' x_2 x_3$	$011d$	$R_{31} = \{6, 7\}$	$T_{131} = \{6\}$	Input x_1 of gate 3
	$x_1 x_2' x_3$	$101d$	$R_{32} = \{10, 11\}$	$T_{132} = \{10, 11\}$	Input x_2 of gate 3
	$x_1 x_2 x_3'$	$110d$	$R_{33} = \{12, 13\}$	$T_{132} = \{12\}$	Input x_3 of gate 3

Step 1 From the prime implicants x_2x_4, $x_3'x_4$, and $x_1x_2x_3$, whose binary-number representation are $d1d1$, $dd01$, and $111d$, respectively, we can find all input combinations (minterms) for which $f = 1$.

$$d1d1 = \{0101, \quad 0111, \quad 1101, \quad 1111\} = \{5, 7, 13, 15\} = S_1$$

$$dd01 = \{0001, \quad 0101, \quad 0101, \quad 1101\} = \{1, 5, 9, 13\} = S_2$$

$$111d - \{1110, \quad 1111\} - \{14, 15\} = S_3$$

We then use Eq. (7.4.3) to obtain the s-a-0 tests, as shown in Table 7.4.1.

Step 2 The adjacent product terms of the prime implicants x_2x_4, $x_3'x_4$, and $x_1x_2x_3$ are found to be $x_2'x_4$, x_2x_4', x_3x_4, $x_3'x_4'$, $x_1'x_2x_3$, $x_1x_2'x_3$, and $x_1x_2x_3'$, which are shown in the second column of Table 7.4.2. As before, from the adjacent product terms, we obtain all input combinations for which adjacent product terms equal 1. They are tabulated in the fourth column of the table. By applying Eq. (7.4.4), the s-a-1 tests are obtained and shown in the fifth column of the table. For example, the s-a-1 tests for detecting input the x_2 s-a-1 fault of gate 1 is obtained by

$$T_{111} = R_{11} - \sum_{k=1}^{p} S_k = \{1, 3, 9, 11\} - \{1, 5, 7, 9, 13, 14, 15\}$$

$$= \{3, 11\}$$

The rest of the T_{1ij} are obtained similarly.

An s-a-1 fault-detection table constructed from T_{1ij} is shown in Table 7.4.3. It is observed that tests 6 and 12 are essential and must be included in the test set. How-

TABLE 7.4.3   s-a-1 Fault-Detection Table

Gate and Input / Test	Gate 1		Gate 2		Gate 3		
	x_2	x_4 ✓	x_3'	x_4 ✓	x_1 ✓	x_2	x_3 ✓
0			1				
3	1		1				
4		1		1			
*6		1			1		
8				1			
10						1	
11	1		1			1	
*12		1		1			1

ever, these two tests also detect the s-a-1 faults of x_4 of gates 1 and 2. The remaining gate inputs, x_2 of gate 1, x_3' of gate 2, and x_2 of gate 3, can be detected by test 11. Therefore, $T_1 = \{6, 11, 12\}$.

Step 3 Combining T_0 and T_1, we obtain the minimal complete test set $T = \{7, 1 \text{ or } 9, 14, 6, 11, 12\}$, which is exactly the same as obtained previously.

Exercise 7.4

1. Find a minimal complete test set for detecting any fault in the minimal two-level AND–OR circuit in Fig. P7.4.1 using the Karnaugh map method.

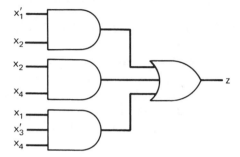

Fig. P7.4.1

2. Modify the Karnaugh map method so that it will work for the minimal two-level OR–AND circuit. Apply this modified version to the circuit of Fig. P7.4.2.

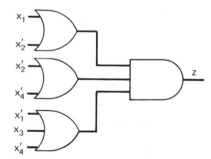

Fig. P7.4.2

3. Modify the Karnaugh map method so that it will work for the minimal two-level NAND circuit. Apply this modified version to the circuit of Fig. P7.4.3.

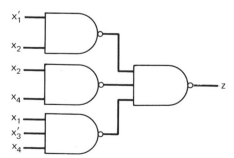

Fig. P7.4.3

4. Modify the Karnaugh map method so that it will work for the minimal two-level NOR circuit. Apply this modified version to the circuit of Fig. P7.4.4.

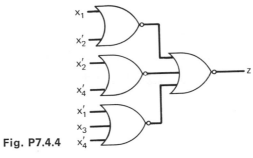

Fig. P7.4.4

5. Apply the Karnaugh map method to derive a minimal complete test set for each of the multilevel circuits of Fig. P7.4.5.

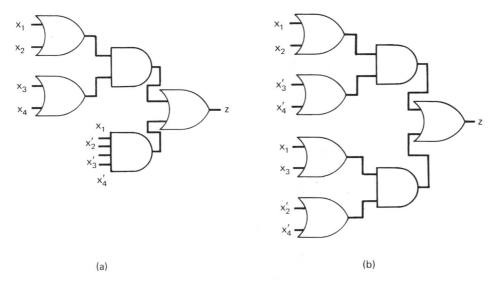

(a)   (b)

Fig. P7.4.5

6. Use the tabular method to derive a minimal complete test set for detecting any faults in the circuit of Fig. P7.4.6.

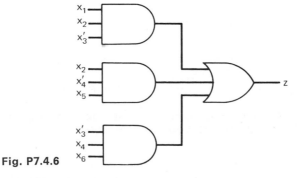

Fig. P7.4.6

7. Derive a minimal complete fault-detection experiment for checking the true/complement, zero/one device SN74H87, whose circuit diagram was shown in Fig. 2.4.2.

8. Derive a minimal complete fault-detection experiment for checking the dual 4-line-to-1-line data selector/multiplexers SN74153 circuit of Fig. P7.4.8.

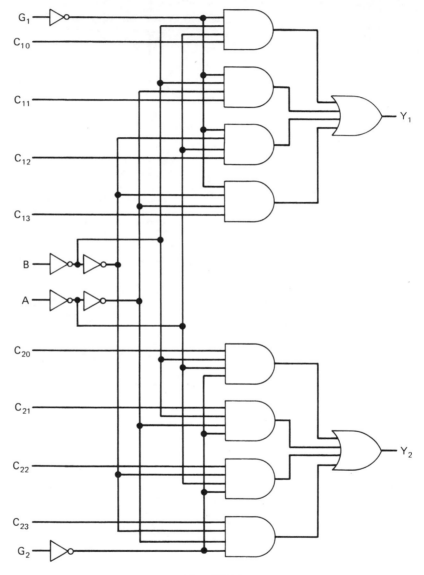

Fig. P7.4.8

9. Derive a minimal complete fault-detection experiment for checking the dual 2-line-to-4-line decoder SN74155 circuit of Fig. P7.4.9.

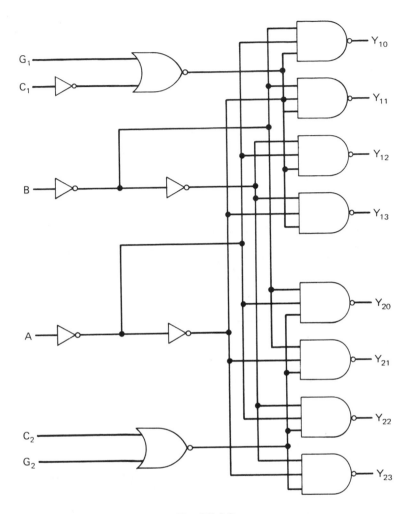

Fig. P7.4.9

7.5 Multilevel-Circuit Fault Detection

Multilevel AND–OR Circuits The minimum sum-of-products or the minimum product-of-sums form of a Boolean expression is not always the most economical to implement. Consequently, a common term or terms is often factored, with the result that a cheaper realization is obtained. AND–OR circuits that realize factored forms are multilevel AND–OR circuits. The purpose of this subsection is to extend the fault-detection methods described above for two-level AND–OR circuits to this class of multilevel circuits. The circuits considered are again assumed to be irredunant.

Let us consider the simple three-level AND–OR circuit of Fig. 7.5.1(a), whose

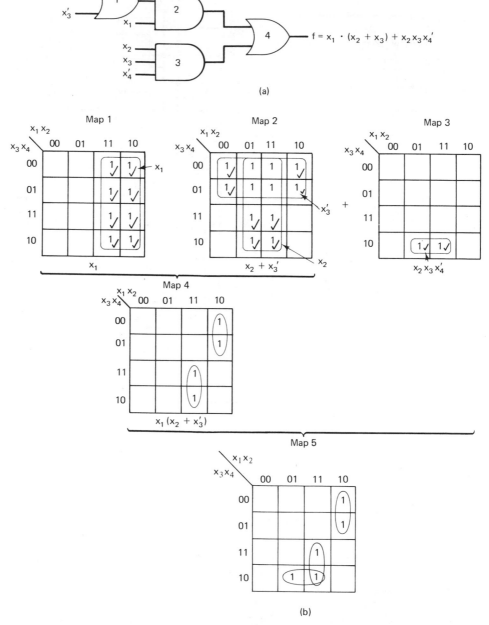

Fig. 7.5.1   (a) Multilevel AND-OR circuit; (b) maps for finding s-a-0 tests.

output function is

$$f(x_1, x_2, x_3, x_4) = x_1 \cdot (x_2 + x_3') + x_2 x_3 x_4' \qquad (7.5.1)$$

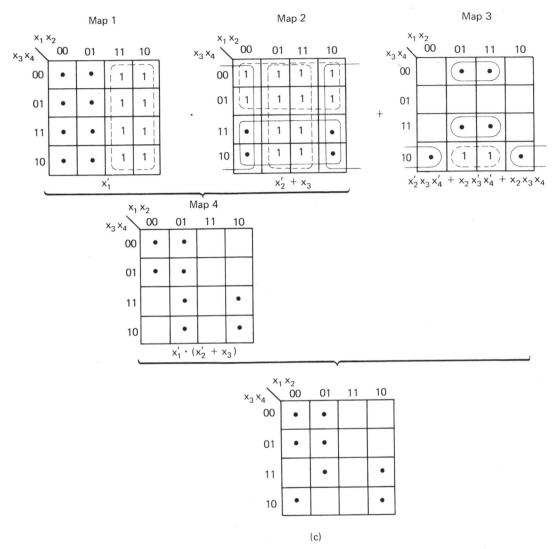

(c)

Fig. 7.5.1 (continued)   (c) maps for finding s-a-1 tests.

Let

$$f_1(x_1, x_2, x_3, x_4) = x_1 \qquad (7.5.2)$$

$$f_2(x_1, x_2, x_3, x_4) = x_2 + x_3' \qquad (7.5.3)$$

$$f_3(x_1, x_2, x_3, x_4) = x_2 x_3 x_4' \qquad (7.5.4)$$

As the first step we find the s-a-0 and s-a-1 tests for each individual subfunction f_1, f_2, and f_3 by using either the map method or the tabular method when the number of variables is large. For this example, since there are four variables, it is convenient to use the map method. The true subcubes of the three subfunctions f_1, f_2, and f_3 are shown in the three Karnaugh maps in Fig. 7.5.1(b). By taking the AND operation of the first two maps and then the OR operation of the resultant map with the third map, we obtain the tests 6, 8, 9, and 15. It is seen that tests 6 and 15 are essential and all the primary inputs s-a-0 fault tests can be covered by a set of three tests {6, 8 or 9, 15}.

Obtaining s-a-1 tests for this circuit is slightly more difficult. The adjacent subcubes of each prime implicant of the three subfunctions are shown in Fig. 7.5.1(c), and the s-a-1 tests for the inputs of each subcircuit are indicated by dark dots. First let us consider the construction of s-a-1 tests for testing inputs of the subnetwork that realizes the function $f_{12} = f_1 \cdot f_2 = (x_1) \cdot (x_2 + x_3')$. This subcircuit may be viewed as a two-level OR–AND circuit. An s-a-1 test for such a circuit may be obtained from the s-a-1 tests of its subcircuits. Any s-a-1 test of one of its subcircuits, for which the output of the other subcircuit is 1, is an s-a-1 test for the two subcircuits combined. The complete s-a-1 test set for testing input s-a-1 faults of these two subcircuits combined is shown in map 4, which is to be ORed with map 3 to obtain s-a-1 tests for the complete circuit. A cell in the resultant map (map 5) is a "dark dot" if (1) at least one of the corresponding cells of the component maps is a "dark dot" and (2) the values of the two subfunctions f_{12} and f_3 for this input combination must not be 1. The final complete input s-a-1 tests are shown in map 5, from which an s-a-1 fault-detection table may be constructed and is shown in Table 7.5.1. We see that tests 2, 4, and 7 are the tests that must be selected. The s-a-1 faults of inputs x_2 and x_3' of

TABLE 7.5.1   Fault-Detection Table

	Gate 1		Gate 2	Gate 3		
	x_2	x_3	x_1	x_2	x_3	x_4'
s-a-0 tests						
*6				1	1	1
8		1	1			
9		1	1			
*15	1		1			
s-a-1 tests						
0			1			
1			1			
*2				1		
*4			1		1	
5			1			
*7			1			1
10	1	1				
11	1	1				

gate 1 may be detected by either test 10 or test 11. The minimal complete s-a-1 test set for the circuit is {2, 4, 7, 10, or 11}. Thus the minimal complete experiment for detecting s-a-0 and s-a-1 faults in the circuit of Fig. 7.5.1(a) consists of the following seven tests: {6, 8 or 9, 15, 2, 4, 7, 10 or 11}.

ENF–Karnaugh Map Method The problem of detecting faults in a general combinational circuit is considerably more complicated than in the case of two-level AND–OR circuits described in Section 7.4. Except in the case where each gate has only a fan-out of 1 (tree-like circuits), and those multilevel AND–OR circuits discussed in the previous subsection, it is not true that testing only the inputs will always detect all the faults within the circuit. If we try to apply an experiment designed for a two-level circuit to a logically equivalent multilevel circuit in which two or more paths emanate from a certain gate and reconverge at a level closer to the output terminal, we shall find that a fault in one path may not always be detectable if the other path is faultless. We shall be concerned with procedures for the detection of single faults This limitation does not, of course, exclude the detection of most double and other multiple faults, but it emphasizes that only single faults will be detected in all cases, while some multiple faults may not be detected.

The design of fault-detection experiments for multilevel combinational circuits that will be presented here is based on the ideas of path sensitization (Section 7.2) and equivalent normal form (ENF) (Section 7.3). The general fault-detection technique using ENF introduced by Armstrong (reference 7) is rather complicated. As described in Section 7.3, it involves separate computations for s-a-0 and s-a-1 faults, and it depends on the order of testing the literals. The order is determined by assigning a score to each of the inputs of the ENF, and, as is often the case, different scoring procedures yield different experiments.

In this section a Karnaugh map technique for deriving a fault-detection experiment for multilevel circuits will be presented which will give the same result as that obtained by the ENF method. It is a much simpler technique than the ENF method, as it does not use a scoring technique and is without the complemented ENF.

Fault-detection tests for equivalent normal forms are easily constructed, owing to the simplicity and uniformity of their structure. All that is needed is to select a set of literals whose paths contain every connection in the corresponding hypothetical circuit, and *to find a set of tests that check at least one appearance of each of these literals for s-a-0 and s-a-1 faults*. This set of tests clearly detects any s-a-0 or s-a-1 faults in the ENF. Note that it is sufficient to test just a single appearance of each literal. Armstrong proved that a set of fault-detection tests devised for the ENF is also a valid set of tests for the original circuit. Thus the problem of designing experiments for multilevel circuits is equivalent to the design of experiments for the two-level equivalent normal forms.

An ENF corresponds to a two-level AND–OR circuit. Consequently, the techniques developed above for two-level circuits can now be applied, with several modifications, to the ENF's.

1. Any literal that appears in two or more terms in the ENF need only be tested once.

2. When there are two literals $(x_i)_\alpha$ and $(x_i)_\beta$ associated with the same input (*not* variable) but different paths in the ENF, it means that in the original circuit the signal x_i diverges into two different paths α and β and then recoverges. There are four possible cases:

(a) The literals are both complemented or both uncomplemented and appear in different terms of the ENF. This corresponds to the case where both paths α and β have either an even or an odd number of inversions. (An inversion is obtained by a NOT, a NAND, and a NOR gate.) The literals associated with these two paths must then have identical polarities, that is, both complemented or both uncomplemented. When these literals appear in different terms in the ENF, *they may be tested simultaneously for s-a-1 faults.* In this case the output will change incorrectly to 1, regardless of whether the fault is common to both paths or is just one of the paths and will be detected. However, *simultaneous s-a-0 tests should be avoided because they may leave some faults undetected.* The correctly operating path will provide the required 1 output, and a fault in the second path will not be detected.

(b) One of the literals is complemented and the other is not, and they appear in different terms of the ENF. This corresponds to the case where one path has an even number of inversions and the other has an odd number of inversions. The literals in the ENF associated with these paths must be complementary. Suppose now that we test one of these literals for an s-a-0 fault. Then if the input combination is such that the other literal is tested for s-a-1 fault, the output will be 1 and the fault will go undetected. Hence *no input combination should be selected so as to test two complementary literals simultaneously for s-a-0 and s-a-1 faults.*

(c) The literals are both complemented or both uncomplemented and they appear in the same term of the ENF. In this case *the literals cannot be tested for s-a-1 faults individually but can be tested simultaneously. Of course, they can be tested individually for s-a-0 faults. Simultaneous s-a-0 tests should be avoided* for the reason stated in (a).

(d) One of the literals is primed and the other is not and they both appear in the same term of the ENF. In this case no test can be found.

For easiness of applying these principles and rules they are summarized as follows:

RULE 1

Any literal that appears in two or more terms in the ENF need only be tested once. Thus the complete s-a-0 and s-a-1 test sets for this literal are the unions of all s-a-0 tests and s-a-1 tests for each appearance of the literal in the ENF, respectively. s-a-0 tests and s-a-1 tests for appearance of the literal are obtained based on the following rules:

(a) Individual and simultaneous testing of two or more appearances of the literal for s-a-0 and s-a-1 faults is valid.

(b) Simultaneous testing of one appearance of the literal for s-a-0 faults and the other for s-a-1 faults is invalid.

(c) Simultaneous s-a-0 testing for the same literal contained in different terms is valid, as is simultaneous s-a-1 testing for the same literal contained in different terms.

RULE 2

When two literals associated with the same input but different paths are both uncomplemented [$(x_i)_\alpha$ and $(x_i)_\beta$] or both complemented [$(x_i')_\alpha$ and $(x_i')_\beta$] and appear in different terms in the ENF:

(a)   Individual s-a-0 and s-a-1 tests for each literal is valid.

(b)   Simultaneous testing of the two literals for s-a-1 faults is valid.

(c)   Simultaneous testing of the two literals for s-a-1 faults should be avoided.

(d)   Simultaneous testing of the two literals in complementary forms [i.e., $(x_i)_\alpha$ and $(x_i')_\beta$ or $(x_i')_\alpha$ and $(x_i)_\beta$] for s-a-0 and s-a-1 faults is invalid.

RULE 3

When two literals associated with the same input but different paths are complementary to each other and appear in different forms in the ENF, s-a-1 tests for an appearance of the literal are obtained based on the following rules:

(a)   Individual and simultaneous testing of two or more appearances of the literal for s-a-0 and s-a-1 faults is valid.

(b)   Simultaneous testing of one appearance of the literal for s-a-0 faults and the other appearance for s-a-1 faults is invalid.

(c)   Simultaneous s-a-0 testing for the same literal contained in different terms is valid, as is simultaneous s-a-1 testing for the same literal contained in different terms.

RULE 4

When two literals associated with the same input but different paths are both uncomplemented or both complemented and appear in the same term in the ENF:

(a)   Individual s-a-0 testing for the literals is valid.

(b)   Individual s-a-1 testing for the literals is invalid.

(c)   Simultaneous s-a-0 testing for the literals should be avoided.

(d)   Simultaneous s-a-1 testing for the literals is valid.

RULE 5

When two complementary literals associated with the same original circuit input appear in the same term in the ENF, no test (either s-a-0 or s-a-1) can be found for them or for any other literals of the term. Hence the term should be discarded completely.

This method for obtaining s-a-0 and s-a-1 tests for any general combinational circuits may be systematically described as follows:

Step 1   Find the equivalent normal form of a given multiple circuit as described in Section 7.3.

Step 2 Draw the map of the ENF of the circuits and label all subcubes in the map. (The map is needed here only for clarity. Actually, the adjacencies can be determined in a systematic manner without the use of map, as described in Section 7.4.)

Step 3 Construct the testing table. Each distinct literal of the ENF appears once as a column heading. Literals corresponding to the same circuit input are grouped together. (This will later help to ensure that no simultaneous s-a-0 and s-a-1 tests are made on complementary inputs.) All possible s-a-0 and s-a-1 tests are made row headings.

Step 4 Complete the testing table. A literal can be tested for s-a-0 faults in any one of the true subcubes in which it is contained. It can be tested for s-a-1 faults by means of a subcube adjacent to any one of the true subcubes.

Step 5 Find an s-a-0 cover and an s-a-1 cover.

This method is illustrated by the following examples.

Example 7.5.1

The ENF for the circuit of Fig. 7.5.2(a), after all the terms containing one or more products of complementary literals associated with the same variable but different paths (there are 14 such terms by rule 5), is

$$z = (x_1')_{acf}(x_1')_{adf}(x_2')_{acf}(x_2')_{adf}(x_3')_{bf}(x_4')_{ef} + (x_1)_{bf}(x_2)_{ef}(x_3)_{cf}(x_4)_{df}$$

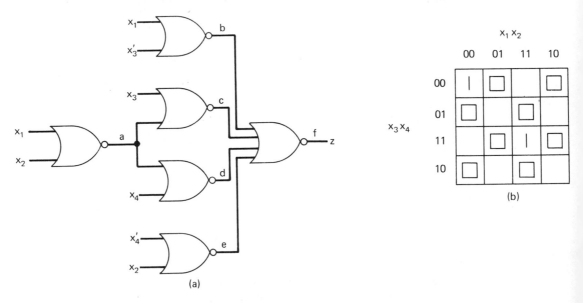

Fig. 7.5.2 (a) Multilevel NOR circuit; (b) true subcubes and adjacent subcubes.

Again, to simplify the notation, let $acf \longrightarrow \alpha$, $adf \longrightarrow \beta$, $bf \longrightarrow \gamma$, $ef \longrightarrow \delta$, $cf \longrightarrow \epsilon$, $df \longrightarrow \sigma$. Then

$$z = x'_{1\alpha}x'_{1\beta}x'_{2\alpha}x'_{2\beta}x'_{3\gamma}x'_{4\delta} + x_{1\gamma}x_{2\delta}x_{3\epsilon}x_{4\sigma}$$

From rule 4(b), the literals $x'_{1\alpha}$ and $x'_{1\beta}$ in the first term cannot be tested for s-a-1 individually, nor can the literals $x'_{2\alpha}$ and $x'_{2\beta}$. However, by rule 4(d), simultaneous s-a-1 testing for them is valid. Figure 7.5.2(b) shows the true subcubes and the adjacent subcubes of the prime implicants of z. There are two true subcubes {0} and {15}, which will detect all s-a-0 faults for all 10 literals. Note that individual s-a-0 tests for literals $x'_{1\alpha}$ and $x'_{1\beta}$ and for literals $x'_{2\alpha}$ and $x'_{2\beta}$ are valid [rule 4(a)]. The s-a-1 tests for these 10 literals are the eight disjoint adjacent subcubes: 1, 2, 4, 7, 8, 11, 13, and 14. For this example the construction of testing table is not needed, and the (minimal) fault-detection experiment is {0, 15, 1, 2, 4, 7, 8, 11, 13, 14}.

Example 7.5.2

The ENF for the circuit of Fig. 7.2.1 was found to be [Eq. (7.3.1)]

$$h = (x_1)_{afh}(x_2)_{bdfh} + (x'_2)_{begh}(x'_3)_{cgh}$$

In order to simplify the notation, let us replace the subscript sequences as follows: $afh \longrightarrow \alpha$, $bdfh \longrightarrow \beta$, $begh \longrightarrow \gamma$, and $cgh \longrightarrow \delta$. Then we write

$$h = x_{1\alpha}x_{2\beta} + x'_{2\gamma}x_{3\delta}$$

and its true subcubes map and adjacent subcubes map are, respectively, shown in Fig. 7.5.3. The literals associated with each original circuit input x_i (see Fig. 7.5.1) are first grouped and the s-a-0 and s-a-1 tests for each group of literals associated with the same input are considered separately.

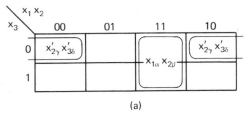

(a)

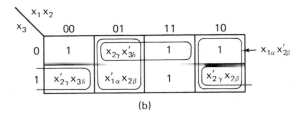

(b)

Fig. 7.5.3 (a) True subcubes; (b) adjacent subcubes of the ENF of the circuit of Fig. 7.2.1.

(1) Literal $x_{1\alpha}$. Since $x_{1\alpha}$ is the only literal associated with x_1 and is contained in only one subcube, the s-a-0 and s-a-1 tests are readily found to be

$$T_0(x_{1\alpha}) = \{6, 7\}, \qquad T_1(x_{1\alpha}) = \{2, 3\}$$

(2) Literals $x_{2\beta}$ and $x'_{2\gamma}$. There are two literals associated with original circuit input x_2, $x_{2\beta}$, and $x'_{2\gamma}$, which are complementary to each other and are contained in different terms of the ENF. Since by rule 3(c) no input combination should be selected so as to test $x_{2\beta}$ and $x'_{2\gamma}$ simultaneously for s-a-0 and s-a-1 faults, input combination 6 (s-a-0 test for $x_{2\beta}$ and s-a-1 test for $x'_{2\gamma}$) and input combination 4 (s-a-0 test for $x'_{2\gamma}$ and s-a-1 test for $x_{2\beta}$) should not be selected for testing literals $x_{2\beta}$ and $x'_{2\gamma}$. Thus

$$T_0(x_{2\beta}) = \{7\}, \qquad T_0(x'_{2\gamma}) = \{0\}$$
$$T_1(x_{2\beta}) = \{5\}, \qquad T_1(x'_{2\gamma}) = \{2\}$$

(3) Literal $x'_{3\delta}$. Again, $x'_{3\delta}$ is the only literal associated with x_3 and is contained in only one true subcube. Thus the s-a-0 and s-a-1 tests for $x'_{3\delta}$ can be obtained in a straightforward manner.

$$T_0(x'_{3\delta}) = \{0, 4\}, \qquad T_1(x'_{3\delta}) = \{1, 5\}$$

A testing table may now be constructed and is shown in Table 7.5.2. It is observed that tests 0 and 7 are essential s-a-0 tests and constitute an s-a-0, and tests 2 and 5 are essential s-a-1 tests and constitute an s-a-1 cover. Therefore, a minimal complete test set for testing any s-a-0 and s-a-1 faults is the circuit of Fig. 7.2.1(a). This is the same result obtained in Example 7.3.1 using the ENF method.

TABLE 7.5.2 Testing Table for Example 7.5.3

	$x_{1\alpha}$	$x_{2\beta}$	$x'_{2\gamma}$	$x'_{3\delta}$
s-a-0 test				
*0			1	1
4				1
6	1			
*7	1	1		
s-a-1 tests				
1				1
*2	1		1	
3	1			
*5		1		1

Example 7.5.3

As a third example, consider the circuit of Fig. 7.5.3(a), which has four input variables, x_1, x_2, x_3, and x_4. The minimal complete test set for this circuit was obtained in Example

7.3.2 using a scoring technique and the complemented ENF. Here we would like to construct it using the map technique. The ENF for this circuit was found to be

$$z = (x_1')_{247}(x_2)_{47} + (x_2)_{47}(x_3')_{1247}(x_4)_{1247} + (x_1)_{2567}(x_3)_{12567}(x_2')_{367}$$
$$+ (x_1)_{2567}(x_3)_{12567}(x_4')_{367} + (x_1)_{2567}(x_4')_{12567}(x_2')_{367} + (x_1)_{2567}(x_4')_{12567}(x_4')_{367}$$

To simplify the notation, let us replace the subscript sequences as follows: $247 \longrightarrow \alpha$, $47 \longrightarrow \beta$, $1247 \longrightarrow \gamma$, $2567 \longrightarrow \delta$, $367 \longrightarrow \epsilon$, $12567 \longrightarrow \lambda$. The above equation can now be written

$$z = x_{1\alpha}'x_{2\beta} + x_{2\beta}x_{3\gamma}'x_{4\gamma} + x_{1\alpha}x_{2\epsilon}'x_{3\lambda} + x_{1\delta}x_{3\lambda}x_{4\epsilon}' + x_{1\delta}x_{2\epsilon}'x_{4\lambda}' + x_{1\delta}x_{4\lambda}'x_{4\epsilon}'$$

The literals of z of the ENF are grouped into six groups:

$$\{x_{1\alpha}', x_{1\delta}\}, \{x_{3\lambda}, x_{3\gamma}'\}, \{x_{4\lambda}', x_{4\gamma}\}, \{x_{2\beta}\}, \{x_{2\epsilon}'\}, \{x_{4\epsilon}'\}$$

The literals in each group are associated with the same original circuit input of the circuit of Fig. 7.3.2.

The ENF map of z is shown in Fig. 7.5.4(a). The s-a-0 and s-a-1 tests for each of the literals are constructed as follows:

1. *Literals $x_{1\alpha}'$ and $x_{1\delta}$.* Since $x_{1\alpha}'$ and $x_{1\delta}$ are complementary to each other and are contained in different terms, by rule 3(c) any input combinations that simultaneously test s-a-0 and s-a-1 faults of the two literals cannot be selected. Tests 4 and 6 cannot be selected because they are s-a-0 tests for $x_{1\alpha}'$ and also s-a-1 tests for $x_{1\delta}$. The s-a-1 tests are determined by observing that 15 is the only adjacent cell for which $z = 0$ and in which the value of x_1 is complementary to its value in the subcube being tested.

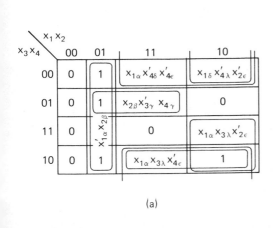

		$x_{1\alpha}'x_{1\delta}$	$x_{3\lambda}x_{3\gamma}'$	$x_{4\lambda}'x_{4\gamma}$	$x_{2\beta}$	$x_{2\epsilon}'$	$x_{4\epsilon}'$
a-a-1 tests	0	1			1		
	1				1		
	2	1			1		
	3	1			1		
	*9		1	1	1		
	*15	1		1		1	1
a-a-0 tests	4				1		
	5				1		
	6				1		
	*7	1			1		
	*8		1		1		
	10		1				
	*11	1	1			1	
	12						1
	*13			1	1	1	
	14						1

(b)

Fig. 7.5.4 (a) ENF map; (b) testing table of Example 7.5.4.

The literal $x_{1\delta}$ is contained in four subcubes. It may be tested in any one of these subcubes. Again, rule 3(c), tests 12 and 14, are invalid as s-a-0 tests for $x_{1\alpha}$ because they are also s-a-1 tests for $x'_{1\alpha}$ (i.e., they are adjacent to $x'_{1\alpha}x_{2\beta}$); the valid s-a-0 tests for $x_{1\delta}$ are therefore tests 8, 10, and 11. Note that tests 8 and 10 are valid [by rule 1(c)], although they belong to more than one subcube. The s-a-1 tests for $x_{1\delta}$ are those adjacent cells in which $z = 0$ and the value of x_1 is complementary. Clearly, 0, 2, and 3 satisfy these conditions. Hence

$$T_0(x'_{1\alpha}) = \{7\}, \qquad T_0(x_{1\delta}) = \{8, 10, 11\}$$
$$T_1(x'_{1\alpha}) = \{15\}, \qquad T_1(x_{1\delta}) = \{0, 2, 3\}$$

2. *Literals $x'_{3\gamma}$ and $x_{3\lambda}$.* From the ENF map we see once again that there exist no tests which test these two literals simultaneously; hence the construction of s-a-0 and s-a-1 tests for one literal is completely independent of that for the other. $x'_{3\gamma}$ is contained in only one subcube. Thus its s-a-0 test is 13 and its s-a-1 test (where the value of x_3 is complementary) is 15. The literal $x_{3\lambda}$ is contained in two subcubes. The only s-a-0 test is 11. Tests 10 and 14 are invalid since they belong to true subcubes that do not contain $x_{3\lambda}$. The s-a-1 test is 9. In summary, the following test sets are found.

$$T_0(x'_{3\gamma}) = \{13\}, \qquad T_0(x_{3\lambda}) = \{11\}$$
$$T_1(x'_{3\gamma}) = \{15\}, \qquad T_1(x_{3\lambda}) = \{9\}$$

3. *Literals $x'_{4\lambda}$ and $x_{4\gamma}$.* The literal $x'_{4\lambda}$ is contained in two terms (true subcubes). One of these subcubes is $x_{1\delta}x'_{4\lambda}x'_{4\epsilon}$. But because this term contains two entries of the same variable (i.e., $x'_{4\lambda}$), which correspond to different inputs in the original circuit, none of the x'_4 inputs in this term can be tested for s-a-1 faults. Test 15 is invalid because it tests only the "untestable" $x_{1\delta}x'_{4\lambda}x'_{4\epsilon}$. Hence the s-a-1 test is 9. The candidates for the s-a-0 test for $x'_{4\lambda}$ are 8, 10, 12, and 14. Tests 10 and 14 are contained in other true subcubes, and test 12 is adjacent to subcube $x_{2\beta}x'_{3\gamma}x_{4\gamma}$ and will simultaneously test $x_{4\gamma}$ for s-a-1, so they are invalid. Thus the s-a-0 test is 8. The only s-a-0 test for $x_{4\gamma}$ is 13, and there are no s-a-1 tests, since there are no adjacent cells in which $z = 0$ and the value of x_4 is complementary. To summarize, we have

$$T_0(x'_{4\lambda}) = \{8\}, \qquad T_0(x_{4\gamma}) = \{13\}$$
$$T_1(x'_{4\lambda}) = \{9\}, \qquad T_1(x_{4\gamma}) = \phi$$

4. *Literal $x_{2\beta}$.* Literal $x_{2\beta}$ can be tested in two subcubes. There are five s-a-0 tests, 4, 5, 6, 7, and 13. There are five s-a-1 tests, 0, 1, 2, 3, and 9.

5. *Literal $x'_{2\epsilon}$.* There are two subcubes containing this literal. The only s-a-0 test is 11. The s-a-1 test is 15.

6. *Literal $x'_{4\epsilon}$.* There are two s-a-0 tests, 12 and 14. The only s-a-1 test is 15. Test 9 is invalid as an s-a-1 test, since $x'_{4\epsilon}$ cannot be tested for s-a-1 in subcube $x_{1\delta}x'_{4\lambda}x'_{4\epsilon}$, only in subcube $x_{1\delta}x_{3\lambda}x'_{4\epsilon}$.

After the complete s-a-0 and s-a-1 test sets for each literal are found, we construct the testing table [Fig. 7.5.4(b)], from which it is seen that tests 9 and 15 are the essential s-a-1 tests, and tests 7, 8, 11, and 13 are the essential s-a-0 tests. After the s-a-1 essential tests which cover columns $x'_{1\alpha}, x_{3\lambda}, x'_{4\lambda}, x_{2\beta}, x'_{2\epsilon}$, and $x'_{4\epsilon}$ are selected, left uncovered is only $x_{1\delta}$, which can be covered by 0, 2, or 3. Similarly, we find that the s-a-0 essential tests plus one additional test (which can be either 12 or 14) will cover every column of the table. Hence the complete test set T is

$$T = \{0 \text{ or } 2 \text{ or } 3, 9, 15, 7, 8, 11, 13, 12 \text{ or } 14\}$$

It can be easily shown that this experiment is identical with the one obtained in Example 7.3.3. However, it has been obtained without a scoring technique and without using the complemented ENF. This method and the scoring technique obtain the same result, but the former is less complicated and easier to comprehend.

As a final remark, this method, like the ENF method, does not guarantee minimal experiments, nor is there a guarantee that a set of sensitized paths can be found for every circuit.

Exercise 7.5

1. Use the ENF–Karnaugh map method to determine a near-minimal complete test set for the circuit of Fig. P7.3.3 and compare the result with that obtained previously.

7.6 Boolean Difference Method

The need for a conceptually simple and straightforward ways of deriving test sequences for combinational circuits is the impetus behind the Boolean difference methods. Boolean difference is defined as being the exclusive-or operation between two Boolean functions, one representing the normal circuit and the other representing the faulty circuit. Thus if the Boolean difference is a 1, a fault is indicated.

Assume that there is a switching function that has one output F and n inputs $x_1, x_2, \ldots, x_n$, so $F(X) = F(x_1, x_2, \ldots, x_n)$. If one of the inputs to the switching function was in error, say input x_i, then the output would be $F(x_1, \ldots, x_i', \ldots, x_n)$. To analyze the action of the circuit when an error occurs, it is desirable to know under what circumstances the two outputs are the same. For this purpose, we define the following function.

DEFINITION 7.6.1

$$\frac{dF(X)}{dx_i} = F(x_1, \ldots, x_i, \ldots, x_n) \oplus F(x_1, \ldots, x_i', \ldots, x_n)$$

The function $dF(X)/dx_i$ is called the *Boolean difference* of $F(X)$ with respect to x_i.

One may ask: Why do we use the derivative notation to represent the Boolean difference? Recall that the meaning of the derivative of a real continuous function $F(x)$ of a real continuous variable x may simply be stated as: the rate of change of $F(x)$ with respect to an infinitesimal change in x. In switching algebra, both F and x can only take on values 0 and 1. Thus, if the value of x is 0(1), it can only change to 1(0). This change can be described by: x changes to x'. The change of x is therefore always 1, which may be seen from the following expression:

$$\Delta x = x \oplus x' = 1$$

Now let us examine the change of a binary function $F(x_1, \ldots, x_i, \ldots, x_n)$ due to the change of a binary variable x_i ($\Delta x_i = 1$). The functions F before and after the change of value of x_i may be represented by $F(x_1, \ldots, x_i, \ldots, x_n)$ and $F(x_1, \ldots, x'_i, \ldots, x_n)$, which may have the following four possible cases:

F before the change $F(x_1, \ldots, x_i, \ldots, x_n)$	F after the change $F(x_1, \ldots, x'_i, \ldots, x_n)$	$\Delta F = F(x_1, \ldots, x_i, \ldots, x_n)$ $\oplus F(x_1, \ldots, x'_i, \ldots, x_n)$	Δx_i	$\dfrac{\Delta F}{\Delta x_i}$ or $\dfrac{dF(X)}{dx_i}$
0	0	0	1	0
1	1	0	1	0
0	1	1	1	1
1	0	1	1	1

It is seen that when $dF(X)/dx_i = 1$ ($dF(X)/dx_i = 0$), it means that the value of $F(X)$ has (has not) changed with x changed to x'. Our interest here is to find input combinations (tests) for detecting detectable faults on a wire x_i of a combinational circuit such that whenever x_i changes to x'_i (caused by a fault on wire x_i), $F(x_1, \ldots, x'_i, \ldots, x_n)$ will immediately be different from $F(x_1, \ldots, x_i, \ldots, x_n)$. In other words, we are interested in finding input combinations for each fault occurring on wire x_i under investigation such that $dF(X)/dx_i = 1$. Before showing how a fault-detection test may be conveniently derived by using the Boolean difference, we first present some useful properties of the Boolean difference.

Operation Properties of Boolean Differences Based on Definition 7.6.1, a set of important operation properties can be derived as follows:

PROPERTY 1

$$\frac{d\overline{F(X)}}{dx_i} = \frac{dF(X)}{dx_i}$$

PROPERTY 2

$$\frac{dF(X)}{dx_i} = \frac{dF(X)}{dx'_i}$$

PROPERTY 3

$$\frac{d}{dx_i}\frac{dF(X)}{dx_j} = \frac{d}{dx_j}\frac{dF(X)}{dx_i}$$

PROPERTY 4

$$\frac{d[F(X)G(X)]}{dx_i} = F(X)\frac{dG(X)}{dx_i} \oplus G(X)\frac{dF(X)}{dx_i} \oplus \frac{dF(X)}{dx_i}\frac{dG(X)}{dx_i}$$

PROPERTY 5

$$\frac{d[F(X) + G(X)]}{dx_i} = \overline{F(X)}\frac{dG(X)}{dx_i} \oplus \overline{G(X)}\frac{dF(X)}{dx_i} \oplus \frac{dF(X)}{dx_i}\frac{dG(X)}{dx_i}$$

PROPERTY 6

$$\frac{d[F(X) \oplus G(X)]}{dx_i} = \frac{dF(X)}{dx_i} \oplus \frac{dG(X)}{dx_i}$$

The bar in $\overline{F(X)}$ denotes the complement of $F(X)$.

These properties can be derived in a straightforward manner and are useful in computing the Boolean difference. The derivations are left to the reader (problem 1).

The most important property of the Boolean difference is that it is equal to 1 when the outputs are different for normal and erroneous settings of input x_i, and equal to 0 if the logic output is the same for both normal and erroneous settings of input x_i. This is the basis for the use of the Boolean difference in the analysis of logic faults. The following definition relates directly to the problem of fault detection.

DEFINITION 7.6.2

A Boolean function $F(X)$ is said to be *independent of* x_i if and only if $F(X)$ is logically invariant under complementation of x_i [i.e., if $F(x_1, \ldots, x_i, \ldots, x_n) = F(x_1, \ldots, x'_i, \ldots, x_n)$].

If we consider $F(X)$ as an output function of a combinational circuit, then we say that $F(X)$ is independent of variable x_i if and only if for any values of the other variables, the output $F(X)$ is independent of the value of x_i. This implies a very important point, that a fault in x_i will not affect the final output $F(X)$.

In the following theorem, the independence relationship can be easily described using the Boolean difference.

THEOREM 7.6.1

A necessary and sufficient condition that a function $F(X)$ be independent of x_i is that $dF(X)/dx_i = 0$.

Proof: By Definition 7.6.2 and the condition $F \oplus F = 0$, the theorem immediately follows. ∎

Based on the result of Theorem 7.6.1, we now have additional operation properties to add to the original set.

PROPERTY 7

$$\frac{dF(X)}{dx_i} = 0 \qquad \text{[if } F(X) \text{ is independent of } x_i]$$

PROPERTY 8

$$\frac{dF(X)}{dx_i} = 1 \qquad \text{[if } F(X) \text{ depends only on } x_i]$$

PROPERTY 9

$$\frac{d[F(X)G(X)]}{dx_i} = F(X)\frac{dG(X)}{dx_i} \qquad \text{[if } F(X) \text{ is independent of } x_i]$$

PROPERTY 10

$$\frac{d[F(X) + G(X)]}{dx_i} = \overline{F(X)}\frac{dG(X)}{dx_i} \qquad \text{[if } F(X) \text{ is independent of } x_i]$$

The derivations can again be easily obtained and thus are left to the reader (problem 1).

Six Methods for Obtaining a Boolean Difference of Switching Function There are at least six methods that can be used to obtain the Boolean difference of a function. One of them is the one presented above, the analytic method using properties 1–10, termed method 1. Method 2 is based on the following theorem.

THEOREM 7.6.2

$$\frac{dF(X)}{dx_i} = F(x_1, \ldots, x_{i-1}, 1, x_{i+1}, \ldots, x_n) \oplus F(x_1, \ldots, x_{i-1}, 0, x_{i+1}, \ldots, x_n)$$

$$\text{for any } x_i, \quad 1 \leq i \leq n$$

Proof:

$$\frac{dF(X)}{dx_i} = [x_i F(x_1, \ldots, x_{i-1}, 1, x_{i+1}, \ldots, x_n) \oplus x_i' F(x_1, \ldots, x_{i-1}, 0, x_{i+1}, \ldots, x_n)]$$

$$\oplus [x_i' F(x_1, \ldots, x_{i-1}, 1, x_{i+1}, \ldots, x_n) \oplus x_i F(x_1, \ldots, x_{i-1}, 0, x_{i+1}, \ldots, x_n)]$$

$$= (x_i \oplus x_i')[F(x_1, \ldots, x_{i-1}, 0, x_{i+1}, \ldots, x_n)]$$

$$\oplus (x_i' \oplus x_i)[F(x_1, \ldots, x_{i-1}, 1, x_{i+1}, \ldots, x_n)]$$

$$= F(x_1, \ldots, x_{i-1}, 1, x_{i+1}, \ldots, x_n) \oplus F(x_1, \ldots, x_{i-1}, 0, x_{i+1}, \ldots, x_n) \quad \blacksquare$$

Method 3 is based on the following theorem.

THEOREM 7.6.3

If $F(X)$ is expressed in the form

$$F(X) = A(X) + x_i \cdot B(X) + x_i' C(X)$$

where $A(X)$, $B(X)$, and $C(X)$ are not functions of x_i, then

$$\frac{dF(X)}{dx_i} = [B(X) \oplus C(X)]\overline{A(X)}$$

The proof of this theorem is left to the reader as an exercise.

In addition to the above three analytical methods, there are three map methods. By Definition 7.6.1, $dF(X)/dx_i = F(x_1, \ldots, x_i, \ldots, x_n) \oplus F(x_1, \ldots, x_i', \ldots, x_n)$, we can construct two Karnaugh maps, one for $F(x_1, \ldots, x_i, \ldots, x_n)$ and the other for $F(x_1, \ldots, x_i', \ldots, x_n)$. The maps are then added using modulo-2 addition in the corresponding entries. The resultant map is the map for $dF(X)/dx_i$ and can be simplified directly. This method will be referred to as method 4.

Another map method, method 5, is derived based on the expression of Boolean functions $F(X)$ in terms of a set of prime implicants summed by exclusive–or operators. The resultant equation is called the exclusive–or form. This prime exclusive–or form of $F(X)$ can be obtained easily by first writing $F(X)$ on the Karnaugh map and then regrouping the map entries. The grouping rule is to allow every entry to be used only for an odd number of times. The steps toward finding $dF(X)/dx$ are summarized as follows:

1. Convert $F(X)$ into its prime exclusive–or form by grouping every entry of $F(X)$ on the map an odd number of times.

2. Apply properties 1 through 6 to the prime exclusive–or form of $F(X)$; the result of $dF(X)/dx_i$ can be readily obtained.

The six method, which is probably the most convenient one, is to obtain the Boolean difference dF/dx_i by rotating the Karnaugh map of F about the axis of x_i using the following two rules.

RULE 1:

A 1 (minterm) is generated and added to the Karnaugh map of F if (1) it is a minor image of a 1 on the map with respect to the axis x_i, and (2) the cell where the 1 is to be added is not occupied by a 1 of the function.

RULE 2:

A 1 (minterm) of the function is deleted from the map if it is a minor image of another 1 of the function with respect to the axis x_i.
The map so obtained is the Karnaugh map of the Boolean difference dF/dx_i, from which the minimized dF/dx_i can be readily obtained.

The following example illustrate these six methods for obtaining the Boolean difference of a switching function.

Example 7.6.1

Find the Boolean difference of $F(X) = x_1x_2 + x_3$ with respect to x_1.

METHOD 1

$$\frac{dF}{dx_1} = \frac{d(x_3 + x_1x_2)}{dx_1} = x_3' \frac{d(x_1x_2)}{dx_1} \oplus (x_1x_2)' \frac{dx_3}{dx_1} \oplus \frac{dx_3}{dx_1} \cdot \frac{d(x_1x_2)}{dx_1} \qquad \text{(by property 5)}$$

$$= x_3' \frac{d(x_1x_2)}{dx_1} \qquad \text{(by property 7)}$$

$$= x_3' \left(x_1 \frac{dx_2}{dx_1} \oplus x_2 \frac{dx_1}{dx_1} + \frac{dx_1}{dx_1} \frac{dx_2}{dx_1} \right) \qquad \text{(by property 4)}$$

$$= x_3'x_2 \qquad \text{(by properties 7 and 8)}$$

METHOD 2

$$\frac{dF}{dx_1} = (1 \cdot x_2 + x_3) \oplus (0 \cdot x_2 + x_3) = (x_2 + x_3) \oplus x_3$$

$$= (x_2 + x_3)x_3' + (x_2 + x_3)'x_3 = x_2 x_3'$$

METHOD 3

It is seen that $A(X) = x_3$, $B(X) = x_2$, and $C(X) = 0$. The $dF(x)/dx_1$ using method 3 is

$$\frac{dF(X)}{dx_1} = \overline{A(X)}[B(X) \oplus C(X)] = x_3'(x_2 \oplus 0) = x_3' x_2$$

METHOD 4

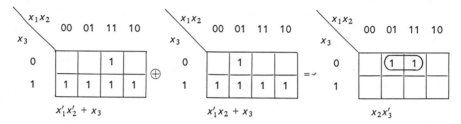

$$x_1' x_2' + x_3 \qquad x_1' x_2 + x_3 \qquad x_2 x_3'$$

METHOD 5

From the Karnaugh map of $F(X) = x_1 x_2 + x_3$ (see Fig. 7.6.1(a)), one prime EXCLU-SIVE–OR form of $F(X)$ is

$$F(X) = x_1 x_2 x_3' \oplus x_3$$

It is easy to verify that $x_1 x_2 + x_3 = x_1 x_2 x_3' \oplus x_3$. Using property 6 we obtain

$$\frac{dF(X)}{dx_1} = \frac{d(x_1 x_2 x_3')}{dx_1} \oplus \frac{dx_3}{dx_1} = \frac{d(x_1 x_2 x_3')}{dx_1} = x_2 x_3'$$

METHOD 6

Referring to Fig. 7.6.1(b), one minterm (minterm 2, indicated by a **1**) is added and four minterms (minterms 1, 3, 5, and 7, indicated by $\mathit{1}$) are deleted; thus the minimized $dF/dx_1 = x_2 x_3'$.

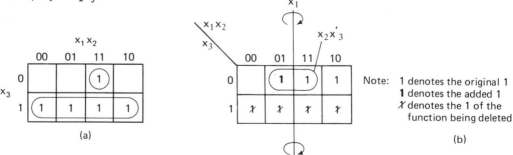

Note: 1 denotes the original 1
        1 denotes the added 1
        X denotes the 1 of the
           function being deleted

Fig. 7.6.1   (a) Prime EXLUSIVE-OR form of $F(X) = x_1 x_2 + x_3$;
(b) rotation of the Karnaugh map of F about x_1.

Derivation of a Complete Fault-Detection Test Set for a Single Fault   To derive a test of a fault on line j, line j is "cut" and is considered as applied a "pseudo-input" x_j. The primary output z is then expressed in terms of the primary input variables $x_1, \ldots, x_n$, and this pseudo-input x_j [i.e., $z_{x_j} = F(x_1, \ldots, x_n, x_j)$]. It is noted that the actual value on j depends on the values $x_1, x_2, \ldots, x_n$, and we let this dependence relation be denoted by the function $x_j(x_1, \ldots, x_n)$. In order to find a test of the s-a-0 (or s-a-1) fault on line j, we must specify an input pattern applied to the primary input terminals such that the value of f depends on the value of x_j and that $X_j = 1$ (or 0) under the fault-free condition. Let a_i be the binary value of the primary input variable x_i. Then it is obvious that $(a_1, \ldots, a_n)$ is a test of the s-a-0 fault on line j if and only if

$$X_j(x_1, \ldots, x_n) \frac{dF(x_1, \ldots, x_n, x_j)}{dx_j} \bigg|_{a_1, \ldots, a_n} = 1 \qquad (7.6.1)$$

Similarly, $(a_1, \ldots, a_n)$ is a test of the s-a-1 fault on line j if and only if

$$\overline{X_j(x_1, \ldots, x_n)} \frac{dF(x_1, \ldots, x_n, x_j)}{dx_j} \bigg|_{a_1, \ldots, a_n} = 1 \qquad (7.6.2)$$

Thus we have the following theorem:

THEOREM 7.6.4

Let $x_i^0 = x_i'$ and $x_i^1 = x_i$. Let a_i be the binary value of input variable x_i. Then $(a_1, \ldots, a_n)$ is a test of the s-a-0 (or s-a-1) fault on line j if and only if $x_1^{a_1} \ldots x_n^{a_n}$ is a minterm in the canonical sum-of-products form of

$$X_j(x_1, \ldots, x_n) \frac{dF(x_1, \ldots, x_n, x_j)}{dx_j} \quad \left[\text{or } \overline{X_j(x_1, \ldots, x_n)} \frac{dF(x_1, \ldots, x_n, x_j)}{dx_j} \right]$$

Following the above theorem, we have:

COROLLARY 7.6.1

The s-a-0 (or s-a-1) fault on line j is undetectable if and only if

$$X_j(x_1, \ldots, x_n) \frac{dF(x_1, \ldots, x_n, x_j)}{dx_j} = 0 \quad \left[\text{or } \overline{X_j(x_1, \ldots, x_n)} \frac{dF(x_1, \ldots, x_n, x_j)}{dx_j} = 0 \right]$$

Example 7.6.2

Consider the circuit of Fig. 7.6.2. By using the truth-table method, we could determine that the faults e_0 and h_1 are undetectable. Now we would like to use the Boolean difference method described in Corollary 7.6.1.

The output functions expressed in terms of primary input variables and the pseudo-inputs e and h are

$$F(x_3, x_4, e) = (e + x_2) + x_3(x_3 + x_4)$$

and

$$F(x_1, x_2, h) = (x_1 x_2 + x_2) + h x_3$$

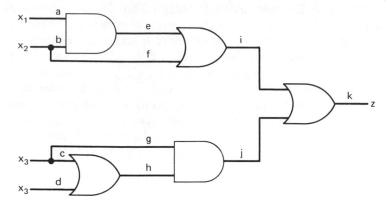

Fig. 7.6.2 Circuit of Example 7.6.2.

respectively, and the Boolean differences with respect to e and h are

$$\frac{dF(x_3, x_4, e)}{de} = [(1 + x_2) + x_3] \oplus [(0 + x_2) + x_3] = x_2' x_3'$$

and

$$\frac{dF(x_1, x_2, h)}{dh} = (x_2 + 1 \cdot x_3) \oplus (x_2 + 0 \cdot x_3) = x_2' x_3$$

It is easy to see that $X_e = x_1 x_2$ and $\bar{X}_h = x_3' x_4'$. We find that

$$X_e \frac{dF(x_3, x_4, e)}{de} = 0 \quad \text{and} \quad \bar{X}_h \frac{dF(x_1, x_2, h)}{dh} = 0$$

According to Corollary 7.6.1, e_0 and h_1 are undetectable faults. We also find that

$$\bar{X}_e \frac{dF(x_3, x_4, e)}{de} = (x_1 x_2)' x_2' x_3' = x_1' x_2' x_3'$$

and

$$X_h \frac{dF(x_1, x_2, h)}{dh} = (x_3 + x_4) x_2' x_3 = x_2' x_3 + x_2' x_3 x_4$$

which indicate that faults e_1 and h_0 are detectable and the complete test sets for detecting e_1 and h_0 are {0000 and 0001} and {0010, 0011, 1010, 1011}, respectively.

Derivation of Complete Fault-Detection Tests for Multiple Faults The derivation of tests for detecting multiple faults are derived from multiple Boolean differences. First let us derive the Boolean difference of a function $F(x_1, \ldots, x_i, \ldots, x_j, \ldots, x_n)$ simultaneously with respect to two variables x_i and x_j, which we call a double Boolean difference.

DEFINITION 7.6.3

$$\frac{dF(X)}{d(x_i x_j)} = F(x_1, \ldots, x_i, \ldots, x_j, \ldots, x_n) \oplus F(x_1, \ldots, x_i', \ldots, x_j', \ldots, x_n)$$

Note that

$$\frac{dF(X)}{d(x_i x_j)} \neq \frac{d}{dx_i} \frac{dF(X)}{dx_j}$$

Definition 7.6.4 defines independence as applied to the double Boolean difference.

DEFINITION 7.6.4

A Boolean function $F(X)$ is said to be *independent of two variables x_i and x_j* if and only if $F(X)$ is logically invariant under complementation of x_i and x_j [i.e., if $F(x_1, \ldots, x_i, \ldots, x_j, \ldots, x_n) = F(x_1, \ldots, x_i', \ldots, x_j', \ldots, x_n)$].

Now from Definition 7.6.3, we have Theorem 7.6.5.

THEOREM 7.6.5

A necessary and sufficient condition that a function $F(X)$ be independent of variables x_i and x_j is that $dF/d(x_i, x_j) = 0$.

Note that Definition 7.6.4 and Theorem 7.6.5 refer strictly to those cases when a multiple fault x_i and x_j occurs simultaneously. It differs from the convenient definition of multiple fault detection. In the latter case, not only do faults in x_i and x_j occur at the same time, but cases of fault in either x_i or x_j are also included. The following cases may be of interest.

CASE 1

$$\frac{dF(X)}{d(x_i x_j)} = F(x_1, \ldots, x_i, \ldots, x_j, \ldots, x_n) \oplus F(x_1, \ldots, x_i', \ldots, x_j', \ldots, x_n)$$

Case 1 is the Boolean difference for studying the effect of multiple faults on a circuit.

CASE 2

$$\frac{dF(X)}{d(x_i \oplus x_j)} = \frac{dF(X)}{dx_i} + \frac{dF(X)}{dx_j}$$

Case 2 is the Boolean difference for studying the effect of a fault in either x_i or x_j.

CASE 3

$$\frac{dF(X)}{d(x_i + x_j)} = \frac{dF(X)}{d(x_i \oplus x_j)} + \frac{dF(X)}{d(x_i x_j)}$$

Case 3 is the Boolean difference for studying a fault in either x_i or x_j or both.

All the preceding double Boolean difference are readily extended to higher-order multiple Boolean differences.

CASE 1'

$$\frac{dF(X)}{d(x_i x_1 \ldots x_k)} = F(x_1, \ldots, x_i, \ldots, x_j, \ldots x_k, \ldots, x_n)$$
$$\oplus F(x_1, \ldots, x_i', \ldots, x_j', \ldots, x_k', \ldots, x_n)$$

CASE 2′

$$\frac{dF(X)}{d(x_i \oplus x_j \oplus \ldots \oplus x_k)} = \frac{dF(X)}{dx_i} + \frac{dF(X)}{dx_j} + \ldots + \frac{dF(X)}{dx_k}$$

CASE 3′

$$\frac{dF(X)}{d(x_i + x_j + \ldots + x_k)} = \frac{dF(X)}{d(x_i \oplus x_j \oplus \ldots \oplus x_k)} + \frac{dF(X)}{d(x_i x_j \ldots x_k)}$$

Parallel to Theorem 7.6.4 and Corollary 7.6.1, we have the following theorem and corollary for test derivation for detecting multiple faults.

THEOREM 7.6.6

Let $x_i^0 = x_i'$ and $x_i^1 = x_i$. Let a_i be the binary value of input variable x_i. Then $(a_1, \ldots, a_n)$ is a test of the s-a-0 (or s-a-1) faults on lines $i, j, \ldots, k$ simultaneously if and only if $x_1^{a_1} \ldots x_n^{a_n}$ is a minterm in the standard sum of

$$X_{i,j,\ldots,k}(x_1, \ldots, x_n) \frac{dF(X)}{d(x_i, x_j, \ldots, x_k)} \left[\text{or } \overline{X_{ij\ldots k}(x_1, \ldots, x_n)} \frac{dF(X)}{d(x_i x_j \ldots x_k)} \right]$$

COROLLARY 7.6.2

The s-a-0 (s-a-1) faults simultaneously occurring on lines $i, j, \ldots,$ and k are undetectable if and only if

$$X_{ij\ldots k}(x_1, \ldots, x_n) \frac{dF(X)}{d(x_i x_j \ldots x_k)} = 0 \left[\text{or } \overline{X_{ij\ldots k}(x_1, \ldots, x_n)} \frac{dF(X)}{d(x_i x_j \ldots x_k)} = 0 \right]$$

It is important to point out that

$$\frac{dF(X)}{d(x_i x_j)} \neq$$
$$F(x_1, \ldots, x_i = 1, \ldots, x_j = 1, \ldots, x_n) \oplus F(x_1, \ldots, x_i = 0, \ldots, x_j = 0, \ldots, x_n)$$

This may be seen from the following example.

Example 7.6.3

Consider the OR gate of Fig. 7.6.3. The output function is $F(x_1, x_2, x_3, x_4) = x_1 + x_2 + x_3 + x_4$. Suppose that we want to derive a test for detecting the multiple fault x_1 s-a-0

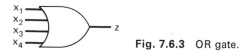

Fig. 7.6.3 OR gate.

and x_2 s-a-0 simultaneously. Then we need first to find the second-order Boolean difference with respect to x_1 and x_2:

$$\frac{dF}{d(x_1 x_2)} = (x_1 + x_2 + x_3 + x_4) \oplus (x_1' + x_2' + x_3 + x_4)$$

$$= x_1 x_2 x_3' x_4' + x_1' x_2' x_3' x_4'$$

The test for detecting the fault x_1 s-a-0 and x_2 s-a-0 may be obtained from

$$X_{x_1 x_2} \cdot \frac{dF}{d(x_1 x_2)} = x_1 x_2 (x_1 x_2 x_3' x_4' + x_1' x_2' x_3' x_4') = x_1 x_2 x_3' x_4'$$

which agrees with our intuition that when the input combination 1100 is applied to the OR gate, the faults x_1 s-a-0 and x_2 s-a-0 may be detected by observing the gate's output. It is fault-free if the output is 1 and has the fault x_1 s-a-0 and x_2 s-a-0 if the output is 0.

Fault-Detection Test-Generation Algorithm Using Boolean Differences

THEOREM 7.6.7

Let f be a switching function realized by a combinational circuit. Let j be an internal line and i be either an internal line or a primary input line of the circuit such that every path from i to the primary output line passes through j. Then

$$\frac{df}{dx_i} = \frac{df}{dx_j}\frac{dx_j}{dx_i} \qquad (7.6.3)$$

where x_i and x_j are the variables representing the logic values on lines i and j, respectively.

Proof: Let $x_1, \ldots, x_n$ be the primary input variables. When all the paths from line i to the primary output line passing through line j, we can write

$$x_j = J(x_1, \ldots, x_n, x_i) \qquad (7.6.4)$$

and

$$
\begin{aligned}
f &= F_j(x_1, \ldots, x_n, x_j) \\
&= F_j[x_1, \ldots, x_n, J(x_1, \ldots, x_n, x_i)] \qquad (7.6.5) \\
&= F_i(x_1, \ldots, x_n, x_i)
\end{aligned}
$$

Consider the following three cases:

CASE 1

Suppose that $dx_j/dx_i = 1$ and $df/dx_j = 1$. Then

$$
\begin{aligned}
\frac{df}{dx_i} &= F_j[x_1, \ldots, x_n, J(x_1, \ldots, x_n, x_i)] \\
&\quad \oplus F_j[x_1, \ldots, x_n, J(x_1, \ldots, x_n, x_i')] \qquad (7.6.6) \\
&= F_j(x_1, \ldots, x_n, \delta) \oplus F_j(x_1, \ldots, x_n, \delta) \\
&= 1
\end{aligned}
$$

CASE 2

Suppose that $dx_j/dx_i = 0$. Then

$$
\begin{aligned}
\frac{df}{dx_i} &= F_j[x_1, \ldots, x_n, J(x_1, \ldots, x_n, x_i)] \\
&\oplus F_j[x_1, \ldots, x_n, J(x_1, \ldots, x_n, x_i')] \\
&= F_j(x_1, \ldots, x_n, \delta) \oplus F_j(x_1, \ldots, x_n, \delta) \\
&= 0
\end{aligned}
\tag{7.6.7}
$$

CASE 3

Suppose that $df/dx_j = 0$. Then

$$
\begin{aligned}
\frac{df}{dx_i} &= F_j[x_1, \ldots, x_n, J(x_1, \ldots, x_n, x_i)] \\
&\oplus F_j[x_1, \ldots, x_n, J(x_1, \ldots, x_n, x_i')] \\
&= F_j(x_1, \ldots, x_n, \delta_1) \oplus F_j(x_1, \ldots, x_n, \delta_2) \\
&= 0
\end{aligned}
\tag{7.6.8}
$$

It is noted that each of δ, δ_1, and δ_2 can be either 0 or 1. Combining the above three cases, we have Eq. (7.6.3). ∎

To avoid ambiguity, we define a line i of a combinational circuit as a *fan-out line* if line i is either a primary input line or a gate-output line such that line i is connected to more than one gate-input terminal. For example, line 2 of the circuit shown in Fig. 7.6.4 is a fan-out line.

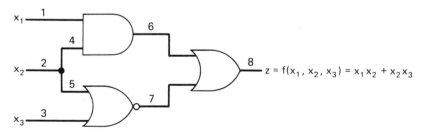

Fig. 7.6.4 Circuit to illustrate the algorithm for the generation of tests based on Boolean differences.

The algorithm for generating the complete test sets of all s-a-0 and s-a-1 single faults using the Boolean difference is described as follows:

Step 1 Number all the lines of the logic circuit under consideration as follows:
(1) Number all the lines using positive integers in ascending order starting with prime inputs from top to bottom.
(2) Draw the circuit diagram in such a way that all the gates are lined up into levels of

**TABLE 7.6.1   Boolean Difference of Six Most Commonly
Used Gates**

Gate	Boolean difference
AND gate NAND gate	$$\frac{dg}{dx_k} = x_1 x_2 \cdots x_{k-1} x_{k+1} \cdots x_m$$
OR gate NOR gate	$$\frac{dg}{dx_k} = x_1' x_2' \cdots x_{k-1}' x_{k+1}' \cdots x_m'$$
NOT gate	$$\frac{dg}{dx_1} = 1$$
EXCLUSIVE-OR gate	$$\frac{dg}{dx_1} = \frac{dg}{dx_2} = 1$$

gates. The NOT gates, if any, are drawn into separate columns from those of other types. But a NOT gate will be treated the same as other types if its input is one of the fan-out branches.

(3) Label the lines from left to right. The last ones to be numbered are the primary output lines.

Step 2   Starting from the line with the smallest number 1 (one of the primary input lines) to the line with the largest number l (the primary output line), where l is the number of lines in the circuit, express the function X_j and its complement $\bar{X}_j$ of each line j in terms of the primary input variables. List X_j and $\bar{X}_j$, $j = 1, 2, \ldots, l$.

Step 3   Find the Boolean differences $df/dx_j = dF(x_1, \ldots, x_n, x_j)/dx_j$ according to (2), where f is the logic function realized by the circuit and j ranges over all fan-out lines.

Step 4   Based on Theorem 7.6.7 and the Boolean differences of various gate outputs with respect to their gate inputs as summarized Table 7.6.1, find the Boolean differences df/dx_j, where j starts from the largest number to the smallest number that is not a fan-out line. List all the df/dx_j, 1, 2, ..., l.

Step 5   Express $X_j(df/dx_j)$, and $\bar{X}_j(df/dx_j)$, $j = 1, 2, \ldots, l$, in terms of the primary input variables. Then the sets of terms of $X_j(df/dx_j)$ and $\bar{X}_j(df/dx_j)$ represent the complete test sets of the s-a-0 and s-a-1 single faults on line j, respectively.

Example 7.6.4

This algorithm is now applied to the circuit shown in Fig. 7.6.4, in which each line is labeled using the rules described in step 1. According to step 2, we obtain Table 7.6.2. It is seen that there is only one fan-out line that is line 2. Following step 3 to find df/dx_2, we obtain

$$\frac{df}{dx_2} = (x_1 \cdot 0 + 0'x_3') \oplus (x_1 \cdot 1 + 1'x_3')$$

$$= x_1 x_3 + x_1' x_3'$$

TABLE 7.6.2 X_j and $\bar{X}_j$ of the Circuit of Fig. 7.6.4

$X_j = x_j$ and $\bar{X}_j = x_j'$, $j = 1, 2, 3$	
$X_4 = x_2$	$\bar{X}_4 = x_2'$
$X_5 = x_2$	$\bar{X}_5 = x_2'$
$X_6 = x_1 x_4 = x_1 x_2$	$\bar{X}_6 = x_1' + x_4' = x_1' + x_2'$
$X_7 = x_3' x_5' = x_2' x_3'$	$\bar{X}_7 = x_3 + x_5 = x_2 + x_3$
$X_8 = x_6 + x_7 = x_1 x_2 + x_2' x_3'$	$\bar{X}_8 = x_6' x_7' = x_1' x_2 + x_1' x_3 + x_2' x_3$

According to step 4, find df/dx_j in the order of $j = 8, 7, \ldots, 2, 1$. All the df/dx_j, $j = 1, 2, \ldots 8$, are listed in Table 7.6.3. According to step 4, the complete test set of the s-a-0 or s-a-1 fault on line j is obtained by taking $X_j(df/dx_j)$ or $\bar{X}_j(df/dx_j)$, respectively.

The complete test sets for all s-a-0 and s-a-1 faults of this circuit are shown in Table 7.6.4.

TABLE 7.6.3 df/dx_j of the Circuit of Fig. 7.6.4

$$\frac{df}{dx_8} = 1 \qquad\qquad \frac{df}{dx_7} = x_6' = x_1' + x_2'$$

$$\frac{df}{dx_6} = x_7' = x_2 + x_3 \qquad\qquad \frac{df}{dx_5} = \frac{df}{dx_7}\frac{dx_7}{dx_5} = x_1'x_3' + x_2'x_3'$$

$$\frac{df}{dx_4} = \frac{df}{dx_6}\frac{dx_6}{dx_4} = x_1x_2 + x_1x_3 \qquad\qquad \frac{df}{dx_3} = \frac{df}{dx_6}\frac{dx_7}{dx_3} = x_2'$$

$$\frac{df}{dx_2} = x_1x_3 + x_1'x_3' \qquad\qquad \frac{df}{dx_1} = \frac{df}{dx_6}\frac{dx_6}{dx_1} = x_2$$

TABLE 7.6.4 Complete Test Sets for Detecting All s-a-0 and s-a-1 Faults of the Circuit of Fig. 7.6.4

$$X_1\frac{df}{dx_1} = x_1x_2 \longrightarrow T(a_0) = \{6,7\} \qquad\qquad \bar{X}_1\frac{df}{dx_1} = x_1'x_2 \longrightarrow T(a_1) = \{2,3\}$$

$$X_2\frac{df}{dx_2} = x_1x_2x_3 + x_1'x_2x_3' \longrightarrow T(b_0) = \{2,7\} \qquad \bar{X}_2\frac{df}{dx_2} = x_1x_2'x_3 + x_1'x_2'x_3' \longrightarrow T(b_1) = \{0,5\}$$

$$X_3\frac{df}{dx_3} = x_2'x_3 \longrightarrow T(c_0) = \{1,5\} \qquad\qquad \bar{X}_3\frac{df}{dx_3} = x_2'x_3' \longrightarrow T(c_1) = \{0,4\}$$

$$X_4\frac{df}{dx_4} = x_1x_2 \longrightarrow T(d_0) = \{6,7\} \qquad\qquad \bar{X}_4\frac{df}{dx_4} = x_1x_2'x_3 \longrightarrow T(d_1) = \{5\}$$

$$X_5\frac{df}{dx_5} = x_1'x_2x_3' \longrightarrow T(e_0) = \{2\} \qquad\qquad \bar{X}_5\frac{df}{dx_5} = x_1'x_2' \longrightarrow T(e_1) = \{0,4\}$$

$$X_6\frac{df}{dx_6} = x_1x_2 \longrightarrow T(f_0) = \{6,7\} \qquad\qquad \bar{X}_6\frac{df}{dx_6} = x_1'x_2 + x_1'x_3 + x_2'x_3 \, T(f_1) = \{1,2,3,5\}$$

$$X_7\frac{df}{dx_7} = x_1'x_2' \longrightarrow T(g_0) = \{0,4\} \qquad\qquad \bar{X}_7\frac{df}{dx_7} = x_1'x_2 + x_1'x_3 + x_2'x_3 \longrightarrow T(g_1) = \{1,2,3,5\}$$

$$X_8\frac{df}{dx_8} = x_1x_2 + x_2'x_3' \longrightarrow T(h_0) = \{0,4,6,7\} \qquad \bar{X}_8\frac{df}{dx_8} = x_1'x_2 + x_1'x_3 + x_2'x_3 \longrightarrow T(h_1) = \{1,2,3,5\}$$

Owing to the fact that the Boolean difference of a function with respect to a variable does not distinguish between the effects of that variable changing from a 0 to a 1, the fault-detection test-generation method using Boolean differences, like many other methods, such as the fault-table method, the path-sensitizing method, the ENF method, and so on, does not give information about the effect of the variable perturbation on the output. But it has been shown that the methods (basically, the methods using the Karnaugh map) described in Sections 7.4 and 7.5 not only generate tests for fault detection but also provide the information about whether the output changes from a 1 to a zero or a zero to a 1. In the following it will be shown that, with slight modification, the fault-detection test-generation method using Boolean differences described above may also provide this information whenever it is needed.

The problem of determining what the correct output should be for a given test can be solved by intersecting the tests for a s-a-1 and a s-a-0 with both the function and its inverse. We now define two equations that define completely the test for an individual error:

$$\nabla_{x_i}F = \left(x_i\frac{dF}{dx_i}\right)[F, \bar{F}] \qquad\qquad (7.6.9)$$

$$\Delta_{x_i}F = \left(x_i'\frac{dF}{dx_i}\right)[F, \bar{F}] \qquad\qquad (7.6.10)$$

In these two equations, the tests for the variables x_i s-a-1 and s-a-0 are intersected with the ordered pair $[F, \bar{F}]$. Equation (7.6.9) gives the tests that will detect a s-a-0 fault in variable x_i, and it partitions these tests into tests corresponding to a 1 or a 0 output for the overall circuit output F. Equation (7.6.10) yields the tests for variable i s-a-1 and partitions them according to their effect on the circuit output. Since each block and line within the total circuit must be tested for a s-a-1 and s-a-0 condition, a complete test for a circuit will consist of one test from Eq. (7.6.9) and one from Eq. (7.6.10), where the index i represents all input lines, internal blocks, and interconnecting lines. All the circuits considered in this section are assumed to have no redundancy.

Example 7.6.5

To illustrate Eqs. (7.6.9) and (7.6.10), let us consider the circuit of Fig. 7.6.4. The complete test sets for detecting all s-a-0 and s-a-1 faults of the circuit by Boolean differences were given in Table 7.6.4, from which we can easily obtain $\nabla_{x_i}F$ and $\Delta_{x_i}F$ for all x_i shown in Table 7.6.5. Take the s-a-0 test for detecting fault a_0, for example, which consists of 110 and 111. From Table 7.6.5 it is seen that $\nabla_{x_i}F = (x_1x_2, 0)$, which means that with the application

TABLE 7.6.5 $\nabla_{x_i}F$ and $\Delta_{x_i}F$ of the Circuit of
Fig. 7.6.4(b)

$\nabla_{x_i}F = \left(x_i\dfrac{dF}{dx_i}\right)\cdot[F, \bar{F}]$	$\Delta_{x_i}F = \left(\bar{x}_i\dfrac{dF}{dx_i}\right)\cdot[F, \bar{F}]$
$\nabla_{x_1}F = (x_1x_2, 0)$	$\Delta_{x_1}F = (0, x_1'x_2)$
$\nabla_{x_2}F = (x_1x_2x_3, x_1'x_2x_3')$	$\Delta_{x_2}F = (x_1'x_2'x_3', x_1x_2'x_3)$
$\nabla_{x_3}F = (0, x_2'x_3)$	$\Delta_{x_3}F = (x_2'x_3', 0)$
$\nabla_{x_4}F = (x_1x_2, 0)$	$\Delta_{x_4}F = (0, x_1x_2'x_3)$
$\nabla_{x_5}F = (0, x_1'x_2x_3')$	$\Delta_{x_5}F = (x_2'x_3', 0)$
$\nabla_{x_6}F = (x_1x_2, 0)$	$\Delta_{x_6}F = (0, x_1'x_2 + x_2'x_3 + x_1'x_3)$
$\nabla_{x_7}F = (x_2'x_3', 0)$	$\Delta_{x_7}F = (0, x_1'x_2 + x_2'x_3 + x_1'x_3)$
$\nabla_{x_8}F = (x_1x_2 + x_2'x_3', 0)$	$\Delta_{x_8}F = (0, x_1'x_2 + x_2'x_3 + x_1'x_3)$

of input combination 110 or 111, the output is 1 if the circuit does not have any fault and is 0 if the circuit has fault a_0. The reader can easily verify this from the circuit. Similarly, for the s-a-1 tests 010 and 011 for detecting fault a_1, the $\Delta_{x_i}F = (0, x_1'x_2)$ indicates that with the application of input combination 010 or 011, the output is 0 if the circuit does not have any fault and is 1 if the circuit has fault a_1.

As a final remark, the Boolean-difference method presented in this section can easily be extended to fault detection of multioutput circuits (problems 11 and 12).

Exercise 7.6

1. Supply the proofs for properties 1–10 of the Boolean difference.

2. Let $F(X) = \prod_{k=1}^{m} G_k(X)$, where $X = (x_1, \ldots, x_i, \ldots, x_n)$.

Prove that

$$\frac{dF(X)}{dx_i} = \left[\prod_{k=1}^{m} G_k(X) + \prod_{k=1}^{m} G_k(X^*)\right] \cdot \sum_{k=1}^{m} \frac{dG_k(X)}{dx_i}$$

where $X^* = (x_1, \ldots, x_i', \ldots, x_n)$.

3. Let $F(X) = \sum_{k=1}^{m} G_k(X)$. Prove that

$$\frac{dF(X)}{dx_i} = \left[\prod_{k=1}^{m} \overline{G_k(X)} + \prod_{k=1}^{m} \overline{G_k(X^*)}\right] \cdot \sum_{k=1}^{m} \frac{dG_k(X)}{dx_i}$$

where X and X^* are defined as in problem 2.

4. Find $dF(X)/dx_1$ for
   (a) $F(X) = x_1 x_2 + x_2 x_3 + x_1 x_3$
   (b) $F(X) = (x_1 + x_2)(x_2 + x_3)(x_1 + x_3)$
   using the six methods given in this section.

5. For the same functions of problem 4, find $dF(X)/dx_1$ using the formulas given in problems 2 and 3.

6. Derive a complete fault-detection test set for the following faults of the NAND circuit in Fig. P7.6.6:
   (a) Single fault: line α s-a-0.
   (b) Single fault: line β s-a-1.
   (c) Single fault: line γ s-a-0.
   (d) Multiple fault: line α s-a-0, line β s-a-1.
   (e) Multiple fault: line β s-a-1, line γ s-a-0.

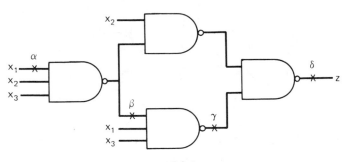

Fig. P7.6.6

7. Find a complete fault-detection test set for the following faults of the NOR circuit of Fig. P7.6.7:
   (a) Single fault: line a s-a-0.
   (b) Multiple fault: line b s-a-1 and line c s-a-0.
   (c) Multiple fault: line d s-a-0, line e s-a-1, and line f s-a-0.

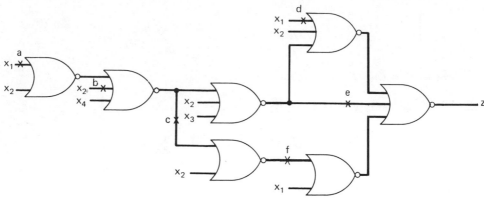

Fig. P7.6.7

8. Construct a complete test set for detecting each single fault in the irredundant tree-like circuit of Fig. P7.6.8 using the method described in this section.

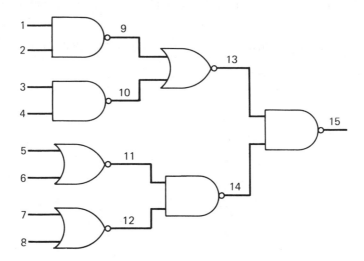

Fig. P7.6.8

9. Find the correct output associated with each test obtained in problem 8.

10. Find a complete fault-detection test for each single fault of the irredundant circuit of Fig. P7.6.10.

11. Consider the fault detection of the multiple-output circuit of Fig. P7.6.11.
    (a) Construct dz_i/dx_j of the circuit.
    (b) Construct the complete sets for detecting s-a-0 and s-a-1 faults on lines 1, 2, 3, and 4.

12. Use the Boolean difference method to construct complete tests for detecting single faults for the irredundant multiple-output circuit of Fig. P7.6.12.

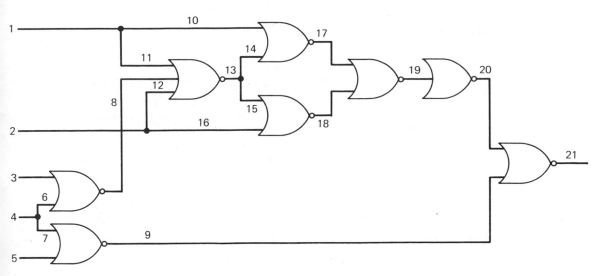

Fig. P7.6.10

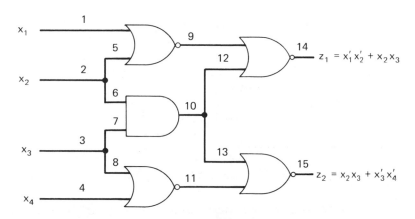

Fig. P7.6.11

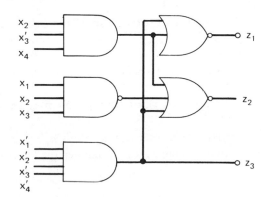

Fig. P7.6.12

7.7 SPOOF Method

In this section an efficient and easy-to-derive method for obtaining tests for detection of single and multiple faults is presented which is based on the use of the *structure-* and *parity-observing output function* (SPOOF). SPOOF provides the information needed for complete analysis of the effects of possible faults on the functional characteristics of a given circuit. The derivation of the SPOOF of a circuit is quite similar to that of the ENF and is best illustrated by use of an example. Consider the circuit of Fig. 7.7.1. The step-by-step derivation of the SPOOF of this circuit is as follows:

$$z = z_h = (z_f z_g)_h = (z_d + z_c)'_{fh}(z_e + z_c)'_{gh} = (z'_{d'}z'_{c'})_{fh}(z'_{e'}z'_{c'})_{gh}$$

$$= z'_{d'fh}z'_{c'fh}z'_{e'gh}z'_{c'gh}$$

$$= z'_{d'gh}(z_a z_b)'_{c'fh}z'_{e'gh}(z_a z_b)'_{c'gh}$$

$$= z'_{d'gh}(z'_{a'c'fh} + z'_{b'c'fh})z'_{e'gh}(z'_{a'c'gh} + z'_{b'c'gh})$$

$$= (x'_3)_{d'fh}[(x'_1)_{a'c'fh} + (x'_2)_{b'c'fh}](x'_4)_{e'gh}[(x'_1)_{a'c'gh} + (x'_2)_{b'c'gh}] \qquad (7.7.1)$$

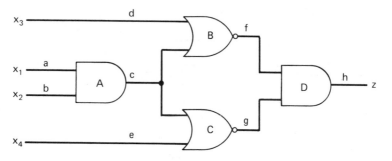

Fig. 7.7.1   Circuit of Example 7.7.1.

It is seen that the essential difference between the SPOOF and the ENF is the presence or absence of primes over each of the subscript symbols of each literal to indicate the parity of the signal.

The application of the SPOOF to the analysis of the effect of any single and multiple s-a-0 and s-a-1 on the output function of a circuit is based on the following algorithm.

Let C and C_F, respectively, denote a fault-free circuit and a faulty circuit with a fault (or a set of faults) F: a_1 s-a-i_1, a_2 s-a-i_2, ..., a_m s-a-i_m, where $i_k \in \{0, 1\}$. The output function $z(C_F)$, or simply $z(F)$, may be determined as follows.

Step 1   Whenever any symbols a_k appear as a subscript to a literal in z, let a_j denote that symbol which designates the closest line to the output terminal of all the lines designated by the symbols a_k. Since the subscript is a sequence of symbols describing a path through the circuit C from a prime input to the circuit output, if any of the a_k appear in a literal's subscript, the symbol designated a_j for that literal must necessarily be unique.

Step 2   Replace the literal by i_j if no prime appears over a_j in the subscript, and by i'_j otherwise.

Step 3   When steps 1 and 2 have been completed for all a_k affected by the fault, remove all the subscripts from the thus-modified SPOOF and simplify the resulting expression by the techniques of Boolean algebra.

The proof of this algorithm may be found in reference 21. The algorithm is illustrated by the following example.

Example 7.7.1

Find the complete test sets for detecting faults for (a) $F_1 = a$ s-a-0; (b) $F_2 = d$ s-a-0 and e s-a-1; and (c) $F_3 = a$ s-a-1, c s-a-0, and f s-a-1, in the circuit of Fig. 7.7.1.

From Eq. (7.7.1) and the algorithm just described, we find that

(a) $z(F_1) = (x'_3)_{a'fh}[1 + (x'_2)_{b'c'fh}](x'_4)_{e'gh}[1 + (x'_2)_{b'c'gh}]$

$\quad = x'_3(1 + x'_2)x'_4(1 + x'_2) = x'_3x'_4$

(b) $z(F_2) = 1 \cdot [(x'_1)_{a'c'fh} + (x'_2)_{b'c'fh}] \cdot 0 \cdot [(x'_1)_{a'c'gh} + (x'_2)_{b'c'gh}]$

$\quad = 1 \cdot (x'_1 + x'_2) \cdot 0 \cdot (x_1 + x_2) = 0$

(c) $z(F_3) = 1 \cdot (1 + 1)(x'_4)_{e'gh}(1 + 1) = x'_4$

Under the normal condition, the output function is

$$z = (x'_1 + x'_2)x'_3x'_4$$

The complete test sets for detecting these faults can be found by

$$T(F_1) = (x'_1 + x'_2)x'_3x'_4 \oplus x'_3x'_4 = x_1x_2x'_3x'_4 \longrightarrow T(F_1) = \{1100\}$$

$$T(F_2) = (x'_1 + x'_2)x'_3x'_4 \oplus 0 = (x'_1 + x'_2)x'_3x'_4 \longrightarrow T(F_2) = \{0000, 0100, 1000\}$$

$$T(F_3) = (x'_1 + x'_2)x'_3x'_4 \oplus x'_4 = x_1x_2x'_4 + x_3x'_4 \longrightarrow T(F_3) = \{1100, 1110, 0010,$$
$$0110, 1010\}$$

As a final remark, recall that if both a variable and its complement are contained in a term in the ENF, that term may be discarded (see the fourth characteristic of the ENF described in Section 7.3). Although the SPOOF greatly resembles the ENF, this rule is not applicable to the SPOOF. As a matter of fact, no terms of the SPOOF may be discarded under any circumstances.

Exercise 7.7

1. For the circuit in Fig. P7.7.1, find the complete set of tests for detecting the following faults:
    (a) Single fault α s-a-0.
    (b) Multiple fault α s-a-0 and β s-a-1.
    (c) Multiple fault α s-a-0, β s-a-1, γ s-a-0, and δ s-a-1.
    Use (1) the path-sensitizing method, (2) the ENF method, and (3) the SPOOF method.

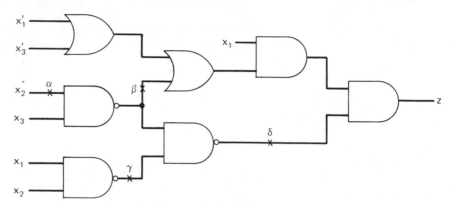

Fig. P7.7.1

2. For the following combinational circuit, find the complete set of tests to detect the multiple fault, all five lines α, β, γ, δ, and ϵ being s-a-0 using the SPOOF method.

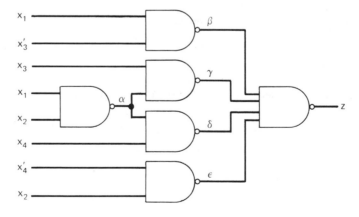

Fig. P7.7.2

3. (a) The normal circuit of Fig. P7.7.3 is a realization of the EXCLUSIVE–OR gate. Suppose that, owing to the presence of single faults in the circuit, the output of the circuit is changed from $z = x_1 \oplus x_2$ to $z = x_1'x_2$. Show that the multiple fault line b s-a-1, line c s-a-1, and line e s-a-1 is the only fault that can cause this change.
    (b) Find all the logic faults that can cause the circuit output to become $z = x_1' + x_2'$.

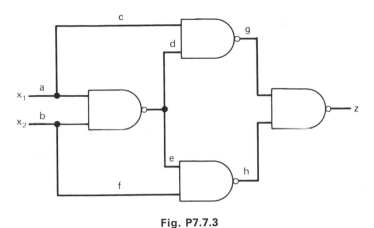

Fig. P7.7.3

Bibliographical Remarks

Three books (references 1–3) on fault detection and diagnosis in digital circuits and systems and two *IEEE Transactions on Computers* special issues on fault-tolerant computing (references 30 and 31) are excellent references for most material covered in this chapter. Readers interested in fault testing and diagnosis in combinational circuits are advised to read them. In recent years interest in this subject has been growing, which results in the publication of a large number of technical papers and the presentation of many papers at conferences. We list here only those which are referred to in the text.

The fault-table method is extensively treated by Kautz (reference 4), Chang (reference 5), and Hadlock (reference 6) .The path-sensitizing and ENF methods are contained in the article by Armstrong (reference 7). The Karnaugh map and tabular method was discovered by Kohavi and Spires (reference 8) and Bearnson and Carroll (reference 9) independently. The ENF–Karnaugh map method is also due to Kohavi and Spires (reference 8). Many techniques for deriving and selecting single-fault diagnostic tests can be found in references 10–16. The problem of fault equivalence in combinational circuits was studied by McCluskey and Clegg (reference 17). Two excellent papers on the analysis of multiple faults in combinational circuits are those of Gault et al. (reference 18) and Bossen and Hong (reference 19). References 20 and 21 are the original papers of the SPOOF method, and references 22–25 are references for the Boolean difference method. A computer-oriented fault-detection test generation algorithm known as the *D*-algorithm which is widely used in industry for large combinational circuits can be found in references 26–29. This algorithm is derived based on the *D*-Calculus (reference 26) and has been implemented in APL language (references 28 and 29). The latest versions of the computer programs are called DALG-II (*D*-Algorithm Version II) and TEST–DETECT (reference 29). An application of the *D*-algorithm to the fault location in large combinational circuits was investigated by Su and Cho (reference 30).

References

Books

1. H. Y. CHANG, E. G. MANNING, and G. METZE, *Fault Diagnosis of Digital Systems*, Wiley, New York, 1970.

2. A. D. FRIEDMAN and P. R. MENON, *Fault Detection in Digital Circuits*, Prentice-Hall, Englewood Cliffs, N.J., 1970.

3. F. F. SELLERS, M. Y. HSIAO, and L. W. BEARSON, *Error Detecting Logic*, McGraw-Hill, New York, 1968.

Papers

4. W. H. KAUTZ, "Fault Testing and Diagnosis in Combinational Digital Circuits," *IEEE Trans. Computers*, Vol. C-17, Apr. 1968, pp. 352–366.

5. H. Y. CHANG, "An Algorithm for Selecting an Optimum Set of Diagnostic Tests," *IEEE Trans. Electron. Computers*, Vol. EC-14, Oct. 1965, pp. 706–711.

6. F. HADLOCK, "On Finding a Minimal Set of Diagnostic Tests," *IEEE Trans Electron. Computers*, Vol. EC-16, Oct. 1967, pp. 674–675.

7. D. B. ARMSTRONG, "On Finding a Nearly Minimal Set of Fault Detection Tests for Combinational Logic Nets," *IEEE Trans. Electron. Computers*, Vol. EC-15, Feb. 1966, pp. 66–73.

8. Z. KOHAVI and D. A. SPIRES, "Designing Sets of Fault-Detection Tests for Combinational Logic Circuits," *IEEE Trans. Computers*, Vol. C-20, No. 12, 1971, pp. 1463–1469.

9. L. W. BEARNSON and C. C. CARROLL, "On the Design of Minimum Length Fault Tests for Combinational Circuits," *IEEE Trans. Computers*, Vol. C-20, No. 11, 1971, pp. 1353–1356.

10. R. BETANCOURT, "Derivation of Minimum Test Sets for Unate Logic Circuits," *IEEE Trans. Computers*, Vol. C-20, No. 11, 1971.

11. G. D. HORNBUCKLE and R. N. SPAN, "Diagnosis of Single-Gate Failures in Combinational Circuits," *IEEE Trans. Computers*, Vol. C-18, Mar. 1969, pp. 216–220.

12. T. J. POWELL, "A Procedure for Selecting Diagnostic Tests," *IEEE Trans. Computers*, Vol. C-18, Feb. 1969, pp. 168–175.

13. H. Y. CHANG, "A Distinguishability Criterion for Selecting Efficient Diagnostic Tests," in *1968 Spring Joint Computer Conf.*, *AFIPS Conf. Proc.*, Vol. 32, Thompson, Washington, D.C., 1968, pp. 529–534.

14. A. D. FRIEDMAN, "Fault Detection in Redundant Circuits," *IEEE Trans. Electron. Computers*, Vol. EC-16, Feb. 1967, pp. 99–100.

15. D. R. SHERTZ and G. A. METZE, "On the Indistinguishability of Faults in Digital Systems," in *Proc. 6th Ann. Allerton Conf. Circuit System Theory*, 1968, pp. 752–760.

16. J. F. POAGE, "Derivation of Optimum Tests to Detect Faults in Combinational Circuits," *Proc. Symp. Math. Theory Automata*, Polytechnic Press, New York, 1963, pp. 483–528.

17. E. J. McCLUSKEY and F. W. CLEGG, "Fault Equivalence in Combinational Networks," *IEEE Trans. Computers*, Vol. C-20, Nov. 1971, pp. 1286–1293.

18. J. W. GAULT, J. P. ROBINSON, and S. M. REDDY, "Multiple Fault Detection in Combinational Networks," *IEEE Trans. Computers*, Vol. C-21, Jan. 1972, pp. 31–36.

19. D. C. BOSSEN and S. J. HONG, "Cause–Effect Analysis for Multiple Fault Detection in Combinational Circuits," *IEEE Trans. Computers*, Vol. C-20, Nov. 1971.

20. F. W. CLEGG, "Use of SPOOF's in the Analysis of Faulty Logic Networks," *IEEE Trans. Computers*, Vol. C-22, Mar. 1973, pp. 229–234.

21. F. W. CLEGG, "The SPOOF—A New Technique for Analyzing the Effects of Faults on Logic Networks," Stanford Electronic Laboratory, Stanford, Calif., Rept. SU-SEL-70-073, IEEE Computer Soc. Repository R-71-105, 1970.

22. F. F. SELLERS, JR., M. Y. HSIAO, and L. W. BEARNSON, "Analyzing Errors with the Boolean Difference," *IEEE Trans. Computers*, Vol. C-17, July 1968, pp. 676–683.

23. S. S. YAU and Y. S. TANG, "An Efficient Algorithm for Generating Complete Test Sets for Combinational Logic Circuits," *IEEE Trans. Computers*, Vol. C-20, Nov. 1971, pp. 1245–1251.

24. G. R. SMITH and S. S. YAU, "A Programmed Fault-Detection Algorithm for Combinational Switching Networks," *Proc. Natl. Electron. Conf.*, Vol. 25, Dec. 1969, pp. 668–673.

25. P. N. MARINOS, "Derivation of Minimal Complete Sets of Test-Input Sequences Using Boolean Differences," *IEEE Trans. Computers*, Vol. C-20, Jan. 1971, pp. 25–32.

26. J. P. ROTH, "Diagnosis of Automata Failures: A Calculus and a Method," *IBM J. Res. Develop.*, Vol. 10, July 1966, pp. 278–291.

27. P. R. SCHNEIDER, "The Necessity to Examine D-Chains in Diagnostic Test Generation—An Example," *IBM J. Res. Develop.*, Vol. 11, Jan. 1967, p. 114.

28. J. P. ROTH, W. G. BOURICIUS, and P. R. SCHNEIDER, "Programmed Algorithms to Compute Tests to Detect and Distinguish Between Failures in Logic Circuits," *IEEE Trans. Electron. Computers*, Vol. EC-16, Oct. 1967, pp. 567–580.

29. W. G. BOURICIUS, E. P. HSIEH, G. R. PUTZOLU, J. P. ROTH, P. R. SCHNEIDER, and C.-J. TAN, "Algorithms for Detection of Faults in Logic Circuits," *IEEE Trans. Computers*, Vol. C-20, Nov. 1971, pp. 1258–1264.

30. S. Y. H. SU and Y.-C. CHO, "A New Approach to the Fault Location of Combinational Circuits," *IEEE Trans. on Computers*, Vol. C-21, Jan. 1972, pp. 21–36.

31. Special issue on Fault-Tolerant Computing, *IEEE Trans. Computers*, Vol. C-20, Nov. 1971.

32. Special issue on Fault-Tolerant Computing, *IEEE Trans. Computers*, Vol. C-22, Mar. 1973.

Fault Detection and Location
in Sequential Circuits

The objective of this chapter is to introduce methods for designing a fault-detection experiment for (synchronous) sequential circuits or machines. The fault testing and diagnosis in sequential circuits is complicated by the presence of memory. There have developed two distinctly different approaches to the problem of fault detection in sequential circuits. The first approach is called the *circuit-test approach*, which assumes that the experimenter has an exact knowledge of the circuit realization and each fault that can occur. The second approach is callled the *transition-checking approach*, which assumes no knowledge whatsoever of the circuit realization but does assume knowledge of the desired transition.

Section 8.1 presents a method of deriving optimal (shortest) experiment for permanent single and multiple s-a-0 and s-a-1 fault detection and location in any sequential circuit. The following three sections are devoted to a discussion of the design of fault-detection experiment using the transition-checking approach. This method will detect, in addition to the class of stuck-at faults, faults that cannot result from a single s-a-1 or s-a-0 fault in the circuit and the types of faults that may occur in the circuit which are not known. Hence this approach protects against a large class of fault. The machines under test are assumed to be strongly-connected, minimal and diagnosable, with no growing states.

The fault-detection checking experiment can be logically divided into three phases: (1) the initialization phase, (2) the state identification phase, and (3) transition verification phase. These tasks cannot be accomplished without using special test sequences. The distinguishing (diagnosing) sequence is for the initial state identification discussed in Section 8.2; the homing sequence and the synchronizing sequence are for the final state identification presented in Section 8.3. Finally, a procedure for designing a fault-detection checking experiment is presented in Section 8.4.

Although both approaches can be applied to fault detection for sequential integrated circuits, the circuit-test approach is recommended for relatively small circuits (SSI and MSI circuits) with circuit diagram known, whereas the transition-checking approach is attractive for relatively large circuits (MSI and LSI circuits), particularly when the circuit diagram is not available to the experimenter.

8.1 Circuit-Test Approach

In Chapter 7 we discussed the problem of fault detection and location in combinational circuits. In this chapter we shall be concerned with the problem of design of optimal (shortest) fault-detection experiment for sequential circuits. The derivation of an optimal test to detect faults in sequential circuits is complicated by the presence of memory.

Quite a few methods that use the circuit-test approach are known to us. The one chosen to be presented here guarantees an optimal test for both fault detection and fault location in any synchronous sequential circuit. By "optimal test" we mean the shortest sequence of symbols guaranteed to detect any (detectable) permanent fault F_i that transforms the transition table T_0 of the normal machine M_0 into another transition table, T_i, which may have *more or fewer states than* M_0.

The following assumptions are made about the machine under test and the faults to be detected. It is assumed that:

1. The machine is minimized and the circuitry is known.
2. The faults are exclusively permanent logical s-a-0 and s-a-1 faults, which may be simple or multiple.
3. A reset signal is available that forces the machine into a unique initial or starting state, q_0. Moreover, if we are testing for a set of m permanent faults, it is assumed that the reset signal forces *all $m + 1$ machines $M_0, M_1, \ldots, M_m$* into the same initial state q_0. Besides, it is assumed that a permanent fault cannot create any new memory elements. This means that if the number of states n_0 of the normal machine is bounded by 2^s, where s is the number of memory elements, then the numbers of states $n_1, n_2, \ldots, n_m$ of the faulty machines are also bounded by 2^s. In other words, the number of memory elements remains unchanged for any permanent detectable faults under consideration.

We begin this investigation with the derivation of tests (experiments) for fault detection.

8.1.1 Fault Detection

A general procedure for deriving the shortest sequence of input symbols for detecting permanent faults in synchronous sequential circuits is best illustrated through use of an example.

Example 8.1.1

Consider the DFF realization of the serial binary adder shown in Fig. 8.1.1. Suppose that an experiment is to be derived for detecting the following faults:

1. s-a-0 fault at location a.
2. s-a-1 fault at location b.
3. s-a-0 fault at location c.
4. s-a-0 fault at location d.

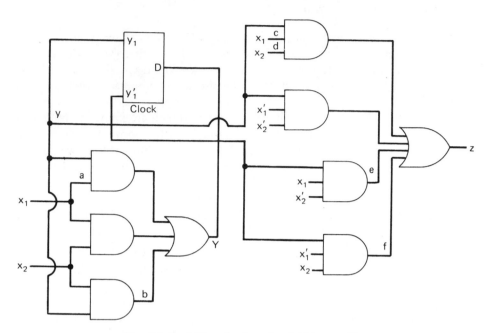

Fig. 8.1.1 DFF realization of serial binary adder.

5. s-a-1 fault at location e.
6. s-a-1 fault at location f.

For convenience, we shall denote these six faults, respectively, by $a_0, b_0, c_0, d_0, e_1,$ and f_1. The transition table of the normal binary adder M_0, denoted by T_0, is shown in Table 8.1.2(a), and the next-state function and output function of M_0 are

$$Y = z_1x_2 + x_1y + x_2y \qquad (8.1.1)$$

$$z = x_1x_2y + x_1'x_2'y + x_1x_2'y' + x_1'x_2y' \qquad (8.1.2)$$

The initial state of the machine is assumed to be q_0 or $y = 0$.

A procedure for obtaining the shortest testing sequence of inputs to detect permanent faults in a sequential machine is as follows:

1. Find the functions Y and z of the faulty machines. The functions Y and z of the six faulty machines are shown in Table 8.1.1.

2. Construct the transition tables of the faulty machines in the form shown in Table 8.1.2.

Before proceeding to the third step, we need the following definitions:

DEFINITION 8.1.1

Let T_0 be the transition table of a normal machine M_0 and T_i be the transition table of a faulty machine M_i of M_0 due to the fault F_i. F_i is *undetectable* if T_0 and T_i are the same.

TABLE 8.1.1 Functions Y and z of the Six Faulty Machines

Fault	Y	z
No fault	$Y = x_1x_2 + x_1y + x_2y$	$z = x_1x_2y + x_1'x_2'y + x_1x_2'y' + x_1'x_2y'$
a_0	$Y = x_1x_2 + x_2y$	$z = x_1x_2y + x_1'x_2'y + x_1x_2'y' + x_1'x_2y'$
b_1	$Y = 1$	$z = x_1x_2y + x_1'x_2'y + x_1x_2'y' + x_1'x_2y'$
c_0	$Y = x_1x_2 + x_1y + x_2y$	$z = x_1'x_2'y + x_1x_2'y' + x_1'x_2y'$
d_0	$Y = x_1x_2 + x_1y + x_2y$	$z = x_1'x_2'y + x_1x_2'y' + x_1'x_2y'$
e_1	$Y = x_1x_2 + x_1y + x_2y$	$z = 1$
f_1	$Y = x_1x_2 + x_1y + x_2y$	$z = 1$

TABLE 8.1.2 Transition Tables of the Normal Machine and the Six Faulty Machines

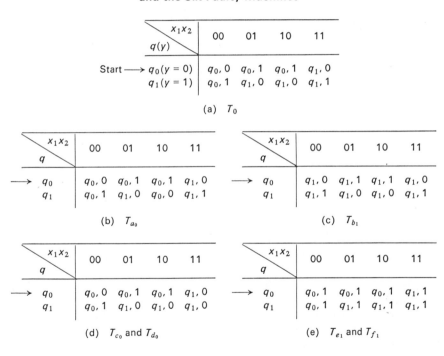

$q(y)$ x_1x_2	00	01	10	11
Start ⟶ $q_0(y=0)$	$q_0,0$	$q_0,1$	$q_0,1$	$q_1,0$
$q_1(y=1)$	$q_0,1$	$q_1,0$	$q_1,0$	$q_1,1$

(a) T_0

q x_1x_2	00	01	10	11
⟶ q_0	$q_0,0$	$q_0,1$	$q_0,1$	$q_1,0$
q_1	$q_0,1$	$q_1,0$	$q_0,0$	$q_1,1$

(b) T_{a_0}

q x_1x_2	00	01	10	11
⟶ q_0	$q_1,0$	$q_1,1$	$q_1,1$	$q_1,0$
q_1	$q_1,1$	$q_1,0$	$q_1,0$	$q_1,1$

(c) T_{b_1}

q x_1x_2	00	01	10	11
⟶ q_0	$q_0,0$	$q_0,1$	$q_0,1$	$q_1,0$
q_1	$q_0,1$	$q_1,0$	$q_1,0$	$q_1,0$

(d) T_{c_0} and T_{d_0}

q x_1x_2	00	01	10	11
⟶ q_0	$q_0,1$	$q_0,1$	$q_0,1$	$q_1,1$
q_1	$q_0,1$	$q_1,1$	$q_1,1$	$q_1,1$

(e) T_{e_1} and T_{f_1}

DEFINITION 8.1.2

Let F_1 and F_2 be two faults of a sequential machine M_0 and T_1 and T_2 the transition tables of the corresponding faulty machines M_1 and M_2. F_1 and F_2 are said to be *indistinguishable* if T_1 and T_2 are the same.

3. Delete the undetectable faults and combine the indistinguishable faults, if any. In this example, all the faults are detectable, since $T_{a_0}, \ldots, T_{f_1}$ are not the same as T_0, which implies that $M_{a_0}, \ldots, M_{f_1}$ are not equivalent to M_0. However, faults c_0 and d_0 are

indistinguishable, and so are faults e_1 and f_1. We can therefore combine the four faults into two distinguishable faults, c_0 and e_1.

4. Construct the *product table* $T_0 \times T_{F_i}$, denoted PT_{F_i}, where F_i denotes a fault, for all the faults under consideration.

(a) The construction of a product table starts with one row corresponding to both machines in their initial states (q_0q_0).

(b) Entries of a row are filled from information contained in the component tables.

(c) New rows are added to the product table when new product states appear as entries.

(d) If the outputs from the two machines differ for an entry, the entry is marked by a cross and a new row is *not* added for the product state.

(e) Construction of the table is complete when there is a row for every product state appearing as an entry that is not marked with a cross.

For example, the product table $T_0 \times T_{a_0}$ is constructed as follows:

x_1x_2 \\ q	00	01	10	11
$\rightarrow q_0$	$q_0, 0$	$q_0, 1$	$q_0, 1$	$q_1, 0$
q_1	$q_0, 1$	$q_1, 0$	$q_1, 0$	$q_1, 1$

$\times$

x_1x_2 \\ q	00	01	10	11
$\rightarrow q_0$	$q_0, 0$	$q_0, 1$	$q_0, 1$	$q_1, 0$
q_1	$q_0, 1$	$q_1, 0$	$q_0, 0$	$q_1, 1$

$=$

x_1x_2 \\ q_iq_j	00	01	10	11
$\rightarrow q_0q_0$	$q_0q_0, 00$	$q_0q_0, 11$	$q_0q_0, 11$	$q_1q_1, 00$
q_1q_1	$q_0q_0, 11$	$q_1q_1, 00$	$q_1q_0, 00$	$q_1q_1, 11$
q_1q_0	$q_0q_0, 10$	$q_1q_0, 01$	$q_1q_0, 01$	$q_1q_1, 10$

PT_{a_0}

$=$

x_1x_2 \\ q_p	00 01 10 11
$\rightarrow A$	A A A B
B	A B C B
C	X X X X

PT_{a_0}

Starting in state q_0q_0 of PT_{a_0}, transitions caused by each input symbol of a sequence are specified as long as a symbol does not call for a transition marked with a cross. In other words, transitions are specified as long as a sequence *does not* detect a fault. As soon as a sequence requires a transition marked with a cross, the fault is detected. Subsequent symbols of a sequence make no difference, so transitions need not be specified.

The major purpose of a product table is to denote a set of sequences. The internal states of component machines and their outputs are of secondary importance. Hence the product table PT_{a_0} can be trimmed down as shown in the final table at the right. The product tables of the remaining faults are shown in Table 8.1.3. To distinguish the faults, we replace the crosses by the faults they denote, $a_0, b_1, c_0,$ and e_1.

5. Construct the sequential fault table. The sequential fault table is the product of the product tables, which is constructed in a very similar way to the way we constructed the product table from a normal transition table and a faulty transition table.

(a) The construction of the sequential fault table starts with one row corresponding to the set of initial product states ($ADGI$).

(b) Entries in a row are filled from information contained in the component product tables.

TABLE 8.1.3 Product Tables of Example 8.1.1

x_1x_2 / q_p	00	01	10	11
→ A	A	A	A	B
B	A	B	C	B
C	a_0	a_0	a_0	a_0

(a) PT_{a_0}

x_1x_2 / q_p	00	01	10	11
→ D	F	F	F	E
E	F	E	E	E
F	b_1	b_1	b_1	b_1

(b) PT_{b_1}

x_1x_2 / q_p	00	01	10	11
→ G	G	G	G	H
H	G	H	H	c_0

(c) PT_{c_0}

x_1x_2 / q_p	00	01	10	11
→ I	e_1	I	I	e_1

(d) PT_{e_1}

(c) New rows are added to the sequential fault table when new sets of product states appear as entries. If a fault indicator F_i appears in an entry, it is regarded as a don't care.

(d) Construction of the sequential fault table is complete when there is a row for every product set of product states appearing in entries of the table.

The sequential fault table for the example is shown in Table 8.1.4(a).

6. Simplify the sequential fault table prior to the derivation of tests. Now suppose that all letters are removed from the sequential fault table of Table 8.1.4(a), leaving only the fault indicators a_0, b_1, c_0, and e_1. Furthermore, if the table is redrawn as shown in Table 8.1.4(b), it is seen that for every cross in column a_0, there is a cross in column e_1. This is also true for every cross of columns b_1 and c_0. This indicates that fault e_1 need not be considered since any sequence that detects fault a_0, b_1, or c_0 will also detect fault e_1. Therefore, fault e_1

TABLE 8.1.4 Sequential Fault Table of Example 8.1.1

x_1x_2 / q_{pp}	00	01	10	11
→ ADGI	$AFGe_1$	AFGI	AFGI	$BEHe_1$
$AFGe_1$	Ab_1Ge_1	Ab_1Ge_1	Ab_1Ge_1	Bb_1He_1
AFGI	Ab_1Ge_1	Ab_1GI	Ab_1GI	Ab_1He_1
$BEHe_1$	$AFGe_1$	$BEHe_1$	$CEHe_1$	BEc_0e_1
Ab_1Ge_1	Ab_1Ge_1	Ab_1Ge_1	Ab_1Ge_1	Bb_1He_1
Bb_1He_1	Ab_1Ge_1	Bb_1He_1	Cb_1He_1	$Bb_1c_0e_1$
⌐→ $CEHe_1$	a_0FGe_1	a_0EHe_1	a_0EHe_1	$a_0Ec_0e_1$

Since all the product states of the four product tables appear in the four columns under a_0, b_1, c_0, and e_1, the table is complete.

(a)

TABLE 8.1.4   continued

q_{pp}		Fault Indicators			
		a_0	b_1	c_0	e_1
00	$ADGI$				X
	$AFGe_1$		X		X
	$AFGI$		X		X
	$BEHe_1$				X
	Ab_1Ge_1		X		X
	Bb_1He_1		X		X
	$CEHe_1$	X			X
01	$ADGI$				
	$AFGe_1$		X		X
	$AFGI$		X		
	$BEHe_1$				X
	Ab_1Ge_1		X		X
	Bb_1He_1		X		X
	$CEHe_1$	X			X
10	$ADGI$				
	$AFGe_1$		X		X
	$AFGI$		X		
	$BEHe_1$				X
	Ab_1Ge_1		X		X
	Bb_1He_1		X		X
	$CEHe_1$	X			X
11	$ADGI$				X
	$AFGe_1$		X		X
	$AFGI$		X		X
	$BEHe_1$			X	X
	Ab_1Ge_1		X		X
	Bb_1He_1		X	X	X
	$CEHe_1$	X		X	X

(b)

**TABLE 8.1.5   Simplified Sequential Fault Table
of Example 8.1.1**

q_{pp} \\ x_1x_2	00	01	10	11
→ ADG	AFG	AFG	AFG	BEH
AFG	Ab_1G	Ab_1G	Ab_1G	Bb_1H
AFG	Ab_1G	Ab_1G	Ab_1G	Ab_1H
BEH	AFG	BEH	CEH	BEc_0
Ab_1G	Ab_1G	Ab_1G	Ab_1G	Bb_1H
Bb_1H	Ab_1G	Bb_1H	Cb_1H	Bb_1c_0
CEH	a_0FG	a_0EH	a_0EH	a_0Ec_0

can be eliminated from the sequential fault table of Table 8.1.4(a). The simplified sequential fault table is shown in Table 8.1.5.

The optimum test sequences for detecting faults is derived from the fault-detection tree, which is defined as follows:

DEFINITION 8.1.3

A fault-detection tree is a tree-like connected graph. It starts with the set of initial product states (ADG). The tree is generated by application of inputs to obtain its next states and by repeating this process.

DEFINITION 8.1.4

A fault-detection tree is terminated if all its branches are terminated by the following two rules:

RULE 1

The kth-level branch b becomes terminal if the set of product states of the kth-level leaf is identical to those of one of its preceding leaves.

RULE 2

All the branches of the kth level are terminated if:

(a)   All the fault indicators appear in one of the kth-level leaves.
(b)   The occurrence is the first time.

7.   Obtain the optimum test sequences by constructing a terminated fault-detection tree. The terminated fault-detection tree for detecting the faults a_0, b_1, and c_0 is shown in Fig. 8.1.2. The tree is terminated at the fourth level. Three optimum test sequences are found:

$$(00), (11), (10), (11)$$
$$(01), (11), (10), (11)$$

and

$$(10), (11), (10), (11)$$

Since they are of the same length, we can use any of them to detect the faults. The output sequences of the normal machine and the faulty machines are given in Table 8.1.6. It is seen that all the output sequences of the faulty machines are different from the output sequence of the normal machine, as expected. As far as fault detection is concerned, the problem is successfully solved.

8.1.2   Fault Location

After having discussed the fault-detection problem, we now turn to the fault-location problem. If the output sequences of the faulty machines obtained from the fault-detection test are *distinct*, we not only can tell whether the machine has any of the prescribed faults, but we can also determine which fault occurred from the output sequence obtained. In this case no fault-location problem need be considered. How-

Level

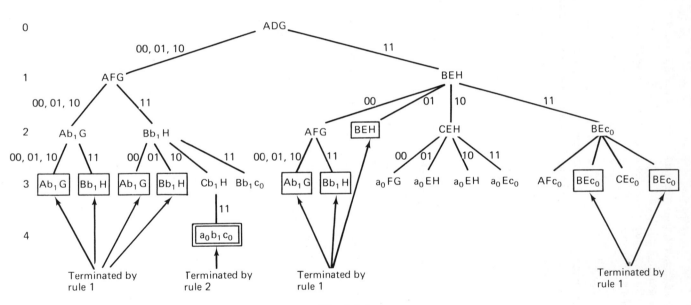

Fig. 8.1.2

**TABLE 8.1.6   Output Sequences of the Normal and
Faulty Versions for the Three Optimum Test Sequences**

Response to (00), (11), (10), (11)					Fault Detection	Fault Location
0	0	0	1	}	Correct one	No fault
0	0	0	0			$a_0, c_0,$ or d_0
0	1	0	1		Indicate the existence	b_1
0	0	0	0		of a fault	$a_0, c_0,$ or d_0
1	1	1	1			e_1 or f_1

Response to (01), (11), (10), (11)					Fault Detection	Fault Location
1	0	0	1	}	Correct one	No fault
1	0	0	0			$a_0, c_0,$ or d_0
1	1	0	1		Indicate the existence	b_1
1	0	0	0		of a fault	$a_0, c_0,$ or d_0
1	1	1	1			e_1 or f_1

Response to (10), (11), (10), (11)					Fault Detection	Fault Location
1	0	0	1	}	Correct one	No fault
1	0	0	0			$a_0, c_0,$ or d_0
1	1	0	1		Indicate the existence	b_1
1	0	0	0		of a fault	$a_0, c_0,$ or d_0
1	1	1	1			e_1 or f_1

ever, in the case where some or all of the output sequences of the faulty machines obtained from the fault-detection test are not distinct, additional tests are needed to identify the fault.

The procedure for deriving the optimum test sequence to locate the faults that cannot be distinguished by the fault-detection test is very similar to the one for fault detection. Recall that the product table of a faulty machine is obtained from the normal transition table and the transition table of the faulty machine. In deriving an optimum test sequence for locating faults, or distinguishing faults, the product tables are obtained among the transition tables of faulty machines whose faults are distinguishable but whose responses to the optimum fault-detection test sequence are indistinguishable. For example, in Table 8.1.6, for all three optimum test sequences, the output sequences of the faulty machines M_{a_0} and M_{c_0} are identical, yet faults a_0 and c_0 are distinguishable faults since the transition tables of the faulty machines M_{a_0} and M_{c_0} are not equivalent. In order to be able to tell which fault actually occurs, additional tests must be performed. The procedure for deriving the optimum test sequence to distinguish or locate these faults is described next.

Suppose that we want to distinguish faults a_0 and c_0 of Example 8.1.1. The general procedure for deriving such a sequence is as follows. Let $F_1, F_2, \ldots, F_l$ be l classes of distinguishable faults and $F_i = \{F_{i1}, \ldots, F_{im_i}\}$ $i = 1, 2, \ldots, l$, where the output sequences of the faulty machines $M_{F_{ij}}$ of the same class are assumed to be identical.

1.   For each class F_i of faults, construct the product tables $T_{F_{ij}} \times T_{F_{ik}}$ for *every pair of faults*. In our example we have only one class of such faults, and in this class

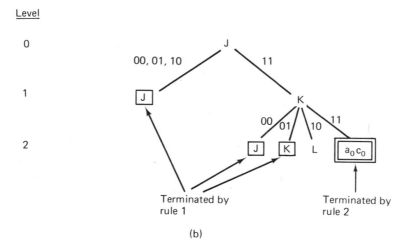

q_p \\ $x_1 x_2$	00	01	10	11
J →	J	J	J	K
K	J	K	L	$a_0 c_0$
L	$a_0 c_0$	$a_0 c_0$	$a_0 c_0$	$a_0 c_0$

(a)

Level

0

00, 01, 10   J   11

1   J   K

00 01 10   11

2   J   K   L $a_0 c_0$

Terminated by rule 1   Terminated by rule 2

(b)

Response to (11), (11)	Fault location
0 0	a_0
0 1	c_0

(c)

Fig. 8.1.3 (a) Fault-location product table; (b) fault-location tree; (c) output sequences of faulty machines M_{a_0} and M_{c_0} upon application of input sequence (11), (11).

there are only two faults, a_0 and c_0. One product table must be constructed, and it is shown in Fig. 8.1.3(a).

2. The rest will be the same as steps 5–7 of the procedure described in Section 8.1.1. Since in this example we have only one product table, no simplification is needed to be considered. The fault-location tree of this example is shown in Fig. 8.1.3(b), from which the optimum test is found, which is (11), (11), as shown in Fig. 8.1.3(c).

Example 8.1.2

Consider the flip-flop realization of the binary serial adder of Fig. 8.1.2, which is redrawn in Fig. 8.1.4. The initial state of the circuit is assumed to be $y = 0$. Suppose that the following multiple faults are to be detected:

1. s-a-1 fault at locations a and b occurred simultaneously.
2. s-a-0 fault at locations c and d occurred simultaneously.
3. s-a-1 fault at location e and s-a-0 fault at location f occurred simultaneously.

We shall denote the faults by α, β, and γ, respectively, and the corresponding faulty machines by M_α, M_β, and M_γ. The S, R, Y, and z of the normal binary adder M_0 and faulty machines M_α, M_β, and M_γ are given in Table 8.1.7.

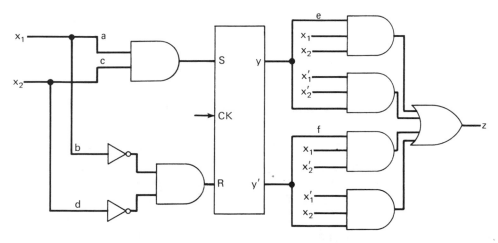

Fig. 8.1.4 Flip-flop realization of binary serial adder.

TABLE 8.1.7 *S, R, Y,* and *z* of the Normal and Faulty Machines of Example 8.1.2

Machine	S	R	$Y = S + R'y$	z
M_0	$x_1 x_2$	$x_1' x_2'$	$Y = x_1 x_2 + x_1 y + x_2 y$	$z = x_1 x_2 y + x_1' x_2' y + x_1 x_2' y' + x_1' x_2 y'$
M_α	x_2	0	$Y = x_2 + y$	$z = x_1 x_2 y + x_1' x_2' y + x_1 x_2' y + x_1' x_2 y'$
M_β	0	x_1'	$Y = x_1 y$	$z = x_1 x_2 y + x_1' x_2' y + x_1 x_2' y + x_1' x_2 y'$
M_γ	$x_1 x_2$	$x_1' x_2'$	$Y = x_1 x_2 + x_1 y + x_2 y$	$z = x_1 x_2 + x_1' x_2' y + x_1' x_2 y'$

Table 8.1.8 shows the transition tables of the normal and faulty machines. From Table 8.1.8 we construct the product tables of the faulty machines, which are shown in Table 8.1.9. The sequential fault tables are shown in Table 8.1.10(a) and (b), from which we find that fault γ can always be detected if fault β is detected, since wherever a row is marked by a cross in column β, there is also a cross in the row under column γ. In other words, fault γ is always detected whenever fault β is detected. Hence fault γ

TABLE 8.1.8   Transition Tables of the Normal and Faulty Machines

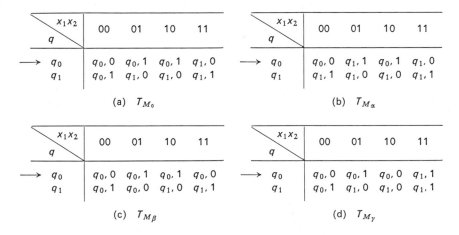

q \ x_1x_2	00	01	10	11
→ q_0	$q_0, 0$	$q_0, 1$	$q_0, 1$	$q_1, 0$
q_1	$q_0, 1$	$q_1, 0$	$q_1, 0$	$q_1, 1$

(a) T_{M_0}

q \ x_1x_2	00	01	10	11
→ q_0	$q_0, 0$	$q_1, 1$	$q_0, 1$	$q_1, 0$
q_1	$q_1, 1$	$q_1, 0$	$q_1, 0$	$q_1, 1$

(b) T_{M_α}

q \ x_1x_2	00	01	10	11
→ q_0	$q_0, 0$	$q_0, 1$	$q_0, 1$	$q_0, 0$
q_1	$q_0, 1$	$q_0, 0$	$q_1, 0$	$q_1, 1$

(c) T_{M_β}

q \ x_1x_2	00	01	10	11
→ q_0	$q_0, 0$	$q_0, 1$	$q_0, 0$	$q_1, 1$
q_1	$q_0, 1$	$q_1, 0$	$q_1, 0$	$q_1, 1$

(d) T_{M_γ}

TABLE 8.1.9   Product Tables of the Faulty Machines

q_p \ x_1x_2	00	01	10	11
→ A	A	C	A	B
B	C	B	B	B
C	α	α	α	α

(a) PT_{M_α}

q_p \ x_1x_2	00	01	10	11
→ D	D	D	D	E
E	β	β	β	β

(b) PT_{M_β}

q_p \ x_1x_2	00	01	10	11
→ F	F	F	F	γ

(c) PT_{M_γ}

TABLE 8.1.10 Sequential Fault Tables

x_1x_2 / q_{pp}	00	01	10	11
→ ADF	ADF	CDF	ADγ	BEγ
CDF	αDF	αDF	αDγ	αEγ
BE	C$\beta\gamma$	B$\beta\gamma$	B$\beta\gamma$	B$\beta\gamma$

(a)

q_{pp}		α	β	γ
00	ADF			
	CDF	x		
	BE		x	
01	ADF		x	
	CDF	x		
	BE		x	x
10	ADF			x
	CDF	x		x
	BE		x	x
11	ADF			x
	CDF	x		x
	BE		x	x

(b)

x_1x_2 / q_{pp}	00	01	10	11
→ AD	AD	CD	AD	BE
CD	αD	αD	αD	αE
BE	Cβ	Bβ	Bβ	Bβ

(c)

need not be considered. Eliminating the columns for γ in Table 8.1.10(a) leaves the simplified sequential fault table shown in Table 8.1.10(c). The fault-detection tree for detecting faults α and β is shown in Fig. 8.1.5, from which the following optimum test sequences are derived:

$$(01), (11), (00, 01, 10, \text{ or } 11)$$

$$(11), (00), (00, 01, 10, \text{ or } 11)$$

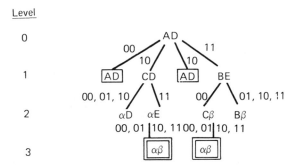

Fig. 8.1.5 Fault-detection tree.

The output sequences of the normal and faulty machines for the optimum test sequences are given in Table 8.1.11. It is seen that by using the test sequence (01), (11), (00, 01, 10, or 11),

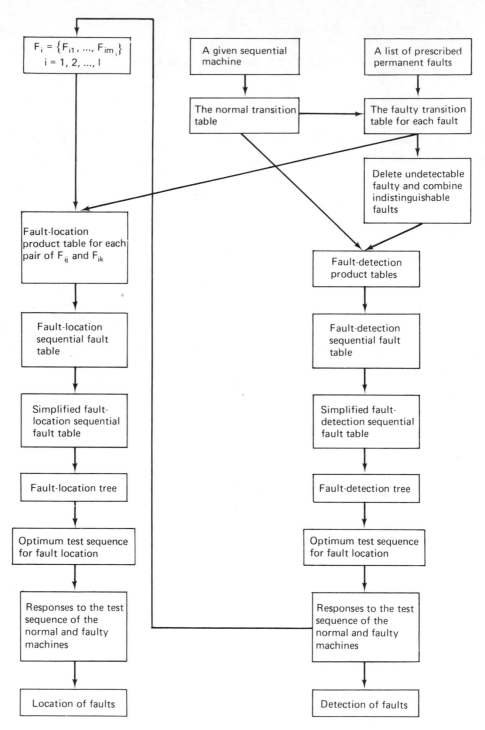

Fig. 8.1.6 Block diagram for showing fault detection and location in synchronous circuits or machines.

TABLE 8.1.11   Output Sequences of the Normal and Faulty Machines for the Optimum Test Sequences

	Fault Detection	Fault Location
Response to (01), (11), (00, 01, 10, or 11)		
1, 0, (1, 0, 0, or 1)	Correct one	No fault
1, 1, (1, 0, 0, or 1) ⎫	Indicate the	⎧ α or γ
1, 0, (0, 1, 1, or 0) ⎬	existence of a	⎨ β
1, 1, (1, 0, 0, or 1) ⎭	fault	⎩ α or γ
Response to (11), (00), (00, 01, 10, or 11)		
0, 1, (0, 1, 1, or 0)	Correct one	No fault
0, 1, (1, 0, 0, or 1) ⎫	Indicate the	⎧ α
0, 0, (0, 1, 1, or 0) ⎬	existence of a	⎨ β
1, 1, (0, 1, 1, or 0) ⎭	fault	⎩ γ

the two output sequences, rows 2 and 4, are identical, which means that faults α and γ are indistinguishable by this sequence. But if we use the sequence (11), (00), (00, 01, 10, or 11) for the test, from the output sequences we shall not only detect the three faults α, β, and γ, but locate them as well.

This method is applicable to any multiple-output sequential circuit or machine (problem 6). The fault-detection and fault-location procedures described above are summarized in the block diagram shown in Fig. 8.1.6.

It is worth noting that the procedure described in this section is applicable to

1. Multiple fault detection.
2. Circuits with multiple outputs.
3. The situation in which some faults in the machine have created new states which were not in the normal machine or have caused some of the states of the normal machine to vanish.

In the following sections we shall introduce the other approach to the fault-detection problem, the transition-checking approach. Before introducing the design of a fault-detection experiment or checking experiment using this approach, we must first study the initial- and final-state identification problems, which are discussed in the next two sections.

Exercise 8.1

1. Consider the following faults of the DFF sequential circuit of Fig. P8.1.1: (1) a_0, (2) b_0, (3) c_0, (4) d_0, (5) e_1, (6) f_1, (7) g_1, (8) h_1, where $a, b, \ldots, h$ denote the location of the faults in the circuit, and the subscripts 0 and 1 denote s-a-0 and s-a-1 faults, respectively.
    (a) Determine the undetectable faults.
    (b) Determine the indistinguishable faults.
    (c) Find the optimum test sequence to detect these faults.
    (d) Find the optimum test sequence to locate them, if required.

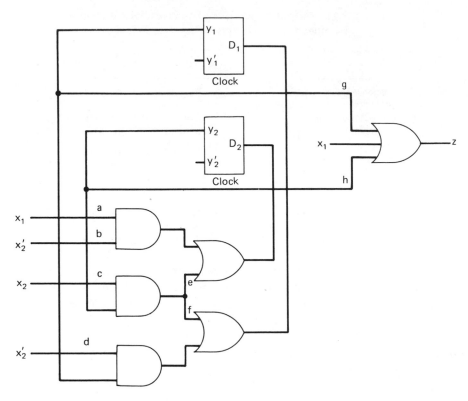

Fig. P8.1.1

2. Detect and locate all possible s-a-0 ans s-a-1 faults for the circuits of Fig. P8.1.2. Assume that the circuits are initially at the state $y_1 = 0$ and $y_2 = 0$.

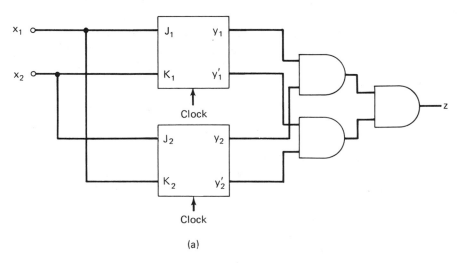

(a)

Fig. P8.1.2

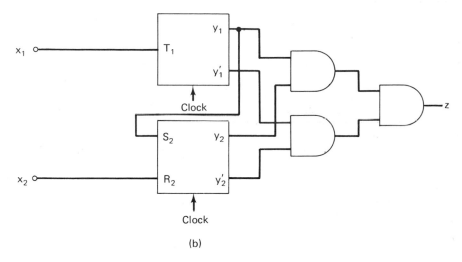

(b)

Fig. P8.1.2 (continued)

3. Consider the following faults of the circuit of Fig. P8.1.3:
   (1) All possible single s-a-0 faults occurred at the input of each of the AND gates.
   (2) All possible single s-a-1 faults occurred at the onput of each of the OR gates.
   Repeat problem 1(a), (b), (c), and (d).

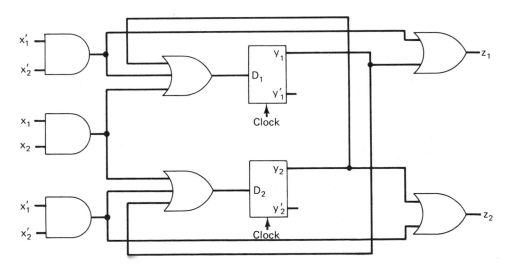

Fig. P8.1.3

4. Locate the s-a-0 faults in the three modules of the flip-flop circuit of Fig. P8.1.4.

5. Suppose that the binary serial adder is realized by a circuit, and four faults of the circuit denoted by α, β, γ, and δ are considered. The transition tables of the normal and faulty circuits are shown in Table P8.1.5.

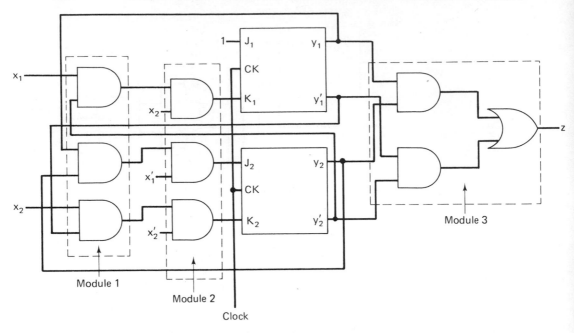

Fig. P8.1.4

TABLE P8.1.5

x_1x_2 $\rightarrow q$	00	01	10	11
$\rightarrow q_0$	$q_0, 0$	$q_0, 1$	$q_0, 1$	$q_1, 0$
q_1	$q_0, 1$	$q_1, 0$	$q_1, 0$	$q_1, 1$

(a) T_{M_0}

x_1x_2 $\rightarrow q$	00	01	10	11
$\rightarrow q_0$	$q_0, 0$	$q_0, 0$	$q_0, 0$	$q_1, 0$
q_1	$q_0, 1$	$q_1, 0$	$q_1, 0$	$q_1, 1$

(b) T_{M_α}

x_1x_2 $\rightarrow q$	00	01	10	11
$\rightarrow q_0$	$q_0, 1$	$q_0, 1$	$q_0, 1$	$q_1, 1$
q_1	$q_0, 1$	$q_1, 0$	$q_1, 0$	$q_1, 1$

(c) T_{M_β}

x_1x_2 $\rightarrow q$	00	01	10	11
$\rightarrow q_0$	$q_2, 0$	$q_0, 1$	$q_0, 1$	$q_1, 0$
q_1	$q_2, 0$	$q_1, 0$	$q_1, 0$	$q_1, 1$
q_2	$q_2, 0$	$q_2, 0$	$q_2, 0$	$q_2, 0$

(d) T_{M_γ}

x_1x_2 $\rightarrow q$	00	01	10	11
$\rightarrow q_0$	$q_0, 0$	$q_0, 1$	$q_0, 1$	$q_2, 0$
q_1	$q_0, 1$	$q_1, 0$	$q_1, 0$	$q_2, 0$
q_2	$q_2, 0$	$q_2, 0$	$q_2, 0$	$q_2, 0$

(e) T_{M_δ}

(a) Show that the optimum test sequences for detecting these four faults are:
   (1) (01 or 10), (11), (00);
   (2) (11), (00), (01 or 10)

(b) Show that the optimum test sequences for locating faults γ and δ are:

(1) (00), (01 or 10)

(2) (11), (11)

6. Consider a single-input and multiple-output sequential circuit whose normal transition table is described in Table P8.1.6(a). The tables of five faults of this machine, represented by α, β, γ, δ, and ϵ, are shown in Table P8.1.6(b)–(f).

TABLE P8.1.6

q \ x	0	1
A	C, 00	B, 01
→ B	A, 10	B, 10
C	A, 10	B, 01

Next state, $z_1 z_2$

(a) T_{M_0}

q \ x	0	1
A	C, 00	B, 01
→ B	A, 10	B, 01
C	A, 10	D, 00
D	D, 00	D, 00

(d) T_{M_α}

q \ x	0	1
A	C, 00	B, 01
→ B	A, 10	B, 01
C	A, 10	B, 01

(b) T_{M_β}

q \ x	0	1
A	D, 10	B, 01
→ B	A, 10	B, 01
C	A, 10	B, 01
D	D, 10	B, 01

(e) T_{M_γ}

q \ x	0	1
A	C, 00	B, 01
→ B	A, 10	B, 01
C	D, 00	B, 01
D	D, 00	D, 00

(c) T_{M_δ}

q \ x	0	1
A	D, 10	B, 01
→ B	A, 10	B, 01
C	A, 10	B, 01
D	D, 10	B, 01

(f) T_{M_ϵ}

(a) Find the optimum fault-detection test sequence.
(b) If one of the two outputs, say z_1, is used to derive the fault-detection test, find the optimum fault-detection test sequence.
(c) Show that the length of the test sequence obtained in (a) is shorter than that of the test sequence obtained in (b).

8.2   Initial-State Identification

The state and machine identification problems are usually solved by performing a specially designed experiment on the machine; based on its result, we draw our conclusion. First, we make precise what we mean by an experiment.

DEFINITION 8.2.1

An *experiment* is the process of applying a sequence of input symbols to the input terminals of a machine and recording the resulting output sequences obtained at the output terminals. An experiment is called *simple* and *multiple* if it requires one copy of the machine and more than one copy of the machine to perform it, respectively.

8.2.1   Successor Tree

The root of state and machine identification by experiment is based on the *successor tree*. A successor tree is a tree-like connected graph. We shall illustrate the procedure of constructing the successor tree of a sequential machine by an example.

Consider a (minimal) machine M_1 whose transition table is described in Table 8.2.1. The successor tree may be constructed in a straightforward manner following the following two steps:

TABLE 8.2.1   Transition Table of a Minimal Machine M_1

x \ q	a	b	c	a	b	c
1	5	2	5	1	1	0
2	1	3	2	1	1	0
3	2	3	6	0	0	1
4	6	3	2	0	0	1
5	1	2	1	0	1	0
6	2	6	6	1	0	1

1.   Apply each input symbol to the states of the machine (or a subset of Q), each of which develops a node that contains the next states obtained from the state-transition function of the machine.

2.   Group all the next states in one group if the corresponding outputs are identical; otherwise, arrange them in different groups.

The set of states $\{q_{i_1}, q_{i_2}, \ldots, q_{i_m}\}$, one of which is, to the experimenter's knowledge, the initial state of machine M, called the *initial uncertainty of M*, denoted $U_0(M)$. The states in $U_0(M)$ are called the *initial uncertainty states*. Both the initial- and final-state identification problems are trivial when $U_0(M)$ is a singleton (i.e., when $m = 1$). We shall therefore concentrate our attention on cases in which $m \geq 2$. The successor tree T_1 for machine M_1 and the initial uncertainty $U_0(M) = \{1, 2, 3, 4, 5, 6\}$ is shown in Fig. 8.2.1, where only two levels are developed. The reader can continue such a process

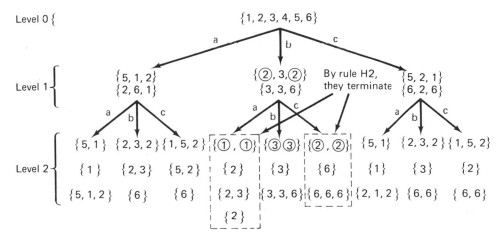

Fig. 8.2.1 Successor tree T_1 of machine M_1.

to any level he wants. The sets of states

$$\{5, 1, 2\}, \quad \{2, 3, 2\}, \quad \{5, 2, 1\}$$
$$\{2, 6, 1\}, \quad \{3, 3, 6\}, \quad \{6, 2, 6\}$$

obtained from U_0 with the applications of a, b, and c, respectively, are called first-level *uncertainty vectors* (UV); the second-, third-, . . . , level uncertainty vectors are defined similarly. The uncertainty vector resulted from the application of the input sequence $\tilde{x}_i$ to U_0 is called the $\tilde{x}_i$-*successor of* U_0. The states in any of the uncertainty vectors are called *successor states*.

The successor tree T_2 for the subset $Q_1 = \{1, 2, 3\}$ is given in Fig. 8.2.2. This corresponds to the case where $Q_1 \subset Q$ is known to contain the initial state, so we do not need to include the state of $Q - Q_1$ in the initial uncertainty in costructing the successor tree of the machine for identifying its initial state.

From the above two examples of the successor trees, we observe that the following important properties of a successor tree:

PROPERTY 1

If, at the kth level, there are two identical states appearing in one group, such as the 2, 2 states in $\{2, 3, 2\}$, the 3, 3 states in $\{3, 3, 6\}$, and the 6, 6 states in $\{6, 2, 6\}$ at the first level of tree T_1, their next states from the $(k + 1)$th-level on will always be identical and in the same group. A UV containing such a group of states is called a *homogeneous uncertainty vector*. For example, the next states of the two states 2 in $\{2, 3, 2\}$ at the

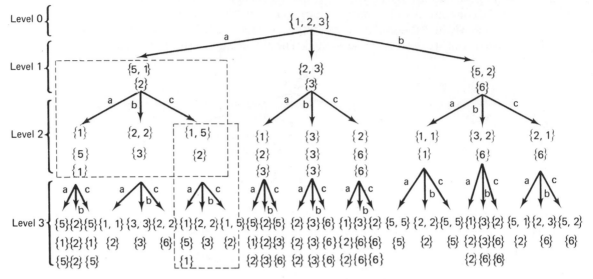

Fig. 8.2.2 Successor tree T_2 for the subset $Q_1 = \{1, 2, 3\}$ of machine M_1.

first level of tree T_1 are $\{1, 1\}$, $\{3, 3\}$, and $\{2, 2\}$, as indicated in Fig. 8.2.1. Thus the two UV's ($\{2, 3, 2\}\{3, 3, 6\}$) and ($\{5, 2, 1\}\{6, 2, 6\}$) are homogeneous uncertainty vectors.

PROPERTY 2

If a kth-level UV is identical to one of *its* preceding UV's [e.g., the second-level UV $\{1, 5\}$, $\{2\}$ of tree T_2 is identical to the first-level UV, as indicated in Fig. 8.2.2], the UV's developed from them will also be identical.

PROPERTY 3

A UV that contains only singleton groups is called a *simple uncertainty vector*. Let V_D be a simple uncertainty vector of a successor tree and P_D be the path from the initial state uncertainty vector to the UV V_D. All the output sequences of the machine to the sequences $\tilde{x}_d$ of the input symbols corresponding to the path P_D for the states in the initial-state uncertainty vector are *distinct*. For example, the UV $\{1\}\{5\}\{1\}$ of tree T_2 is a simple UV. The output sequences of the states 1, 2, and 3 of machine M_1 to the input sequence $\tilde{x}_d = aa$ corresponding to the path

$$\{1, 2, 3\} \xrightarrow{a} \{5, 1\}\{2\} \xrightarrow{a} \{1\}\{5\}\{1\}$$

are shown in Table 8.2.2.

PROPERTY 4

Let V_H be a UV that contains only singleton groups or groups consisting of one state with duplicates, and P_H be the path from the initial-state uncertainty to the UV

TABLE 8.2.2 Output Sequences of States 1, 2, 3 of Machine M_1

Initial State	Input	
	a	*aa*
1	1	10
2	1	11
3	0	01
	not distinct	distinct

V_H. Then, regardless of the initial state of the machine, the output sequences to the sequence $\tilde{x}_k$ of input symbols corresponding to the path P_H completely determine the final state of the machine.

8.2.2 Distinguishing (Diagnosing) Sequence

Based on the above four properties of the successor tree, we can derive effective experiments for solving the three state and machine identification problems stated previously. The problem of identifying the initial state of a sequential machine can be solved by constructing a *distinguishing sequence*.

DEFINITION 8.2.2

An input sequence $\tilde{x}_d$ is said to be a *distinguishing sequence* (DS) or *diagnosing sequence* for a machine $M = (\Sigma, Q, Z, \delta, \lambda)$ if the output response of M to $\tilde{x}_d$ is different for each initial state.

A distinguishing sequence is obtained from the distinguishing tree, which is a *terminated* successor tree obtained from a successor tree by stipulating the following termination rules:

RULE D1

A kth-level branch b becomes terminal if the kth-level UV associated with b is identical to a UV of a level less than k. When two or more UV's of kth level are identical, all except one of them may be deleted.

This rule is derived from property 2. Property 2 indicates that if such a situation arises, the UV's developed from them will also be identical. Hence rule D1 is to eliminate the unnecessary effort in the process of finding a solution to the problem.

RULE D2

A kth-level branch b becomes terminal if the kth-level UV associated with b is a homogeneous UV.

Clearly, this rule is based on property 1—that if such a situation occurs, we will not be able to distinguish the initial state by observing the output sequences of the machine; thus we should abandon this part of the successor tree by terminating it.

The most important termination rule is the third one.

RULE D3

A kth-level branch b becomes terminal whenever (the first time such situation occurs) the kth-level UV associated with b is a simple UV. When such UV occurs, the complete procedure may be terminated. The DS is the input sequence corresponding to the path from the initial uncertainty to this simple UV.

In accordance with property 3 of the successor tree, the initial state of the machine may be identified by observing the output sequence of the machine to this DS. This is illustrated by the following example.

Example 8.2.1

Consider machine M_2 in Table 8.2.3, with its associated distinguishing tree in Fig. 8.2.3. Note that in this distinguishing tree, we have, in addition to the uncertainty vectors, included

TABLE 8.2.3 Machine M_2

$\dfrac{x}{q}$	0	1
A	$D, 0$	$C, 0$
B	$C, 1$	$D, 1$
C	$B, 1$	$A, 0$
D	$A, 0$	$A, 0$

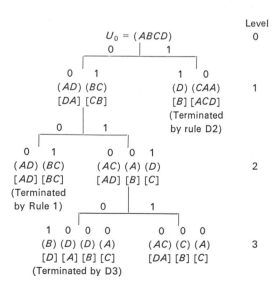

Fig. 8.2.3 Distinguishing tree of machine M_2.

the output symbol and the predecessor state associated with each component and each state, respectively, of every uncertainty vector. For example, the first-level uncertainty vector

$$0 \quad 1$$
$$(AD)\ (BC)\dagger$$
$$[DA]\ [CB]$$

indicates that states A and D in (AD) are obtained from states D and A in U_0, respectively, and states B and C in (BC) are obtained from states C and B in U_0, respectively. The 0 and 1 above the two components (AD) and (BC) of the uncertainty vector indicate their corresponding outputs. The third-level simple uncertainty vector

$$1 \quad 0 \quad 0 \quad 0$$
$$(B)\ (D)\ (D)\ (A)$$
$$[D]\ [A]\ [B]\ [C]$$

for example, indicates that the states B, D, D, and A in (B), (D), (D), and (A) are the successors of states D, A, B, and C, respectively, of the initial uncertainty U_0.

The node associated with the homogeneous uncertainty vector $(D)(CAA)$ in level 1 is terminated by rule D2, since no further input can split the component (AA), and it cannot be determined whether the initial states was C or D. The uncertainty vector $(AD)(BC)$ is ended by rule D1 because if a path describing a DS reaches a simple uncertainty vector from level 2, a shorter path must reach a simple uncertainty vector from the node in level 1. The uncertainty vector $(B)(D)(D)(A)$ in level 3 ends by rule D3, and the path leading to it corresponding to a DS (i.e., $\tilde{x}_d = 010$). The responses of M_2 to $\tilde{x}_d = 010$ and the initial and final states indicated by them are as follows:

Response $\tilde{z}$ to $\tilde{x}_d = 010$	Initial State Indicated by $\tilde{x}_d$	Final State Indicated by $\tilde{x}_d$
000	A	D
100	B	D
110	C	A
001	D	B

Hence the initial state is uniquely identified by the output sequence of the machine to $\tilde{x}_d$. Because their corresponding outputs and corresponding predecessors in U_0 are recorded on the tree while it is constructed, the response $\tilde{z}$ to $\tilde{x}_d = 010$ for each state in U_0 and the initial and final states indicated by it can be read out directly from the successor tree of Fig. 8.2.3 without referring to the transition table of the machine. Note that the state-group output and the predecessor information are not necessary for the uncertainty vectors that are terminated by rules D1 and D2, since they will not be used. The following problems about the distinguishing sequence are of our interest.

1. Find the class of machines that does not possess a distinguishing sequence.
2. For the class of machines that possess distinguishing sequences:

†From now on, for notational convenience, we shall represent an uncertainty vector such as $(\{A, D\}\{B, C\})$ simply by $(AD)(BC)$.

(a)   Find the greatest lower bound.

(b)   Find the least upper bound.

We begin our investigation with the first problem.

DEFINITION 8.2.3

A machine $M = (\Sigma, Q, Z, \delta, \lambda)$ is said to be x_i-*mergeable* if and only if for each input symbol $x_i \in \Sigma$ there exists at least two states q_{i_1} and q_{i_2} such that

$$\delta(q_{i_1}, x_i) = \delta(q_{i_2}, x_i)$$

and

$$\lambda(q_{i_1}, x_i) = \lambda(q_{i_2}, x_i)$$

Otherwise, it is called *non-x_i-mergeable*.

An example of an x_i-mergeable machine is given in Table 8.2.4. In this example, states A and B are merged into C under input a, and states B and C are merged into

TABLE 8.2.4   Example of x_i-Mergeable
Machine

x \ q	a	b
A	C, 0	A, 0
B	C, 0	B, 1
C	A, 1	B, 1

state B under input b. Thus this machine is x_i-mergeable. States 1 and 2 are merged into state 3 under input a. States 2 and 3 are merged into state 2 under input b.

THEOREM 8.2.1

The class of x_i-mergeable sequential machine does not possess a distinguishing sequence.

Proof: Since every UV in the first level of the distinguishing tree of an x_i-mergeable machine is homogeneous UV, by rule D2, all the branches terminate at the first level. Hence an x_i-mergeable machine does not possess a DS. ∎

We now investigate the greatest lower bound and the least upper bound on the length of a DS when it does exist. It will be seen in Section 8.4 that the length of a checking experiment that uses a distinguishing sequence is heavily influenced by the length of the distinguishing sequence used. First we present the greatest lower bound on the length of a distinguishing sequence of a sequential machine.

DEFINITION 8.2.4

An *n*-state, *m*-input *p*-output machine will be denoted (n, m, p)-*machine*.

THEOREM 8.2.2

The greatest lower bound on the length of a distinguishing sequence (GLB_{DS}) of an (n, m, p)-machine is $\lceil \log_p n \rceil$.

Proof: To distinguish *n* states, the relation $p^l \geq n$ must hold, where *l* denotes the number of levels of the distinguishing tree, which also denotes the length of a DS. Since *l* must be an integer, $l \geq \lceil \log_p n \rceil$. Hence $\text{GLB}_{\text{DS}} = \lceil \log_p n \rceil$. ■

To distinguish the four states of M_2, we found a DS $\tilde{x}_d = 010$ which is of length 3 and $|\tilde{x}_d| < \lceil \log_2 4 \rceil = 2$. It agrees with Theorem 8.2.2.

Next, we present a necessary condition for a sequential machine to possess a DS with length equal to GLB_{DS}. Before presenting this condition, we will show by an example that such machines exist.

Example 8.2.2

Consider the machine M_3 (Table 8.2.5), whose distinguishing tree is given in Fig. 8.2.4. It can be seen from the tree that M_3 possesses two distinguishing sequences of length 2 ($= \lceil \log_2 4 \rceil$) (i.e., $\tilde{x}_{d1} = 01$ and $\tilde{x}_{d2} = 00$).

TABLE 8.2.5 Machine M_3

q \ x	0	1
A	D, 1	C, 0
B	B, 0	D, 1
C	A, 1	B, 1
D	A, 0	C, 1

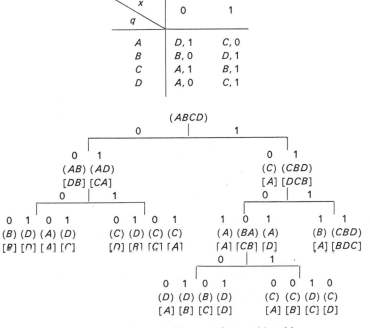

Fig. 8.2.4 Distinguishing tree for machine M_3.

From Theorem 8.2.2 and Example 8.2.2, we immediately obtain the following necessary condition for a machine having a minimum-length DS.

THEOREM 8.2.3

An (n, m, p)-machine will possess a minimum-length distinguishing sequence (a distinguishing sequence whose length is equal to $\mathrm{GLB_{DS}}$) only if it has at least one input symbol which will cause the states in the initial uncertainty to be partitioned with respect to their corresponding outputs into p components.

The proof of this theorem is obvious and is omitted.

In practice, it is more important for us to know the least upper bound on the length of DS of a machine than to know the greatest lower bound. First we need the following definitions.

DEFINITION 8.2.5

Let π_1 and π_2 be two partitions of the set of the states of a sequential machine. A partition π_1 is said to be a *proper refinement* of a partition π_2 if and only if $\pi_1 \cdot \pi_2 = \pi_1$ and $\pi_1 \neq \pi_2$.

DEFINITION 8.2.6

The partition that partitions the set of states of a machine into singleton groups is called the *O-partition*.

An upper bound on the length of a distinguishing sequence is stated below.

THEOREM 8.2.4

The class of minimal, non-x_i-mergeable (n, m, p)-machines possess at least one distinguishing sequence which is no longer than $n(n-1)/2$ input symbols.

Proof: Let M be a minimal, non-x_i-mergeable (n, m, p)-machine. We begin with the initial-state uncertainty $U_0 = (q_1, q_2, \ldots, q_n)$ and attempt to form the distinguishing tree. For each positive integer k, define a relation R_k to exist between any two states q_i and q_j of M if and only if q_i is k-indistinguishable from q_j. Indistinguishability is an equivalence relation and therefore defines a partition π_k on the state set Q of M. The blocks of this partition are the initial uncertainty states associated with the components of the resulting uncertainty vector. The non-x_i-mergeability hypothesis guarantees that each of the successor states will not merge to form any homogeneous uncertainty vector in the resultant uncertainty vector.

If π_k is not the O-partition, then π_{k+1} is a proper refinement of π_k. To show this, note that if any two states are $(k+1)$-indistinguishable, then they are certainly k-indistinguishable, and hence at least $\pi_{k+1} \leq \pi_k$. To show that $\pi_{k+1} < \pi_k$, pick any two states q_i and q_j which are k-indistinguishable. Since by hypothesis the machine is minimized, choose the shortest sequence of inputs that will distinguish the states.

If the sequence is of length t, consider the $(t-k-1)$ successors of states q_i and q_j. These successors are $(k+1)$-distinguishable. The sequence that distinguishes them is the rest of the above sequence. The $(t-k-1)$-successors are not distinguishable by any sequence of length k, as that would contradict the minimality of the chosen sequence. Thus $\pi_{k+1} < \pi_k$.

π_1 partitions Q into at least two blocks because, if not, every state would have the same output for all single inputs, contradicting the assumption that machine M is minimized. It is known from the argument concerning proper refinements of π_1 that at least two of the states in U_0 in the same block of π_1 are partitioned into seperate blocks by a sequence of length 2; thus $\pi_2 < \pi_1$, and the maximum number of states in any block of π_2 is $n - 2$. An inductive proof given below shows that at most $n(n - 1)/2$ input symbols will be required to reach the O-partition. The inductive hypothesis is that at most $k(k + 1)/2$ inputs will be required to reduce the maximum number of states in any block of π_k to $n - k$.

The hypothesis was shown above to be true for $k = 1$ and $k = 2$. Assume that the statement is true for $k - 1$, i.e., that the maximum number of states in any block of π_{k-1} is $n - k + 1$, and that at most $(k - 1)k/2$ inputs are needed to arrive at this partition. From the proper refinement argument above, it is known that at most k more inputs are required to form the π_k partition such that $\pi_k < \pi_{k-1}$, and therefore the maximum number of states in any block of π_k is $n - k$. Hence π_k can be reached by a sequence of inputs whose total length is no greater than $k + k(k - 1)/2 = k(k + 1)/2$ symbols. The O-partition must be reached when $k \leq n - 1$, as then the maximum number of states in any block of π_{n-1} is $n - (n - 1) = 1$. By letting $k = n - 1$, one can see that the O-partition $= \pi_{n-1} < \pi_{n-2} < \ldots < \pi_1 < \pi_0 = U^0$ can be reached with at most $n(n - 1)/2$ input symbols, where the O-partition corresponds to a simple uncertainty vector. ∎

The non-x_i-mergeability structure of a state table provides a useful existence test to determine that a given machine possesses a distinguishing sequence. The next corollary is especially interesting in that it is possible to construct a non-x_i-mergeable machine for which the length of a minimal-length distinguishing sequence actually attains the bound of $n(n - 1)/2$ symbols.

COROLLARY 8.2.1

The bound of $n(n - 1)/2$ symbols on the length of a minimal-length distinguishing sequence for the class of minimal, non-x_i-mergeable machine is the least upper bound.

Proof: The length of $n(n - 1)/2$ symbols was shown in Theorem 8.2.4 to be an upper bound. Example machine M_4 of Table 8.2.6 demonstrates that $n(n - 1)/2$ is the least upper bound. ∎

TABLE 8.2.6   Machine M_4

x / q	0	1
A	$A, 0$	$B, 0$
B	$C, 1$	$A, 0$
C	$B, 1$	$C, 0$

The distinguishing tree of machine M_4 is given in Fig. 8.2.5, which shows that for $n = 3$, $n(n - 1)/2 = 3$ and $|\tilde{x}_d| = 3$, with $\tilde{x}_d = 010$.

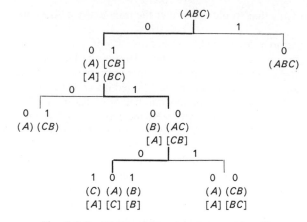

Fig. 8.2.5 Distinguishing tree for machine M_4.

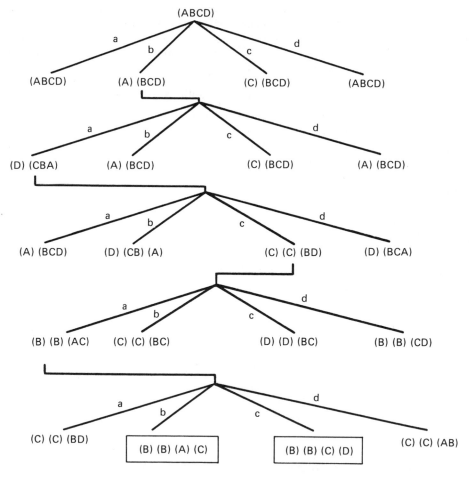

Fig. 8.2.6 Partial DS tree for machine M_5.

The search for a four-state machine for which the bound $n(n-1)/2$ is reached was not successful. However, the bound of $[n(n-1)/2] - 1$ for $n = 4$ was reached with machine M_5 of Table 8.2.7 as DS $x_d = bacab$ with length 5 is found. The pertinent

TABLE 8.2.7 Machine M_5

q \ x	a	b	c	d
A	D 0	A, 0	C, 0	A, 1
B	C, 0	B, 1	B, 1	C, 1
C	B, 0	C, 1	D, 1	B, 1
D	A, 0	D, 1	C, 1	D, 1

parts of the DS tree for M_5 are given in Fig. 8.2.6 to show the flow through the machine.

8.3 Final-State Identification

8.3.1 Homing Sequence

By slightly modifying the procedure for identifying the initial state of a sequential machine described above, we can design an experiment for identifying the final state of a sequential machine without knowing its present (initial) state. The final state of a sequential machine is identified by a homing sequence, which is defined as follows.

DEFINITION 8.3.1

An input sequence $\tilde{x}_h$ is said to be a *homing sequence* (HS) for a machine $M = (\Sigma, Q, Z, \delta, \lambda)$ if its final state can be determined uniquely from the response to $\tilde{x}_h$ irrespective of the actual initial state of the machine.

A homing sequence is derived from a *homing tree*, which is another terminated successor tree. The termination rules for constructing a homing tree are very similar to those for constructing a distinguishing tree. They are:

RULE H1

The first termination rule is the same as rule D1, to avoid unnecessary duplication of the uncertainty vectors and branches in the successor tree.

RULE H2

This is the main termination rule. A kth-level branch b becomes terminal whenever (the first time such situation occurs) the kth-level uncertainty vector contains only singleton groups and groups consisting of one state with duplicates.

In accordance with property 4 of the successor tree, when a branch is terminated by rule H2, a homing sequence is formed. Hence the entire procedure may also be terminated.

Suppose that we want to identify the final state of machine M_1 after applying a sequence of input symbols. A close observation of the successor tree of Fig. 8.2.1 indicates that it is in fact a homing tree, since the two second-level UV's [(1, 1), (2), (6, 6), (2)) and ((2, 2), (6), (6, 6, 6)] both satisfy the condition stated in rule H2; thus the procedure is terminated at the second level. The two output sequences corresponding to the input sequences ba and bc are given in Table 8.3.1. For each input sequence, the output sequence *uniquely* determines the final state of the machine.

TABLE 8.3.1 Response of Machine M_1 to the Input
Sequences *ba* and *bc,* with Respect to Each of Its
States as the Initial State

Initial State	Input		Final State	Initial State	Input		Final State
	b	*ba*			*b*	*bc*	
1	1	11	1	1	1	10	2
2	1	10	2	2	1	11	6
3	0	00	2	3	0	01	6
4	0	00	2	4	0	01	6
5	1	11	1	5	1	10	2
6	0	01	2	6	0	01	6

The procedures for identifying the initial and final states of a sequential machine may be represented by a flow chart, as shown in Fig. 8.3.1.

Only some machines have a distinguishing sequence, but every machine possesses a homing sequence. The following theorem gives the upper bound on the length of a homing sequence.

THEOREM 8.3.1

For every minimal n-state machine M, there always exists a preset homing sequence whose length $\leq (n - 1)^2$.

Proof: Let the initial-state uncertainty be $(q_1, q_2, \ldots, q_n)$. Since M is minimized, by Corollary 5.2.2, for every pair of states q_i, q_j there exists an experiment of length $n - 1$ or shorter which distinguishes q_i from q_j. Let us denote this experiment $\tilde{x}_i$. Starting with the initial-state uncertainty, the application of the sequence $\tilde{x}_1$, which distinguishes between some pair of states in M, yields the $\tilde{x}_1$-successor uncertainty vector, which contains at least two components. Next, select any two states in one component and apply the appropriate sequence $\tilde{x}_2$, which distinguishes between them. The $\tilde{x}_1\tilde{x}_2$-successor uncertainty vector contains at least three components. In a similar manner we obtain the $\tilde{x}_1\tilde{x}_2 \ldots \tilde{x}_{n-1}$-successor vector, which consists of n components each of which contains only one state. Therefore, the sequence $\tilde{x}_1\tilde{x}_2 \ldots \tilde{x}_{n-1}$ is a homing sequence. Since the length of each $\tilde{x}_i$ experiment is at most $n - 1$, the total length of a homing sequence cannot exceed

$$\underbrace{(n - 1) + (n - 1) + \ldots + (n - 1)}_{n - 1 \text{ times}} = (n - 1)^2 \quad \blacksquare$$

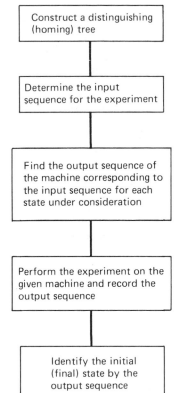

Fig. 8.3.1 Procedure for identifying the initial and final states of a sequential machine.

Note that the bound on the length of a homing sequence in Theorem 8.3.1 is not the least upper bound. It can be shown that the length of the homing sequence need not exceed $1/2n(n-1)$.*

8.3.2 Synchronizing Sequence

A synchronizing sequence is a special case of a homing sequence.

DEFINITION 8.3.2

A *synchronizing sequence* of a machine M is a sequence $\tilde{x}_s$ of inputs that take M to a specified final state *regardless of the output or the initial state of the machine.*

Since a synchronizing sequence is a special case of a homing sequence, it can be constructed from a successor tree using the two termination rules, rules H1′ and H2′, which are modifications of rules H1 and H2, respectively.

*For proof, see reference 12.

RULE H1′

A kth-level branch b becomes terminal whenever the same states, irrespective of the number of their appearances and the groups they belong to, in the UV associated with b have appeared in one of the UV's of a level less than k. If there are two more UV's at the kth level containing the same states, all except one of them may be terminated.

RULE H2′

A kth-level branch b becomes terminal whenever all the groups of the kth level UV associated with b have the same state.

For example, the successor tree of the machine M_6 of Table 8.3.2 is constructed in Fig. 8.3.2. At level 1, the UV $(BCDA)$ is terminated by rule H1′, which is obvious.

TABLE 8.3.2　Machine M_6

$\dfrac{x}{q}$	0	1
A	$D, 1$	$B, 0$
B	$D, 1$	$C, 0$
C	$A, 0$	$D, 0$
D	$B, 0$	$A, 0$

At level 3, the UV $(BB)(DD)$ is terminated by rule H1′, which is again obvious. At level 4, there are four UV's, which are terminated by rule H1′. Take the fourth-level UV $(DD)(AA)$, for instance. The reason it is terminated by rule H1′ is that it contains the two states A and D, which are the same states contained in the third-level UV (D) $(A)(DD)$. All the branches of level less than 5 are terminated by rule H1′ except the one with uncertainty vector (D) (D) (DD), which has one common state D in all its groups, hence is terminated by rule H2′, and a synchronizing sequence $\tilde{x}_s = 01010$ of this machine is thus found.

From the definition of a synchronizing sequence, we find that it can be obtained by a simpler process. Observe that a synchronizing sequence of a machine takes it to a specified final state (in the above example, this state was D) *regardless of the output or the initial state of the machine*. We therefore define a synchronizing uncertainty vector as follows.

DEFINITION 8.3.3

A *synchronizing uncertainty vector* is an uncertainty vector with following modifications:

1.　All the groups (components) of states are combined into one group.
2.　All the duplicates of the states are deleted.

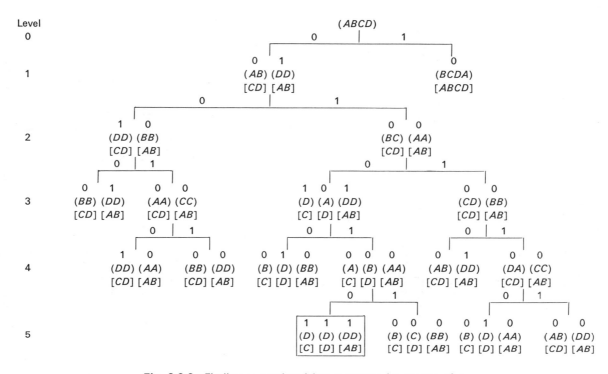

Fig. 8.3.2 Finding a synchronizing sequence by means of a homing tree.

In terms of synchronizing uncertainty vectors, rules H1' and H2' become:

RULE S1

A kth-level branch b becomes terminal whenever the synchronizing uncertainty vector associated with b is identical to a synchronizing uncertainty vector at a level lower than k. When two or more identical synchronizing uncertainty vectors occur at the kth level, all but one of them may be terminated.

RULE S2

A kth-level branch b becomes terminal whenever the synchronizing uncertainty vector is a singleton.

The tree constructed using synchronozing uncertainty vectors and terminated by rules S1 and S2 is called a *synchronizing tree*. The synchronozing tree of machine M_{14} is depicted in Fig. 8.3.3. Comparing this tree with that of Fig. 8.3.2, it is seen that this tree is much simpler, thus easy to construct.

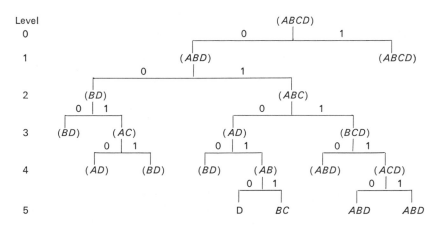

Fig. 8.3.3 Synchronizing tree for machine M_6.

The synchronozing tree for M_6 is drawn in Fig. 8.3.3, where it is seen that $\tilde{x}_s = 01010$ is a SS that will leave the machine in state D.

Mathematically, rules S1 and S2 can be conveniently expressed in terms of the next-state functions. Define

$$\delta(Q, x) = \{q_i \,|\, q_i = \delta(q_j, x) \text{ for all } q_j \in Q\}$$

Then $\delta(Q, \tilde{x}_s)$, where $\tilde{x}_s$ is a synchronizing sequence, can be thought of as an *absorption mapping* of the state set Q onto a single state q_k (i.e., $\delta : Q \xrightarrow{\tilde{x}_s} q_k$). The two termination rules can be expressed in terms of the state transition functions as: A kth-level branch b becomes terminal whenever one of the following two rules is satisfied:

RULE S1'

$\delta(Q, \tilde{x}) = \delta(Q, \tilde{x}')$, where $|x'| < |\tilde{x}| = k$.

RULE S2'

$\delta(Q, \tilde{x})$ is a singleton where $|\tilde{x}| = k$.

The symbol $|\quad|$ in $|\tilde{x}|$ denotes the length of sequence $\tilde{x}$.

Thus it is seen that different paths through the synchronizing tree correspond to the possible input sequences that might be applied to the machine. Each node then represents the image set of the transition map $\delta(Q, \tilde{x})$, where $\tilde{x}$ is the input sequence corresponding to the branches leading to that node. A tree path terminated by rule S2' corresponds to a minimal-length synchronozing sequence for machine M. There may be more than one minimal synchronizing sequence which takes the machine to different final states.

An upper bound on the height of the synchronizing tree can be obtained from the next theorem.

THEOREM 8.3.2

If an n-state machine M possesses a synchronizing sequence $\tilde{x}_s$, then its length is no greater than $2^n - (n + 1)$ input symbols.

Proof: In the first level there must exist an input symbol such that the n-state machine will merge to a state set containing at most $n - 1$ elements; otherwise, the experiment ends, by rule S2'. This input symbol will be the first element in the $\tilde{x}_s$ string. It is possible that the $(n - 1)$-element state set can be permuted into each of the $\binom{n}{n-1}$ distinct combinations before a state reduction must occur at which the state uncertainty contains $n - 2$ elements. Thus far $\tilde{x}_s$ could be of length $\leq \binom{n}{n-1}$. In a similar manner, the $(n - 2)$-element state set could be permuted $\binom{n}{n-2}$ times before terminating by rule S1'.

This process can continue until finally a single state must be reached in which none of the $\binom{n}{1}$ permutations would be allowed, because the tree would end, by rule S2'. Such a state is known as a binomial-coefficients sum:

$$2^n = \sum_{i=1}^{n} \binom{n}{i} = 1 + n + \sum_{i=2}^{n} \binom{n}{i}$$

Hence the longest possible tree path must end by rule S1' at $2^n - (n + 1)$ steps. ∎

For $n \geq 8$ a better bound is given in the following.

THEOREM 8.3.3

If a synchronozing sequence for an n-state machine M exists, then its length is at most $n(n - 1)^2/2$. This bound is derived by considering the pairwise absorption of states.

A comparison of the two bounds for $1 \leq n \leq 9$ is as follows:

n	$2^n - (n + 1)$	$\dfrac{n(n - 1)^2}{2}$
1	0	0
2	1	1
3	4	6
4	11	18
5	26	40
6	57	75
7	120	126
8	247	196
9	502	288

By the use of L'Hospital's rule, one can easily see that

$$\lim_{n \to \infty} \frac{2^n}{n^3} \longrightarrow \infty$$

and thus the exponential bound 2^n is not as tight as the cubic bound n^3 for $n \geq 8$.

There exist one-, two-, and three-state machines for which the bound $2^n - (n + 1)$ is attainable, as can be demonstrated by machines M_7, M_8 and M_9, whose state tables are given in Tables 8.3.3, 8.3.4, and 8.3.5, respectively.

TABLE 8.3.3 Machine M_7 (Trivial)

q \ x	Λ
A	A

$\tilde{x}_s = \Lambda \quad |\tilde{x}_s| = 0$

TABLE 8.3.4 Machine M_8

q \ x	0	1
A	A	B
B	B	B

$\tilde{x}_s = 1, \quad |\tilde{x}_s| = 1$

TABLE 8.3.5 Machine M_9

q \ x	0	1
A	A	B
B	C	C
C	C	A

$\tilde{x}_s = 0110, \quad |\tilde{x}_s| = 4$

The attempt to design a two-input, four-state machine for which the length of a minimal synchronizing sequence reaches the bound of 11 was unsuccessful. The four-

TABLE 8.3.6 Machine M_{10}

q \ x	0	1
A	A	B
B	C	C
C	C	D
D	D	A

state machine M_{10} given in Table 8.3.6 has the minimum-length state set of $\tilde{x}_s = 011101110$ whose length is $\tilde{x}_s = 9$. The synchronizing tree of M_{10} is given in Fig. 8.3.4.

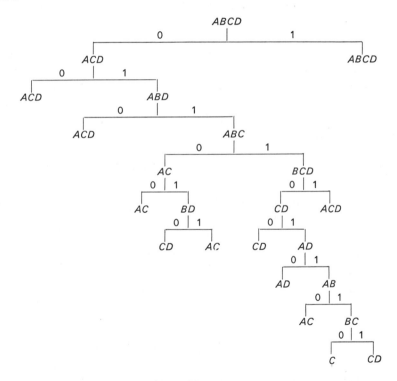

Fig. 8.3.4 Synchronizing tree for machine M_{10}.

Based upon the fact that the state-set mapping is fixed for a given machine, it is conjectured that a better upper bound exists for the length of a synchronizing sequence which is a function of the number of states n, and the number of inputs m, i.e.,

$$|\tilde{x}_s| \leq f(n, m)$$

The two previous bounds were both indepedent of m.

In order that a machine possess a synchronizing sequence, an easily detectable condition is required of its state table.

THEOREM 8.3.4

A necessary condition for a machine M to possess a synchronizing sequence is that there exists some input symbol x_k such that

$$\delta(q_i, x_k) = \delta(q_j, x_k), \qquad \text{where } i \neq j, 1 \leq i, j \leq n, n \geq 2$$

Proof: If the condition is not satisfied, it will not be possible to obtain the first-level reduction from n to $n - 1$ states in the state uncertainty, and hence the n states in the state uncertainty will simply permute. ■

8.3.3 Machine Identification by Means of State Identification

Just as state equivalence can be extended to machine equivalence, so can state identification. The problem of machine identification that we shall be concerned with is that given a machine and several machine descriptions, we are asked which of these descriptions (assume that one of them is the real description) describes the machine. We shall assume that the machines described by the descriptions do not have equivalent states up to isomorphism.

Suppose that we are given a machine and two machine descriptions M_{11} and M_{12} of Tables 8.3.7 and 8.3.8. Determine which of the two describes the machine.

TABLE 8.3.7 Machine M_{11}

q \ x	a	b	a	b
A	C	C	0	1
B	C	B	1	1
C	B	A	0	1

TABLE 8.3.8 Machine M_{12}

q' \ x	a	b	a	b
A'	B'	C'	0	1
B'	B'	B'	1	1
C'	C'	A'	1	1

The homing tree on the set $Q_T = \{A, B, C, A', B', C'\}$ of the states of the two machines is shown in Fig. 8.3.5. The tree is terminated at the fourth level. Since there are fourth-level UV's (one is sufficient) containing groups of $\{C\}\{B'\}\{C\}\{C\}\{B'\}\{C'\}$ and groups of $\{B\}\{B'\}\{B\}\{B\}\{B'\}\{B'\}$, both satisfy the condition of the termination rule, rule H2. The input sequences corresponding to these two terminations are

aaba   and   *abaa*

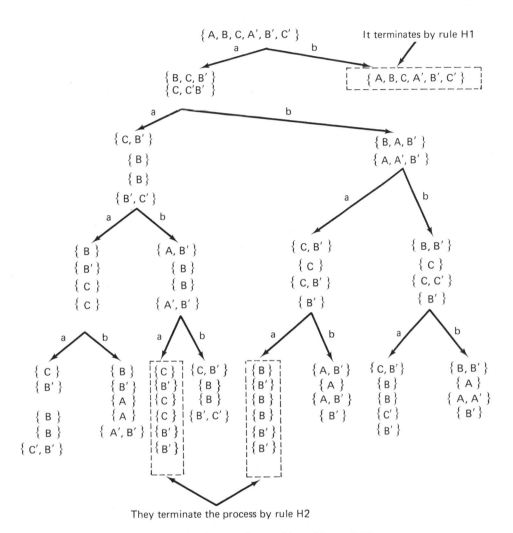

Fig. 8.3.5 Homing tree for machines M_{11} and M_{12}.

Both are the shortest length for an experiment concerned with identifying the machine. The output sequences for the sequence *abaa* and the final states that they identify are given in Table 8.3.9.

It is interesting to note that in this example, the homing sequences *aabа* and *abaa* found are not only the shortest homing sequences but also the shortest distinguishing sequences and the shortest homing sequences as well. Therefore, from this experiment, not only the given machine is identified to be M_{11} or M_{12}, the initial state and the specified final state of the machine are known also.

Next, we present an upper bound on the length of an experiment to distinguish

TABLE 8.3.9   Output Sequences Corresponding to the
Input Sequence abaa and the States They Identify

Machine	Initial State	Output Sequences to Input Sequence abaa				Final State
		a	b	a	a	
M_{11}	A	0	1	0	0	B
	B	1	1	0	0	B
	C	0	1	1	0	B
M_{12}	A'	0	1	1	1	B'
	B'	1	1	1	1	B'
	C'	1	1	0	1	B'

an (n, m, p)-machine M_a from all other (n, m, p)-machines via a simple state-identification experiment. First we prove the following lemma.

LEMMA 8.3.1

Let M_a and M_b be two (n, m, p)-machines. If state q_i of M_a can be distinguished from state q_j of M_b, there exists such an experiment of length $\leq 2n - 1$.

Proof: Define a $(2n, m, p)$-machine $M_a + M_b$ with states 1 to n defining the transitions of M_a and states $n + 1$ to $2n$ defining the transitions of M_b. By Corollary 5.2.2 we can distinguish the state of $M_a + M_b$ by an experiment of length $\leq 2n - 1$, and state q_i of M_a can be distinguished from state q_j of M_b by this experiment.   ∎

THEOREM 8.3.5

There exists a simple experiment of length $< n^2(np)^{nm}$ which distinguishes a strongly-connected minimal Mealy (n, m, p)-machine M_a from all other (n, m, p)-machines. If M_a is a Moore machine, the bound for the length of the experiment becomes $n^2[n^{mn}p^n]$.

Proof: First, suppose that M_a is a Mealy machine. Then there are $(np)^{nm}$ possible Mealy (n, m, p)-machines, since these include machines isomorphic to M_a obtained by permuting the state labels $(np)^{nm}$. Define the composite machine M for all of these as in Theorem 8.2.3 and the subsets of states of M obtained from each of the original machines. By Theorem 8.2.4 there exists an experiment of length $\leq n(n - 1)/2$ which will distinguish the state of each subset. Thus all but one of the members of each subset can be eliminated by an experiment of length $L_1 \leq [n(n - 1)/2](np)^{nm}$. Considering pairs of machines containing M_a and each of the remaining $(np)^{nm} - 1$ machines, it follows from Lemma 8.3.1 that a sequence of length $L_2 \leq (2n - 1)[(np)^{nm} - 1]$ is sufficient to determine whether machine C is in one of the states of M_a. Thus the entire experiment to distinguish M_a is of length L:

$$L = L_1 + L_2 < \left[\frac{n(n - 1)}{2} + (2n - L)\right](np)^{np}$$

$$\leq n^2(np)^{nm}$$

If M_a is a Moore machine, there are $n^{nm}p^n$ possible Moore (n, m, p)-machines. Therefore, $L < n^2(n^{nm}p^n)$. ∎

This upper bound is very high. In Section 8.4 we shall present another approach to the problem, machine identification via a checking experiment, by which the length of the experiment will be greatly shortened.

Exercise 8.3

1. Design a homing experiment for the machine shown in Fig. P8.3.1.

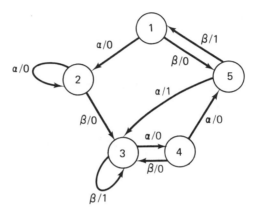

Fig. P8.3.1

2. Given the machine in Fig. P8.3.2 design a homing experiment that will pass it into state 2.

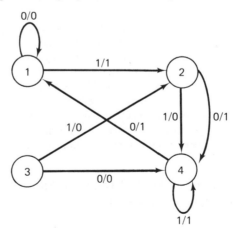

Fig. P8.3.2

3. Let M be a machine in which every pair of states is L-distinguishable. Show that the homing problem for M with n states can always be solved by applying an input sequence of length l, where $l \leq L(n - 1)$.

4. Using the result of problem 3, show that the homing problem for an n-state minimal machine M is always solvable by an input sequence of length l, where $l \leq (n - 1)^2$.

5. Show that every machine that can be realized by a circuit of the form in Fig. P8.3.5 has a synchronous sequence. Find such a sequence and specify its length.

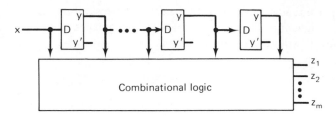

Fig. P8.3.5

6. Consider the five-state machine whose transition diagram is given in Fig. P8.3.6.

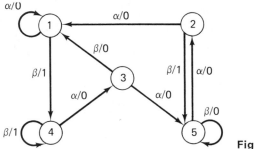

Fig. P8.3.6

(a) Is there a distinguishing sequence for the five-state machine with the initial uncertainty vector $U_0 = \{2, 3, 4, 5\}$?
(b) If so, find a minimal distinguishing sequence.
(c) Is the minimal distinguishing sequence unique?
(d) If not, what is another minimal distinguishing sequence?
(e) Is this machine minimal?
(f) Minimize this machine if it is not already minimal, and give a minimal homing sequence for the minimized machine.
(g) Show that this machine has a synchronizing sequence.
(h) Suppose that it is necessary to synchronize this machine to state 1 with a minimum number of input symbols. Find this sequence.

8.4 Design of Fault-Detection Experiment
for Diagnosable Machines

In the previous two sections we introduced methods for deriving minimum-length experiments to identify the initial state and the final state of a machine. In this section we shall apply them to the design of a fault-detection experiment or a checking experiment for a machine.

The problem of designing fault-detection experiments is actually a restricted problem of machine identification. An experimenter is supplied with a machine and its transition table. His task is to determine from terminal experiments whether the given table accurately describes the behavior of the machine, that is, if the actual machine is isomorphic to the one described by the transition table. If at any time during the execution of the experiment the machine being tested produces a response that is different from that of the correctly operating machine, the test can be stopped, and we would conclude that the circuit definitely is faulty. The subtlest part of checking experiment design is to make sure that only the fault-free circuit or its isomorphisms will respond correctly to the experiment.

In Section 8.1 a fault-detection method using the circuit-checking approach was presented that will detect any single or multiple permanent logical faults (s-a-0 and s-a-1 faults) in a synchronous sequential circuit. The method presented in this section will use the transition-checking approach, which will detect faults that cannot result from any s-a-0 or s-a-1 fault in the circuit, realizing the specified flow table and the types of faults that may occur in the circuit, which are not known. Hence this approach will detect a much larger class of faults.

The multiple experiment may be used for state and machine identification experiments, as described in the previous sections, where the machine was assumed to be fault-free. But it is almost never used for fault-detection experiments because of the cost of replicated equipment and the impossibility of assuring that the copies fail in the same manner and in the same state. For this reason we shall consider only the simple experiment. We shall also restrict our study to strongly connected, completely specified, and minimized machines.

DEFINITION 8.4.1

A *diagnosable sequential machine* is one that possesses one or more distinguishing sequences.

As mentioned before, the class of machines that naturally possess distinguishing sequences is quite small. A way of making a machine to have a distinguishing sequence is to embed a given machine in a larger machine which has a distinguishing sequence in such a manner that the given machine still maintains its original input/output relationship. Two basic techniques have been applied to constructing a machine that contains the given machine as a submachine. They are:

1. The addition of input terminals and associated logic.
2. The addition of output terminals and associated logic.

For example, consider machine M_{13}, shown in Table 8.4.1(a). Machine M_{13} does not possess any distinguishing sequence, so it is not distinguishable. An easy and direct way to make it diagnosable is to add one (in general, we add as many as needed) additional output terminal, denoted by z_t, which will be used only for diagnosing purpose. One such machine is given in Table 8.4.1(b). Denote this machine by M'_{13}. The original machine M_{13} is now embedded in M'_{13}. When the output terminal z_t is ignored during the regular operation, the two machines M_{13} and M'_{13} are identical in performance.

**TABLE 8.4.1 Example Illustrating Techniques Used to
Make a Nondiagnosable Machine Diagnosable**

x $\diagdown$ q	$x = 0$	$x = 1$
A	B, 1	B, 0
B	A, 0	A, 1
C	B, 1	D, 1
D	B, 1	C, 0

Next state, z

(a) M_{13}

x $\diagdown$ q	$x = 0$	$x = 1$
A	B, 10	B, 00
B	A, 01	A, 11
C	B, 10	D, 10
D	B, 11	C, 11

Next state, zz_t

z_t is used only for diagnosing purposes

(b) M'_{13}

xx_t $\diagdown$ q	$xx_t = 00$	$xx_t = 10$	$x_t = 1$ ($x = 0$ or 1)
A	B, 1	B, 0	11
B	A, 0	A, 1	10
C	B, 1	D, 1	01
D	B, 1	C, 0	00

Next state, z

x_t is used only for identifying the initial state

(c) M''_{13}

With the aid of this additional output terminal z_t, M'_{13} becomes diagnosable. Clearly, this machine now has at least one distinguishing sequence, which is $\tilde{x}_d = 1$.

As an alternative, we can add, instead of output terminals, input terminals to the given machine. An example of such a machine is shown in Table 8.4.1(c). Denote this machine by M''_{13}. Again, M''_{13} has at least one distinguishing sequence, $\tilde{x}_d = (x_1 = 1$ and $x_t = 1)$; hence M''_{13} is diagnosable. It should be mentioned again that the additional input terminal x_t has no effects whatsoever on the machine's normal performance and is used only for diagnosing purpose.

We shall assume that in addition to the above assumptions about the machine under investigation in this section, it is also diagnosable, and thus a simple fault-detection experiment can be designed.

A fault-detection experiment for a (n, m, p)-machine can be divided into three phases:

1. *Initialization phase.* During the initialization phase the machine to be tested must be brought to a specific state from which the second phase will begin. Generally, the initialization phase is *adaptive*, in that a homing sequence is applied to the machine, and observation of its response enables identification of the machine's final state. A transfer sequence is then applied to take the machine to the desired starting state. The initialization phase can be *preset* if the machine should possess a synchronizing sequence.

2. *State-identification phase.* During the state-identification phase, a distinguishing sequence is repeatedly applied to the machine to see whether it has n different responses to the distinguishing sequence, indicating n distinct states. This is a present experiment.

3. *Transition-verification phase.* During this phase the machine is made to go through all possible transitions. This is also a present experiment.

Although the three phase must be considered as logically distinct by the experimenter, it is important to point out that in actual practice it is not necessary at all to separate them. On the contrary, in the sequel it will be seen that in designing a checking experiment, we should try to mix up and overlap the subsequences used for state identification and transition verification whenever possible to shorten the total length of the checking experiment.

The procedure for designing a checking experiment for a machine is outlined as follows. The experiment must begin with a homing sequence to determine the machine's present state. A transfer sequence is applied to bring the machine to the starting state of the state-identification sequence. The distinguishing sequence is applied between appropriate transfer sequences to display the response of each state to the DS and to ascertain the DS successor of each state. *This successor will aid in the transition-verification phase of the experiment.* This is explained as follows. If our checking sequence contains the two subsequences

$$\tilde{x}_d \tilde{x}^t \tilde{x}_d \qquad \tilde{x}_d \tilde{x}^t \tilde{x}_a \tilde{x}_d$$
$$\text{and}$$
$$\tilde{z}_i \tilde{z}^t \tilde{z}_j \qquad \tilde{z}_i \tilde{z}^t \tilde{z}_a \tilde{z}_k$$

where $\tilde{z}_i$ is the response of M to $\tilde{x}_d$ in initial state q_i, we can conclude from the outputs that at the beginning of *both* subsequences, the machine was in state i. From the first subsequence we can conclude that the input sequence $\tilde{x}_d \tilde{x}^t$ takes the machine from state q_i to state q_j, and from the second subsequence we can conclude that the input sequence $\tilde{x}_d \tilde{x}^t \tilde{x}_a$ takes the machine from state q_i to state q_k. Thus it is evident that the next state and the output of the machine with present state q_j and input $\tilde{x}_a$ are

$$\delta(q_j, \tilde{x}_a) = q_k \quad \text{and} \quad \lambda(q_j, \tilde{x}_a) = \tilde{z}_a$$

The state-identification sequence must display n different responses to a *fixed* DS for an n-state machine. Finally, the machine is made to go through each of its $m \times n$ transitions with the state before and after each transition identified in terms of the DS used.

As mentioned at the beginning of this section, it is not necessary that the responses of each state appear during the state-identification phase, but the state-identification and transition-verification phases may be mixed up. In fact, different subsequences of the checking experiment can overlap each other. As an example of what is meant by overlapping subsequences, consider the two subsequences

$$\tilde{x}_d \tilde{x}^{t_1} \tilde{x}_a \tilde{x}_d \qquad \tilde{x}_d \tilde{x}^{t_2} \tilde{x}_d$$
$$\text{and}$$
$$\tilde{z}_i \tilde{z}^{t_1} \tilde{z}_a \tilde{z}_k \qquad \tilde{z}_k \tilde{z}^{t_2} \tilde{z}_j$$

They are overlapped in the single sequence

$$\tilde{x}_d \tilde{x}^{t_1} \tilde{x}_a \tilde{x}_d \tilde{x}^{t_2} \tilde{x}_d$$
$$\tilde{z}_i \tilde{z}^{t_1} \tilde{z}_a \tilde{z}_k \tilde{z}^{t_2} \tilde{z}_j$$

The design of a fault-detection experiment for a sequential machine possessing a distinguishing sequence is given below.

The design of a fault-detection experiment for a diagnosable machine is best illustrated by use of examples. The first example illustrates the fact that when a machine possesses a synchronizing sequence, the fault-detection experiment can be designed by an entirely present experiment. The example also illustrates that when a machine possesses a repeated symbol-distinguishing sequence and other distinguishing sequences, the fault-detection experiment using the repeated symbol-distinguishing sequence in general will be substantially shorter, because it allows overlapping portions of such a distinguishing sequence.

The second example presents the design of fault-detection experiment of a machine that does not possess a synchronizing sequence. By Theorem 8.3.1, we can always find a preset homing sequence for it with length $\leq (n-1)^2$. When a homing sequence is used to bring the machine to a specific state, an adaptive experiment must be employed.

It is easy to show that if a machine M displays n different responses to a *fixed sequence*, M must have at least n different states. But if n different responses resulted from the application of several different sequences to the machine, we *cannot* guarantee that the machine has at least n states. Our third example demonstrates the validity of this statement.

Example 8.4.1

Machine M_{14} (Table 8.4.2) is a minimal, strongly-connected machine that possesses two distinguishing sequences and one synchronizing sequence, thus making it possible to

TABLE 8.4.2   Machine M_{14}

x / q	0	1
A	$B, 0$	$A, 1$
B	$C, 0$	$A, 1$
C	$A, 1$	$B, 0$

construct an entirely preset checking experiment. If the machine did not possess a synchronizing sequence, it would still have a homing sequence, as guaranteed by Theorem 8.3.1. The DS and SS trees for machine M_{13} are given in Figs. 8.4.1 and 8.4.2, respectively. The two distinguishing sequences are 01 and 00. Notice that the latter is a repeated symbol-distinguishing sequence. The responses of the machine to these two distinguishing sequences are given in Table 8.4.3. The design of experiments based on the distinguishing sequence 01 is presented first.

The experiment will begin with the SS = 11 to bring the machine into the known initial state A. The DS = 01 will be used to determine that three different states do, in fact, exist. Application of 01 will give the output 01 leaving the machine in state A. Applying 01 again determines that the final state is indeed A when the output 01 is again produced.

The transfer sequence $\bar{x}_i/\bar{z}_i = 01/11$ should bring the machine to state B. Applying 01

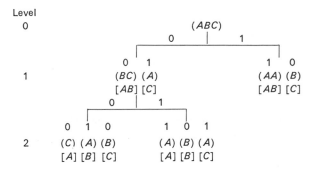

Fig. 8.4.1 Distinguishing tree for machine M_{13}.

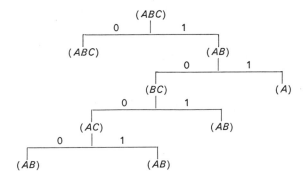

Fig. 8.4.2 Synchronizing tree for machine M_{14}.

TABLE 8.4.3 Responses of Machine M_{14} to the DS
$\tilde{x}_d = 01$ and the RSDS $\tilde{x}_d = 00$

Initial State	Response to 01	Final State		Initial State	Response to 00	Final State
A	01	A		A	00	C
B	00	B		B	01	A
C	11	A		C	10	B

(a) Response to the DS = 01	(b) Response to the RSDS = 00

should produce $\tilde{z} = 00$ with the machine stopped in state B. This end state is verified with another 01 input with the correct 00 output. Next transfer the machine from state B to C with the input/output sequence 0/0 and apply 01 to yield 11, demonstrating that the state after the transfer was C. Another input of 01, to give $\tilde{x} = 01$, verifies that the 01-successor of state C was A and leaves the machine in state A, as was already shown. The sequence and known states look as follows.

$\tilde{x}$	1 1 0 1 0 1 0 0 1 0 1 0 0 1 0 1
State	A A A B B B C A A
$\tilde{z}$	0 1 0 1 0 0 0 0 0 1 1 0 1

Although the states thus far identified should be labeled A', B', and C', the correct machine labels are used because the machines would have to be isomorphic when the experiment is successfully completed.

Observe that the overlapped subsequences

$$
\begin{array}{ccc}
01 & & 01 \\
A & A & A \\
01 & & 01
\end{array}
\Longrightarrow
\begin{array}{c}
\tilde{x}_d \tilde{x}^t \tilde{x}_d \\
\tilde{z}_A \tilde{z}^t \tilde{z}_A
\end{array}
$$

where $\tilde{x}^t = \Lambda$ and $\tilde{z}^t = \Lambda$ (Λ denotes the empty sequence), and

$$
\begin{array}{cccc}
01 & 0 & & 01 \\
A & A & & B \\
01 & 0 & & 00
\end{array}
\Longrightarrow
\begin{array}{c}
\tilde{x}_d \tilde{x}^t \tilde{x}_a \tilde{x}_d \\
\tilde{z}_A \tilde{z}^t \tilde{z}_a \tilde{z}_B
\end{array}
$$

where $\tilde{x}^t = \tilde{z}^t = \Lambda$ and $x_a/z_a = 0/0$, jointly imply that

$$
\delta(A, 0) = B \quad \text{and} \quad \lambda(A, 0) = 0
$$

Similarly, the overlapped subsequences

$$
\begin{array}{ccc}
01 & & 01 \\
B & B & B \\
00 & & 00
\end{array}
\quad \text{and} \quad
\begin{array}{cccc}
01 & 0 & & 01 \\
B & B & & C \\
00 & 0 & & 11
\end{array}
$$

jointly imply that

$$
\delta(B, 0) = C \quad \text{and} \quad \lambda(B, 0) = 0
$$

At this point we can fill in the following table:

q \ x	0	1
A	$B, 0$	
B	$C, 0$	
C		

The states with the circles can be filled in in the above state-identification sequence.

At this point the 0 transition from A to B and from B to C has been checked, so the states with the circles can be filled in.

$\tilde{x}$	1 1 0 1 0 1 0 0 1 0 1 0 0 1 0 1
State	$A\;\textcircled{B}\;A\;\textcircled{B}\;A\;\;B\;\textcircled{C}\;B\;\textcircled{C}\;B\;\;C\;\;\;\;\;A\;\textcircled{B}\;A$
$\tilde{z}$	0 1 0 1 0 0 0 0 0 0 1 1 0 1

An examination of the $\tilde{x}/\tilde{z}$ sequence thus far reveals that the transition from B to A under a 1 and from C to B under a 1 have also been checked. To check the transition from A under 1, apply the sequence 101 to get 101. State C under an input of 0 must still be checked. Apply the transfer sequence 00 with the output 00 to transfer from A to C. The resultant state will be C, as the transition A to B under 0 and B to C under 0 have already been checked. Now apply 001 to get 101. The complete checking sequence with all the missing states can now be filled in:

$\tilde{x}$	1 1 0 1 0 1 0 0 1 0 1 0 0 1 0 1 ┊ 1 0 1 0 0 0 0 1
State	$A\;\textcircled{B}\;A\;\textcircled{B}\;A\;\;B\;\textcircled{C}\;B\;\textcircled{C}\;B\;\;C\;\textcircled{A}\;A\;\textcircled{B}\;A$ ┊ $A\;\;B\;\;A\;\;B\;\;C\;\;A\;\;B\;\;A$
$\tilde{z}$	0 1 0 1 0 0 0 0 0 0 1 1 1 1 ┊ 1 0 1 0 0 1 0 1

state identification plus transition verification	transition verification

The total length of the checking experiment is 24 symbols. It is apparent that a checking experiment is not unique. A shorter experiment could have been designed using $DS = 00$ because the end states, which are the 00 successors, could have been chained together for the state-identification phase. This experiment would have begun with the sequence

$\tilde{x}$	1 1 0 0 0 0 0 0 0 0
State	$A\;\;\;\;\;C\;\;\;\;\;B\;\;\;\;\;A\;\;\;\;\;C$ + transition checks
$\tilde{z}$	0 0 1 0 0 1 0 0

Example 8.4.2

Consider the minimal and strongly-connected machine M_{15} described in Table 8.4.4. The successor tree of this machine is shown in Fig. 8.4.3, from which it can be seen that the tree has a homing sequence $\tilde{x}_h = 0$ but no synchronizing sequence. Therefore, we need to have an adaptive experiment to bring the machines to a specific state. Suppose that the preset part

TABLE 8.4.4 Machine M_{15}

q \ x	0	1
1	4, 1	5, 0
2	3, 0	5, 1
3	3, 0	2, 1
4	3, 0	3, 1
5	3, 0	1, 0

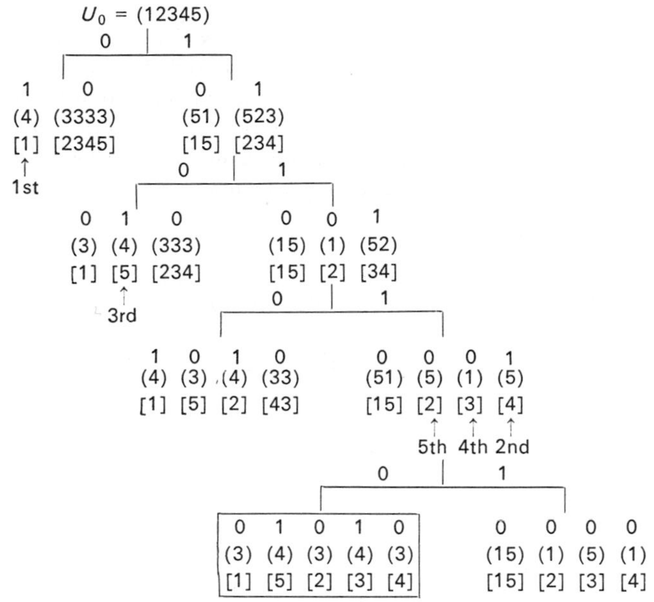

Fig. 8.4.3 Successor tree for machine M_{15}.

of the experiment is designed so that state 1 is the initial state, to which it is therefore neces-sary to transfer the machine. To this end we apply the homing sequence 0 and observe the following response:

  1. If the response is 1, the machine is in state 4. Apply the transfer sequence $T(4, 1)$ = 1111 to transfer the machine to state 1.

  2. If it is 0, the machine is in state 3. Apply the transfer sequence $T(3, 1) = 111$ to trans-fer the machine to state 1.

From Fig. 8.4.3 we find that the shortest distinguishing sequence this machine has is $\tilde{x}_d = 1110$. The five states of the machine can be identified by a preset experiment as follows:

x	1110 1110 1110 1 1110 11 1110 1111
State	1 3 4 3 2 3 5 4
z	0000 1101 1110 1 1000 11 0001 1110

Note that the machine displays five responses to the sequence $\tilde{x}_d = 1110$ (Table 8.4.5). Thus machine M_{15} has at least five states. A preset transition-verification experiment can be designed in the same manner as was done in the previous example; it is left to the reader.

TABLE 8.4.5

Initial State	Response to 1110	Final State
1	0000	3
2	1000	3
3	1101	4
4	1110	3
5	0001	4

Example 8.4.3

Consider the machine of Example 8.4.2. Suppose that after it is brought to the initial state 3, instead of applying the state-identification sequence shown above, we apply three sequences 0, 10, and 111 to the machine to identify its states (Fig. 8.4.3). The states are identified in the order given by the arrows under the initial uncertainty states of the tree in Fig. 8.4.3. Assume that the machine has been subjected to a homing sequence and transferred into starting state 1. State 1 can be identified by applying a 0 and observing a 1 output, because only state 1 produces that output. The machines final state should be 4. State 4 will be identified by applying 111 to M_{15}. The output of $\tilde{z} = 111$ can only be generated by state 4, as shown in the tree. The successor of 4 is state 5, which will be identified next with an input of 10 to produce $\tilde{z} = 01$. The resulting state is 4, which has already been checked. Apply a transfer sequence $x_1/z_1 = 0/0$ to bring M_{15} to state 3. State 3 is uniquely identified by applying 111, with the resulting output 110. The resultant state is 1, which has already been checked, so apply a transfer sequence of 001 to take the machine to state 2. The resulting output sequence is 101. State 2 is the fifth state to be identified, as shown in the distinguishing tree. It is partitioned into a singleton uncertainty with the input/output sequence 111/100.

The state identification phase can be summarized as follows:

x	0	111	10	0	111	001	111
State	①④	⑤	④T③	① T ②	⑤		
$\tilde{z}$	1	111	01	0	110	101	100

The states are circled to differentiate them from the input/output symbol set. The T is used to point out the location of transfer sequences. The above reasoning, used to deduce that M_{15} has five states, is perfectly valid *provided that the machine being tested is operating correctly*. If it is not known whether the machine is fault-free, it is *not* sufficient to conclude that the machine has five states.

The next machine to be considered is M_{16}, which is represented by Table 8.4.6. It has only four stated, but it will be seen to produce the same "correct" output sequence

TABLE 8.4.6   Machine M_{16}

x q	0	1
①③ A	D, 1	B, 1
② B	A, 1	C, 1
⑤ C	D, 1	C, 0
④ D	A, 0	A, 1

for the given input $\tilde{x} = 0111100111001111$ as machine M_{15}. Machines M_{15} and M_{16} thus serve as examples to demonstrate that the same input sequence must be used in the state-identification phase.

The correspondences between the five states of M_{15} and the four states of M_{16} are shown to the left of the first column of M_{16}. They are states 1 and 3 $\approx A$, 2 $\approx B$, 5 $\approx C$, 4 $\approx D$. If machine M_{16} is assumed to start in state A, the following states are visited, producing the same output as machine M_{15}:

Input	0	111	10	0	111	001	111
State	A	D	C	D T A	C T B	C
Output	1	111	01	0	110	101	100

Thus the input test sequence applied to M_{16} with the above expected resulting response is seen to be necessary, *but not sufficient*, to guarantee that M_{16} has five distinct states. In fact, M_{16} has only four states. It must be concluded from this example that a diagnostician has to carefully design a test sequence so that only the correct machine will produce the expected test response.

Exercise 8.4

1. Consider the machine whose transition table is given in Table P8.4.1.
   (a) Show that this machine possesses a distinguishing sequence and thus is diagnosable.
   (b) Show that this machine has three synchronizing sequences: *cab*, *cac*, and *cdc*.
   (c) Design a checking experiment for this machine.

TABLE P8.4.1

x \ q	a	b	c	d
A	A, 0	B, 1	A, 0	D, 1
B	C, 1	A, 0	B, 0	A, 0
C	D, 1	B, 1	A, 1	B, 0
D	B, 0	C, 1	A, 1	C, 1

2. (a) Modify the machine in Table P8.4.2 by adding additional inputs and additional output terminals so that it will have a distinguishing sequence.

TABLE P8.4.2

x \ q	0	1
1	2, 1	1, 1
2	5, 0	1, 1
3	1, 0	5, 1
4	2, 1	4, 1
5	5, 0	4, 1

(b) Find the fault-detection experiments for the diagnosable machines obtained in (a).

3. (a) Describe an experiment that will allow you in general to determine the actual numbering of the input and output terminals of the black box of Fig. P5.2.12. The experiment may be adaptive; that is, you may apply an input sequence and, depending on the response observed, apply another sequence, and so on. Only one copy of the machine is available. It is in an unknown state and cannot be reset to a known or initial state at any time. That is, the only control you have over the machine is at the input terminals. (Do not give specific sequences in this part but describe what your method of attack would be in general for a problem of this type.)

(b) Assuming that the black box is actually in state 3 (as described by Table P5.2.12), use your method to devise an experiment to determine the terminal numbering. Describe all sequences applied and responses.

Bibliographical Remarks

The earliest work on the circuit-testing approach to the fault-detection problem was done by Moore (reference 17). The derivation of optimum test sequences for sequential machines was published by Poage and McCluskey (reference 18). The design of initial and final state-identification experiments was proposed by Ginsburg (reference 8) and Gill (references 3 and 7). Reference 5 contains many original papers on automata theory. The transition-checking approach to the fault detection problem was first introduced by Hennie (reference 10). This subject is extensively treated in references 1–4. The problem of better organizing a checking experiment for the class of reduced, strongly connected sequential machines with distinguishing sequences was attacked by Kime (reference 14), Gönenc (reference 9), and Kohavi and Kohavi (reference 16). For the class of machine that does not have a distinguishing sequence, Kim (reference 13) showed how the addition of a single input column to a machine will enable that machine to be strongly connected and have a distinguishing sequence. Kohavi and Lavalle (reference 15) employed the technique of adding output terminals to construct a definitely diagnosable machine. The problem of designing checking experiments for machines that did not have distinguishing sequences without the necessity of modifying the given circuit to make it possess one was attacked by Hsieh (references 11 and 12).

References

Books

1. H. Y. CHANG, E. G. MANNING, and G. METZE, *Fault Diagnosis of Digital Systems*, Wiley, New York, 1970.

2. A. D. FRIEDMAN and P. R. MENON, *Fault Detection in Digital Circuits*, Prentice-Hall, Englewood Cliffs, N.J., 1971.

3. A. GILL, *Introduction to the Theory of Finite-State Machines*, McGraw-Hill, New York, 1962.

4. A. KOHAVI, *Switching and Finite Automata Theory*, McGraw-Hill, New York, 1970.

5. C. E. SHANNON and J. MCCARTHY (eds.), *Automata Studies*, Princeton University Press, Princeton, N.J., 1956.

Papers

6. H. Y. CHANG, R. C. DORR, and R. A. ELLIOTT, "Logic Simulation and Fault Analysis of a Self-Checking Switching Processor," *Proc. 1972 IEEE–ACM Design Automation Workshop*, June 1972, pp. 128–135.

7. A. GILL, "State Identification Experiments in Finite Automata," *Inform. Control*, Vol. 4, 1961, pp. 132–154.

8. S. GINSBURG, "On the Length of the Smallest Uniform Experiment Which Distinguishes the Terminal States of a Machine," *J. ACM*, Vol. 5, 1958, pp. 266–280.

9. G. GÖNENC, "A Method for the Design of Fault Detection Experiments," *IEEE Trans. Computers*, Vol. C-19, June 1970, pp. 551–558.

10. F. C. HENNIE, "Fault Detecting Experiments for Sequential Circuits," *Proc. Eighth Annual Symposium on Switching Circuit Theory and Logical Design*, 1964, pp. 95–110.

11. E. F. HSIEH, "Optimal Checking Experiments for Sequential Machines," Ph.D. Dissertation, Department of Electrical Engineering, Columbia University, 1969; University Microfilms, Ann Arbor, Mich.

12. E. P. HSIEH, "Checking Experiments for Sequential Machines," *IEEE Trans. Computers*, Vol. C-20, Oct. 1971, pp. 1152–1166.

13. C. R. KIME, "A Failure Detection Method for Sequential Circuits," Department of Electrical Engineering, University of Iowa, *Tech. Rept. 66–13*, 1966.

14. C. R. KIME, "An Organization for Checking Experiments on Sequential Circuits," *IEEE Trans. Electron. Computers*, Vol. EC-15, Feb. 1966, pp. 113–115.

15. Z. KOHAVI and P. Lavallee, "Design of Sequential Machines with Fault Detection Capabilities," *IEEE Trans. Electron. Computers*, Vol. EC-16, Aug. 1967, pp. 473–484.

16. I. KOHAVI and Z. KOHAVI, "Variable Length Distinguishing Sequences and Their Application to the Design of Fault-Detection Experiments," *IEEE Trans. Computers*, Vol. C-17, Aug. 1968, pp. 792–795.

17. E. F. MOORE, "Gedanken-Experiments on Sequential Machines," in *Automata Studies*, Princeton University Press, Princeton, N.J., 1956, pp. 129–153.

18. J. F. POAGE and E. J. McCLUSKEY, "Derivation of Optimum Test Sequences for Sequential Machines," *Proc. Fifth Ann. Symp. Switching Theory and Logical Design*, Princeton University, Nov. 11–13, 1964, pp. 121–132.

9

Logic Design Using Microprocessors

In this chapter, a new method of implementing logic circuits is introduced. Instead of using hard-wired logic (i.e., gates, flip-flops, counters, etc.), this method uses a microprocessor containing a microgrammed read-only memory (ROM). The microprocessor replaces hard-wired logic by storing program sequences in the ROM rather than implementing these sequences with gates, flip-flops, counters, and so on. It can be said that 8 to 16 bits of ROM are the logical equivalent of a single gate. Assuming that on the average an IC contains 10 gates, the number of IC's that are replaced by a single ROM is as follows:

ROM Memory-Sized Bits	Gates Replaced	IC's Replaced
2,048	128–256	13–25
4,096	256–512	25–50
8,192	512–1024	50–100
16,384	1024–2048	100–200

Many engineers have begun to adopt this method, for the following reasons:

1. Manufacturing costs of products can be significantly reduced.
2. Products can get to the market faster, providing a company with the opportunity to increase product sales and market share.
3. Product capability is enhanced, allowing manufacturers to provide customers with better products, which can frequently command a higher price in the marketplace.
4. Development costs and time are reduced.
5. Product reliability is increased, which leads to a corresponding reduction in both service and warranty costs.

This chapter begins with a comparison of three information-handling systems: the conventional computer system, the man–calculator system, and the micropro-grammed computer system, by which the general structure and the operation of a microcomputer is described. Then, a typical microcomputer system for implementing logic using the Inter 4040 microprocessor is presented. Both the hardware and software of this system are discussed. Several examples of the use of this system to implement logic circuits are given.

9.1 Functions of a Computer

This section introduces certain basic computer concepts. It provides background information and definitions that will be useful in later sections.

A typical digital computer consists of the following three parts:

1. A central processor unit (CPU).
2. A memory.
3. Input/output (I/O) ports.

The CPU contains two major parts: an arithmetic/logic unit (which is often simply referred to as the ALU) and a control unit. A block diagram showing these basic elements and the organization of a digital computer is shown in Fig. 9.1.1. A brief description of such a typical computer system is given next.

The memory is composed of storage space for a large number of "words," with each storage space identified by a unique "address." The word stored at a given address might be either computational data or a machine directive (such as add, or read from memory). Two temporary store registers, each capable of containing one word, complete our memory. These registers are designated as "memory address register" (MAR) and "memory data register" (MDR). The MAR contains the binary representation of the address where information is to be read from memory or written (stored) into memory, and the MDR contains the data being exchanged with memory.

Based on its usage, a memory may be referred to as a *program memory* or a *data memory*. The program memory serves primarily as a place to store *instructions*, the coded pieces of data that direct the activities of the CPU. A group of logically related

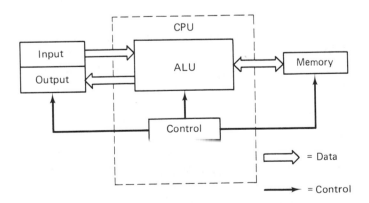

Fig. 9.1.1 Basic elements of a digital computer.

instructions stored in memory is referred to as a *program*. The CPU "reads" each instruction from memory in a logically determinate sequence, and uses it to initiate processing actions. If the program structure is coherent and logical, processing produces intelligible and useful results. The data memory is used to store the data to be manipulated. The CPU can access any data stored in memory, but often the memory is not large enough to store the entire data bank required for a particular application. The program can be resolved by providing the computer with one or more *input ports*. The CPU can address these ports and input the data contained there. The addition of input ports enables the computer to receive information from external equipment (such as a paper tape reader) at high rates of speed and in large volumes.

Almost any computer requires one or more *output ports* that permit the CPU to communicate the result of its processing to the outside word. The output may go to a display, for use by a human operator, to a peripheral device that produces "hard copy," such as a line printer, to a peripheral storage device, such as a magnetic tape unit, or the output may constitute process control signals that direct the operations of another system, such as an automated assembly line. Like input ports, output ports are addressable. The input and output ports together permit the processor to interact with the outside world. Data exchanges take place with external or peripheral devices via an I/O register.

The CPU unifies the system. It controls the functions performed by the other components. The CPU must be able to fetch instructions from memory, decode their binary contents, and execute them. It must also be able to reference memory and I/O ports as necessary in the execution of instructions. In addition, the CPU should be able to recognize and respond to certain external control signals, such as INTERRUPT and STOP requests. The functional units within a CPU that enable it to perform these functions are described below.

Registers   Registers are temporary storage units within the CPU. Some registers, such as the program counter and instruction register, have dedicated uses. Other registers, such as the accumulator, are used for more general purposes.

Accumulator   The accumulator usually stores one of the operands to be manipulated by the ALU. A typical instruction might direct the ALU to add (or to perform similar logical operations upon, e.g., OR) the contents of some other register to the contents of the accumulator and store the result in the accumulator. In general, the accumulator is both a source (operand) and destination (result) register. Often a CPU will include a number of additional general-purpose registers that can be used to store operands or intermediate "scratch-pad" data.

Program Counter (Jumps, Subroutines, and the Stack)   The instructions that make up a program are stored in the system's memory. The central processor examines the contents of the memory to determine what action is appropriate. This means that the processor must know which location contains the next instruction. Each of the locations in memory is numbered, to distinguish it from all other locations in memory. The number that identifies a memory location is called its *address*. The processor maintains a counter that contains the address of the next program instruction. This register is called the *program counter*. The processor updates the program counter by adding "1" to the counter each time it fetches an instruction, so that the program counter is always current. The programmer therefore stores his instructions in numerically adjacent addresses, so that the lower addresses contain the first instructions to be executed and the higher addresses contain later instructions. The only time the programmer may violate this sequential rule is when the last instruction in one block of memory is a *jump* instruction to another block of memory. A jump instruction contains the address of the instruction that is to follow it. The next instruction may be stored in any memory location, as long as the programmed jump specifies the correct address. During the execution of a jump instruction, the processor replaces the contents of its program counter with the address embodied in the jump. Thus the logical continuity of the program is maintained.

A special kind of program jump occurs when the stored program accesses or "branches" to a subroutine. In this kind of jump, the processor is logically required to "remember" the contents of the program counter at the time that the jump occurs. This enables the processor to resume execution of the main program when it is finished with the last instruction of the subroutine. A *subroutine* is a program within a program. Usually it is a general-purpose set of instructions that must be executed repeatedly in the course of a main program. Routines that calculate the square, the sine, or the logarithm of a program variable are good examples of the functions often written as subroutines. Other examples might be programs designed for inputting or outputting data to a particular peripheral device.

The processor has a special way of handling subroutines, in order to ensure an orderly return to the main program. When the processor receives a jump to subroutine instruction, it increments the program counter and stores the counter's contents in a register memory area known as the *stack*. The stack thus saves the address of the instruction to be executed after the subroutine is completed. Then the processor stores the address specified in the subroutine jump in its program counter. The next instruction fetched will therefore be the first step of the subroutine. The last instruction in any subroutine is a branch back. Such an instruction need specify no address. When the processor fetches a branch-back instruction, it simply replaces the current contents of the program counter with the address on the top of the stack. This causes the processor to resume execution of the program at the point immediately following the original branch.

Subroutines are often *nested;* that is, one subroutine will sometimes call a second subroutine. The second may call a third, and so on. This is perfectly acceptable, as long as the processor has enough capacity to store the necessary return addresses, and the logical provision for doing so. In other words, the maximum depth of nesting is determined by the depth of the stack itself. If the stack has space for storing three return addresses, then three levels of subroutines may be accommodated. Processors have different ways of maintaining stacks. Most have facilities for the storage of return addresses built into the processor itself. The integral stack is usually more efficient, since fewer steps are involved in the execution of a call or a return.

Instruction Register and Decoder   Every computer has a *word length* that is characteristic of that machine. A computer's word length is usually determined by the size of its internal storage elements and interconnecting paths (referred to as *buses*); for example, a computer whose registers and buses can store and transfer 4 bits and is referred to as a 4-bit parallel processor. The characteristic 8-bit field is sometimes referred to as a *byte*; a 4-bit field can be referred to as a *nibble*. Each operation that the processor can perform is identified by a unique binary number known as an *instruction code* or *operation code* (OP code). An 8-bit word used as an instruction code can distinguish among 256 alternative actions, more than adequate for most processors.

The processor fetches an instruction in two distinct operations. In the first, it transmits the address in its program counter to the memory. In the second, the memory returns the addressed byte to the processor. The CPU stores this instruction byte in a register known as the *instruction register*, and uses it to direct activities during the remainder of the instruction execution. The mechanism by which the processor translates an instruction code into specific processing actions requires more elaboration than we can afford here. The concept, however, will be intuitively clear to any experienced logic designer. The 8 bits stored in the instruction register can be decoded and used to selectively activate one of a number of output lines, in this case up to 256 lines. Each line represents a set of activities associated with execution of a particular instruction code. The enabled line can be combined coincidentally with selected timing pulses, to develop electrical signals that can then be used to initiate specific actions. This translation of code into action is performed by the *instruction decoder* and by the associated control circuitry.

An 8-bit field is more than sufficient, in most cases, to specify a particular processing action. There are times, however, when execution of the instruction code requires more information than 8 bits can convey. One example of this is when the instruction references a memory location. The basic instruction code identifies the operation to be performed, but it cannot also specify the object address. In a case like this, a two-word instruction must be used. Successive instruction bytes are stored in sequentially adjacent memory locations, and the processor performs two fetches in succession to obtain the full instruction. The first byte retrieved from memory is placed in the processor's instruction register, and the second byte is placed in temporary storage, as appropriate. When the entire instruction is fetched, the processor can proceed to the execution phase.

Address Register(s)   A CPU may use a register or register pair to temporarily store the address of a memory location that is to be accessed for data. If the address register is *programmable* (i.e., if there are instructions that allow the programmer to alter the contents of the register), the program can "build" an address in the address register prior to executing a *memory reference* instruction (i.e., an instruction that reads data from memory, writes data to memory, or operates on data stored in memory).

Arithmetic/Logic Unit (ALU)   By way of analogy, the ALU may be thought of as a super adding machine with its keys commanded automatically by the control signals developed in the instruction decoder and the control circuitry. This is essentially how the first stored-program digital computer was conceived. The ALU naturally bears little resemblance to a desk-top adder. The major difference is that the ALU calculates by creating an electrical analogy rather than by mechanical analogy. Another important difference is that the ALU uses binary techniques, rather than decimal methods, for representing and manipulating numbers. In principle, however, it is convenient to think of the ALU as an electronically controlled calculator.

The ALU must contain an adder that is capable of combining the contents of two registers in accordance with the logic of binary arithmetic. This provision permits the processor to perform arithmetic manipulations on the data it obtains from memory and from its other inputs. Using only the basic adder, a capable programmer can write routines that will subtract, multiply, and divide, giving the machine complete arithmetic capabilities. In practice, however, most ALU's provide other built-in functions, including hardware subtraction, Boolean logic operations, and shift capabilities.

The ALU contains *flag bits* which register certain conditions that arise in the course of arithmetic manipulations. Flags typically include *carry* and *zero*. It is possible to program jumps that are conditionally dependent on the status of one or more flags. For example, the program may be designed to jump to a special routine, if the carry bit is set following an addition instruction. The presence of a carry generally indicates an overflow in the accumulator and sometimes calls for special processing actions.

Control Circuitry   The control circuitry is the primary functional unit within a CPU. Using clock inputs, the control circuitry maintains the proper sequence of events required for any processing task. After an instruction is fetched and decoded, the control circuitry issues the appropriate signals (to units both internal and external to the CPU) for initiating the proper processing action. Often the control circuitry will be capable of responding to external signals, such as an interrupt request. An *interrupt* request will cause the control circuitry to temporarily interrupt main program execution, jump to a special routine to service the interrupting device, then automatically return to the main program. A diagram of a simplified CPU plus a memory is depicted in Fig. 9.1.2. Certain operations are basic to almost any computer. A sound understanding of these basic operations is a necessary prerequisite to examining the specific operations of a particular computer.

Timing   The activities of the central processor are cyclical. The processor fetches an instruction, performs the operations required, fetches the next instruction, and so on, Such an orderly sequence of events requires timing, and the CPU therefore contains a free-running oscillator clock which furnishes the reference for all processor

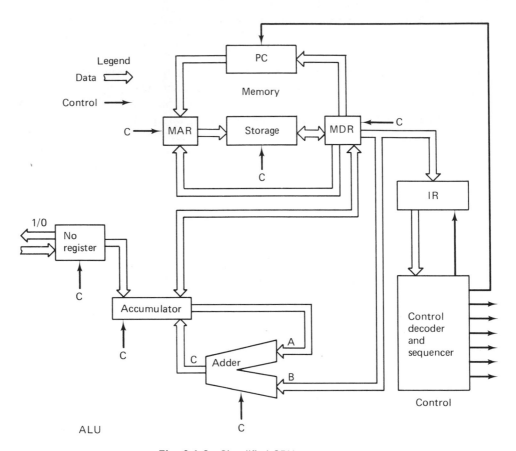

Fig. 9.1.2 Simplified CPU + memory.

actions. The combined fetch and execution of a single instruction is referred to as a *machine cycle*. The portion of a cycle identified with a clearly defined activity is called a *phase*. And the interval between pulses of the timing oscillator is referred to as a *clock period*. As a general rule, one or more clock periods are necessary to the completion of a phase, and there are several phases in a cycle.

Instruction Fetch   The first phase(s) of any machine cycle will be dedicated to fetching the next instruction. The CPU issues a read operation code and the contents of the program counter are sent to program memory, which responds by returning the next instruction word. The first word of the instruction is placed in the instruction register. If the instruction consists of more than one word, additional cycles are required to fetch each word of the instruction. When the entire instruction is present in the CPU, the program counter is incremented (in preparation for the next instruction fetch) and the instruction is decoded. The operation specified in the instruction will be executed in the remaining phases of the machine cycle. The instruction may call for a memory read or write, an input or output, and/or an internal CPU operation, such as a register-to-register transfer or an add-registers operation.

Memory Read   The instruction fetched may then call for data to be read from data memory into the CPU. The CPU issues a read operation code and sends the proper memory address; memory responds by returning the requested word. The data received are placed in the accumulator (not the instruction register).

Memory Write   A program memory write operation is similar to a read except for the direction of data flow. The CPU issues a write operation code, sends the proper memory address, then sends the data word to be written into the addressed memory location.

Input/Output   Input and output operations are similar to memory read and write operations with the exception that a peripheral I/O port is addressed instead of a memory location. The CPU issues the appropriate input or output command, sends the proper device address, and either receives the data being input or sends the data to be output. Data can be input/output in either parallel or serial form. All data within a digital computer are represented in binary-coded form. A binary data word consists of a group of bits; each bit is either a 1 or a 0. *Parallel I/O* consists of transferring all bits in the word at the same time, one bit per line. *Serial I/O* consists of transferring one bit at a time on a single line. Naturally serial I/O is much slower, but it requires considerably less hardware than does parallel I/O.

Interrupts   *Interrupt provisions* are included on many central processors as a means of improving the processor's efficiency. Consider the case of a computer that is processing a large volume of data, portions of which are to be output to a printer. The CPU can output a byte of data within a single machine cycle but it may take the printer the equivalent of many machine cycles to actually print the character specified by the data byte. The CPU will have to remain idle waiting until the printer can accept the next data byte. If an interrupt capability is implemented on the computer, the CPU can output a data byte, then return to data processing. When the printer is ready to accept the next data byte, it can request an interrupt. When the CPU acknowledges the interrupt, it suspends main program execution and automatically branches to a routine that will output the next data byte. After the byte is output, the CPU continues with main program execution. Note that this is, in principle, quite similar to a subroutine call, except that the jump is initiated externally rather than by the program. More complex interrupt structures are possible, in which several interrupting devices share the same processor. Interruptive processing is an important feature that enables maximum utilization of a procesor's capacity.

Having introduced the structure and operations of a digital computer, we now turn to another information-processing system, the man–calculator system shown in Fig. 9.1.3. This system has all the "functional parts" that a digital computer has. The man's fingers represent the input; his eyes, coupled with the calculator's output, represent the output; his brain is the control element, while the calculator electronics function as the ALU; and his brain serves also as the memory.

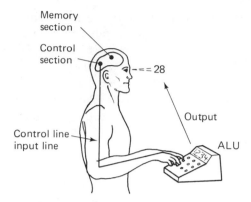

Fig. 9.1.3 Man-calculator system.

Let us examine the sequence of events which occur when our man–calculator solves the problem $2 + 3 = ?$:

1. Brain accesses first number of be added, a 2.
2. Brain orders hand to depress 2 key.
3. Brain identifies addition operation.
4. Brain orders hand to depress + key.
5. Brain accesses second number to be added, a 3.
6. Brain determines that all necessary information has been provided and signals the ALU to complete computation by ordering hand to depress = key.
7. ALU (calculator) makes computation.
8. ALU displays result on readout.
9. Eyes signal brain; brain recognizes this number as the result of the specified calculation.
10. Brain stores result, 5, in a location which it appropriately identifies to itself to facilitate later recall.

Now let us see how a digital computer solves this problem without having any human intervention. The task involved in solving this problem may be described as follows: "Read in a number from the I/O. Store it in memory location 10. Read in another number from the I/O. Store it in memory location 11. Add the two numbers together. Store the result in the memory location 12, print out the output, and halt."

A "program" has been written to execute this task and is stored in consecutive memory locations beginning at 0. The program, written in an artificial symbolic "language," is shown in Tabel 9.1.1. Suppose that our machine is a 8-bit word machine. Each instruction is described by one word. Half of it (4 bits) is used for describing the operation code (OP code) and the other half (4 bits) for describing the address* (ADDR). The above sample problem for example involves five distinct operations:

*Address would be longer in actual design; so would the OP code.

TABLE 9.1.1 Program

Memory location	Instruction (contents)
0	Input to accumulator.
1	Store accumulator at location 10.
2	Input to accumulator.
3	Store accumulator at location 11.
4	Add accumulator at location 10.
5	Store accumulator at location 12.
6	Display accumulator to I/O.
7	Halt.

CLEAR, INPUT, STORE, ADD, and HALT. They are usually abbreviated by three-letter words and coded in binary codes. For example:

Operation	OP code	Meaning
INP	0100	Input a number from I/O device 1 (0001) into the accumulator.
STR	0110	Store the number in the accumulator to a memory location.
ADD	0101	Add the number in a memory location to the number in the accumulator.
DSP	0111	Display the number in the accumulator in I/O device 1.
HLT	0001	Stop operation.

With these OP code's one can rewrite the program in Table 9.1.1 in a binary coded form as shown in Table 9.1.2.

TABLE 9.1.2 Program in Table 9.1.1 Described by Binary Code

	Instruction (contents)	
Memory location	OP code (OPR)	ADDR (OPA)
0000	0100	0001
0001	0110	1010
0010	0100	0001
0011	0110	1011
0100	0101	1010
0101	0110	1100
0110	0111	0001
0111	0001	—

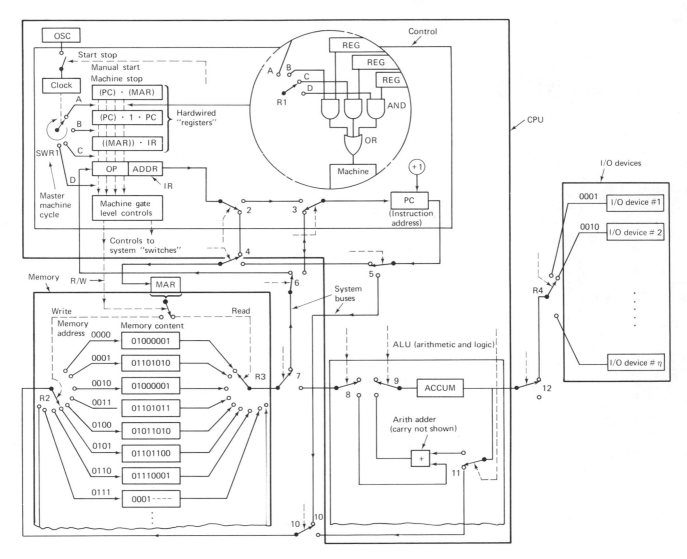

Fig. 9.1.4 Digital computer system architecture for illustrating its operation.

Figure 9.1.4 shows a system architecture diagram of a digital computer, with just enough details of registers and the control of interconnecting paths to indicate how it might operate. In this diagram, mechanical switches are shown for the purpose of illustration, but they are meant to present logic gates. It is seen that the program in

TABLE 9.1.3 Program To Make Computer Do 2 + 3 = 5

Master cycle no.	SW R1	Shorthand description	Comments
		Start	Somebody pushes start button and connects clock.
1	A	(PC) $\longrightarrow$ MAR	PC loads MAR to point at mem. loc. 0000.
	B	(PC) + 1 $\longrightarrow$ PC	PC becomes 0001.
	C	((MAR)) $\longrightarrow$ IR	First instruction (loc. 0000) goes to IR.
	D	(I/O 0001) $\longrightarrow$ ACC	Instruction says: input a number from I/O device 1 into acc.
2	A	(PC) $\longrightarrow$ MAR	MAR points at mem. loc. 0001.
	B	(PC) + 1 $\longrightarrow$ PC	PC becomes 0010.
	C	((MAR)) $\longrightarrow$ IR	Second instruction (loc. 0001) goes to IR.
	D	(ACC) $\longrightarrow$ (1010)	Instruction says: store the contents of acc. in mem. loc. 1010.
3	A	(PC) $\longrightarrow$ MAR	MAR points at mem. loc. 0010.
	B	(PC) + 1 $\longrightarrow$ PC	PC becomes 0011.
	C	((MAR)) $\longrightarrow$ IR	Third instruction (loc. 0010) goes to IR.
	D	(I/O 0001) $\longrightarrow$ ACC	Instruction says: input a number from I/O device 1 into acc.
4	A	(PC) $\longrightarrow$ MAR	MAR points at mem. loc. 0011.
	B	(PC) + 1 $\longrightarrow$ PC	PC becomes 0100.
	C	((MAR)) $\longrightarrow$ IR	Fourth instruction (loc. 0011) goes to IR.
	D	(ACC) $\longrightarrow$ (1011)	Instruction says: store the contents of acc. in mem. loc. 1011.
5	A	(PC) $\longrightarrow$ MAR	MAR points at mem. loc. 0100.
	B	(PC) + 1 $\longrightarrow$ PC	PC becomes 0101.
	C	((MAR)) $\longrightarrow$ IR	Fifth instruction (loc. 0100) goes to IR.
	D	(1010) + (ACC) $\longrightarrow$ ACC	Instruction says: add the contents of mem. loc. 1010 to the contents of the acc.
6	A	(PC) $\longrightarrow$ MAR	MAR points at mem. loc. 0101.
	B	(PC) + 1 $\longrightarrow$ PC	PC becomes 0110.
	C	((MAR)) $\longrightarrow$ IR	Sixth instruction (loc. 0101) goes to IR.
	D	(ACC) $\longrightarrow$ (1100)	Instruction says: store the contents of acc. to mem. loc. 1100.
7	A	(PC) $\longrightarrow$ MAR	MAR points at mem. loc. 0110.
	B	(PC) + 1 $\longrightarrow$ PC	PC becomes 0111.
	C	((MAR)) $\longrightarrow$ IR	Seventh instruction (loc. 0110) goes to IR.
	D	(ACC) $\longrightarrow$ (I/O 0001)	Instruction says: display the contents of acc. to I/O device 1.
8	A	(PC) $\longrightarrow$ MAR	MAR points at mem. loc. 0111.
	B	(PC) + 1 $\longrightarrow$ PC	PC becomes 1000.
	C	((MAR)) $\longrightarrow$ IR	Eighth instruction (loc. 0111) goes to IR.
	D	HALT	Instruction says: disconnect clock.

TIME

TABLE 9.1.4 Register Contents

PC	Acc.	MAR	MDR	I/O Reg.	IR	Memory (read, R; write, W)	State	Operation
0000	?	?	?	?	?	?	Fetch	
0000	?	0000	(0000)	?	?	R	Fetch	Input the number 2 (= 0010_2) into the accumulator from
0001	?	0000	(0000)	0010	(0000)	?	Fetch	I/O device 0001.
0001	0010	0000	(0000)	0010	(0000)	?	Execute	
0001	0010	0001	(0001)	?	(0000)	R	Fetch	Store the number 2 in the accumulator in memory location
0010	0010	0001	(0001)	?	(0001)	?	Fetch	10 (= 1010_2).
0010	0010	1010	0010	0011	(0001)	W	Execute	
0010	0010	0010	(0010)	0011	(0001)	R	Fetch	Input the number 3 (= 0011_2) into the accumulator from
0011	0010	0010	(0010)	0011	(0010)	?	Fetch	I/O device 0001.
0011	0011	0010	(0010)	?	(0010)	?	Execute	
0011	0011	0011	(0011)	?	(0010)	R	Fetch	Store the number 3 in the accumulator in memory location
0100	0011	0011	(0011)	?	(0011)	?	Fetch	11 (= 1011_2).
0100	0011	1011	0011	?	(0011)	W	Execute	
0100	0011	0100	(0100)	?	(0011)	R	Fetch	
0101	0011	0100	(0100)	?	(0100)	?	Fetch	Add the number in memory location 10 to the number in the
0101	0011	1010	(1010)	?	(0100)	R	Execute	acculator.
0101	0101	1010	(1010)	?	(0100)	?	Execute	

Note: Addition has occurred; content of MDR (2) is at adder input B; content of accumulator (3) is at adder input A; result at C replaces previous accumulator contents.

PC	Acc.	MAR	MDR	I/O Reg.	IR	Memory (read, R; write, W)	State	Operation
0101	0101	0101	(0101)	?	(0100)	R	Fetch	Store the contents of the accumulator in memory location
0110	0101	0101	(0101)	?	(0101)	?	Fetch	12 (= 1100_2).
0110	0101	1101	(0101)	?	(0101)	W	Execute	
0110	0101	0101	0110	?	(0100)	R	Fetch	Display the contents of the accumulator on I/O device
0111	0101	0101	0110	?	(0101)	?	Fetch	0001.
0111	0101	0101	(0110)	0101	(0101)	W	Execute	
0111	0101	0110	(0111)	?	(0101)	R	Fetch	
1000	0101	0110	(0111)	?	(0110)	?	Fetch	Disconnect the clock.
1000		HALT					Execute	

TIME ↓

binary form is stored in the memory in locations 0000 to 1000. To implement this program, the PC is set to 0000, which is the address of the first instruction of the program. When the start button is pushed, the clock pulses (dashed lines) cause the master machine cycles switch to rotate clockwise continuously. This triggers the three hard-wired step—A, B, C—that "fetch" instructions into the instruction register so that upon the fourth step, D, the machine will do what you, the user, have programmed it to do. In the first mastercycle step, A, the contents of PC goes to MAR. In the second step, B, the PC is incremented by 1, so it will be ready to point at the next memory location on the next go-round of the master cycle. On the third master-cycle step, C, the MAR causes the instruction to be loaded into the IR ready for execution. The fourth step, D, is the instruction-execution step. The upper or MSB's (most-significant bits) of the instruction in the IR are the OP code and their states are gated to the machine gate-level controls below. There they will control the routing and manipulations of the data. Which data are moved or manipulated is determined by the lower or LSB's (least-significant bits) of the instruction in the IR. This "address" part of the instruction will be routed to the MAR before the OP code is executed, so that the MAR will point to memory locations for data just as it did during master-cycle step C for instructions.

Now let us examine the sequence of events that occur when this automatic computer solves the problem $2 + 3 = ?$ The detailed steps that the program makes the computer do in the computation $2 + 3$ is given in Table 9.1.3. In this table, parentheses around a register means "the contents of that register." For example, (PC) means "the contents of the program counter." Arrows between the expressions indicate movements of data between the registers. Thus (PC) arrow (MAR) means "the contents of the program counter are moved to the memory-address register." The double parentheses around MAR in the third step of the master cycle means "the contents of the memory address that the contents of MAR is pointing at." The sequence of events shown by register contents is given in Table 9.1.4. The operation is complete! No human intervention is required—all operations are automatic.

All computers (processors, CPU's, etc.) operate in a similar manner, regardless of their size or intended purpose. It must be emphasized that many variations are possible within this basic architectural framework. More common variations include highly sophisticated I/O structures (some of which have direct and/or autonomous communication with memory), multiple accumulators for programming flexibility, index registers that allow a memory address to be modified by a computed value, multilevel interrupt capability, and on and on.

9.2 Microprocessors and Microcomputers

Since their inception, digital computers have continuously become more efficient, expanding into new applications with each major technological improment. A time chart of computer technology is shown in Fig. 9.2.1. The advent of minicomputers enabled the inclusion of digital computers as a permanent part of various process

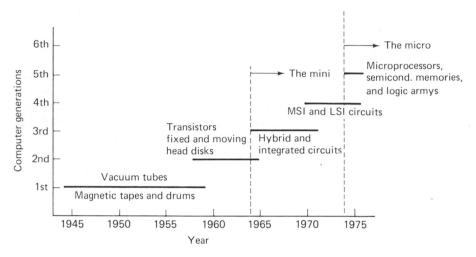

Fig. 9.2.1 Time chart of computer technology.

control systems. Unfortunately, the size and cost of minicomputers in dedicated applications has limited their use. Another approach has been the use of custom-built systems made up of random logic. However, the huge expense and development time involved in the design and debugging of these systems has restricted their use to large-volume applications where the development costs could be spread over a large number of machines.

One of the most exciting architectural concepts to gain popularity in the past few years is that of microprogrammed control. A microprogrammed computer differs from the classical machine presented in the previous section in the control unit implementation. The classical machine has for its control unit an assemblage of logic elements (gates, counters, flip-flops, etc.) interconnected to realize certain combinatorial and sequential Boolean equations. On the other hand, a microprogrammed machine utilizes the concept of a "*computer within a computer.*" That is, the control unit has all the functional elements which comprise a classical computer, including ROM. This microprogrammed control unit is called a *microprocessor*. The purpose of this "inner computer," which is not apparent to the user, is to execute the user's program instructions by executing a series of its own microinstructions, thereby controlling data transfers and all functions from computed results. Herein lies the key distinction: The control signals, hence the very "personality" of the computer, are controlled by computed results. The implication is immediately obvious—by simple changing the stored microprogram which generates the control signals, we may alter the entire complexion of the computer. By altering a few words stored in the ROM, we can cause our computer to behave in an entirely new fashion—to execute a completely different set of instructions, to emulate other computers, to tailor itself to a specified application. It is this capability for "custom tailoring" that allows such a machine to be optimized for a given usage. By so extracting the utmost measure of efficiency, a microprogram-controlled machine is more reliable, less costly, and easier to adapt to any given situation.

The two types of control unit, hard-wired and microprogrammed, are depicted in Fig. 9.2.2. The instructions stored in the ROM are called *microinstructions*, and the program stored in the ROM is called a *microprogram*. In general, the execution of a microinstruction is a two-step process: (1) place a word from the ROM in IR and (2) issue a control pulse and/or update MAR. If no asynchronous transfers are specified, two clock periods are required per microinstruction. By contrast, in a machine with a hard-wired control sequencer, synchronous transfers require only one clock period, and branches within the control sequence required no clock periods. Thus it is safe to say that a machine with a hard-wired sequencer can always be made to operate faster than an apparently identical machine which employs microprogramming. It should, however, be noted that a small-size microprogrammed computer can perform complex computations without requiring the use of a high-level programming language compiler (such as FORTRAN compiler) and any scientific subroutines. For example, suppose we would like to compute $\sin x + \cosh y$ on a microcomputer, in which both $\sin x$ and $\cosh y$ have been microprogrammed† in a ROM. Then the computational complexity of $\sin x + \cosh y$ would be the same as that of $2 + 3$, as described in Table 9.1.3.

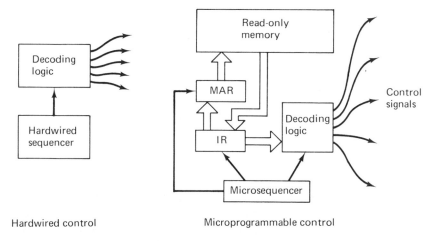

Hardwired control   Microprogrammable control

Fig. 9.2.2 Two types of control units.

The following are some of the characteristics of a microprocessor:

1.   It consists of from 1 to 30 integrated circuits.

2.   It contains: ALU, registers, control, register stack or RAM, data busses, and ROM with microprograms.

3.   It handles shorter words than other computers, usually from 4 to as many as 16 bits.

---

†The technique for storing functions such as sine and hyperbolic cosine functions, in a ROM is first to describe them in the form of a truth table, and then to program the truth table into the ROM (problem 11).

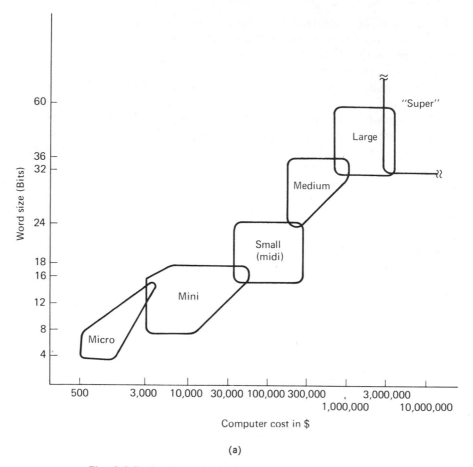

(a)

Fig. 9.2.3   (a) Cost of micro- to super computer systems.

These facts mean that in nearly every type of design, with any complexity at all, microprocessors have the potential for drastically reducing component count and shortening design time. In fact, many people consider microprocessors to represent the long-awaited next generation of digital building blocks, and that microprocessors will provide the best single approach to the system-level digital integrated circuit.

Any computer that uses a microprocessor as its main processor will be considered a *microcomputer*. A microcomputer is generally used for a dedicated task as part of a system. Because of its low cost (see Fig. 9.2.3(a)), the microcomputer beings to replace the minicomputer used in many dedicated systems, such as data-acquisition systems, process control systems, data-processing systems, and remote-terminal control systems (see Fig. 9.2.3). There is a mushrooming variety of microprocessor applications. A few of them are:

    Special (high-level) language computers
    Microprocessor-based instruments

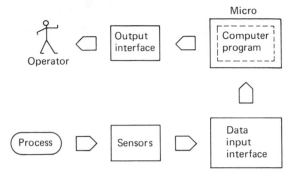

Data acquisition system, limited or no feedback, man-machine interface

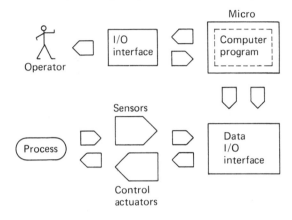

Process control system, large amount of feedback, man-machine interface

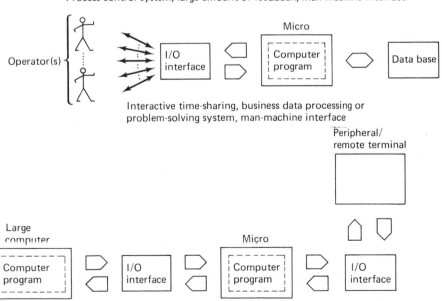

Interactive time-sharing, business data processing or
problem-solving system, man-machine interface

Peripheral or remote terminal control system, machine-machine interface

(b)

Fig. 9.2.3 (b) microcomputer as a system element.

513

TABLE 9.2.1 Available

Manufacturer	Device	Technology	Bit slice	CPU† One chip	CPU† Chip set	Memory family RAM	Memory family ROM PROM
Burroughs	Mini-D	PMOS	No	✓		On Chip	On Chip
Fairchild	F-8	NMOS	No		✓	✓	✓
Semiconductor	PPS-25	PMOS	No		✓	✓	✓
General Instrument	CP-1600	NMOS	No		✓	✓	✓
Intel	4004	PMOS	No		✓	✓	✓
	4040	PMOS	No		✓	✓	✓
	8008	PMOS	No		✓	✓	✓
	8080	NMOS	No		✓	✓	✓
	3000	Schottky Bipolar	Yes		✓		
Intersil	6100	CMOS	No		✓	✓	✓
Monolithic Memories	5701–6701	Schottky Bipolar	Yes		✓	✓	✓
Mostek	5065	PMOS	No		✓	✓	✓
Motorola	M6800	NMOS	No		✓	✓	✓
National	IMP4	PMOS	Yes		✓	✓	✓
	IMP8	PMOS	Yes		✓	✓	✓
	IMP16	PMOS	Yes		✓	✓	✓
	PACE	PMOS	No		✓	✓	✓
RCA	COSMAC	CMOS	No		✓	✓	✓
Raytheon	RP16	ECL Bipolar	Yes		✓	✓	✓
Rockwell	PPS-4	PMOS	No		✓	✓	✓
	PPS-8	PMOS	No		✓	✓	✓
Signetics	2650	NMOS	No		✓	✓	✓
Texas Instruments	TMS 1000	PMOS	No	✓		On Chip	On Chip
	SBP 0400	I²L Bipolar	Yes		✓	✓	✓
Toshiba Transistor Works	TLCS 12	NMOS	No		✓	✓	✓

†CPU includes ALU, controller, and control memory.

Intelligent CRT terminals
Point-of-sale terminals
Computer peripheral controllers
On-board automobile control
Control of automation and continuous processes
Home entertainment and games
Replacing hard-wired logic by a microprocessor

Here, however, we shall be only concerned with this last application, logic design using microprocessors.

For reference convenience, lists of some available microprocessors and micro-computers are given in Tables 9.2.1 and 9.2.2.

Microprocessors

Data word size	Memory capacity	Micropro- grammed	Instructions	Stack	Interrupts	DIP pins	Voltage
8	256	Yes	—	—	No	16	−12, +5
4/8	64K	No	101	64	Multilevel	40	−12, +5
4	12K	No	32	4	No	16/18/ 24/40	−10, +5
16	64K	Yes	68	8	Multilevel	40	−12, +5
4	4K	No	46	3	No	16/24	−10, +5
4	4K	No	60	7	No	24	−10, +5
8	16K	No	48	7	Multilevel	18	−10, +5
8	64K	No	78	RAM	Multilevel	40	−5, +5, −9
2-bit slice	64K	Yes	512	RAM	Multilevel	28	+5
12	4K	No	40+	—	1	40	+5
4-bit slice	64K	No	36	16	1	40	+5
8	32K	No	51+	RAM	3	40	+5, −12
16	64K	Yes	72	RAM	1	40	+5
4	64K	Yes	43+	16	2	24	−12, +5
8	64K	Yes	43+	16	2	24	−12, +5
16	64K	Yes	43+	16	2	24	−12, +5
8/16	64K	Yes	337	10	Multilevel	40	−12, +5
8	64K	No	37	RAM	Multilevel	28/40	+12
4-bit slice	64K	No	32	RAM	2	48	+5
4	12K	No	50	3	No	42	−17
8	12K	No	90+	2	Multilevel	42	−17
8	32K	No	72+	8	Multilevel	40	+5
4	8K	No	43	1	No	28/40	+15
4-bit slice	64K	Yes	512	RAM	Multilevel	40	+4
12	4K	Yes	18	RAM	Multilevel	16/24/ 26/42	−5, +5

9.3 Intel 4040 Microprocessor

Among many microprocessors now available in the market, the two smallest ones, the Intel 4040 and the Rockwell PPS–4, have a special appeal to the logic designer. Since the two systems are similar, we shall use one of them, say Intel 4040, to illustrate the principle and practice of logic design using microprocessors.

To understand the operation of a microprocessor, we need to know its architecture and instruction set. The architecture of a microprocessor deals with defining the nature of the blocks that make up the processor and the way they will work together. The instruction set yields insight into how these basic blocks work together by showing the manufacturer's intended use.

TABLE 9.2.2 Available

Manufacturer and model	Microprocessor			Memory min/max	Instruction size/ data path (bits)
	Manufacturer	Device No.	Chip set		
Applied Computing Technology					
CBC-4N	Intel	4004	Yes	ROM 256–1024 RAM 80–1280	8/4
PPS-4MP	Rockwell	PPS-4	Yes	ROM 4K RAM 1K-4K, PRAM 1K-4K	8/4
Control Logic					
L-Series	Intel	8008	Yes	256–16K Any mix of RAM-ROM	8/8
M-Series	Intel	8080	Yes	256–64K Any mix of RAM-ROM	8/8
Digital Equipment Corp.					
MPS	Intel	8008	Yes	1K–16K Any mix of RAM-ROM	8/8
Intel					
Immy-42	Intel	4004	Yes	ROM 256–1K, RAM 320	8/4
Immy-43	Intel	4040	Yes	ROM 256–1K, RAM 320	8/4
Imm 8	Intel	8008	Yes	256–16K	8/8
Imm 8	Intel	8080	Yes	256–65K	8/8
Intellec 4 MOD-40	Intel	4040	Yes	ROM 256–4K RAM 80–1280	4/8
Intellec 8 MOD-80	Intel	8080	Yes	256–64 Any mix of RAM-ROM	8/8
Intellec MDS 800	Intel	8080	Yes	ROM 256–12K RAM 256–64K	8/8
Motorola					
EXORciser	Motorola	M6800	Yes	ROM 1K-16K RAM 4K-64K	8/8
National Semiconductor					
8 C 200	National Semiconductor	IMP 8	Yes	256 RAM, 0–2048 ROM	8/8
16 C 200	National Semiconductor	IMP 16	Yes	625 RAM, 0–512 ROM	16/16
16 L 304	National Semiconductor	IMP 16	Yes	4K-65K, 0–2048 ROM	16/16
PCS					
Micropac 80	Intel	8080	Yes	5K-64K Min 4K ROM, 1K ROM	8/8
RZE Microcomputers					
Micral 1000	Intel	8008–8080	Yes	256–64K Any mix of RAM-ROM	8/8
Warner & Swasey					
Comstar 4	Intel	4004	Yes	RAM 320–2560×4 ROM 1K-4K×8	8/4

9.3.1 Intel 4040 Architecture

The Intel 4040 is a single-chip 4-bit parallel MOS central processor. It is an enhanced version of the Intel 4004 and retains all the functional capacity of that

Microcomputers

No. of boards (size)	Configuration		Software					
	Console	Mini-based	Assembler	Cross assembler	High-level language	OS	Peripheral interfaces
1 (8×8)	No	No	No	No	No	No	No
Packaged	Yes	No	Yes	No	No	Yes	TTY
3 (2.5×4)	Yes	No	Yes	Yes	No	Yes	TTY, serial, parallel, digital
4 (2.5×4)	Yes	No	Yes	Yes	No	Yes	TTY, serial, parallel, digital
2 (8.5×11)	Yes	No	No	Yes	No	Yes	All DEC peripherals
1 (8×6.18)	Yes	No	Yes	Yes	No	Yes	TTY
1 (8×6.18)	Yes	No	Yes	Yes	No	Yes	TTY
2 (8×6.18)	Yes	No	Yes	Yes	Yes	Yes	TTY
2 (8×6.18)	Yes	No	Yes	Yes	Yes	Yes	TTY
2 (8×6.18)	Yes	No	Yes	Yes	No	Yes	TTY, CRT, floppy printer, card reader
2 (8×6.18)	Yes	No	Yes	Yes	Yes	Yes	TTY, CRT, floppy printer, card reader
2 (8×6.18)	Yes	No	Yes	Yes	Yes	Yes	TTY, CRT, floppy printer, card reader
2 (7×17.50)	Yes	No	Yes	Yes	Yes	Yes	TTY, CRT, floppy printer, card reader
1 (8.5×11)	Yes	No	No	Yes	No	Yes	TTY, card reader
1 (8.5×11)	Yes	No	Yes	Yes	No	Yes	TTY, card reader
Packaged	Yes	No	Yes	Yes	No	Yes	TTY, DMA, card reader
Packaged	Yes	No	Yes	No	No	Yes	TTY, parallel, serial, digital
Packaged	Yes	No	Yes	Yes	Yes	Yes	TTY, floppy, printer, MT, PT, card reader, 6MB disk
3 (6.5×4.5)	Yes	No	Yes	Yes	Yes	Yes	TTY, DAC, ADC, floppy, card reader, tape, printer

device. Here, we shall only present the architecture and instruction set of Intel 4040 (those of Intel 4004 can be found in reference 11). From now on, we shall abbreviate Intel 4040 as 4040.

Chip Pin Description The 4040 is packaged in a 24-pin chip. The pin configuration is shown in Fig. 9.3.1. A brief functional description of each pin is given in the following pin description.

Pin No.	Desig- nation	Description of function
1–4	D_0–D_3	Bidirectional data bus. All address and data communication between the processor and the RAM and ROM chips is handled by way of these four lines.
5	STPA	STOP ACKNOWLEDGE output. This signal acknowledges that the processor has entered the stop mode.
6	STP	STOP input signal. A logic 1 level at this input causes the processor to enter the STOP mode.
7	INT	INTERRUPT input signal. A logic 1 level at this input causes the processor to enter the INTERRUPT mode.
8	INTA	INTERRUPT ACKNOWLEDGE output. This signal acknowledges receipt of an INTERRUPT command and prevents additional INTERRUPTs from entering the processor. INTERRUPT ACKNOWLEDGE remains active until cleared by the new BRANCH BACK and SRC (BBS) instruction.
9	V_{SS}	Circuit GND potential—most positive supply voltage.
10, 11	ϕ_1–ϕ_2	Nonoverlapping clock signals, which determine processor timing.
12	RESET	RESET input. A 1 level applied to this pin clears all flag and status flip-flops and forces the program counter to 0. To completely clear all of the address and index registers, RESET must be applied to 96 clock cycles (12 machine cycles).
13	TEST	TEST input. The logical state of this input can be examined with the JCN instruction.
14	V_{DD}	Main supply voltage to the processor. Value must be V_{SS} − 15.0 V ± 5%.
15	V_{DD2}	Supply voltage for output buffers. May be varied, depending on interface conditions.
16	SYNC	SYNC output. Synchronization signal generated by the processor and sent to ROM and RAM chips. Indicates beginning of instruction cycle.
17–20	CM-RAM$_0$– CM-RAM$_3$	CM-RAM outputs. These outputs act as bank select signals for the 4002 RAM chips in the system.
21	V_{DD1}	Supply voltage for timing circuit. Value must be V_{SS} − 15.0 V ± 5%. Allows low-power standby operation. Only SYNC will be generated when this pin is the only active V_{DD}.
22, 23	CM-ROM$_0$– CM-ROM$_1$	CM-ROM outputs. These outputs act as bank select signals for the ROM chips in the system.
24	CY	CARRY output buffer. The state of the CY flip-flop is presented at this output and is updated at X_1.

Basic Circuit Timing   The basic system timing for the 4040 is shown in Fig. 9.3.2. Two nonoverlapping clock signals, ϕ_1 and ϕ_2, are used to define the basic timing. The start of an instruction cycle is indicated by the SYNC signal, which is generated by the

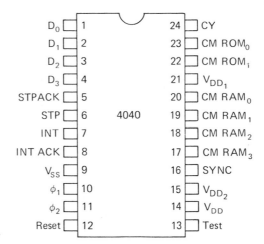

Fig. 9.3.1 4040 pin configuration.

processor and sent to the various ROM and RAM or peripheral chips in the system. An instruction cycle consists of the following operations:

1. The 12-bit content of the program counter is sent out to the ROM chips in three 4-bit groups during A_1, A_2, and A_3.

2. The 8-bit instruction or data from the addressed ROM location is received by the processor at M_1 and M_2, at which time the instruction is decoded.

3. Instruction execution occurs during X_1, X_2, and X_3. Data or address information may be sent to output ports or RAM chips, data may be received from input ports or RAM chips, or data may be operated on within the processor.

Observe that the generation of the CM–RAM signals at M_1 will occur for all single-cycle instructions and for the first cycle of all double-cycle instructions. This feature allows external logic to distinguish between instruction information and address or data at M_1 and M_2 time.

Basic Description of Major Circuit Blocks Figure 9.3.3 is a block diagram of the 4040, indicating the major circuit blocks and their interconnections. The following major functional blocks are contained in the 4040:

1. Address register stack and address incrementer.
2. Index register array.
3. SRC register.
4. Four-bit adder/accumulator.
5. Instruction register/decoder and control logic.
6. Hardware interrupt and stop control.

A brief functional description of each of these major elements is given below.

1. *Address Register Stack and Address Incrementer* The address register is a dynamic RAM array of 8×12 bits operating as a push-down stack. One level of the stack is used to store the effective address, leaving seven levels available for subroutine calls and interrupt processing. The stack address is provided by the effective address counter to the decoder. The contents of the selected address register are stored in the

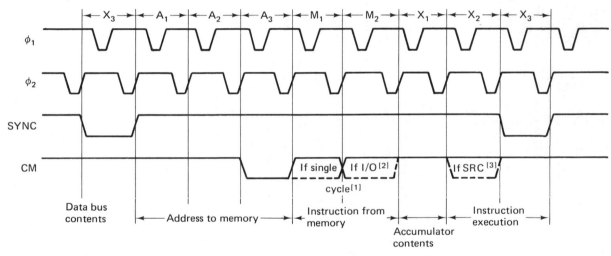

Notes:

1. CM ROM, RAM signals will be present at M_1 for any single cycle instruction or for the first cycle of a double cycle instruction.

2. CM ROM, RAM signals will be present at M_2 for any of the sixteen I/O group instructions.

3. CM ROM, RAM signals will be present at X_2 during execution of an SRC instruction.

Fig. 9.3.2 4040 basic timing diagram. (Courtesy of Intel Corporation.)

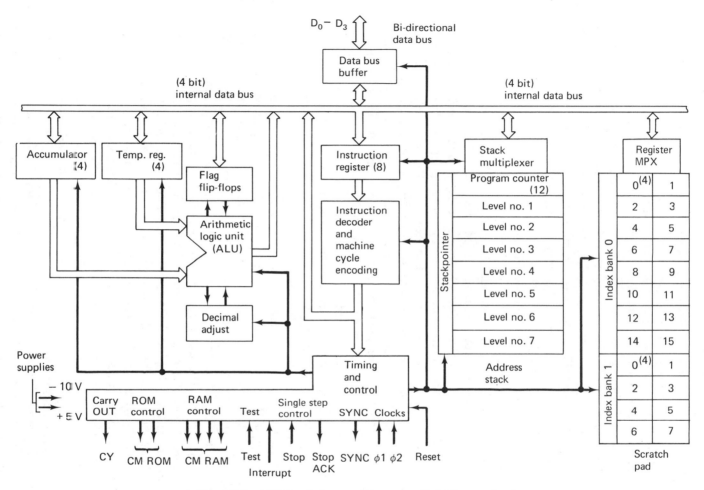

Fig. 9.3.3 4040 block diagram. (Courtesy of Intel Corporation.)

address buffer and multiplexed to the internal bus during A_1, A_2, and A_3 in 4-bit nibbles. The contents of the address buffer are incremented by a 4-bit carry-look-ahead circuit following the outputting of each 4-bit nibble. The incremented value is transferred back to the address buffer and written back into the selected address register.

2. *Index Register Array*   The index register is a dynamic RAM array of 12×8 bits organized as three banks of 4×8 bits. Two of the banks have identical address locations and so must be individually selected with the SB0, SB1 instructions. The third bank is always available for use. Two modes of operation are possible for the index register array. In one mode the array provides 24 directly addressable 4-bit storage locations for intermediate computation or control purposes. In the second mode the array provides 12 pairs of register locations for addressing RAM, ROM, and I/O ports or by storing data fetched from ROM.

Index register addressing is provided by the internal bus to normal read/write operations and by a refresh counter for refresh operation. The address are multiplexed to the array decoder. The content of the selected register is stored in a temporary register and multiplexed to the internal bus. During write operations the internal bus contents are transferred to the temporary register and then to the selected index register.

3. *SRC REGISTER*   The SRC register is an 8-bit dynamic latch which stores the contents of the designated index register pair during the execution of the SRC instruction. This 8-bit value is sent to the ROM and RAM chips as an address for any succeeding I/O instruction (see detailed description in Definition of Instruction Set section). The SRC register is used to hold this value in the case that an interrupt should occur, thus allowing the value to be automatically restored when a return from interrupt is made. The SRC register is not loadable during an interrupt routine. However, an SRC may be executed during an interrupt routine.

4. *Four-Bit Adder/Accumulator*   The 4-bit adder is of the ripple-through carry type. One term of the addition comes from the "adder buffer" register, which communicates with the internal bus on one side and can transfer to the adder data, or $\overline{\text{data}}$. The other term of the addition comes from the accumulator and carry flip-flop. Both data and $\overline{\text{data}}$ can be transferred. The output of the adder is transferred to the accumulator and carry FF. The accumulator is provided with a shifter to implement shift-right and shift-left instructions. The accumulator communicates also with the command register, with special ROMs, with the condition logic, and with the internal bus. The command register holds a 3-bit code used for CM–RAM line switching and 1 bit for CM–ROM switching. The special ROM's perform a code conversion for DAA (decimal-adjust accumulator) and KBP (process keyboard) instructions. The special ROM's communicate with the internal bus. The condition logic senses ADD = 0 and ACC = 0 conditions, the state of the carry FF, and the state of an external signal (TEST) to implement JCN (jump-on condition) and ISZ (increment index register, skip if zero) instructions.

5. *Instruction Register/Decoder and Control Logic*   The instruction register is loaded with the content of the internal bus at M_1 and M_2 during the first instruction cycle through a multiplexer and holds the instruction fetched. The instructions are decoded in the instruction decoder and appropriately gated with timing signals to provide the control signals for the various functional blocks. A single-cycle FF is

reset from one of five double-length instructions. Double-length instructions are instructions that need two system cycles (16 clock cycles) for their execution. A condition FF controls JCN and ISZ instruction and is set by the condition logic.

6. *Interrupt and Stop Control Logic* The 4040 is provided with hardware INTERRUPT and STOP controls which override the normal processor operation. The INTERRUPT logic detects and acknowledges the presence of an INTERRUPT signal and forces the processor to execute a JMS instruction to location 003. The STOP control logic behaves in a similar manner by detecting and acknowledging the presence of a STOP signal. The processor is forced to execute a NOP instruction and will remain in the STOP condition until the STOP signal is removed.

9.3.2 Intel 4040 Instruction Set

The 4040 is functionally compatible with the 4004 and therefore recognizes all 46 instructions valid for the 4004. In addition, the 4040 recognizes 14 new instructions, giving a total of 60 instructions in the set. They can be grouped into four groups, which can be defined as follows:

1. *Machine Instructions* This group of 16 instructions are designated by an OPR code of 0000–1101. There are one-word machine instructions and two-word machine instructions. The one-word instructions are 8 bits wide and require 8 clock periods (one instruction cycle) for execution. The two-word instructions are 16 bits wide and require 16 clock periods (two instruction cycles) for execution. A one-word instruction occupies one location in ROM (each location can hold one 8-bit word), and a two-word instruction occupies two successive locations in ROM. Each instruction word is divided into two 4-bit nibbles. The upper 4 bits is called the *OPR* and contains the operation code. The low 4 bits is called the *OPA* and contains the modifier. The formats of one-word and two-word instructions are shown in Fig. 9.3.4. For a single-word machine instruction the operation code (OPR) contains the code of the operation that is to be performed (add, substract, load, etc.). The modifier (OPA) contains one of four things:

    1. A register address.
    2. A register-pair address.
    3. Four bits of data.
    4. An instruction modifier.

For a two-word machine instruction the first word is similar to a one-word instruction; however, the modifier (OPA) contains one of four things:

    1. A register address.
    2. A register-pair address.
    3. The upper portion of another ROM address.
    4. A condition for jumping.

The second word contains either the middle portion (in OPR) and lower portion (in OPA) of another ROM address or 8 bits of data (the upper 4 bits in OPR and the lower 4 bits in OPA).

The upper 4 bits of instruction (OPR) will always be fetched before the lower 4 bits of instruction (OPA) during M_1 and M_2 times, respectively.

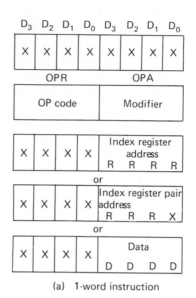

(a)   1-word instruction

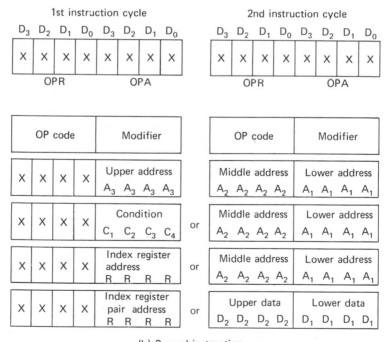

(b) 2-word instruction

Fig. 9.3.4   Machine instruction format.

2. *4040 Group* This group of 14 instructions is designated by an OPR code of 0000 and an OPA code of 0001–1110. These are the new instructions which have been added to the 4040.

3. *I/O Group* Designated by an OPR code of 1110, this group of 16 instructions is used for transferring data between the processor and the RAM chips or I/O circuits.

4. *Accumulator Group* This group of 14 instructions is designated by an OPR code of 1111 and operates only on the accumulator/carry flip-flop, the special ROM's, and the command register.

All the 4040, I/O, and accumulator group instructions are single-word instructions. Their formats are shown in Fig. 9.3.5. Since every instruction is described by two nibbles or four nibbles, and each nibble can be represented by a hexadecimal number, besides the binary-code representation we can also represent an instruction by a hexadecimal code. For example, the HLT (00000001) can be represented by 01_{16}. There are two major advantages of using the hexadecimal-code representation:

1. The representation is short and thus easy to remember.
2. From its first number, we can determine to which instruction group it belongs:

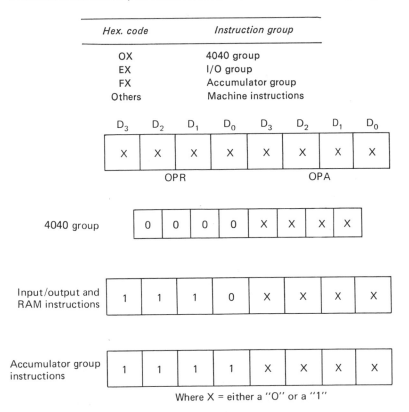

Hex. code	Instruction group
OX	4040 group
EX	I/O group
FX	Accumulator group
Others	Machine instructions

Fig. 9.3.5 4040 I/O and accumulator group instruction formats.

The X represents any one of the 16 digits 0, 1, . . . , F. There is one exceptional case. The instruction 00 (NOP) is an old machine instruction.

The following symbols and abbreviations are commonly used in describing the instructions and their operations:

Symbol and abbreviation	Meaning
P_L	Low-order program counter field (4 bits)
P_M	Middle-order program counter field (4 bits)
P_H	High-order program counter field (4 bits)
A_1	Low-order address bits
A_2	High-order address bits
A_3	Chip select
RRRR	Index register address
RRR	Index register-pair address
P_X^1	Register pairs P_0 through P_7, designated by odd characters 1, 3, 5, 7, 9, B, D, F
P_X^0	Register pairs P_0 through P_7 designated by even characters 0, 2, 4, 6, 8, A, C, E
R_X	Register $0 \longrightarrow F$
D_X	Data
D_1	Data for odd register
D_2	Data for even register
C_X	Jump conditions
SRCR	SRC register
( )	The content of is transferred to
ACC	Accumulator (4 bits)
CY	Carry flip-flop
ACBR	Accumulator buffer register (4 bits)
a_i	Order i content of the accumulator
CM_i	Order i content of the command register
M	RAM main character location
M_{si}	RAM status character i
DB (T)	Data bus content at time T
Stack	The three or seven registers in the address register other than the program counter
CR	Command register
IE	Interrupt enable
RB0	Register bank 0 $RRRR_0$–RRRR, enable
RB1	Register bank 1 $RRRR_0$–RRRR, enable
V	Logical OR
Λ	Logical AND

The complete set of instructions is given in Table 9.3.1. It should be noted that the 4040 instruction set is implemented with the negative logic assignment (i.e., logic 1 = low voltage = negative voltage and logic 0 = high voltage = ground). Table 9.3.2 shows

TABLE 9.3.1 4040 Instruction Description

Mnemonic	OPR D_3 D_2 D_1 D_0	OPA D_3 D_2 D_1 D_0	Hex code	Symbolic representation of the instruction	Description of operation
				Machine Instructions	
NOP	0 0 0 0	0 0 0 0	00	Not applicable	No operation.
JCN	0 0 0 1	C_1 C_2 C_3 C_4	$1C_X$	If $C_1C_2C_3C_4$ is true, $A_2A_2A_2A_2 \longrightarrow P_M$, $A_1A_1A_1A_1 \longrightarrow P_L$, P_H unchanged.	Jump to ROM address $A_2A_2A_2A_2$, $A_1A_1A_1A_1$ (within the same ROM that contains this JCN instruction)
	A_2 A_2 A_2 A_2	A_1 A_1 A_1 A_1	A_2A_1	If $C_1C_2C_3C_4$ is false, $(P_H) \longrightarrow P_H$, $(P_M) \longrightarrow P_M$, $(P_L + 2) \longrightarrow P_L$	if condition $C_1C_2C_3C_4$ is true; otherwise, skip (go to the next instruction in sequence).
FIM	0 0 1 0	R R R 0	$2P_X^0$	$D_2D_2D_2D_2 \longrightarrow RRR0$, $D_1D_1D_1D_1 \longrightarrow RRR1$	Fetch immediate (direct) from ROM data D_2, D_1 to index register-pair
	D_2 D_2 D_2 D_2	D_1 D_1 D_1 D_1	D_2D_1		location RRR.
SRC	0 0 1 0	R R R 1	$2P_X^1$	$(RRR0) \longrightarrow DB(X_2)$, $(RRR1) \longrightarrow DB(X_3)$	Send register control. Send the address (contents of index register pair RRR) to ROM and RAM at X_2 and X_3 time in the Instruction cycle.
FIN	0 0 1 1	R R R 0	$3P_X^0$	$(P_H)(0000)(0001) \longrightarrow ROM$ address, $(OPR) \longrightarrow RRR0$, $(OPA) \longrightarrow RRR1$	Fetch indirect from ROM. Send contents of index register-pair location 0 out as an address. Data fetched are placed into register-pair location RRR.
JIN	0 0 1 1	R R R 1	$3P_X^1$	$(RRR0) \longrightarrow P_M$, $(RRR1) \longrightarrow P_L$; P_H unchanged	Jump indirect. Send contents of register pair RRR out as an address at A_1 and A_2 time in Instruction cycle.
JUN	0 1 0 0	A_3 A_3 A_3 A_3	$4A_3$	$A_1A_1A_1A_1 \longrightarrow P_L$, $A_2A_2A_2A_2 \longrightarrow P_M$, $A_3A_3A_3A_3 \longrightarrow P_H$	Jump unconditional to ROM address A_3, A_2, A_1.
	A_2 A_2 A_2 A_2	A_1 A_1 A_1 A_1	A_2A_1		
JMS	0 1 0 1	A_3 A_3 A_3 A_3	$5A_3$	$(P_H, P_M, P_L + 2) \longrightarrow Stack$, $A_1A_1A_1A_1 \longrightarrow P_L$, $A_2A_2A_2A_2 \longrightarrow P_M$, $A_3A_3A_3A_3 \longrightarrow P_H$	Jump to subroutine ROM address A_3, A_2, A_1, save old address. (Up 1 level in stack.)
	A_2 A_2 A_2 A_2	A_1 A_1 A_1 A_1	A_2A_1		
INC	0 1 1 0	R R R R	$6R_X$	$(RRRR) + 1 \longrightarrow RRRR$	Increment contents of register RRRR.

Table 9.3.1   continued

Mnemonic	OPR D_3 D_2 D_1 D_0	OPA D_3 D_2 D_1 D_0	Hex code	Symbolic representation of the instruction	Description of operation
ISZ	0 1 1 1	R R R R	$7R_X$	$(RRRR) + 1 \longrightarrow RRRR$. If result = 0, $(P_H) \longrightarrow P_H$, $(P_M) \longrightarrow P_M$, $(P_L + 2) \longrightarrow P_L$.	Increment contents of register RRRR. Go to ROM address A_2, A_1 (within the same ROM that contains this ISZ instruction) if result $\neq 0$; otherwise, skip (go to the next instruction in sequence).
	A_2 A_2 A_2 A_2	A_1 A_1 A_1 A_1	A_2A_1	If result $\neq 0$, $(P_H) \longrightarrow P_H$, $A_2A_2A_2A_2 \longrightarrow P_M$, $A_1A_1A_1A_1 \longrightarrow P_L$	
ADD	1 0 0 0	R R R R	$8R_X$	$(RRRR) + (ACC) + (CY) \longrightarrow ACC, CY$	Add contents of register RRRR to accumulator with carry.
SUB	1 0 0 1	R R R R	$9R_X$	$(ACC) + (RRRR) + (CY) \longrightarrow ACC, CY$	Subtract contents of register RRRR to accumulator with borrow.
LD	1 0 1 0	R R R R	AR_X	$(RRRR) \longrightarrow ACC$	Load contents of register RRRR to accumulator.
XCH	1 0 1 1	R R R R	BR_X	$(ACC) \longrightarrow ACBR$, $(RRRR) \longrightarrow ACC$, $(ACBR) \longrightarrow RRRR$	Exchange contents of index register RRRR and accumulator.
BBL	1 1 1 1	D D D D	CD_X	$(Stack) \longrightarrow P_L, P_M, P_H$; $DDDD \longrightarrow ACC$	Branch back (down 1 level in stack) and load data DDDD to accumulator.
LDM	1 1 0 1	D D D D	DD_X	$DDDD \longrightarrow ACC$	Load data DDDD to accumulator.
				New 4040 Instructions	
HLT	0 0 0 0	0 0 0 1	01	$1 \longrightarrow HALT$, $1 \longrightarrow STOP$	Halt—inhibit program counter and data buffers.
BBS	0 0 0 0	0 0 1 0	02	$(Stack) \longrightarrow P_L, P_M, P_H$, $SRCRO \longrightarrow DB(X2)$, $SRCR1 \longrightarrow DB(X3)$	Branch back from Interrupt and restore the previous SRC. The program counter and send register control are restored to their pre-interrupt value.
LCR	0 0 0 0	0 0 1 1	03	$(CR) \longrightarrow ACC$	The contents of the COMMAND REGISTER are transferred to the accumulator.
OR4	0 0 0 0	0 1 0 0	04	$(RRRR_4) \vee (ACC) \longrightarrow ACC$	The 4-bit contents of register 4 are logically OR-ed with the ACCUM.

									Hex	Operation	Description
OR5	0	0	0	0	0	1	0	1	05	$(RRRR_5) \vee (ACC) \longrightarrow ACC$	The 4-bit contents of index register 5 are logically OR-ed with the accumulator.
AN6	0	0	0	0	0	1	1	0	06	$(RRRR_6) \wedge (ACC) \longrightarrow ACC$	The 4-bit contents of index register 6 are logically AND-ed with the accumulator.
AN7	0	0	0	0	0	1	1	1	07	$(RRRR_7) \wedge (ACC) \longrightarrow ACC$	The 4 bit contents of index register 7 are logically AND-ed with the accumulator.
DB0	0	0	0	0	1	0	0	0	08	Enable $\longrightarrow$ CM-ROM$_0$	DESIGNATE ROM BANK 0. CM-ROM$_0$ becomes enabled.
DB1	0	0	0	0	1	0	0	1	09	Enable $\longrightarrow$ CM-ROM$_1$	DESIGNATE ROM BANK 1. CM-ROM$_1$ becomes enabled.
SB0	0	0	0	0	1	0	1	0	0A	$1 \longrightarrow RB0, 0 \longrightarrow RB1$	SELECT INDEX REGISTER BANK 0. The index registers 0–7.
SB1	0	0	0	0	1	0	1	1	0B	$0 \longrightarrow RB0, 1 \longrightarrow RB1$	SELECT INDEX REGISTER BANK 1. The index registers 0*–7*.
EIN	0	0	0	0	1	1	0	0	0C	$1 \longrightarrow IE$	ENABLE INTERRUPT.
DIN	0	0	0	0	1	1	0	1	0D	$0 \longrightarrow IE$	DISABLE INTERRUPT.
RPM	0	0	0	0	1	1	1	0	0E	$(1111)(SRC) \longrightarrow$ ROM/RAM address, $(DDDD) \longrightarrow ACC$	READ PROGRAM MEMORY.

Input/Output and RAM Instructions†

									Hex	Operation	Description
WRM	1	1	1	0	0	0	0	0	E0	$(ACC) \longrightarrow M$	Write the contents of the accumulator into the previously selected RAM main memory character.
WMP	1	1	1	0	0	0	0	1	E1	$(ACC) \longrightarrow$ RAM output register	Write the contents of the accumulator into the previously selected RAM output port.
WRR	1	1	1	0	0	0	1	0	E2	$(ACC) \longrightarrow$ ROM output lines	Write the contents of the accumulator into the previously selected ROM output port (I/O lines).

†The RAM's and ROM's operated on in the I/O and RAM instructions have been previously selected by the last SRC instruction executed.

Table 9.3.1 continued

Mnemonic	OPR D_3 D_2 D_1 D_0	OPA D_3 D_2 D_1 D_0	Hex code	Symbolic representation of the instruction	Description of operation
WPM	1 1 1 0	0 0 1 1	E3	$(ACC) \longrightarrow RAM$	Write the contents of the accumulator into the previously selected half byte of read/write program memory (for use with 4008/4009 only).
WRϕ	1 1 1 0	0 1 0 0	E4	$(ACC) \longrightarrow M_{s0}$	Write the contents of the accumulator into the previously selected RAM status character 0.
WR1	1 1 1 0	0 1 0 1	E5	$(ACC) \longrightarrow M_{s1}$	Write the contents of the accumulator into the previously selected RAM status character 1.
WR2	1 1 1 0	0 1 1 0	E6	$(ACC) \longrightarrow M_{s2}$	Write the contents of the accumulator into the previously selected RAM status character 2.
WR3	1 1 1 0	0 1 1 1	E7	$(ACC) \longrightarrow M_{s3}$	Write the contents of the accumulator into the previously selected RAM status character 3.
SBM	1 1 1 0	1 0 0 0	E8	$(\overline{M}) + (ACC) + (\overline{CY}) \longrightarrow ACC, CY$	Subtract the previously selected RAM main memory character from accumulator with borrow.
RDM	1 1 1 0	1 0 0 1	E9	$(M) \longrightarrow ACC$	Read the previously selected RAM main memory character into the accumulator.
RDR	1 1 1 0	1 0 1 0	EA	$(ROM\ input\ lines) \longrightarrow ACC$	Read the contents of the previously selected ROM input port into the accumulator (I/O lines).
ADM	1 1 1 0	1 0 1 1	EB	$(M) + (ACC) + (CY) \longrightarrow ACC, CY$	Add the previously selected RAM main memory character to accumulator with carry.
RDϕ	1 1 1 0	1 1 0 0	EC	$(M_{s0}) \longrightarrow ACC$	Read the previously selected RAM status character 0 into accumulator.

RD1	1	1	1	0	1	1	0	1	ED	$(M_{s1}) \longrightarrow ACC$	Read the previously selected RAM status character 1 into accumulator.
RD2	1	1	1	0	1	1	1	0	EE	$(M_{s2}) \longrightarrow ACC$	Read the previously selected RAM status character 2 into accumulator.
RD3	1	1	1	0	1	1	1	1	EF	$(M_{s3}) \longrightarrow ACC$	Read the previously selected RAM status character 3 into accumulator.

Accumulator Group Instructions

CLB	1	1	1	1	0	0	0	0	F0	$0 \longrightarrow ACC, 0 \longrightarrow CY$	Clear both accumulator and carry.
CLC	1	1	1	1	0	0	0	1	F1	$0 \longrightarrow CY$	Clear carry.
IAC	1	1	1	1	0	0	1	0	F2	$(ACC) + 1 \longrightarrow ACC$	Increment accumulator.
CMC	1	1	1	1	0	0	1	1	F3	$(\overline{CY}) \longrightarrow CY$	Complement carry.
CMA	1	1	1	1	0	1	0	0	F4	$\bar{a}_3 \bar{a}_2 \bar{a}_1 \bar{a}_0 \longrightarrow ACC$	Complement accumulator.
RAL	1	1	1	1	0	1	0	1	F5	$C_0 \longrightarrow a_0, a_i \longrightarrow a_{i-1}, a_3 \longrightarrow CY$	Rotate left (accumulator and carry).
RAR	1	1	1	1	0	1	1	0	F6	$a_0 \longrightarrow CY, a_i \longrightarrow a_{i-1}, C_0 \longrightarrow a_3$	Rotate right (accumulator and carry).
TCC	1	1	1	1	0	1	1	1	F7	$0 \longrightarrow ACC, (CY) \longrightarrow a_0, 0 \longrightarrow CY$	Transmit carry to accumulator and clear carry.
DAC	-	1	1	1	1	0	0	0	F8	$(ACC) - 1 \longrightarrow ACC$	Decrement accumulator.
TCS	-	1	1	1	1	0	0	1	F9	$1001 \longrightarrow ACC$ if $(CY) \longrightarrow 0, 1010 \longrightarrow ACC$ if $(CY) = 1, 0 \longrightarrow CY$	Transfer carry subtract and clear carry.
STC	1	1	1	1	1	0	1	0	FA	$1 \longrightarrow CY$	Set carry.
DAA	1	1	1	1	1	0	1	1	FB	$(ACC) + 0000$ or $0110 \longrightarrow ACC$	Decimal-adjust accumulator.
KBP	1	1	1	1	1	1	0	0	FC	$(ACC) \longrightarrow KBP\ ROM \longrightarrow ACC$	Keyboard process. Converts the contents of the accumulator from a 1-out-of-4 code to a binary code.
DCL	1	1	1	1	1	1	0	1	FD	$a_0 \longrightarrow CM_0, a_1 \longrightarrow CM_1, a_2 \longrightarrow CM_2$	Designate command line.

the C_X condition table for JCN instruction. From Table 9.3.1, observe that:

OPA pattern	Operations indicated
0XXX	WRITE operations.
1XXX	READ operations.
X1YY	The instruction applies to one of the four RAM status characters (indicated by YY).

and so on. The clues you get from the bit patterns are often less ambiguous than the verbal descriptions that accompany the instruction.

TABLE 9.3.2 C_X Condition Table for JCN Instruction

JCN HEX	C_X mnemonic	C_8	C_4	C_2	C_1	Invert jump condition Jump if accumulator = 0 Jump if carry bit = 1 Jump if test input = 0 (high)
10		0	0	0	0	No operation.
11	TO	0	0	0	1	Jump if test = 0 (high).
12	C1	0	0	1	0	Jump if CY = 1.
13	TO+C1	0	0	1	1	Jump if test = 0 or CY = 1.
14	AO	0	1	0	0	Jump if AC = 0.
15	TO+AO	0	1	0	1	Jump if test = 0 or AC = 0.
16	C1+AO	0	1	1	0	Jump if CY = 1 or AC = 0.
17	TO+C1+AO	0	1	1	1	Jump if test = 0 or CY = 1 or AC = 0.
18		1	0	0	0	Jump unconditionally.
19	T1	1	0	0	1	Jump if test = 1 (low).
1A	CO	1	0	1	0	Jump if CY = 0.
1B	T1CO	1	0	1	1	Jump if test = 1 and CY = 0.
1C	A1	1	1	0	0	Jump if AC $\neq$ 0.
1D	T1A1	1	1	0	1	Jump if test = 1 and AC $\neq$ 0.
1E	COA1	1	1	1	0	Jump if CY = 0 and AC $\neq$ 0.
1F	T1COA1	1	1	1	1	Jump if test = 1 and CY = 0 and AC $\neq$ 0.

Register-Pair PX Lookup Table

PX	PX 0 RRR 0 FIM	FIN	PX 1 RRR 1 SRC	JIN
P0	20	30	21	31
P1	22	32	23	33
P2	24	34	25	35
P3	26	36	27	37
P4	28	38	29	39
P5	2A	3A	2B	3B
P6	2C	3C	2D	3D
P7	2E	3E	2F	3F

9.4 Procedure for Logic Design Using Microprocessors

Logic diagrams using graphic symbology are the key to visualization and implementation of hard-wired logic designs. The sequential nature of programmed logic does not lend itself to logic diagrams. The visual and verbal aids available to the program logic designer are block diagrams, flow charts, register maps, and coding forms.

The designer begins with a block diagram to make input and output, ROM and RAM register assignments. The problem is flow-charted and detailed assignments are made of registers on register maps. The flow charts are progressively partitioned into more and more detail until each flow symbol can be converted to program instructions on the coding forms. The instructions are first written in mnemonic form for easily verbalizing the solution to the problem. When the complete problem or a workable partition has been solved, the mnemonic instructions are converted to code for the ROM. The code is programmed into the ROM and tested with the hardware on the breadboard system.

Logic design using microprocessors and microprogrammed PROM's consists of the following three steps:

Step 1   Determination of system block diagram and components.
Step 2   Preparation of a program to be programmed in PROM.
Step 3   Program the PROM.

These three steps are described in detail as follows.

Step 1   Determination of system block diagram and components: Figure 9.4.1 shows a block diagram of a typical system using a 4040 microprocessor for logic design. In this system, we may use up to 16 RAM's and 32 ROM's. The number of input ports and the number of output ports should be equal to the number of ROM's used in the system. All the system components (4040CPU microprocessor, RAM's, and ROM's) are interconnected via the 4-bit CPU bus. The program sequence describing logic design stored in the ROM program memory controls the interaction between the elements connected on the CPU bus.

Step 2   Preparation of a program to be programmed in PROM: The preparation of a program for PROM involves (1) assigning registers (index registers inside CPU and RAM registers) for specific tasks, (2) describing program logic by flow charts, and (3) coding instructions in hexadecimal numbers.

1. *Register Assignments*   Register assignments are valuable for visualizing register storage allocation. Figure 9.4.2 illustrates assignments for both the index registers and the RAM registers. The 4040 provides an additional bank of eight registers, making a total of twenty-four 4-bit registers. These eight registers are referenced in identically the same manner as registers 0 through 7. Thus there are two

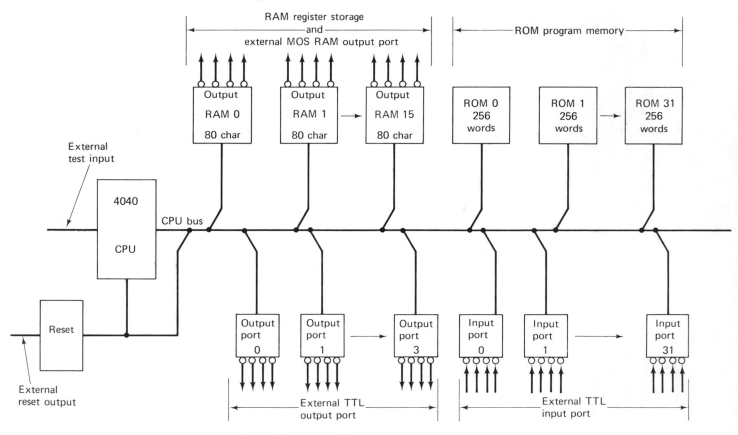

Note: All buses are 4-bit

Fig. 9.4.1 Block diagram of a typical system for logic design.

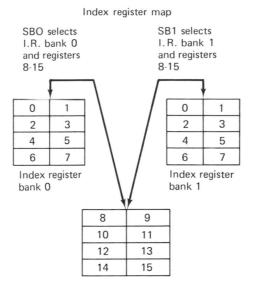

Index register map

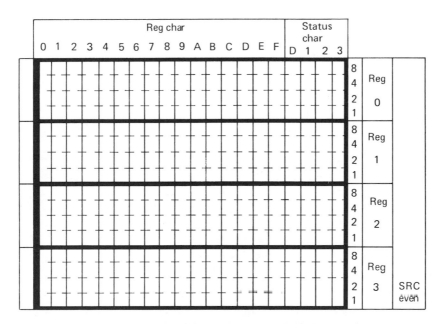

RAM register map

Fig. 9.4.2 Register maps.

register 0's, two register 1's, and so on. Two instruction, SB0 and SB1 are provided which enable the user to choose which of the two banks of registers are active at any instant. Registers 8 through 15 are unique and are always available.

The RAM register assignment shows all the bits available in one 4002 device. The organization is four registers of twenty 4-bit characters addressable by SRC. Each register contains 16 characters addressable by SRC and 4 characters addressable by individual instructions. When using register assignments, it is helpful to write an abbreviated mnemonic on the map to verbalize its assignment. A mnemonic is written for each register used in a routine to show which registers have been used and what they are used for. A convenient place for recording register assignments is on the document containing the flow chart.

2. *Flow Charts* The sequential nature of programmed logic fits directly into the graphic representation provided by flow charts. Programmed logic being sequential with only yes or no decisions allows for every simple flow-charting procedures. The graphic symbols used in design examples in the research are shown in Fig. 9.4.3. The main symbols used are the rectangle for operations and the diamond for decisions. The rectangle contains an abbreviated question concerning the decision.

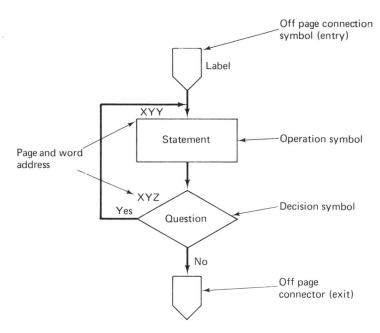

Fig. 9.4.3 Flow chart.

The 4040 CPU has only two decision instructions, JCN and ISZ. Any time that a diamond symbol appears in the flow chart, one of these two instructions must be involved. All other instructions perform operations either alone or in groups and are thus represented by the rectangle. The use of flow charts correlates with the use of logic block diagrams in hardware design. The hardware designer progressively partitions his problem into more and more detailed logic diagrams until each block represents a logic device. The program logic designer uses progressively more detailed flow diagrams, to the point where individual instructions or groups of instructions can be written for each flow symbol.

3. *Hexadecimal Coding Form* The hexadecimal coding form, or some variation of the form, is an absolute necessity for keeping track of the bookkeeping details inherent in programming. In addition, when properly implemented the coding form becomes the program listing which defines how the program accomplishes its task. The program listing in programmed logic is equivalent to a combined logic schematic, wire list, and assembly drawing of a hardwired logic system.

The hexadecimal coding form in Fig. 9.4.4 has seven columns and is divided into two major sections, each serving a distinct requirement. The four rightmost columns constitute the major portion of the form, which is used for mnemonic listing of the program as it is generated. The three leftmost columns constitute the second section, which provides the hexadecimal coding of addresses and instructions in machine language used by the CPU and program memory. The mnemonic section of the form is completed first as the designer sequentially lists the program steps in easy-to-remember mnemonic form. When the designer has solidified the mnemonic listing, the hexadecimal address and instruction codes are assigned. The coding operation of programmed logic is similar to assigning device location, pin numbers, and wire listing in hardware logic. The mnemonic listing of instructions in programmed logic is equivalent to the hardware logic operation of creating a logic schematic diagram where the program designer assembles instructions and subroutines in a mnemonic list and the hardware designer selects gates MSI and LSI.

ADD	OPR	OPA	Label	Mnemonic	Operand	Comment

Fig. 9.4.4 Hex coding form

Column 1 Hexadecimal program memory page and word address. The address has three hexadecimal digits. The leftmost (most-significant) digit addresses the 16 pages of program memory. The other two digits address the 256 words of the page.

Column 2   Hexadecimal instruction word for OPR.

Column 3   Hexadecimal word for OPA.

Column 4   Mnemonic address label used to verbalized the destination of the address control instructions. Address labels in this column must appear only once, with each label having unique spelling. This column is left blank if the line does not require a label.

Column 5   Mnemonic instructions—usually an abbreviation that verbalizes the operation. The second word of double-word instructions does not have a mnemonic and is left blank. The exception is the FIM, where the even register data character is inserted.

Column 6   Mnemonic operand, which can be blank, data constants, instruction modifiers, register designation, register-pair designation, or a source address label. Instructions 0 through 9 and A through D always require operand information. I/O, RAM, and accumulator instructions never have operand information.

Column 7   Written comments defining the purpose of an instruction or a group of instructions.

Step 3   Program the PROM: After a program is prepared, we then proceed to program the PROM. Suppose that a 4702A 256 $\times$ 8-bit electrically programmable and erasable read-only memory is used in the system. The 4702A is programmed by the momentary application of high-amplitude pulses on selected pins of the chip. The 4702A, however, may be cleared by controlled exposure to high-intensity ultraviolet. The device may be reloaded as often as desired, making it particularly suitable for use in program development.

Programming of the 4702A requires a carefully controlled sequence of operations. The safety of the chip demands that both the interelement voltage and the duty cycle of the programming pulses be maintained within specific limits. This ensures against breakdown and overheating. The Intel INTELLEC 4/MOD 40 microcomputer development system has the PROM programmer module, which is designed to automatically provide pulses of the correct level, duration, and frequency. A 4702A that has been programmed previously must be erased prior to reloading. Erasure is accomplished by exposing the silicon die to ultraviolet light.

A 4702A PROM can be programmed using the INTELLEC 4/MOD 40 micro-computer development system. It has two system programs designed to help the user write, debug, and prepare his own programs for execution. They are the system monitor and the assembler. The system monitor contains a teletype I/O routine and service routines which permit:

a. Loading "program RAM" from paper tape
b. Dumping the contents of program RAM
c. Modification of individual RAM instructions
d. Writing program RAM onto paper tape
e. Programming of programmable read only memory (PROM)

The assembler allows user to write the source program in assembly language and then the assembler will convert it into object program which may be directly loaded into PROM's or control program RAM, and then executed.

The procedure for programming a 4702A PROM consists of the following steps:

1. The teletype console, in the off-line mode, is used to prepare a paper tape containing the entire assembly-language listing of the program.

2. With the teletype (or high-speed paper tape reader) on line, and with the system monitor selected, the tape containing the assembler program is run through the teletype's (or the high-speed) tape reader. This loads the assembler into program RAM.

3. The program RAM memory is selected, using the console RAM switch.

4. The tape containing the assembly-language listing is run through the tape reader (teletype or high-speed) for the first pass.

5. The assembly-language listing is run through the reader again. The user can specify pass 2, which will produce a listing of the assembled program or can specify pass 3, which will produce an object program tape. Once the first pass is completed, the second and third passes may be done in any order, or either may be omitted.

6. If pass 3 has been executed (i.e., if an object tape has been punched), the system monitor is again selected, using the console switch, and the object tape is run through the reader to be loaded into program RAM.

7. Program RAM is again selected, and execution may begin.

9.5   Design Example

To illustrate the logic design using the system described in Fig. 9.4.1 and the procedure described in the previous section, a design of a real-time 12-hour digital clock is presented. The design consists of two parts: finding a counting technique to convert the CPU cycles into $H_2H_1 : M_2M_1 : S_2S_1$ data, and the display of the data (second, minute, and hour), which are shown in Fig. 9.5.1. The display of the outputs can be accomplished by using ready-made MSI TTL chips such as SN 7448 and FND 70 seven-segment LED display. Since this is relatively easy to accomplish, the output-display design part will not be discussed.

The block diagram of the MCS-40 system in Fig. 9.5.1 is shown in Fig. 9.5.2, in which S1 and S2 represent the two digits of the second count, M1 and M2 for the minute count, and H1 and H2 for the hour count. In this system, the following components are used:

1. 4201: clock generator
2. 4289: standard memory interface.
3. 4702A: 2,048-bite erasable and electrically reprogrammable read-only memory.
4. 4002: 320-bit RAM and 4-bit output port.
5. 3205: 1-out-of-8 binary decoder.
6. DM 7098: tristate buffer.
7. 3404: 6-bit latch.

The complete descriptions of these devices can be found from reference 15.

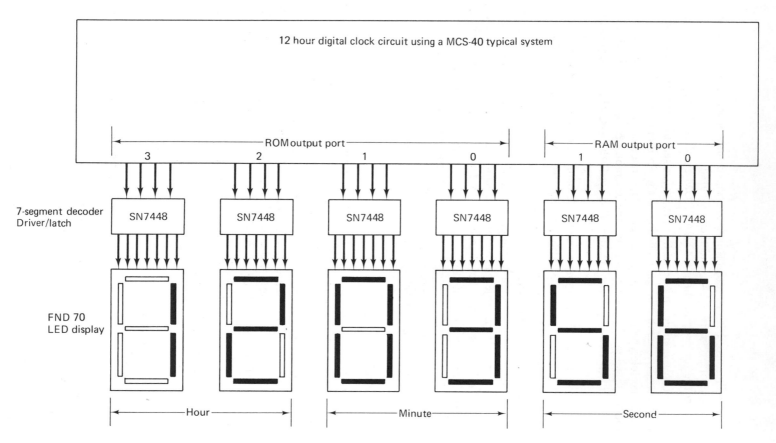

Fig. 9.5.1 Design of 12-hour digital clock including displays of the second, minute, and hour.

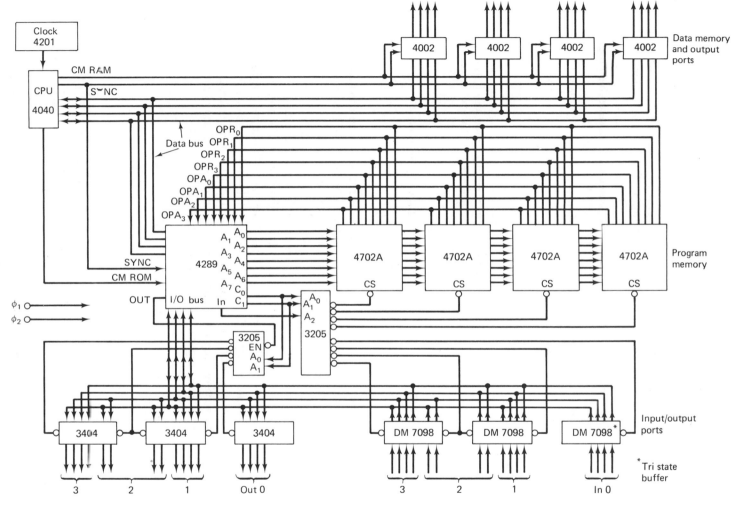

Fig. 9.5.2 MCS-40 typical system. (Courtesy of Intel Corporation.)

Two major programming techniques that are used to implement a 12-hour digital clock are described below.

1. *Programming Technique for Achieving Time Delays*
   a. Achieving time delays using index registers. Although the easiest and simplest technique to achieve an amount of time delay required by a design is to execute a number of non-operative instructions such as the NOP, any significant time delay would soon use up considerable program memory space, which is not economical. A more efficient technique for achieving time delay is to use the ISZ instruction (see Table 9.3.1). For example:

Label	OPR	OPA	Comment
LOOP	ISZ	A LOOP	Increment the content of index register A by 1. If result $\neq 0$, go to LOOP; otherwise, go to the next instruction in sequence.

The amount of time delay τ resulted from this program is

$$\tau = n \cdot T \qquad (9.5.1)$$

where n denotes the number of ISZ instruction being executed and T denotes the time for executing a two-word instruction. It is clear that the n depends on the initial setting of the register. For instance, if register A is initially set to 0_{16} ($= 0000_2$) and $T = 32$ μs, then n will be 16 and the delay τ will be 512 μs. As another example, suppose that the initial setting of register A is 7; then the corresponding value of n will be 9_{16} (see Table 4-1) and the τ will be 288 μs. Note that with one index register, the maximum time delay can be achieved by this technique is $16 \times T$. However, this technique can be extended to using multiple index registers to achieve longer time delay. For instance, the following example uses four index registers, designated A, B, C, and D:

```
LOOP    ISZ     A
                LOOP
        ISZ     B
                LOOP
        ISZ     C
                LOOP
        ISZ     D
                LOOP
```

Let $\sigma_A, \sigma_B, \sigma_C, \ldots, \sigma_D$ be the initial setting of registers A, B, C, and D, respectively. The number of ISZ instructions being executive for a given set of σ_A, σ_B, σ_C, and σ_D is given by

$$n(\sigma_A, \sigma_B, \sigma_C, \sigma_D) = \underbrace{(16 - \sigma_A)}_{n_A(\sigma_A)} + \underbrace{16(15 - \sigma_B) + (16 - \sigma_B)}_{n_B(\sigma_B)}$$

$$+ \underbrace{16^2(15 - \sigma_C) + 16(15 - \sigma_C) + (16 - \sigma_C)}_{n_C(\sigma_C)} \qquad (9.5.2)$$

$$+ \underbrace{16^3(15 - \sigma_D) + 16^2(15 - \sigma_D) + 16(15 - \sigma_D) + (16 - \sigma_D)}_{n_D(\sigma_D)}$$

It is easy to generalize this formula for the case where the number of index registers is more than four. Note that n_A, n_B, n_C, and n_D, as defined in Eq. (9.5.2), are functions of σ_A, σ_B, σ_C, and σ_D, respectively. For reference convenience, the values of $n_A(\sigma_A)$, $n_B(\sigma_B)$, $n_C(\sigma_C)$, and $n_D(\sigma_D)$, for σ_A σ_B, σ_C, and σ_D equal to 0 to are tabulated in Table 9.5.1.

TABLE 9.5.1   Values of $n_A(\sigma_A)$, $n_B(\sigma_B)$, $n_C(\sigma_C)$, and $n_D(\sigma_D)$

	Reg. A		Reg. B		Reg. C		Reg. D
σ_A	n_A	σ_B	n_B	σ_C	n_C	σ_D	n_D
F	1	F	1	F	1	F	1
E	2	E	18	E	274	E	4,370
D	3	D	35	D	547	D	8,739
C	4	C	52	C	820	C	13,108
B	5	B	69	B	1,093	B	17,477
A	6	A	86	A	1,366	A	21,846
9	7	9	103	9	1,639	9	26,215
8	8	8	120	8	1,912	8	30,584
7	9	7	137	7	2,185	7	34,953
6	10	6	154	6	2,458	6	39,322
5	11	5	171	5	2,731	5	43,691
4	12	4	188	4	3,004	4	48,060
3	13	3	205	3	3,277	3	52,429
2	14	2	222	2	3,550	2	56,798
1	15	1	239	1	3,823	1	61,167
0	16	0	256	0	4,096	0	65,536

b. Achieving a specific time delay. From Table 9.5.1 and Eq. (9.5.1), we can easily achieve a time delay of a specific value. For example, suppose that we want to have a 1-s time delay. Assume that $T = 32$ μs. From Eq. (9.5.1),

$$n = \frac{1 \text{ s}}{32 \ \mu s} = 31{,}250 \qquad (9.5.3)$$

Equation (9.5.2) indicates that

$$n = n_A + n_B + n_C + n_D \qquad (9.5.4)$$

From the fourth column of Table 9.5.1, we find that the largest number that is less that 31,250 is 30,584. Thus

$$n_D = 30,584$$

From Eqs. (9.5.3) and (9.5.4),

$$n_A + n_B + n_C = 666$$

In a similar manner, n_C, n_B, and n_A can be found. They are

$$n_C = 547$$
$$n_B = 18$$
$$n_A = 1$$

2. *Programming Technique for Constructing Counters*   Three types of counters —modulo-10, module-6, and modulo-12—are needed to implement the second, minute, and hour counts.

  a. Modulo-10 counter.   Two modulo-10 counters are used for counting the least-significant digits of the minute and second counts. The 4-bit output port 0 via ROM and the output port 0 via RAM are programmed by the instructions WRR (write the contents of accumulator into the previously selected ROM output port I/O lines) and WMP (write the contents of accumulator into the previously selected RAM output port) to provide the counting sequences M1 and S1 from 0000_2 up to 1001_2 in the following sequence:

Binary count	S1, M1
0000	0
0001	1
0010	2
0011	3
0100	4
0101	5
0110	6
0111	7
1000	8
1001	9

b. Modulo-6 counter.   Two modulo-6 counters are used for counting the most-significant digits of the second and minute counts. These counters output their counting sequences S2 and M2 from 0000_2 up to 0110_2, respectively, at the 4-bit output port 1 via RAM and the output port 1 ROM:

Binary count	S2, M2
0000	0
0001	1
0010	2
0011	3
0100	4
0101	5

c. Modulo-12 counter.   A modulo-12 counter is used for counting the two digits of the hour count. The output port 2 via ROM is programmed by the instruction WRR to give the counting for H1:

Binary count	H1
0001	1
0010	2
0011	3
0100	4
0101	5
0110	6
0111	7
1000	8
1001	9
0000	0
0001	1
0010	2

For the most-significant digit of hour conting, output port 3 via ROM is programmed by the instruction WRR to count 0000 with the exception that it counts 0001 when the number at the output port 2 via ROM is over 1001 as described in the H1 counting sequence. The flow chart of the complete design of a 12-hour digital clock is presented in Fig. 9.5.3. The index register assignment and the program for programming the PROM 4702A for implementing the real-time clock is given in Table 9.5.2.

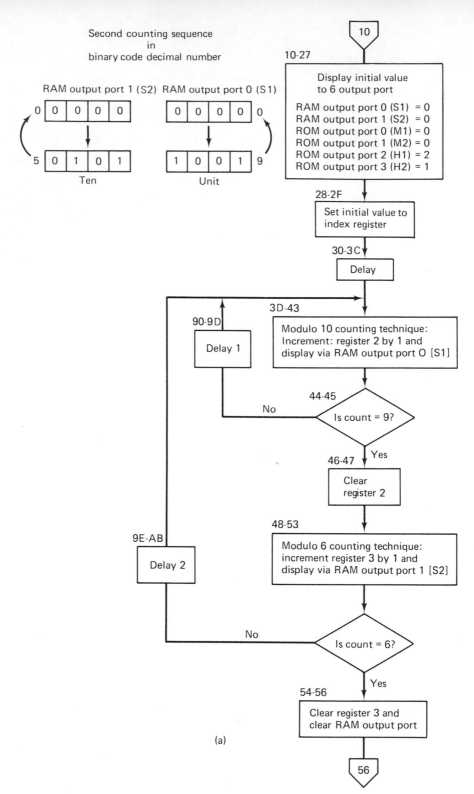

Fig. 9.5.3 Flow chart of the complete design of a 12 hour digital clock.

Minute counting sequence in.
binary code decimal number

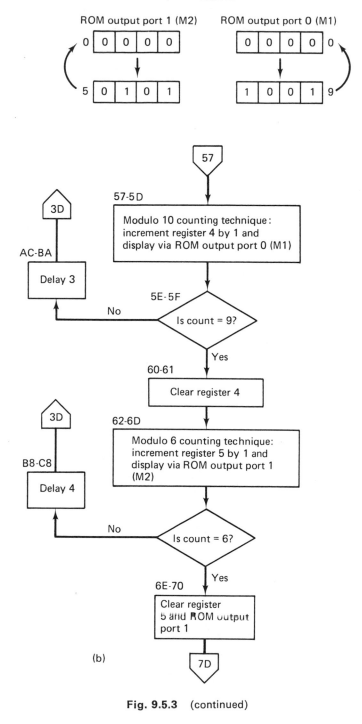

(b)

Fig. 9.5.3   (continued)

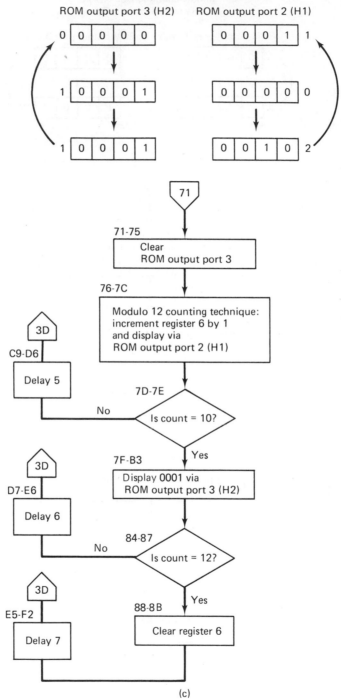

Hour counting sequence in
Binary code decimal number

ROM output port 3 (H2)   ROM output port 2 (H1)

| 0 | 0 | 0 | 0 | 0 | | 0 | 0 | 0 | 1 | 1 |

| 1 | 0 | 0 | 0 | 1 | | 0 | 0 | 0 | 0 | 0 |

| 1 | 0 | 0 | 0 | 1 | | 0 | 0 | 1 | 0 | 2 |

71

71-75
Clear
ROM output port 3

76-7C
Modulo 12 counting technique:
increment register 6 by 1
and display via
ROM output port 2 (H1)

3D

C9-D6
Delay 5

7D-7E
Is count = 10?   No

Yes

3D

7F-B3
Display 0001 via
ROM output port 3 (H2)

D7-E6
Delay 6

84-87
Is count = 12?   No

Yes

3D

E5-F2
Delay 7

88-8B
Clear register 6

(c)

Fig. 9.5.3   (continued)

548

Table 9.5.2 Index register assignment and the program for programming the PROM 4702A for implementing the real-time clock

```
┌──────────────────────────────────────────────────┐
│         12-HOUR-DIGITAL-CLOCK PROGRAM IN           │
│     MACHINE-LANGUAGE HEXADECIMAL NUMBER            │
│            AND ASSEMBLY LANGUAGE                    │
└──────────────────────────────────────────────────┘
```

	HOUR			MINUTE			SECOND		
	H2	H1		M2	M1		S2	S1	} output
	▭	▭		▭	▭		▭	▭	} port
	ROM p. 3	ROM p. 2		ROM p. 1	ROM p. 0		RAM chip 1	RAM chip 0	} assignment

Index Register Assignment

0	← FIM and SRC →		1
2	INC S1	INC S2	3
4	INC M1	INC M2	5
6	INC H1		7
8	0100 (6)	1100 (12)	9
A	Delay	Delay	B
C	Delay	Delay	D
E			F

Machine language in hexadecimal number			Assembly language			
Address	OPR	OPA	Label	Mnemonic	Operand	Comment
00	0	0		NOP		
01	0	0		NOP		
02	0	0		NOP		
03						
06						
05						No operation
06						
07						
08						
09						
0A						

Table 9.5.2 continued

Machine language in hexadecimal number			Assembly language			
Address	OPR	OPA	Label	Mnemonic	Operand	Comment
0B						
0C						No operation
0D						
0E	0	0		NOP		
0F	0	0		NOP		
10	2	1	START	SRC	0	Write to chip 0 RAM output port with 0
11	E	1		WMP		[0] S1.
12	2	0		FIM	0	
13	4	0		4	0	Write to chip 1 RAM output port with 0
14	2	1		SRC	0	[0] S2.
15	E	1		WMP		
16	2	0		FIM	0	
17	0	0		0	0	Write to page 0 c/o ROM output port with 0
18	2	1		SRC	0	[0] M1.
19	E	2		WRR		
1A	2	0		FIM	0	
1B	1	0		1	0	Write to page 1 c/o ROM output port with 0
1C	2	1		SRC	0	[0] M2.
1D	E	2		WRR		
1E	2	0		FIM	0	
1F	2	0		2	0	Write to page 2 c/o ROM output port with 0
20	2	1		SRC	0	
21	D	2		LDM	2	[2] H1.
22	E	2		WRR		
23	2	0		FIM	0	
24	3	0		3	0	Write to page 3 c/o ROM output port with 1
25	2	1		SRC	0	
26	D	1		LDM	1	[1] H2.
27	E	2		WRR		
28	D	6	Set index reg.	LDM	6	
29	B	8		XCH	8	Set index reg. 8 = 6 ⎱ for JCN
2A	D	C		LDM	C	Set index reg. 9 = 12 ⎰
2B	B	9		XCH	9	Set index reg. C = D ⎱ for time
2C	D	D		LDM	D	Set index reg. D = 8 ⎰ delay
2D	B	C		XCH	C	
2E	D	8		LDM	8	
2F	B	D		XCH	D	

Table 9.5.2 continued

Machine language in hexadecimal number			Assembly language			
Address	OPR	OPA	Label	Mnemonic	Operand	Comment
30	D	8	DELAY	LDM	8	
31	B	A		XCH	A	Set index reg. A = 8 ⎱ for time
32	D	B		LDM	B	Set index reg. B = B ⎰ delay
33	B	B		XCH	B	31,250 − 42 = 31,208 instructions
34	F	0		CLB		(1 s needs 31,250 instructions).
35	7	A	LOOP	ISZ	A	
36	3	5			LOOP	Time delay for 31,208 ISZ
37	7	B		ISZ	B	instructions.
38	3	5			LOOP	
39	7	C		ISZ	C	
3A	3	5			LOOP	
3B	7	D		ISZ	D	
3C	3	5			LOOP	
*3D	2	0	INC S1	FIM	0	
3E	0	0		0	0	Write out at S1 with 1, 2, 3, . . . ,
3F	2	1		SRC	0	9, 0 every 1 s.
40	6	2		INC	2	
41	A	2		LD	2	
42	F	B		DAA		
43	E	1		WMP		
44	1	A		JCN	CY = 0	If acc < 0, cy. bit = 0, keep
						increase sec.
45	9	0			DELAY 1	If acc ≧ 10, cy. bit = 1, go to
						N.I. (next instruction).
46	F	0		CLB		Reset IR 2 = 0.
47	B	2		XCH	2	
48	2	0	INC S2	FIM	0	
49	4	0		4	0	
4A	2	1		SRC	0	Write out at S2 with 1, 2, 3, 4, 5
4B	6	3		INC	3	every 10 s.
4C	A	3		LD	3	
4D	9	8		SUB	8	
4E	1	4		JCN	AC = 0	
4F	5	4			SET 1	
50	A	3		LD	3	
51	E	1		WMP		
52	4	0		JUN	page 0	
53	9	E		DELAY	2	
54	E	1	SET 1	WMP		When increase up to 6, write out
55	F	0		CLB		at S2 with 0 and go to N.I.
56	B	3		XCH	3	

Table 9.5.2 continued

Machine language in hexadecimal number			Assembly language			
Address	OPR	OPA	Label	Mnemonic	Operand	Comment
57	2	0	INCM1	FIM	0	
58	0	0		0	0	Write out at M1 with 1, 2, 3, . . . ,
59	2	1		SRC	0	9, 0 every 60 s.
5A	6	4		INC	4	
5B	A	4		LD	4	
5C	F	B		DAA		
5D	E	2		WRR		
5E	1	A		JCN	CY = 0	If acc < 0, cy. bit = 0, keep increase sec.
5F	A	C		DELAY	3	If acc ≧ 10, cy. bit = 1, go to N.I.
60	F	0		CLB		
61	B	4		XCH	4	Reset IR 4 = 0.
62	2	0	INCM2	FIM	0	
63	1	0		1	0	
64	2	1		SRC	0	Write out at M2 with 1, 2, 3, 4, 5,
65	6	5		INC	5	6 every 10 min.
66	A	5		LD	5	
67	9	8		SUB	8	
68	1	4		JCN	AC = 0	
69	6	E		Set	2	
6A	A	5		LD	5	
6B	E	2		WRR		
6C	4	0		JUN	page 0	
6D	B	B		DELAY	4	
6E	E	2	SET 2	WRR		When increase up to 6,
6F	F	0		CLB		write out at M2 with 0
70	B	5		XCH	5	and go to N.I.
71	2	0	SET H2 = 0	FIM	0	
72	3	0		3	0	Change output H2 from 1 to 0.
73	2	1		SRC	0	
74	F	0		CLB		
75	E	2		WRR		
76	2	0	INCH 1	FIM	0	
77	2	0		2	0	Write out at H1 with 1, 2, 3, 4,
78	2	1		SRC	0	. . . , 9, 0, 1, 2 every 60 min.
79	6	6		INC	6	
7A	A	6		LD	6	
7B	F	B		DAA		
7E	E	2		WRR		
7D	1	A		JCN	CY = 0	If acc < 10, cy. bit = 0, keep increase sec.
7E	C	9		DELAY	5	If acc ≧ 10, cy. bit = 1, go to N.I.

Table 9.5.2 continued

Address	OPR	OPA	Label	Mnemonic	Operand	Comment
	Machine language in hexadecimal number			*Assembly language*		
7F	2	0		FIM	0	
80	3	0		3	0	Write out at H2 with 1 every
81	2	1		SRC	0	10 hr.
82	F	7		TCC		
83	E	2		WRR		
84	A	6		LD	6	Check if IR 6, increase to 12.
85	9	9		SUB	9	If IR 6 < 12, keep increase sec.
86	1	C		JCN	AC $\neq$ 0	If IR 6 = 12, go to N.T.
87	D	7		DELAY	6	
88	F	0		CLB		Clear IR 6 = 0.
89	B	6		XCH	6	Go back to increase sec.
8A	4	0		JUN		
8B	E	5		DELAY	7	
8C	0	0		NQP		
8D	0	0		NOP		
8E	0	0		NOP		
8F	0	0		NOP		
90	D	F	DELAY 1	LDM	F	
91	B	A		XCH	A	Time delay for (31,250 − 15) =
92	D	9		LDM	9	312,35 ISZ instructions.
93	B	B		XCH	B	
94	7	A	LOOP 1	ISZ	A	
95	9	4			LOOP 1	
96	7	B		ISZ	B	
97	9	4			LOOP 1	
98	7	C		ISZ	C	
99	9	4			LOOP 1	
9A	7	D		ISZ	D	
9B	9	4			LOOP 1	
9C	4	0		JUN	page 0	
9D	3	D			INC S1	
9E	D	C	DELAY 2	LDM	C	
9F	B	A		XCH	A	
A0	D	A		LDM	A	Time delay for (31,250 − 29) =
A1	B	B		XCH	B	31,221 ISZ instructions.
A2	7	A	LOOP 2	ISZ	A	
A3	A	2			LOOP 2	
A4	7	B		ISZ	B	
A5	A	2			LOOP 2	
A6	7	C		ISZ	C	
A7	A	2			LOOP 2	
A8	7	D		ISZ	D	
A9	A	2			LOOP 2	
AA	4	0		JUN	page 0	
AB	3	D			INCS 1	

Table 9.5.2   continued

Machine language in hexadecimal number			Assembly language			
Address	OPR	OPA	Label	Mnemonic	Operand	Comment
AC	D	7	DELAY 3	LDM	7	
AD	B	A		XCH	A	
AE	D	B		LDM	B	Time delay for (31,250 − 41) =
AF	B	B		XCH	B	31,209 ISZ instructions.
B0	7	A	LOOP 3	ISZ	A	
B1	B	0			LOOP 3	
B2	7	B		ISZ	B	
B3	B	0			LOOP 3	
B4	7	C		ISZ	C	
B5	B	0			LOOP 3	
B6	7	D		ISZ	D	
B7	B	0			LOOP 3	
B8	4	0		JUN	page 0	
B9	3	D		INCS	1	
BA	0	0				
BB	D	4	DELAY 4	LDM	4	
BC	B	A		XCH	A	
BD	D	C		LDM	C	Time delay for (31,250 − 55) =
BE	B	B		XCH	B	31,195 ISZ instructions.
BF	7	A	LOOP 4	ISZ	A	
C0	B	F			LOOP 4	
C1	7	B		ISZ	B	
C2	B	F			LOOP 4	
C3	7	C		ISZ	C	
C4	B	F			LOOP 4	
C5	7	D		ISZ	D	
C6	B	F			LOOP 4	
C7	4	0		JUN	page 0	
C8	3	D		INCS	1	
C9	D	4	DELAY 5	LDM	4	
CA	B	A		XCH	A	
CB	D	D		LDM	D	Time delay for (31,250 − 72) =
CC	B	B		XCH	B	31,178 ISZ instructions.
CD	7	A	LOOP 5	ISZ	A	
CE	C	D			LOOP 5	
CF	7	B		ISZ	B	
D0	C	D			LOOP 5	
D1	7	C		ISZ	C	
D2	C	D			LOOP 5	
D3	7	D		ISZ	D	
D4	C	D			LOOP 5	
D5	4	0		JUN	page 0	
D6	3	D		INCS	1	

Table 9.5.2   continued

Machine language in hexadecimal number			Assembly language			
Address	OPR	OPA	Label	Mnemonic	Operand	Comment
D7	D	D	DELAY 6	LDM	D	
D8	B	A		XCH	A	
D9	D	D		LDM	D	Time delay for (31,250 − 81) =
DA	B	B		XCH	B	31,169 ISZ instructions.
DB	7	A	LOOP 6	ISZ	A	
DC	D	B			LOOP 6	
DD	7	B		ISZ	B	
DE	D	B			LOOP 6	
DF	7	C		ISZ	C	
E0	D	B			LOOP 6	
E1	7	D		ISZ	D	
E2	D	B			LOOP 6	
E3	4	0		JUN	page 0	
E4	3	D		INCS 1		
E5	D	0	DELAY 7	LDM	0	
E6	B	A		XCH	A	
E7	D	E		LDM	E	Time delay for (31,250 − 85) =
E8	B	B		XCH	B	31,165 ISZ instructions.
E9	7	A	LOOP 7	ISZ	A	
EA	E	9			LOOP 7	
EB	7	B		ISZ	B	
EC	E	9			LOOP 7	
ED	7	C		ISZ	C	
EE	E	9			LOOP 7	
EF	7	D		ISZ	D	
F0	E	9			LOOP 7	
F1	4	0		JUN	page 0	
F2	3	D		INCS1		
F3				END		
F4						
F5						
F6						
F7						
F8						
F9						
FA						
FB				End of page 0.		

Exercise 9.5

1. The implementation of logic gates is faciliated on the 4040 by the instructions AN6 and AN7 (which AND-registers 6 and 7, respectively, with the accumulator) and OR4

and OR5 (which OR registers 4 and 5, respectively, with the accumulator). Write a program in 4040 assembly language to implement the AND–OR circuit of Fig. P9.5.1.

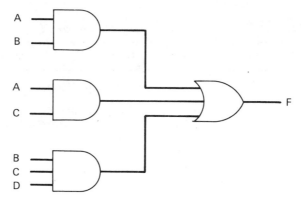

Fig. P9.5.1

2. Write a program to implement each of the following gates:
   (a) NOT gate
   (b) NAND gate
   (c) NOR gate
   (d) XOR gate

3. (a) Use the results of problem 2 to write a program to implement the combinational circuit in Fig. P9.5.3.

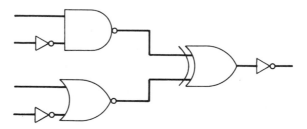

Fig. P9.5.3

   (b) Write a program to implement the 4-bit parallel adder in Fig. 2.2.3(b).

4. Write a program to implement the following:
   (a) True/complement and zero/one device in Fig. 3.4.2.
   (b) Four-bit carry look-ahead adder in Fig. 2.3.1.
   (c) Adders/subtractors of Fig. 2.4.3.

5. Write a program to implement the following:
   (a) SRFF
   (b) JKFF
   (c) DFF
   (d) TFF

6. Use the results of problem 5 to write a program to implement the sequential circuit in Fig. P9.5.6. It is assumed that x_1 and x_2 will never be simultaneously equal to 1.

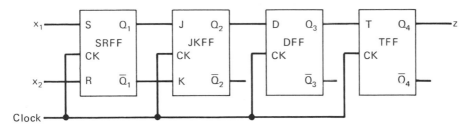

Fig. P9.5.6

7. Write a program to implement each of the shift registers of Fig. 4.4.1.

8. Write a program to implement the counter of Fig. 4.4.4.

9. Write computer programs to implement the multipliers in Fig. P4.4.1.

10. Write a program to implement the binary multiplier in Fig. P.4.4.2.

11. (a) Describe the sine function using a 4-bit by 4-bit truth table.
    (b) Program the truth table into a 4702A PROM.

12. Write a program to test (a) an IC gate chip, (b) an IC flip-flop chip, and (c) an IC RAM memory chip, that you are most familiar with. Print out "good" and "bad" if they pass and fail the test, respectively.

13. Design a traffic-signal controller using a MCS-40 microcomputer system. The inputs are in the form of count of passing autos, and the like; the outputs are to control signal lights and to notify a central location of the existence of congestion. It is desirable that the controller be able to execute sequence on command to expedite the flow of emergency vechicles.

14. Many recently developed diversions have been based about microprocessor systems (e.g., table tennis). A relatively straight forward demonstration would be the implementation of a two-dimensional tic-tac-toe game. Design such a system using the 4040 microprocessor.

Bibliographical Remarks

The material presented in this chapter is new and can be found in the following references.

1. R. H. CUSHMAN, "Don't Overlook the 4-Bit Microprocessor: They're Here and They're Cheap," *EDN*, Vol. 19, No. 4, 1974, pp. 44–50.

2. R. H. CUSHMAN, "Intel 8080: The First of the Second-Generation Microprocessors," *EDN*, Vol. 19, No. 9, 1974, pp. 30–36.

3. R. H. CUSHMAN, "Microprocessors Are Rapidly Gaining on Minicomputers," *EDN*, Vol. 19, No. 10, 1974, pp. 16–19.

4. R. H. CUSHMAN, "Understand the 8-Bit Microprocessor: You'll See a Lot of It," *EDN*, Vol. 19, No. 2, 1974, pp. 48–54.

5. R. H. CUSHMAN, "What Can You Do with a Microprocessor?" *EDN*, Vol. 19, No. 6, 1974, pp. 42–48.

6. R. H. CUSHMAN, "Microprocessors Are Changing Your Future. Are You Prepared?" *EDN*, Vol. 18, No. 21, 1973, pp. 26–32.

7. R. H. CUSHMAN, "Understanding the Microprocessor Is No Trivial Task," *EDN*, Vol. 18, No. 22, 1973, pp. 42–49.

8. R. FAGGIN and others, "MCS-4—A LSI Microcomputer System," *IEEE Region 6 Conference Record*, 1972, pp. 1–6.

9. C. C. FOSTER, "Something New—The Intel MCS-4 Microcomputer Set," *Computer Architecture News*, Vol. 1, No. 2, 1972, pp. 16–17.

10. B. GLADSTONE, "Designing with Microprocessors Instead of Wired Logic Asks More of Designers," *Electronics*, Vol. 46, No. 21, 1973, pp. 91–104.

11. Intel MCS-40, *User's Manual for Logic Designers*, 1974.

12. Intellec 4/MOD 40, *Microcomputer System Operator's Manual*, 1974.

13. Intel 4004 and 4040, *Assembly Language Programming Manual*, 1975.

14. Intellec 4/MOD 40, *Microcomputer Development System Reference Manual*, 1975.

15. *Intel Data Catalog*, 1975.

Semiconductor Memories

A.1 Semiconductor Arrays Get Bigger and Cheaper

Over the past few years the semiconductor memory device has grown from a temporary storage device of 8 bits to a main memory storage device of 4,000 (4k) bits. Even now research is going on for a storage device four magnitudes larger than the 4K device, such as INTEL 2416 16K bit CCD (charged-coupled device) serial memory. Not only are the storage capacities ever-enlarging, but the power per bit is dropping fast. As technology develops, these memory devices—with increased access speed and cost reductions—are now making this area of progress very interesting.

From the beginning the ferrite magnetic-core memory has been used for the main memory banks of computer systems, all the while competing with such other magnetic devices as plated wire, ferrite sheets, and a ribbon-like device called the twister. The core memory has also been reduced from its original price and is now only about 0.6 cent per bit. Now since it is possible to produce semiconductor memory circuits for this price or less, there is a new problem for memory-core manufacturers. Since the development rate is very high in semiconductor memories, it is only a matter of time until no new core memory systems are manufactured, not only for reasons of cost but for other reasons as well.

The mass memory market is probably the next area to be served by semiconductor memories. The downtime of most computer systems is usually caused by a mechanical malfunction rather than a electrical malfunction. Because of this, more and more designers are doing all they can to become as independent as possible of mechnical devices. This was clearly shown when CRT (cathode-ray tube) terminals replaced some teletype machines. The mass storage devices currently in use are magnetic drums, discs, tapes, and paper or card punches. All these devices depend to a high degree on a mechnical part of the device. If the semiconductor memory device can be implemented into the average computer system, the reliability of the system will be greatly enhanced. One big disadvantage of this scheme is the volatility of the all-electronic semiconductor memory device system (such as CCD memories). However, this is changing. For example, CMOS memories can be made nonvolatile if battery power is made available. Other nonvolatile memories such as nonvolatile magnetic bubble memories are in the development stage.

Another application for the semiconductor memory chip is in the replacement of complex logic circuitry or in the decoding and encoding of data lines. This application is also used in fixed program storage control under a turn-key environment. These uses are discussed further in Section A.3.

A.1.1   Three Categories of Storage Elements

The three categories of storage elements that will be discussed are random-access storage, sequentially accessed storage, and read-only random-access storage. Random-access storage involves retrieving or storing any particular bit of information in approximately the same length of time independent of the address of the item. Sequentially accessed storage has a wide range of time for retrieving or storing any particular bit of information that is dependent on the address of the item. Read-only random-access storage is used in special applications and are just what the term "read–only" indicates. Some special exceptions to this will be discussed later.

The random-access storage semiconductor device has undergone a vast amount of research. In the past the cost of the main memory of these computers was about 30 to 50 per cent of the total cost. With the development of MSI and LSI circuits for use in the CPU's of computer systems, the total cost of these computer systems has been reduced drastically. Now the cost of the main memory is about 50 to 95 per cent of the total cost. Core memory has been used for the main memory of computers for many years, and, if a manufacturer can find a suitable replacement for it, a large market will be opened up. The random-access storage semiconductor device is trying for a piece of this market.

Sequentially accessed storage devices provide data with varying time delays from a reference point, depending on the address of the data referenced. This characteristic usually defines the applications for this device. The semiconductor sequentially accessed memory device is a direct replacement for storage delay lines. While the average access time for any particular bit of information is very long compared to the random-access storage semiconductor device, as previously described, this device does have its own set of features. Two of these features, low cost per bit and low power per bit, make this device attractive as the semiconductor device to eventually replace the current mechanical–electrical mass storage devices used. Read-only random-access storage devices are relatively new to the memory division of logic circuits. In the past the functions currently provided by these ROM's were provided by diode matrix networks. The current ROM's can output as many as eight unique lines for 2,048 different addresses. The ROM has made possible fixed-program-storage computer systems for special applications. Other applications are in the decode circuitry, code conversion, and tabular implementation of complex functions (such as binary multiplication), with ROM's replacing popular logic networks.

A.1.2 Bipolar Family

When digital logic first began appearing in mass as integrated circuits, the transistor–transistor logic family was introduced. This was an obvious carryover from the discrete-logic-component days, when transistors were used throughout computer systems. In the early days, there was one flip-flop per integrated circuit. This flip-flop could be connected to the circuit as a storage cell for a RAM or as an element of a shift register.

The term "yield" will now be used to indicate the percentage of acceptable memory chip produced by a particular process. Silicon real estate, area per bit, and the complexity of the process all affect the manufacturing yield. The design goal is to achieve the greatest circuit density possible on a given silicon area, in accordance with good engineering practice and overall product reliability. The fabrication of the bipolar device is a most complex process, involving as many as 30 process steps. More information on this process can be found in Section A.4.

The main member of the bipolar family is the TTL standard circuitry that employs epitaxial material, plus buried collector regions in the epitaxial material. The electrical isolation is achieved with p-diffused regions. The TTL circuitry usually consists of a bistable flip-flop with one of two transistors in saturation. In order to switch the flip-flop, the transistor that is saturated and the transistor that is not saturated must exchange their state of saturation. Since one transistor is always turned on, a large amount of current will be used and a finite length of time is required to switch logic levels.

The fast member of the bipolar family is the ECL (emitter-coupled logic) memory device, which is similar to the TTL except that the emphasis for ECL is on speed, dictating low stray capacitance, close component spacing, and careful control of base–emitter input characteristics. Since it is a time-consuming process to bring a transistor out of saturation, the ECL fast-access-time performance is achieved by preventing the transistor in the storage cells from entering into the saturation region in contrast with

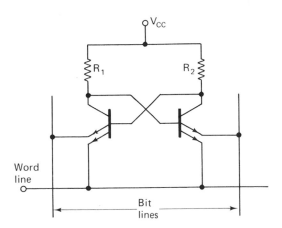

Fig. A.1.1 Bipolar RAM memory cell using multiple-emitter structures.

TTL. Furthermore, the OR–NOR decode ECL circuitry, with its multiple base–emitter structures, consumes more silicon area than the TTL multiple-emitter structures within a single base region of a TTL NAND decoder (see Figs. A.1.1, A.1.2, and A.1.3 for examples of this). Notice the lower values of R_L for the ECL memory cell; this causes the circuit to be fast but also increases the current consumed.

There are two other members of the bipolar family, which are not as widely

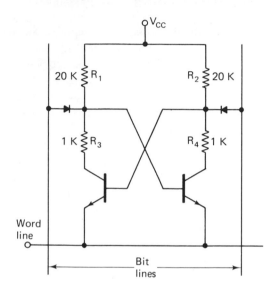

Fig. A.1.2 Bipolar RAM memory cell using diode coupling.

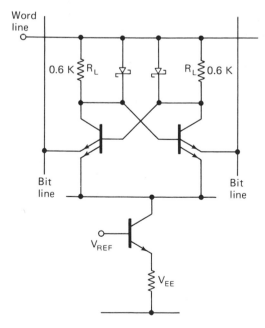

Fig. A.1.3 ECL form of memory cell. Two diodes as shown (sometimes Schottky) are usually used to limit saturation. Schottkys will actually keep the transistors out of saturation and in the active region.

known as TTL and ECL devices. One of these members is called *collector diffusion isolation* (CDI) and is designed to simplify and minimize the number of bipolar circuit processing steps. The CDI structure differs from TTL in two major ways: (1) it uses an epitaxial layer for the base structure, thereby eliminating the base-diffusion step; and (2) it combines the deep collector contact and isolation step. The simplicity of the CDI process causes other problems to develop, such as: (1) the collector–base breakdown voltage is lower; (2) the reverse current gain is higher; (3) the collector-junction capacitance is increased; and (4) the process has less control over base width; therefore, this method is incapable of high-speed performance, but it is inexpensive.

Isoplanar is a manufacturing technique and is not unique to any device family. The goal of isoplanar is to increase circuit density by compacting component areas. In the isoplanar process, the isolation regions consist of selectively grown oxide formed only to the depth of the buried collector. The depletion layers that accompany diffusion isolation are thereby eliminated, and the total chip area is reduced by up to 40 per cent. Also, an epitaxial base may be used and the process then requires only five photomasking steps. These improved process fabrications should result in improved yields.

A.1.3   MOS Family

The MOS arrays, used for memory storage devices with their low power dissipation and high packed density, are finding a place in main memory banks and in read-only and associative memories while yielding to bipolars only when speed is essential. This is the semiconductor memory technology that is the most glamorous. The main types of MOS devices are: (1) *p*-channel; (2) *n*-channel; (3) CMOS; (4) SOS; (5) CCM; and (6) ion implantion.

The MOS fabrication process involves less photolithographic and diffusion steps than those required for bipolar. As stated earlier, the more steps, the lower the yield. Furthermore, the starting material for MOS circuits is less expensive than for bipolar circuits. In addition, the circuit density of MOS can be compacted by a factor of 5 and thus results in a lower cost per bit for MOS as compared to bipolar. The MOS fabrication process will be discussed in Section A.4.

Gate control of the conducting channel of an MOS field-effect device is achieved by means of a metallic or doped polysilicon gate isolated from the channel by an insulating layer of silicon dioxide. The gate controls an electric field in the channel region, which in turn controls conductance. Gate input impedance is capacitative rather than resistive. Depending on the current-flow-process mechanism, an MOS device may have *p*- or *n*-channel configuration and operate in the depletion or enhancement mode.

The predominant type of MOS device is the *p*-channel enhancement mode (*p*-MOS) unit, in which holes are the vehicle of current flow. While this device is easy to produce, inexpensive, and reliable, it is relatively slow and limited in packing density. Because electron mobility is greater than hole mobility, *n*-channel transistors are two to three times faster than *p*-channel. Until recently, *n*-channel MOS devices have had various problems in the manufacturing processes; now manufacturers are using the *n*-channel MOS technology to produce some of the new large-capacity medium-speed memory devices.

MOS storage devices have two basic circuit arrangements. The two arrangements are a static storage cell and a dynamic storage cell. The static storage cell is similar to the bistable flip-flop. A particular MOS transistor is turned on or off depending on the bit being stored in the cell. An MOS cell can also be used as the basis for a dynamic memory system. In the operation of this device, data are written in by charging or not charging a capacitor. Reading the dynamic MOS memory cell entails sensing the presence or absence of a charge. Since there is junction leakage in the transistor, periodic refreshing is necessary to keep the capacitor charged or not charged, as the case may be.

Another technology that has received considerable attention is that of complementary MOS, or CMOS. This technology combines *p*-channel and *n*-channel transistors on the same substrate. Each storage cell in a CMOS memory device has an *n*-channel transistor, which is normally the driver device, and a *p*-channel transistor, which is the load. Figure A.1.4 shows a typical CMOS memory cell. Only one of these transistors can be conducting at a time, except during a switching operation; and the nonlinearity of the series combination is greater than that of either transistor working by itself. CMOS circuitry, being static in nature, exhibits extremely low standby power and high-speed operation.

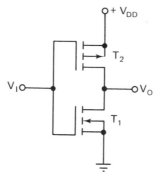

Fig. A.1.4 Typical CMOS memory cell.

Unfortunately, the CMOS memory device requires a very complex fabrication process. Since both *n*-channel and *p*-channel technology are used, CMOS requires two additional diffusions and photomasking steps as compared to a standard single-channel MOS. In addition, *n*- and *p*-channel devices must have electrical isolation as separate entities. Considerable chip area is consumed as a result of these situations, with the conclusion of not being able to increase the CMOS memory capacity with as much ease as with other MOS devices. Because of this capacity limitation, other processes are usually used for manufacturing memory devices.

Another member of the MOS family is the silicon-on-sapphire (SOS) storage device. This new technique, in which a very thin layer of single-crystal silicon is deposited on an insulating substrate of synthetically grown sapphire, will result in the

MOS transistors' capacitances of the source and the drain being reduced by a factor of 20 over standard MOS transistors made in bulk silicon material. The aluminum interconnections, since they are isolated from ground by the thick slab of sapphire, are also free of parasitic capacitance. The combined result of these low-capacitance values produces a memory device that has high speed *and* power dissipation.

By using SOS technology, shift registers with a very high clock rate, as well as RAM's with a 50-ns access time with low power, are possible. The capacity of these devices is presently limited to 256 bits. Also, at the present time, SOS structures are more expensive than bulk devices, as well as more complicated to manufacture, and will be used only where the additional cost is justified. As the cost differential is narrowed with improved manufacturing techniques, SOS devices should have an important place in the integrated-circuit memory industry.

Probably the most economic mass storage device to come to the attention of the designer is the charge-coupled device (CCD). The CCD is a relatively new concept with regard to semiconductor storage devices. These devices have no preprepared *p-n* junction, and the circuitry does not require load devices or flip-flop configurations. It is basically an analog shift register or just a linear array of closely spaced MOS capacitors. Minority-carrier charges are stored in potential wells created at the silicon surface and are transported along the surface by moving the potential wells.

Clearly, 100 per cent of the charge cannot move instantaneously from one potential well to another; therefore, the frequency must be comparatively low, about 500 kHz, and the fraction of the charge left behind or the loss per transfer determines how many transfers can be made before the signal is distorted. At the present time, charge-transfer efficiencies of 99.99 per cent have been achieved, giving way for some shift registers of about 1,000 bits. When the transfer improves, very large memories will be possible.

The ion-implantation process can be used to adjust the threshold-voltage values in the gate region after the formation of source and drain areas. The ions can be either the *p*- or the *n*-type doping variety. This method may be applied to *p*-MOS, *n*-MOS, or CMOS devices. An application of this process is the forming of *p*-channel depletion-mode loads for *p*-channel enhancement-mode drivers all on the same chip, which is similar to CMOS because the circuitry is insensitive to power-supply variations, the internal voltage swings are almost equal to the supply voltage, and the operation takes place from a single voltage power supply. The speed and power dissipated will be improved with this method. A drawback of this process is that an expensive ion accelerator is required, and, in addition, the technique is difficult to adapt to multiwafer batch processing, even though some manufacturers are now producing this memory device.

A.1.4 Comparison of Technology Characteristics

There are many characteristics by which a semiconductor memory device may be evaluated. For any particular application, one or more of these characteristics will become a dominate feature of the device. The seven most important characteristics are: economy, low power dissipation per bit, speed, small area per bit, nondestructive readout, nonvolatility of the memory cell, and last, and of great importance, reliability.

In general the digital design engineer will first consider cost, speed, and power dissipation. After a basic device type has been selected, the engineer will then determine a device that fits his other requirements. High speed is attainable with an ECL storage device, but we sacrifice high power consumption and cost. A p-MOS device will cost less and consume less power, but the speed will be greatly reduced. The speed can be improved by using n-MOS or CMOS, but at the same time the cost will greatly increase. SOS will have 50-ns access time and low power dissipation, but again the cost increases.

The reliability of the device and the area per bit of the device are two parameters that the semiconductor engineer is always worried about. If the device is not reliable, the manufacturer will definitely not make any money. The area per bit parameter determines a number of things. It not only determines the capacity for a particular device, but it determines the number of chips a manufacturer can get from a 3-inch slice during the manufacturing process. This parameter also interacts with the reliability and the yield of the product. Naturally the larger the total area of the chip, the more likely the device will not work to start with, and also the more likely the device will fail later. For example, a simple bipolar cell has up to eight transistors, but a CCD cell only has up to four capacitors. This is why the capacity of an MOS device is between 1,000 and 16,000, and the capacity of a CCD device will have a capacity of 16,000 bits or greater.

Yield is not one of the seven most important characteristics to the applications engineer, but it is of great importance to the semiconductor process engineer. The yield must be good for a product to be sold with a profit. This parameter alone dictates when a new semiconductor device will be introduced. The more complicated the process and the larger the area per bit, the lower the yield will be. Sometimes the process will be made more complex in order to reduce the chip area. These two characteristics must be adjusted to have the best yield.

All the processes mentioned in Sections A.1.2. and A.1.3 are giving good yields, with the exception of the bipolar isoplanor, the MOS SOS, and MOS CCD process. Semiconductor manufacturers should have these process yield problems worked out within the next few years. Rapid change is the byword for the evolution of the semiconductor memory-device process and design. With the competition at a peak level, a new device introduced last year, and used this year, will be obsolete next year. For example, look at the manufacturers of the 4K RAM.

A.2 Random-Access Memories

As stated previously, the memory of a computer system is becoming the focal point of computer research. The 750-ns memory-access-time barrier can now be passed with the advent of high-speed semiconductor storage device. These RAM's are now utilized in the computer system not only as scratch pads and temporary buffers, but also as main-frame memories. The magentic core has played this role for so long that the user of a computer often makes reference to his main-frame memory as "core" even if his memory bank has no core elements.

Along with increased access speed, the semiconductor storage devices can be used for large memory capacities and be very compact. This feature alone enables the computer manufacturer to build complete computer systems in smaller and smaller packages. Now with the arrival of the one-chip computer, a 6- by 8-inch printed circuit board may make up the entire computer system, including CPU, memory, and input/ontput ports.

A.2.1 Bipolar Random-Access Memories

Bipolar RAM's have been used for some time in scratch-pad applications for the register array in the main CPU. These RAM's are extremely fast, with an access time of 10 to 200 nsec. Until the last few years bipolar RAM's had a limited capacity of about 64 bits. Now the SN74200, a 256-bit RAM, has been accepted, and manufacturers are currently introducing 1,024-bit RAM's (SN74S209) with comparable access time.

Speed is of primary interest for bipolar RAM's. Power dissipation and the physical size of the storage cell come second. These last two parameters are not independent of each other. Because the integrated circuit package can only dissipate a limited amount of power, the cell capacity must be limited for overall power dissipation considerations.

The standard storage cell for the bipolar RAM is the bistable flip-flop (Fig. A.2.1) consisting of two cross-coupled inverters. Some modifications to the basic cell are the multiple-emitter TTL–RAM cell (Fig. A.2.2), the diode coupled cell (Fig. A.2.3), and the multiple-emitter ECL–RAM cell (Fig. A.2.4).

The widely used TTL–RAM is read by raising the word line, thereby transfering current from the word line to the bit line of the turned transistor. A current-sensing amplifier will then detect the memory state of the cell selected. By lowering the voltage

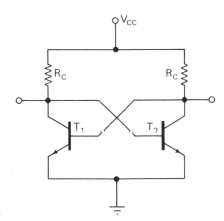

Fig. A.2.1 Bistable flip-flop circuit.

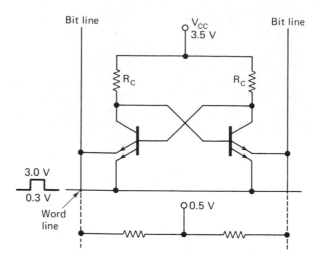

Fig. A.2.2   Multiple-emitter RAM memory cell (TTL).

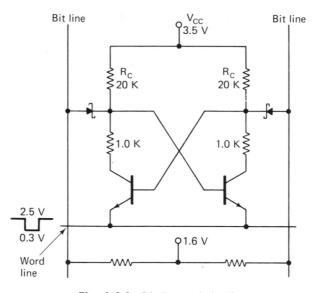

Fig. A.2.3   Diode coupled cell.

on one bit line, the transistor with the lowest emitter voltage will be turned on and the data will be stored in the memory cell when the word line selected is lowered. The R_c of the cell must be relatively low for fast switching speeds, and thus the cell itself will consume quite a bit of power.

The diode-coupled cell is particularly suited for the Schottky TTL process because the two additional gating diodes can be of the Schottky variety. Since R_c does not

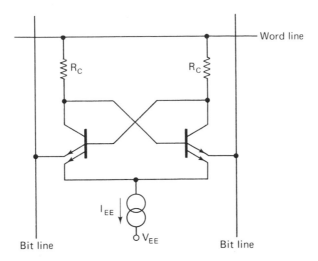

Fig. A.2.4 Multiple-emitter RAM memory cell (ECL).

now affect the access speed of the device, it can be increased to reduce the power consumption of the cell. The voltage swings of the word line during read/write operations and the bit line during write operations is smaller than the previous TTL–RAM circuit. This circuit has a lower power consumption and is used in the SN74200 bipolar RAM.

The ECL–RAM is similar to the TTL–RAM except that the transistors are never driven into saturation. Also, when a read operation is selected, a large sense current is available on one bit line for the sense-amplifier. This cell is the fastest cell of the group under study. It is, however, somewhat difficult to read because it generally presents a low (< 0.5 V) differential voltage to the sense-amplifier.

A.2.2 MOS Random-Access Memories

The initial form of the MOS RAM cell is similar to the bistable flip-flop circuitry with load resistors being replaced with MOS transistors (Fig. A.2.5). This difference alone increases the circuit density and produces higher yield. As stated previously, MOS memory cells have two basic configurations. These are the dynamic and the static modes of operation. Also, the newer MOS devices have TTL compatible inputs and outputs, as well as static-charge protection.

The static RAM cell has progressed from an eight-transistor cell (Fig. A.2.6) to a six-transistor cell (Fig. A.2.7). The operation of the eight-transistor cell is similar to the bipolar cell, and the operation of the six-transistor cell is also similar, with the exception that address circuitry and the bit lines have been combined. The two cross-coupled transistors, T_1 and T_2, serve as the storage elements, while T_5 and T_6 are transistors that connect or isolate a given cell from the sense-bit lines. The word-

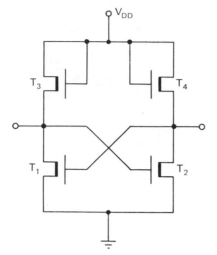

Fig. A.2.5 MOS bistable flip-flop.

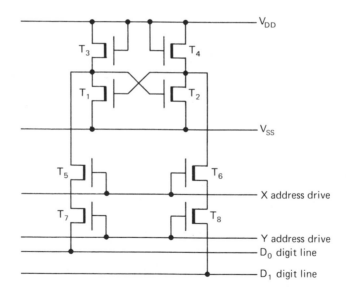

Fig. A.2.6 8-transistor static RAM MOS cell.

select line controls the on–off condition of T_5 and T_6. To write or read with this cell, a particular column-select line and a particular word-select line are enabled, thereby selecting the cell. For a read operation, a current will be or will not be sensed by the sense amplifier after cell selection. For a write operation, a low level is set on one of the two bit lines, with the other bit line left on high. At this time T_1 or T_2 will turn on and the other transistor of the pair will turn off, and the memory state will be locked in at the completion of the write operation when the word-select line is returned to V_{ss}.

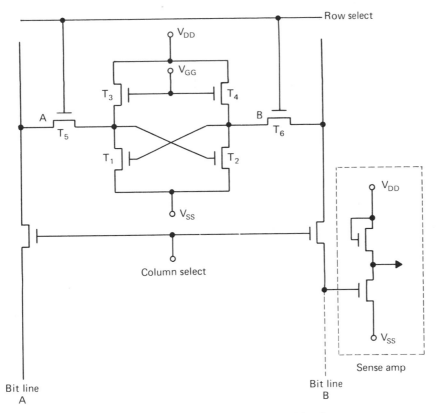

Fig. A.2.7 6-transistor static MOS RAM cell.

Figure A.2.8 shows a typical word-select circuitry for selecting a particular memory cell.

Since stable states exist unless purposely disturbed, the static cell is latched and does not require a refresh cycle or any periodic clocks. But, since the circuit requires at least six transistors per cell, a large-capacity memory device is not feasible. For primarily this reason, the dynamic memory cell was introduced.

Circuitry can now be manufactured to lower the number of transistors used per memory cell. The state-of-the-art MOS storage cell has only one transistor and one capacitor. This is a remarkable improvement over the six-transistor static MOS–RAM. This one-transistor cell is used as a dynamic memory cell. That is, data are written in by charging or not charging a capacitor, depending on the value of the data. By sensing the presence or absence of the charge, data may be read out. Since there is junction leakage in the transistor, periodic refreshing is required to keep the data valid. The refresh period varies with temperature and is usually required every 2 ms. Figure A.2.9 shows a single-transistor dynamic-memory cell. The features and drawbacks of this cell will be discussed in Section A.2.3.

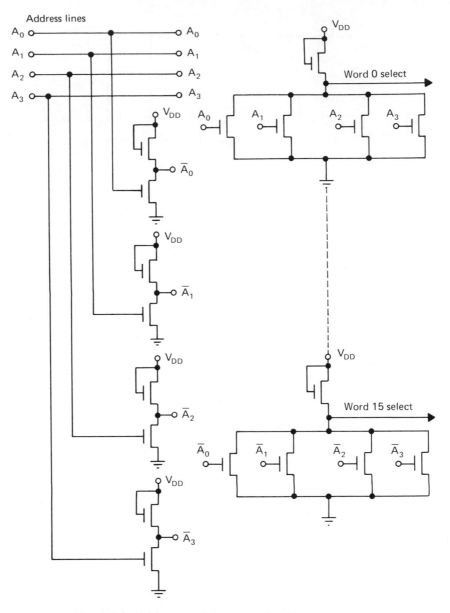

Fig. A.2.8 Address word decoder with 4 binary inputs.

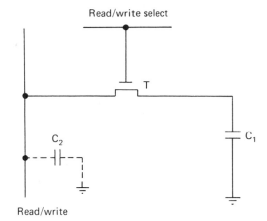

Fig. A.2.9 1-transistor dynamic MOS RAM cell.

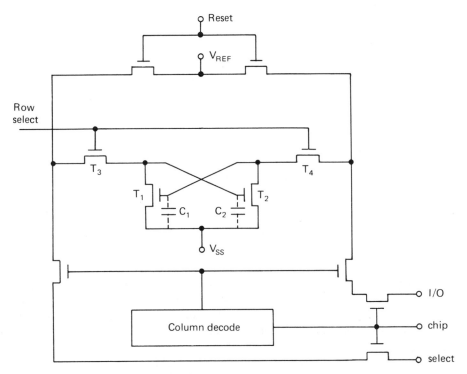

Fig. A.2.10 4-transistor dynamic MOS RAM cell with overhead circuitry.

Before the one-transistor memory cell was produced in high quantities, the four-transistor (Fig. A.2.10) and the three-transistor (Fig. A.2.11) dynamic memory cells were widely used. The feature of the four-transistor cell is that it is very simple to design external refresh circuits. The three-transistor cell is used in the 1103 RAM. This

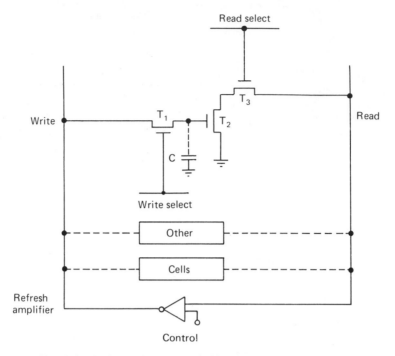

Fig. A.2.11   3-transistor dynamic MOS RAM cell with overhead circuitry.

cell also employs simple refresh circuits, but unfortunately the read/write address pulse shapes are critical to the unit working properly. All of these dynamic RAM's have an access time of about 300 ns, with lower power consumption than comparable static MOS RAM's.

A.2.3   4K Random-Access Memories

A relatively new semiconductor memory chip, the 4K RAM device, is manufactured by a number of companies. These new memories offer the systems designer a wide variety to choose from, in contrast with the 1.024-bit MOS RAM's, which were mostly derivations of the three-transistor-per-cell 1103. 4K RAM's were in the development phase for a number of years, and each manufacturer therefore has his own idea of the features the 4K RAM should have. Nearly all 4K RAM's have the following characteristics: (1) n-channel silicon gate; (2) dynamic; (3) TTL compatibility; (4) 2- to 4-ms refresh period; (5) 16- or 22-pin package; and (6) operating power of about 300 mW.

Some of the parameters that vary with the manufacturer are: (1) access time (from 180 to 450 ns); (2) power-supply voltages; (3) standby power (from 0.1 to 40 mW); (4) clock requirements; and, most important, (5) transistors per cell (1 or 3). Also, Mostek has a 4K RAM that comes in a 16-pin package. At first this may seem impossible, since 12 pins are used for the address with 6 or more pins being used for power supplies, data in and out, and control lines. Mostek requires the 12-bit address to be entered in two 6-bit bytes. During the first 150 ns the first six bits are latched into the chip, by the row address strobe. Approximately 150 ns later in the memory cycle, the second six bits are latched into the chip, by the column address strobe.

Most manufacturers are using the standard three-transistor dynamic memory cell. As a sesult of the amount of research that has already been done on this cell, they feel relatively safe, from a yield standpoint, to manufacture it. At least two manufacturers, Texas Instruments and Mostek Corporation, are using the one-transistor dynamic memory cell. They have overcome some large problems in using this cell. The cell's major problem is that of obtaining a large-enough voltage swing to read information out of the cell easily. This problem is due to the large ratio of bit-sense-line capacitance to the cell-storage capacitance, and thus in the readout of the original voltage stored in the cell, a reduction of this voltage will occur. Evidently this problem has been solved because all the leading semiconductor memory manufacturers indicate that their yields of the 4K RAM's are now good.

A.2.4 Outlook for the Future

The outlook for both bipolar and MOS RAM's looks very bright. The main goal for bipolar research is to manufacture a memory device that not only has a very fast access time but also has low power consumption and is economical. Design goals for MOS memory devices are to improve access time and bit capacity per chip; 16K MOS storage devices are now under development.

The standby power consumption of the bipolar memory device is being lowered with additional development, to power-down large portions of a particular memory bank when not in use. Bit-density improvement is very important because it would solve a number of problems. This would result in a bipolar 4,096-bit RAM. Other circuit designs are being investigated to replace the bistable flip-flop. These are: the two-diode array; the two-transistor array with avalanche breakdown; and the Wiedmann–Berger memory cell. New fabrication processes (isoplanar and CDI) are now being developed, as described earlier.

The guiding principle of MOS RAM design is that of economy. The initial cost of the device is important, but the cost of the support circuitry is equally important. The address-line decoding circuitry has been incorporated on the memory chip, and probably in the future the dynamic cell will operate with an internal refresh circuit. The system designer need not concern himself with the details of cell refresh as such. New MOS fabrication techniques, such as using an electron beam during the photolithographic process, have been developed. The yield of these techniques still needs to be greatly improved for them to become favorable.

A.3 Read-Only Memories

Read-only memories are intended to store nonchanging information. Although some ROM's can be written in or altered, the write operation is complex as compared to a read operation. During normal use they are considered fixed in their memory content. Either the bipolar or MOS fabrication process may be used during the manufacturing of a ROM. Of course the access time and power consumption depends on which process is used.

Since the basic circuit of a ROM memory cell is very simple, a high-capacity chip of 16K bits is currently being sold. The ROM is manufactured in a variety of sizes, because each designer has a unique use for the ROM and usually does not need a very large capacity chip. Two used for the ROM are: (1) as a substitute for a combinatorial logic network, and (2) as a device to store a fixed program for a computer to read out.

Normally the bit pattern will be custom-made at the semiconductor manufacturer's location for a particular customer or for a commonly used logic decode purpose. An example of the previous application is a character dot generator used in CRT display circuits. Some ROM's are made so as to be field-programmable by the user. Of course, this is more expensive for large quantities, but it is cheaper for small-quantity deliveries.

A.3.1   Field-Programmable Read-Only Memories

The field-programmable ROM, or p/ROM, is fabricated using fusible metal parts throughout the circuit. In this way the connections to the individual emitters for each of the memory cells can be open-circuited in a controllable way and the memory may be programmed after packaging. Only bipolar ROM's are used at this time. Programming this ROM is very simple and does not require complex equipment.

A few problems occur with the p/ROM; one concerns factory testing. This problem is similar to that of testing flashbulbs in that until the ROM is programmed, nobody can be sure that the circuitry will operate satisfactorially. Also, once the ROM is programmed, it cannot be reprogrammed, so programming mistakes will result in wasted units. There does exist a reprogramable ROM, but it uses circuitry different from the p/ROM and is more expensive to operate.

A.3.2   Reprogrammable Read-Only Memories

The main problem with a large ROM is that it is a one-time programmable device. That is, once a bit pattern has been installed, the device is thrown away if the bit pattern must be changed. Electrically alterable ROM's, reprogrammable ROM's, have now been designed just for this reason. These devices are field-programmed, and if the need arrises, they can be field-programmed again and again. While this memory device is called a reprogrammable ROM, in truth it is a RAM with a very slow and complex write operation. A definite requirement of this device is that it be nonvolitile. This is one of the few cases in the semiconductor memory field in which a nonvolatile device is encountered.

The best known and most readily available device of this nature is an arrangement that typically uses an array of floating-gate p-channel MOS transistors. This

ROM is made ready for programming by exposure to ultraviolet radiation, which discharges the gates. A tape is next placed over the device's window and standard programming is started. A high voltage between the source and the drain removes positive carriers from the floating gate, and the isolated negative carriers on the gate create a conducting channel.

Another semiconductor storage device that offers the nonvolatile feature is the MNOS, metal–nitride–oxide–silicon, device. Binary data are stored by shifting the threshold of a MNOS transistor to a high or a low value. The threshold shift is caused by the trapping of charge in deep-energy traps in the specially prepared gate insulator. Once the threshold is set, data are stored without further consumption of power, the readout is nondestructive, and the data may be electrically reset. In its present form, the MNOS memory system is medium-speed, nonvolatile, extremely low powered, and expensively priced at 2.5 cents per bit in 1,000-unit quantities.

The access time of the reprogrammable ROM is of the order of 1 μs (for example, INTEL 8702A 2K(256 × 8) reprogrammable ROM has access time spread at 1.3 μs. Reprogrammable ROM's are great devices to use in the design phase, while the bit pattern of a ROM is continuously changing.

A.3.3 Read-Only Memories as Combinational Logic Circuits

The ROM has opened up a wide range of applications for memory devices to take the place of logic circuits. These include code conversion, table look-up, arithmetic logic, control logic, and character generation.

An example of the use of a ROM for a logic circuit is demonstrated in a code-conversion application. A particular bit code is translated into another bit code. The contents of a memory location is the translated code for the particular bit code used for the address of this location. The bit size does not necessarily have to be the same for the address and the bit length of the contents. A character generator will give out 35 bits of information for a 7-bit address code.

ROM's are very much in use to replace the control logic in small to medium-sized computer control sections. Some minicomputers have as many as 256 by 64-bit ROM arrays. A fast multiply can be accomplished using ROM's manufactured for this purpose. The 74187 is an example of this product. Needless to say, the ROM has a definite use in today's computer systems.

A.3.4 Fixed Program Storage

During the past few years the cost of a small computer has dropped significantly. With this have come many additional applications. Small computers are usually programmed for a particular function, as contrasted with the larger computer, which is programmed for a large variety of general applications. Because the same program is always in the memory of these computers, it would simplify things if the program were locked into the memory. A great deal of time could be saved each time the computer system was powered up if the program did not have to be loaded into memory. This may be accomplished by use of large ROM's. Such a system is called a *turn-key system*, because once the power is turned on, the system is "off and running."

With the advent of the one-chip CPU (central processing unit) many small computer systems can be used to replace complex logic circuitry hardware. All these applications require fixed program storage, which the ROM can provide. These fixed-program-storage ROM's are also used for part of the memory in large-scale memory systems. This area of memory has special software routines that are used quite often by the program. This ROM use for the storage of software instructions to be executed by the computer hardware is called *computer "firmware."*

A.4 Fabrication Processes

A.4.1 Bipolar Fabrication Process

Bipolar fabrication is more complicated than MOS fabrication. Epitaxial silicon is used in the first process step. In the epitaxial materials preparation, *n*-type single-crystalline silicon is grown on a *p*-type silicon substrate at a temperature of 1200°C. *N*-type impurity dopant is added to the gas flow, and a 0.2-mil-thick sheet of *n*-type silicon is formed on top of the *p*-type substrate. Prior to the epitaxial disposition, regions of heavy *n*-type dopant are diffused into the substrate. These epitaxial slices are a factor of 3 to 5 times more expensive than the silicon starting material used in MOS. The first process step consists in thermally oxidizing these preprepared epitaxial slices of several mils thickness. Then, employing photolithographic techniques, holes are etched in the oxide in preparation for isolation diffusion. In the next step, the photolithographic process is used to form holes in the oxide where *p*-base regions or *p*-resistor regions are desired. After another diffusion is completed and is followed by suitable oxidation, a photolithographic process is used to cut holes in the oxide, this time for n^+-emitter structures and n^+-collector contact regions. Diffusion of the n^+-emitter structures is performed in a furnace that contains phosphorus. Finally, contact holes are etched in the oxide and aluminum is deposited on the slice. The metal is then selectively removed in a photomasking step to define the desired interconnection pattern. Usually a protective glass layer is deposited on the completed wafer with contact holes cut in the glass layer to bond connecting wires between the chip and package terminals.

A.4.2 MOS Fabrication Process

The starting material for *p*-channel MOS processing is *n*-type silicon having about 10^{15} donor atoms per cubic centimeter. The round crystal slice is 3 inches in diameter and 10 mils in thickness. First, a uniform layer of silicon dioxide is formed over the entire surface of the slice at a temperature of 1,000°C for about 1 hour.

Next, the oxide layer is selectively removed wherever the source and drain regions are to be formed by a process of photomasking and oxide etching, which is repeated a number of times during one processing sequence. The slice is next subjected to an atmosphere of acceptor impurity atoms, again at a high temperature. During the subsequent oxidation step, the boron diffuses into the lattice, doping the crystal. Where the silicon is masked with oxide, the impurity atoms are not able to enter the lattice, and the conductivity of the substrate will remain as it was originally.

In the next step additional oxide is grown over the entire slice. Growth of a layer this thick takes an appreciable length of time. It is during this time that the boron diffuses into the substrate to produce the source and drain regions. Another photo-masking operation opens regions where contacts are made to the source and drain regions and where the thin dielectric layer of SiO_2 is formed under the gate. The oxide is completely removed in the gate region and then re-formed by an additional thermal oxidation step.

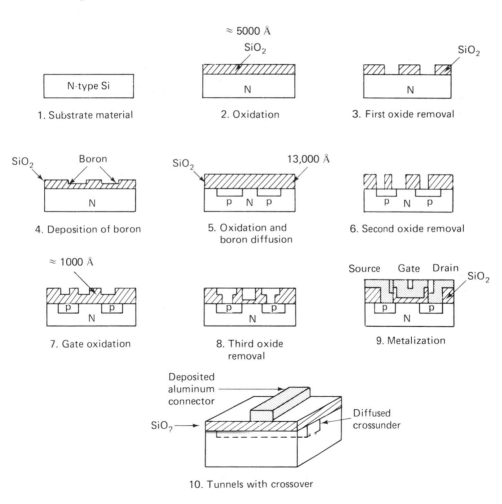

Fig. A.4.1

Another photomasking operation is required to protect the gate oxide while the oxide in the contact holes is removed. Next, to interconnect the components, a layer of aluminum is deposited over the entire slice and then selectively removed. This film of aluminum forms the connections to the source and drain regions, the gate electrode, and the interconnections from device to device. It is difficult to design complex circuits with only a single layer of interconnecting metal. A way around this is to use diffused cross-unders, as shown in Fig. A.4.1.

References

1. G. Luecke, J. P. Mize, and W. N. Carr, *Semiconductor Memory Design and Applications*, McGraw-Hill, New York, 1973.

2. W. N., Carr and J. P. Mize, *MOS/LSI Design and Application*, McGraw-Hill, New York, 1972.

3. R. G. Hibberd, *Integrated Circuits*, McGraw-Hill, New York, 1969.

4. W. R. Runyan, *Silicon Semiconductor Technology*, McGraw-Hill, New York, 1965.

5. *COS/MOS Integrated Circuits Manual*, RCA Corporation, 1971.

6. *McMOS Motorola Handbook*, Motorola, Inc., 1974.

7. *The TTL Data Book*, Fairchild Semiconductor 1974.

8. *MOS*, Signetics Corporation, 1974.

9. *The TTL Data Book*, Texas Instruments, 1973.

10. *The Semiconductor Memory Data Book*, Texas Instruments, 1975.

11. *Intel Data Catalog*, 1975.

Laboratory Experiments

Experiment 1
Static Characteristics of IC Gates

Equipment

4 Power supply 5V, (500 mA)
1 Voltmeter,
1 Ammeter (mA)

Components

3 SN 7400
1 Bread board
1 Nose plier (small)
1 Cutter (small)
1 10K Potentiometer, 1/2 watt (±10%)

1. Measure and plot the voltage transfer characteristic of a quad 2-input NAND gate in TTL (SN 7400). Vary the input voltage continuously from 0 to 5 volts and measure the output voltage. Do this with the gate unloaded and loaded with rated number of gate loads. Indicate $(V_{OH})_{min}$ and $(V_{OL})_{max}$.

2. Using the circuits in Figs. B.1.1 and B.1.2 plot the $I_{OH} - V_{OH}$ and $I_{OL} - V_{OL}$ characteristics of the 2-input NAND gate. Indicate $(I_{OH})_{max}$ and $(I_{OL})_{max}$ and also find the output impedance Z_{OH} and output impedance Z_{OL}.

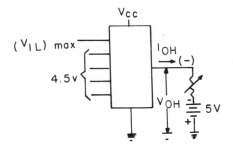

Fig. B.1.1 $V_{IH}, V_{IL}, V_{OH},$ and I_{OH}.

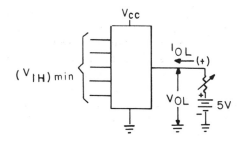

Fig. B.1.2 $V_{IH}, V_{IL},$ and V_{OL}.

3. Using the circuit in Fig. B.1.3, plot the $I_i - V_i$ characteristic. Determine the current $(I_i)_{max}$ at which $V_i = (V_{OL})_{max}$.

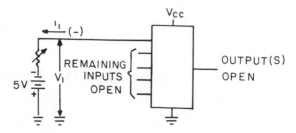

Fig. B.1.3 V_I.

4. Determine the fanout of the gate. The fanout η of the gate is computed by

$$\eta = \frac{(I_{OL})_{\max}}{(I_{iL})_{\max}}$$

Verify your result.

5. Measure the power consumption of a quad 2-input NAND in TTL SN 7400 for various combinations of inputs. What would you consider a reasonable average power rating *per gate*, when used as a part of a large system of many gates?

Experiment 2
Dynamic Characteristics of IC Gates

Equipment	*Components*
1 Dualtrace oscilloscope	1 SN 7400
(at least 50 MHZ band width)	1 SN 74106
1 Power Supply 5V	1 CD 4013AE
1 Voltmeter	5 Resistors 1K, 1/4 watt, 10%
1 Pulse generator (1 ηs rise time)	1 Bread board
	1 Nose plier
	1 Cutter

In Exp. 1 we looked at the static characteristics of gates. These are very important, but since gates are rarely used in static situations, these static characteristics are not sufficient. In most situations we wish to operate logic at the maximum possible speed, therefore a good understanding of dynamic characteristics is even more important.

The two most important dynamic characteristics are *speed* and *delay*. Speed refers to the maximum rate at which we can make changes at the input, while delay refers to the time required for a change at the input to be seen at the output. While these two characteristics are distinct, they are not independent of one another. For example, we might expect problems if we make a second change in an input while the output is still responding to the first change.

The problem of characterizing speed and delay in terms of specific measurements is somewhat different for logic than for linear circuits because the output waveforms are not necessarily the same as the inputs. For example, the input and output rise times may be quite different, raising problems as to what points in the waveforms you use in measuring the delay.

1. Speed

With reference to Fig. B.2.1, we will characterize speed as the duration of the minimum pulse at A such that the pulse at B_1 fails to reach the specified minimum $V_{H_{in}}$, that is, the minimum voltage that the manufacturer guarantees will be interpreted as a high input by succeeding stages.

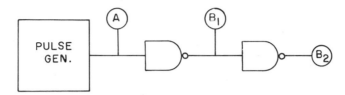

Fig. B.2.1

(a) Measure the speed as defined above for TTL SN 7400. For pulse widths of 200 nsec, and minimum pulse width, sketch the pulse generator waveform and the waveforms at the outputs of each gate.

(b) Repeat the above measurements with the eight gates of two quad 2-input NAND gate connected in parallel to point B_1. Do this first with all unused inputs tied high, then tied low.

2. Delay

Because input and output waveforms are not the same, it is necessary to specify the points at which delay is to be measured. Since gates switch at a certain voltage levels, it is standard to specify delay from the time the input passes through the nominal switching level (1.5v for TTL) to the time the output passed through the same level. The delay may depend on whether the gate is turning on or turning off, so we specify two delay times, t_{pd+} for a positive-going output transition, t_{pd-} for a negative-going transition, as shown in Fig. B.2.2.

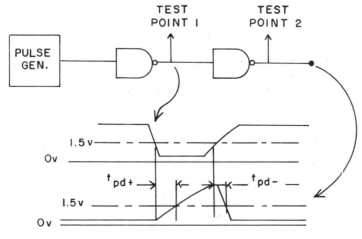

Fig. B.2.2

(a) Using the setup shown in Fig. B.2.2, measure the delays for TTL SN 7400, with a single gate as load. Do this with all unused inputs of the gate under test tied high, then tied low.

(b) Repeat 1 with the gates loaded with their rated loads of 8 gates.

3. Repeat 1 and 2 for a TTL J-K negative-edge-triggered flip-flop with preset and clear, say, SN 74106.

4. Repeat 1 and 2 for a CMOS flip-flop CD4013AE.

Experiment 3
Combinational Logic Design Using SSI IC Chips

Equipment	Components	
1 Voltmeter	1 SN7400	4 Resistors 1K, 1/4 watt, 10%
1 Power supply 5V	1 SN7402	1 Bread board
	1 SN7454	1 Nose plier
	4 LEDs	1 Cutter

1. Construct an "AND" gate, an "OR" gate, a "NOR" gate, an EXCLUSIVE OR gate using "NAND" gates only and study their operations.

2. The Boolean expression $F(A, B, C) = AB + AC$ is in a sum-of-products form and can be realized using a two-level AND/OR representation. Using DeMorgan's theorem, obtain a NAND-NAND realization of the expression, build the circuit and verify its operation by completing the truth table shown below.

A	B	C	F
0	0	0	
0	0	1	
0	1	0	
0	1	1	
1	0	0	
1	0	1	
1	1	0	
1	1	1	

3. A two-level NOR-NOR realization of some expression, $F(A, B, C)$, is shown in Fig. B.3.1. Determine the Boolean expression and express it in the sum-of-products form. Build the circuit shown in Fig. B.3.1 and verify its operation by completing the truth table.

A	B	C	F
0	0	0	
0	0	1	
0	1	0	
0	1	1	
1	0	0	
1	0	1	
1	1	0	
1	1	1	

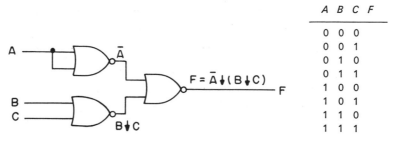

Fig. B.3.1

4. Minimize the function $f(A, B, C) = \sum (0, 4, 5, 7)$ using a Karnaugh map and determine minimum NAND-NAND realization of the function. Build the circuit and verify its operation.

5. Find the minimum sum-of-products form of the function $f(A, B, C) = (A + B + \bar{C})$ $(A + \bar{B} + \bar{C})(\bar{A} + B + C)$.

Experiment 4
MSI Combinational IC Chips

Equipment	*Components*	
1 Power supply 5V	2 SN 7400	13 1K, 1/4 W., 10%
1 Voltmeter	1 SN 7410	1 SN 7480
	1 SN 7432	1 SN 7487
	10 LEDs	1 SN 74184

1. A 1-bit binary full adder has three inputs and two outputs as shown in Fig. B.4.1. It is capable of adding two 1-bit numbers, A and B plus a carry C_{in} from a prior addition. The two outputs are the sum S and the carry C_{out}. The latter is used as an input to the next 1-bit binary full adder.

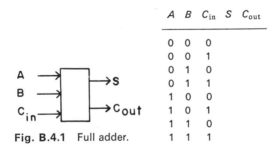

A	B	C_{in}	S	C_{out}
0	0	0		
0	0	1		
0	1	0		
0	1	1		
1	0	0		
1	0	1		
1	1	0		
1	1	1		

Fig. B.4.1 Full adder.

The truth table for a full adder is shown above. You are provided with two input, three input Nand gates, and two input OR gate. Verify your result with SN 7480.

2. (a) Design a logic circuit which will satisfy the truth table in Fig. B.4.2

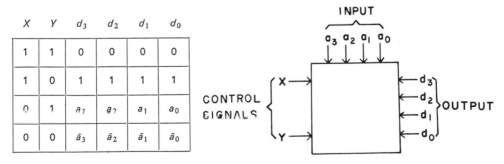

X	Y	d_3	d_2	d_1	d_0
1	1	0	0	0	0
1	0	1	1	1	1
0	1	a_3	a_2	a_1	a_0
0	0	$\bar{a}_3$	$\bar{a}_2$	$\bar{a}_1$	$\bar{a}_0$

Fig. B.4.2 True-complement zero-one circuit

(b) Verify you result with SN 7487.

3. Construct the circuit as shown in Fig. B.4.3. What function does this circuit perform?

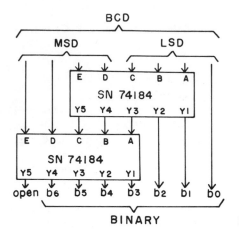

Fig. B.4.3

Experiment 5
Design of a Clock

Equipment

1 Oscilloscope
1 Power supply 5V
1 Voltmeter

Components

2 SN 7492   8 500 Ω resistors 1/4w., 10%
1 SN 7495   1 Dual in Line IC socket
1 SN 7446   3 Printed circuit-board [276–024 Radio Shack]
1 SN 7417   1 Metal cabinet, cat. no. 270–236 [Radio Shack]
1 DL 707   2 Banana Jacks 274–724 [Radio Shack]
4 LEDs   1 3/4″ Knob (274–415)
1 NE 555   2 Nuts, Screws, 1 1/4″

1. Construct the circuit shown in Fig. B.5.1.

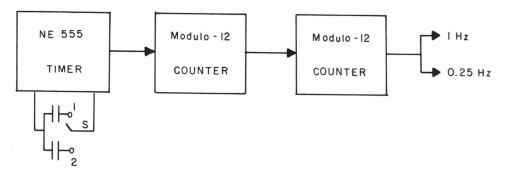

Fig. B.5.1

(a) First, switch the switch S to position 1 and measure the frequency of the output.
(b) Next, swtich the switch S to position 2 and measure the frequency of the output.

2. Design an up/down (0–7) synchronous counter using JK flip-flops. The counter should be designed such that there is one input, IN, which determines the mode of operation. If IN = 0, the counter increments and if IN = 1, the counter decrements. Build the counter and verify its operation.

3. A shift register is basically a series of flip-flops connected such that the state of each flip-flop is transferred to the next at each clock pulse. A shift register may shift right or shift left. Using a Lamp Set, observe the operation (left and right shift) of a shift register (SN 7495) and make a timing diagram.

4. A standard synchronous up/down counter, if clocked a sufficient number of times, will reset when the maximum counting sequence has been exceeded, i.e., a 4-bit counter will count to 15 (BCD) and reset. The counting sequence may be modified, however, such that the counter will count to some number n and reset ($n <$ max count). One method which may be used to construct an n-counter is shown below. The counter (SN 74193) increments on each clock pulse and the output is compared (SN 7485) to some number n. When the output of the counter is equal to n, the counter is reset. Design a counter that counts to 10 and resets, and verify its operation. The counter should be preset so that the counting sequence begins at 0000.

Experiment 6
Design of a Digital Clock

Equipment

1 Power Supply
1 Voltmeter

Components

1 SN 74193
1 SN 7485
4 SN 74185
2 SN 7446
2 SN 7417
2 DL 707
14 Resistors 500Ω, 1/4 W., 10%
1 Clock

1. (a) Design an asynchronous (ripple) decade counter using the following procedure. Suppose SN 74106 J-K flip-flops are used in the design.

(i) Find the number n of flip-flops required:

$$2^{n-1} \leq N \leq 2^n$$

where N = counter cycle length. (For our problem, $N = 60$.) If N is not a power of 2, use the next higher power 2.

(ii) Connect all flip-flops as a ripple counter, i.e., all J and K inputs are tied together and applied with a constant logic 1 input and the output Q of $(i - 1)$th J-K flip-flop is connected to the clock of ith J-K flip-flop.

(iii) Find the binary number $N - 1$.

(iv) Connect all flip-flop outputs that are 1 at the count $N - 1$ as inputs to a NAND gate. Also feed the clock pulse to the NAND gate.

(v) Connect the NAND gate output to the preset inputs of all flip-flops for which $Q = 0$ at the count $N - 1$.

(b) The maximum clock frequency for a ripple counter is given by

$$f \leq \frac{1}{N \cdot (T_p) + T_s} = f_{\max} \qquad (B.6.1)$$

where N = number of flip-flop stages
  T_p = propagation delay of one flip-flop
  T_s = strobe time, width of decoded output pulse
Find $f_{\max}$ for this counter.

(c) Design a modulo-60 ($N = 60$) ripple counter and find out $f_{\max}$. Compare your result with that calculated by Eq. (B.6.1).

2. Cascade the clock you built in Exp. 5 with two modulo-60 ripple counters and a modulo-12 counter (SN 7492) to make a 12-hour digital clock. It is required that this clock has second, minute, and hour seven-segment displays.

Experiment 7
Sequential Circuit Design

Equipment	Components
1 Power supply	4 SN 7400
1 Voltmeter	1 SN 7410
	3 SN 7411
	1 SN 7432
	1 SN 74106
	1 Bread board

1. Sequential locks. A particular sequence of numbers is provided as inputs to the lock which "remembers" each number and opens at the termination of the correct sequence. Design a sequential system (combination lock) that has two inputs (x_1, x_2), one output (z) and four stable states (R, E, B, C). The proper input sequence (combination) for a logic "1" output (lock opens) should be 00–01–11 (R-B-C), i.e., the output of all states is *zero* with the exception of state C which has an output of *one*. The system should return to state R (reset) whenever the input is *00*. The system should go to state E (error) whenever the input sequence is incorrect. Additionally, the system should remain in state E (error) until the input is *00* which returns it to the reset state. Construct a state diagram, build the circuit using J-K flip-flops and NAND gates and verify its operation.

2. Serial adder. There are basically two types of adder configuration in digital systems. The first type is a parallel adder which adds two n-bit numbers simultaneously. The second type is a serial adder which adds two n-bit numbers two bits at a time. The serial adder has two inputs (x_1 and x_2) and one output (S). Each present state of the adder corresponds to a carry from the previous addition which is also considered in determining the present output. Complete the transition table shown in Fig. B.7.1, which will be used to design the adder. Realize the transition table using J-K flip-flops and NAND gates and verify its operation.

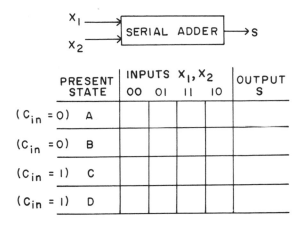

PRESENT STATE		INPUTS x_1, x_2				OUTPUT S
		00	01	11	10	
$(C_{in} = 0)$	A					
$(C_{in} = 0)$	B					
$(C_{in} = 1)$	C					
$(C_{in} = 1)$	D					

Fig. B.7.1   Transition table.

Experiment 8
Random Access Memory (RAM)

Equipment

1 Power supply
2 Voltmeter

Components

1 SN 7489
1 SN 74157
1 SN 7495
4 LED
1 Bread board
4 1K, 1/4w, 10%

1.   Design a counter that will count consecutively from zero to nine and reset at the tenth clock pulse. The counter should increment on each clock pulse (the only input to the system) and display the present count on a LED. A block diagram of the system to be used is shown in Fig. B.8.1

The memory device used is a 16 word-bit RAM (SN 7489). The RAM output is connected to a LED which displays the output. After appropriate output data has been determined, it should be loaded (written) into the RAM.

The purpose of the selector (SN 74157) is to select four of the eight inputs (As or Bs) which are used as address locations. The selector simply connects inputs A or B directly to the output. The selector output may be reset (all zero) by enabling terminal 15.

The shift register (SN 7495) is used in a parallel-in, parallel-out mode. The output of the RAM is tied to the input of the register. When the register is clocked, the data passes to the output and becomes the address for the RAM memory location, thus the previous RAM output becomes the next address.

2.   Design a system that will operate an elevator in a building that has four floors. The system should have two inputs (x_1 and x_2) and four possible states. Each state should have an output such that a LED may be utilized to display the selected floor in BCD form. For purposes of simplification, the system may receive only one set of inputs (x_1 and x_2) at a time. The system

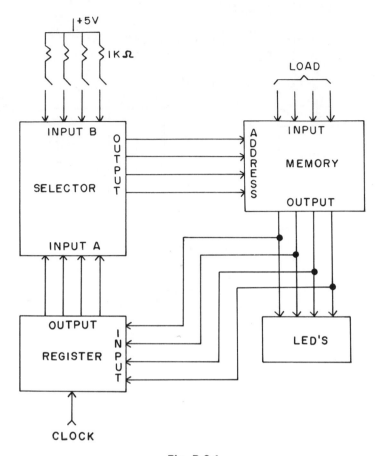

Fig. B.8.1

should be designed such that its next state is determined both by the inputs and its present state, i.e., the RAM address will consist of the two inputs plus two outputs from the previous state of the RAM. The system will be similar to the block diagram shown in Figure B.8.1 with the exception to *two* shift register inputs which will be disconnected from the RAM output and will become x_1 and x_2.

3. Repeat the design of a sequential lock (Exp. 7) using a memory device (SN 7489) instead of JK flip-flops. A block diagram of the system to be designed is shown on page 591. Build the circuit and verify its operation.

4. Repeat the design of a serial adder (Exp. 7) using a memory device (SN 7489) instead of JK flip-flops. Build the circuit and verify its operation.

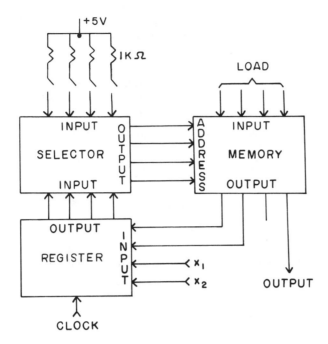

Fig. B.8.2

Experiment 9
Programming PROM

Equipment	*Components*
INTELLEC 4/40 or INTELLEC 8/80 or INTEL MDS-800	4702A or equivalent

1. Program the assembly language program of the 12-hour digital clock of Table 9.5.2 into PROM 4702A.

2. Write a computer program in assembly language to check a simple IC chip, say SN 7400. Print out "good" and "bad" if it passes and fails the test, respectively. Program this program into PROM.

Experiment 10
Logic Design Using Microprocessors

Equipment

3 power supply, +5V, +12V, and −12V
1 voltmeter

Components

1	4040	1	SN 7400
1	4201	6	SN 7448
4	4002	6	FND 70
1	4289	1	14-pin socket
4	4702A		
2	3205		
3	DM 7098		
3	3404		

1. Construct the typical MCS-40 system as shown in Fig. 9.5.2.

2. Build the 12-hour digital clock using the programmed PROM obtained in Part 1 of Exp. 9 using this system.

3. Build an IC tester to test SN 7400 using the programmed PROM obtained in Part 2 of Exp. 9 using this system.

Index